Microbes for Sustainable Development and Bioremediation

Microbes for Sustainable Development and Bioremediation

Edited by
Ram Chandra and R.C. Sobti

CRC Press
Taylor & Francis Group
Boca Raton London New York

CRC Press is an imprint of the
Taylor & Francis Group, an **informa** business

CRC Press
Taylor & Francis Group
6000 Broken Sound Parkway NW, Suite 300
Boca Raton, FL 33487-2742

Visit the Taylor & Francis Web site at
http://www.taylorandfrancis.com

and the CRC Press Web site at
http://www.crcpress.com

Contents

v

Preface

Environment comprises air, water, and soil that support the growth of all living things and provide basic necessity for the all biological activities. But, due to population explosion, indiscriminate utilization of natural resources, production of toxic industrial waste, and deforestation resulted in the challenge of environmental pollution and emergence of life-threatening disease. Consequently, the problem of global warming, climate change, and shrinkage of biodiversity and extinction of numerous important flora and fauna all over the world have been observed by the scientific communities. Since microorganisms and plant play a key role for maintaining the natural resources and balancing the ecosystem, the major role of microorganisms is for regulation of biogeochemical cycle. The microbial world includes bacteria, fungi, viruses, cyanobacteria, algae, and protozoa. Microorganisms, being the pioneer colonizer of this planet and master of the biosphere as they are considered ubiquitous, can exist in most inhospitable habitats with extreme temperature, water and salt stress. The microorganisms are major source for industrial enzyme production, biosurfactant to clean up toxic waste, biobleaching, sewage treatment and oil extraction, biogas, and biofuel production. In the past decades, microbial-based technology (microbial technology) revealed several important knowledge and devices for production of useful products for human life improvement and existing technology for eco-restoration for sustainable development. Keeping in view the global threat of environmental problem for human society, United Nations Organization (UNO) in 2016 in Marrakech, Morocco, declared strategies for development of green technology for sustainable development of environment and human society, which emphasized for the balance utilization of natural resources with help of microbial technology. But unfortunately, several hidden mechanisms of microbial world for maintaining the ecosystem on the earth are still to be revealed. Moreover, useful related important knowledge for development of green technology is away from the researchers, students, and industrialists. Therefore, the proposed book has been focused to imbibe the rare deep knowledge regarding the mechanism of bioremediation and mineralization of various toxic organic and inorganic pollutants for the recycling, which is as a threat to environment. The role of rhizospheric bacteria for maintaining the plant and the agricultural growth to for sustainable development of environment is also an indispensable microbial process of environment. Therefore, this book will provide an opportunity for a wide range of readers, including students, researchers, and consulting professional of biotechnology, microbiology, biochemistry, molecular biology, and environmental science. We gratefully acknowledge the cooperation and support of all contributing author for the publication of this book.

Editors

Dr. Ram Chandra is currently a professor in Department of Microbiology, School for Environmental Sciences, B.B. Ambedkar University (A Central University), Lucknow, India. He has obtained his BSc (Hons.) and MSc from Banaras Hindu University (BHU), Varanasi, India, in 1987. Subsequently, his PhD was awarded in 1994. He has internationally led work on bacterial degradation of chlorolignin compounds for detoxification of pulp paper and mill effluent and molasses melanoid in compounds of distillery waste. He has published more than 160 original research papers in peer-reviewed journal of high-impact, review articles and technical reports with citations 2,700 and h-index 32. He has published six books from CRC Press, USA, and CBS Publisher, India, and more than 30 book chapters on biodegradation and bioremediation of industrial pollutants. He has been awarded Fellow of the Academy of Environmental Biology (FAEB), Fellow of the Association of Microbiologists of India (FAMI), and Fellow of the Biotech Research Society of India (FBRSI). He is also the member in International Academic of Sciences, Indian Science Congress, and the Association for Overseas Technical Scholarship (AOTS), Japan. He has received several honors and awards from the National Agency. He is a reviewer in various international journals of high reputed published from peer-reviewed journals from Elsevier-USA, Springer-USA, Taylor & Francis-USA, and John Wiley & Sons-USA. Prof. Ram Chandra has also trained scientists from Germany and Nigeria under the TWAS-CSIR Fellowship Programme. He has completed more than 25 grant-in-aid projects funded from Ministry of Environment & Forests (MoEF) New Delhi, Department of Science and Technology (DST), India, and Department of Biotechnology (DBT) and Council of Scientific and Industrial Research (CSIR), India.

Dr. R.C. Sobti is currently Professor Emeritus in the Department of Biotechnology, Panjab University, Chandigarh, and former Vice-Chancellor of B.B. Ambedkar University (A Central University), Panjab University, Chandigarh. He has completed his PhD (1974) from Panjab University, Chandigarh, and DSc (2011) from Himachal Pradesh University, India. His main contribution is in cancer biology. He has demonstrated that the regulation of stem cell character under invitro conditions is a function of a morphological assortment of solid lipid nanoparticles (SLNCs). He has published more than 180 research papers, technical reports, and review articles in peer-reviewed journal with more than 2,600 citations. He has published 35 books and completed 23 sponsored research project funded from various agencies. He is, indeed, a polymath, a renowned academician, distinguished scientist, dynamic administrator, and a visionary gifted with an immensely optimistic disposition and integrity of character, words, and action. He has been bestowed with Padma Shri in 2009 for remarkable leadership and extraordinary contribution in higher education. He has been also honored with Bharat Gaurav, lifetime achievement award, at the House of Commons, British Parliament London, United Kingdom. Professor Sobti is a Fellow of the Third World Academy of Sciences (TWAS), National Academy of Sciences, and Indian National Science Academy.

List of Contributors

Abhilash
CSIR—National Metallurgical Laboratory
Jamshedpur, Jharkhand, India

Kaushik Adhikari
Department of Agriculture
University of Arkansas at Pine Bluff
Pine Bluff, Arkansas

Olumuyiwa Samuel Alabi
Department of Pharmaceutical Microbiology
University of Ibadan
Ibadan, Oyo State, Nigeria

Sanjeev Balda
Department of Microbiology
Panjab University
Chandigarh, Punjab, India

Tajudeen Akanji Bamidele
Department of Microbiology
Nigerian Institute of Medical Research
Lagos, Nigeria

Ram Naresh Bharagava
Laboratory for Bioremediation and
    Metagenomics Research (LBMR),
    Department of Microbiology, School of
    Biomedical and Pharmaceutical Sciences
Babasaheb Bhimrao Ambedkar University
    (A Central University)
Lucknow, Uttar Pradesh, India

Muhammad Bilal
School of Life Science and Food Engineering
Huaiyin Institute of Technology
Huaian, China

Olanike Maria Buraimoh
Department of Microbiology
University of Lagos
Lagos, Nigeria

Neena Capalash
Department of Biotechnology
Panjab University
Chandigarh, Punjab, India

Hillol Chakdar
ICAR–National Bureau of Agriculturally
    Important Microorganisms
Kushmaur, Maunath Bhanjan, Uttar Pradesh,
    India

Aneesh Kumar Chandel
Department of Agriculture
University of Arkansas at Pine Bluff
Pine Bluff, Arkansas

Ram Chandra
Department of Microbiology, School for
    Environmental Sciences, Babasaheb
    Bhimrao Ambedkar University
    (A Central University)
Lucknow, Uttar Pradesh, India

Veeranna A. Channashettar
Environmental and Industrial Biotechnology
    Division
The Energy and Resources Institute
New Delhi, India

Hao Chen
Department of Agriculture
University of Arkansas at Pine Bluff
Pine Bluff, Arkansas

Ianny Andrade Cruz
Post-Graduated Program on Process
    Engineering
Institute of Technology and Research
Tiradentes University
Farolândia, Aracaju-Sergipe, Brazil

Subhasis Das
Environmental and Industrial Biotechnology
    Division
The Energy and Resources Institute
New Delhi, India

Shailesh R. Dave
Xavier's Research Foundation
Loyola Centre of Research and
    Development
Ahmedabad, Gujarat, India

Clara Dourado Fernandes
Post-Graduated Program on Process
    Engineering
Institute of Technology and Research
Tiradentes University
Farolândia, Aracaju-Sergipe, Brazil

Silvia Maria Egues Dariva
Post-Graduated Program on Process
    Engineering
Institute of Technology and Research
Tiradentes University
Farolândia, Aracaju-Sergipe, Brazil

Venkata Gadhamshetty
BuG ReMeDEE Consortium
South Dakota School of Mines and Technology
Rapid City, South Dakota
and
Department of Civil and Environmental
    Engineering
South Dakota School of Mines and Technology
Rapid City, South Dakota

Bin Gao
Department of Agricultural and Biological
    Engineering
University of Florida
Gainesville, Florida

Kamini Gautam
Department of Biosciences and Bioengineering
Indian Institute of Technology-Bombay
Powai, Mumbai, Maharashtra, India

Robin Gerlach
Department of Chemical and Biological
    Engineering
Montana State University
Bozeman, Montana

Tanvi Govil
Department of Chemical and Biological
    Engineering
South Dakota School of Mines and Technology
Rapid City, South Dakota
Composite and Nanocomposite Advanced
    Manufacturing Center—Biomaterials
    (CNAM-Bio Center)
Rapid City, South Dakota

Ravi Kr. Gupta
Department of Microbiology, School of
    Biomedical and Pharmaceutical Sciences
Babasaheb Bhimrao Ambedkar University
    (A Central University)
Lucknow, Uttar Pradesh, India

Prakash M. Halami
Department of Food Microbiology
CSIR—Central Food Technology Research
    Institute
Mysore, Karnataka, India

En Huang
Department of Environmental and
    Occupational Health
University of Arkansas for Medical Sciences
Little Rock, Arkansas

Hafiz M. N. Iqbal
Tecnologico de Monterrey, School of
    Engineering and Sciences
Campus Monterrey
Monterrey, Mexico

Yutaka Kawarabayasi
Bioproduction Research Institute
National Institute of Advanced Industrial
    Science and Technology (AIST)
Tsukuba, Ibaraki, Japan

Lee R. Krumholz
Department of Microbiology and Plant Biology
University of Oklahoma
Norman, Oklahoma

Narendra Kumar
Department of Environmental Science (SES),
    School of Environmental Sciences
Babasaheb Bhimrao Ambedkar University
    (A Central University)
Lucknow, Uttar Pradesh, India

Vineet Kumar
Department of Microbiology, School for
    Environmental Sciences, School of
    Biomedical and Pharmaceutical Sciences
Babasaheb Bhimrao Ambedkar University
    (A Central University)
Lucknow, Uttar Pradesh, India

Nanthakumar Kuppanan
Environmental and Industrial Biotechnology
    Division
The Energy and Resources Institute
New Delhi, India

Banwari Lal
Environmental and Industrial Biotechnology
    Division
The Energy and Resources Institute
New Delhi, India

Fangfei Lou
Department of Food Science and Technology
The Ohio State University
Columbus, Ohio

Shahid Mehmood
Bio-X Institute
Key Laboratory for the Genetics of
    Developmental and Neuropsychiatric
    Disorders (Ministry of Education)
Shanghai Jiao Tong University
Shanghai, China

Diego Batista Menezes
Institute of Technology and Research
Tiradentes University
Farolândia, Aracaju-Sergipe, Brazil

Sun Hee Moon
Department of Environmental and
    Occupational Health
University of Arkansas for Medical Sciences
Little Rock, Arkansas

Nandkishor More
Department of Environmental Science (SES),
    School of Environmental Sciences (SES)
Babasaheb Bhimrao Ambedkar University
    (A Central University)
Lucknow, Uttar Pradesh, India

Bhaskar Narayan
Department of Meat and Marine Sciences
CSIR-Central Food Technological Research
    Institute
Mysore, Karnataka, India
and
Food Safety & standards Authority of India
New Delhi, India

Bamidele Tolulope Odumosu
Department of Pharmaceutical
    Microbiology
University of Lagos
Yaba, Lagos, Nigeria

Sunil Pabbi
Division of Microbiology
Centre for Conservation and Utilisation of
    Blue Green Algae
Indian Agricultural Research Institute
New Delhi, India

Prashant S. Phale
Department of Biosciences and
    Bioengineering
Indian Institute of Technology—Bombay
Powai, Mumbai, Maharashtra, India

Hemant J. Purohit
Biotechnology and Genomics
    Division (EBGD)
CSIR-National Environmental Engineering
    Research Institute
Nagpur, Maharashtra, India

Asifa Qureshi
Biotechnology and Genomics
    Division (EBGD)
CSIR-National Environmental Engineering
    Research Institute
Nagpur, Maharashtra, India

Vrinda Ramakrishnan
Department of Microbiology and Fermentation
    Technology
CSIR—Central Food Technological Research
    Institute
Mysore, Karnataka, India

Luiz Fernando Romanholo Ferreira
Post-Graduated Program on Process
    Engineering
Institute of Technology and Research
Tiradentes University
Farolândia, Aracaju-Sergipe, Brazil

David R. Salem
Department of Chemical and Biological
    Engineering
South Dakota School of Mines and Technology
Rapid City, South Dakota
and
Composite and Nanocomposite Advanced
    Manufacturing Center—Biomaterials
    (CNAM-Bio Center)
Rapid City, South Dakota

Dipayan Samanta
Department of Chemical and Biological
    Engineering
South Dakota School of Mines and Technology
Rapid City, South Dakota
and
BuG ReMeDEE Consortium
South Dakota School of Mines and Technology
Rapid City, South Dakota

Rajesh K. Sani
Department of Chemical and Biological
    Engineering
South Dakota School of Mines and Technology
Rapid City, South Dakota
and
Composite and Nanocomposite Advanced
    Manufacturing Center—Biomaterials
    (CNAM-Bio Center)
Rapid City, South Dakota
and
BuG ReMeDEE Consortium
South Dakota School of Mines and Technology
Rapid City, South Dakota

Gaurav Saxena
Laboratory for Bioremediation and
    Metagenomics Research (LBMR),
    Department of Microbiology, School of
    Biomedical and Pharmaceutical Sciences
Babasaheb Bhimrao Ambedkar University
    (A Central University)
Lucknow, Uttar Pradesh, India

Maulin P. Shah
Industrial Waste Water Research Lab,
    Division of Applied & Environmental
    Microbiology
Enviro Technology Limited
Ankleshwar, Gujarat, India

Aarjoo Sharma
Department of Microbiology
Panjab University
Chandigarh, Punjab, India

Amrita Sharma
Department of Biosciences and Bioengineering
Indian Institute of Technology—Bombay
Powai, Mumbai, Maharashtra, India

Hem Chandra Sharma
Department of Agriculture
University of Arkansas at Pine Bluff
Pine Bluff, Arkansas

Prince Sharma
Department of Microbiology
Panjab University
Chandigarh, Punjab, India

Desh Deepak Singh
Department of Biotechnology, BMS-I,
    South campus
Panjab University
Chandigarh, India

Kshitij Singh
Department of Microbiology, School of
    Biomedical and Pharmaceutical Sciences
Babasaheb Bhimrao Ambedkar University
    (A Central University)
Lucknow, Uttar Pradesh, India

Asha B. Sodha
Department of Microbiology & Biotechnology,
    School of Sciences
Gujarat University
Ahmedabad, Gujarat, India

Ranyere Lucena de Sousa
Post-Graduated Program on Process
    Engineering
Institute of Technology and Research
Tiradentes University
Farolândia, Aracaju-Sergipe, Brazil

Hitesh Tikariha
Biotechnology and Genomics Division (EBGD)
CSIR-National Environmental Engineering
    Research Institute
Nagpur, Maharashtra, India

Adeline Su Yien Ting
School of Science
Monash University Malaysia
Bandar Sunway, Selangor Darul Ehsan,
    Malaysia

Devayani R. Tipre
Department of Microbiology & Biotechnology,
    School of Sciences
Gujarat University
Ahmedabad, Gujarat, India

Nádia Hortense Torres
Institute of Technology and Research
Tiradentes University
Farolândia, Aracaju-Sergipe, Brazil

Yago Araújo Vieira
Institute of Technology and Research
Tiradentes University
Farolândia, Aracaju-Sergipe, Brazil

Débora Vilar
Post-Graduated Program on Process
    Engineering
Institute of Technology and Research
Tiradentes University
Farolândia, Aracaju-Sergipe, Brazil

Sangeeta Yadav
Department of Microbiology, School of
    Biomedical and Pharmaceutical Sciences
Babasaheb Bhimrao Ambedkar University
    (A Central University)
Lucknow, Uttar Pradesh, India

Ling Sze Yap
School of Science
Monash University Malaysia
Bandar Sunway, Selangor Darul Ehsan,
    Malaysia

1 Bacterial-Assisted Phytoextraction Mechanism of Heavy Metals by Native Hyperaccumulator Plants from Distillery Waste–Contaminated Site for Eco-restoration

Vineet Kumar and Ram Chandra
Babasaheb Bhimrao Ambedkar University

CONTENTS

1.1 INTRODUCTION

Heavy metals (HMs) forms the main group of inorganic contaminants, and the restoration of sites contaminated such compounds is one of the major challenges for environmentalist (Deng et al. 2004; Tchounwou et al. 2012). Phytoextraction, also called phytoaccumulation, phytoabsorption, or phytosequestration, is one of the most promising and developing phytoremediation technologies applied for the removal of toxic HMs from complex environments (Garbisu and Alkorta 2001; Chandra et al. 2015; Chandra and Kumar 2018). It is an emerging, esthetically pleasing, ecologically nonintrusive, socially accepted, in situ plant-based technology that is being applied for cleanup of toxic metals and cocontaminated (a mixture of inorganic and organic pollutants) and/or polluted sites globally or renders them harmless (Chaney et al. 1997; Chandra and Kumar 2017a; Chandra et al. 2018a). Under normal growing conditions, plants can potentially accumulate certain HMs ions an order of magnitude greater than the surrounding medium (Cunningham et al. 1995; Bhargava et al. 2012). Generally, there are two strategies of phytoextraction: (i) chelate-assisted or induced phytoextraction and (ii) continuous phytoextraction. The chelate-assisted phytoextraction is based on the fact that the application of metal chelators to the soil significantly enhances metal accumulation by plants (Wu et al. 1999; Evangelou et al. 2007), whereas continuous phytoextraction depends on the natural ability of some plants to accumulate, translocate, and resist high amounts of HMs over the complete growth cycle (Garbisu and Alkorta 2001). At the contaminated site, there are many plants that have several strategies to avoid HMs uptake or tolerate the presence of excess HMs in soil, sludge, and/or sediment (Chandra and Kumar 2017a; Chandra et al. 2018c,d). Based upon this, there are three categories of plant i.e. metal excluders, metal indicators, and hyperaccumulators (Baker 1981; Baker and Brooks 1989). Among them, hyperaccumulator plants are the best candidate for metal remediation purpose from the contaminated site, as they can increase internal sequestration, translocation, and accumulation of metals in their harvestable biomass(stem or leaves) (Baker and Brooks 1989; Leitenmaier and Küpper 2013; Chandra et al. 2018b,c). Usually, hyperaccumulators have high accumulation capacity, i.e., the minimum concentration in the shoots of a hyperaccumulator for arsenic (As), lead (Pb), copper (Cu), nickel (Ni), and cobalt (Co) should be greater than 1,000 mg/kg dry mass, and zinc (Zn) and manganese (Mn) should be 10,000 mg/kg, gold (Au) is 1 mg/kg, and Cd is 100 mg/kg, respectively (Baker 1981). The idea of using hyperaccumulator plants to extract and remove metals from contaminated site was first introduced and developed by Chaney (1983). However, the efficiency of metal acquisition and accumulation in plant tissues differed considerably between the elements, plant species, and plant tissues. Four processes are generally believed to be crucial for heavy metal accumulation in plant tissues: uptake of metals by roots, transport of metals from roots to shoot, complexation with chelating molecules, and compartmentalization into the vacuole (McGrath and Zhao 2003; Chandra and Kumar 2018). Although plants suitable for phytoremediation have to be adapted to the polluted environment, the presence of organic pollutants in soil generally reduces plant development and eventually phytoremediation efficacy. As elevated levels of metals are toxic to most plant species apart from hyperaccumulators, leading to impaired metabolism and reduced plant growth, the potential for phytoextraction of HMs is highly restricted and necessitates the development of other phytoremediation strategies for HMs contaminated soils.

The rhizoplane (the part of root remaining in contact with soil, sludge, or water) of growing plant is the most active zone of the presence of various microbial communities. However, the interface between microbes and rhizosphere is considered to greatly influence the growth and survival of plants at contaminated sites (Abou-Shanab et al. 2007, 2008; Kumar and Chandra 2018). The rhizosphere microbial communities includes bacteria, fungi, nematodes, protozoa, algae, and microarthropods (Figure 1.1). Among them, bacteria are the most abundant microorganisms in the rhizosphere, occupying 1 g of soil with up to half a billion individual cells.

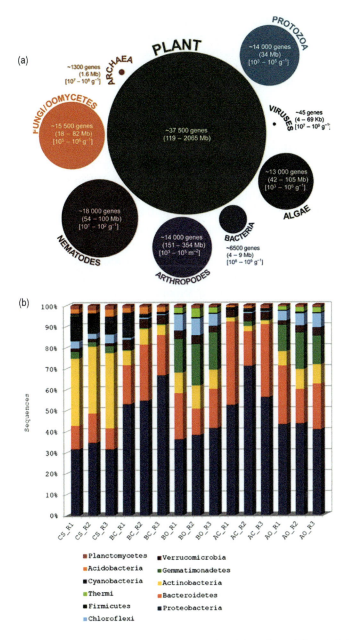

FIGURE 1.1 (a) Overview of microorganisms present in the rhizosphere. The circle's size, except for viruses, is a measure of the average number of genes in the genomes of representative species of each group of organisms; the size (or size range) of their respective genomes is indicated between parentheses. (Reprinted from Mendes et al. 2013 with Permission from Oxford University Press.) (b) Phylum level distribution of microbial population across samples. Phyla represented by at least 0.5% of the total assigned sequences (11 of total 36 phyla) are shown here. CS, clean sandy soil; BC, control oil-contaminated bulk soil without barley plants; BO, barley-planted oil-contaminated rhizosphere soil; AC, control oil-contaminated bulk soil without alfalfa plants; AO, alfalfa-planted oil-contaminated rhizosphere soil. (Reprinted with Permission from Kumar et al. 2018).

At contaminated sites, plants and bacteria coevolved to interact with each other in complex manners, which yielded a vast spectrum of interfaces between the two kingdoms of life. The discovery of rhizosphere bacteria that are heavy metal resistant and able to promote plant growth has raised high hopes for ecologically friendly and cost-effective strategies toward eco-restoration of HMs–polluted sites (Abou-Shanab et al. 2008; Ahemad and Khan 2010a,b; Ahemad 2014a,b). Therefore, alternative phytoremediation methods that exploit rhizosphere bacteria to reduce metal toxicity to plants have been investigated (Abou-Shanab et al. 2007). Rhizobacteria colonize the close vicinity of roots, whereas endophytic bacteria colonize the plant interior without causing pathogenicity to their host plant. Rhizobacteria have been investigated for their plant growth–promoting capacity, and in the last decade, the use of such bacteria to enhance phytoremediation efficiency has been reported (Glick 2003, 2012). When using plants and microbes in combination, the plant provides the habitat as well as nutrients to the associated rhizospheric and endophytic bacterial communities. In return, the bacteria enhance the stress tolerance of the plant or improve plant growth and detoxify the plant environment by degrading the pollutant (Glick 2010, 2014). There are complex and varied interactions between plants and their associated microbes, and these interactions have been extensively studied and used to increase soil fertility, plant development, and phytoremediation of polluted soil and water. Muehe et al. (2015) compared the effects of a "native" and a strongly disturbed (gamma-irradiated) soil microbial communities on Cd and Zn accumulation by the *Arabidopsis halleri* plant in soil microcosm experiments. *A. halleri* accumulated 100% more Cd and 15% more Zn when grown on the untreated than on the gamma-irradiated soil. Pyrosequencing of 16S rRNA gene amplicons of DNA extracted from rhizosphere samples of *A. halleri* showed the higher relative sequence abundance of microbial taxa *Lysobacter, Streptomyces, Agromyces, Nitrospira, Candidatus,* and *Chloracido bacterium* in the rhizospheres of *A. halleri* plants grown on untreated than on gamma-irradiated soil, leading to hypotheses on their potential effect on plant metal uptake. Kumar et al. (2018) used high-throughput sequencing to explore the rhizosphere microbial diversity in the alfalfa and barley-planted oil-contaminated soil samples. The analysis of 16s rRNA sequences showed *Proteobacteria* to be the most enriched (45.9%) followed by *Bacteriodetes* (21.4%) and *Actinobacteria* (10.4%) phyla. The oil-contaminated rhizospheric soil showed enrichment of known oil-degrading genera, such as *Alcanivorax* and *Aequorivita,* later being specifically enriched in the contaminated soil samples planted with barley. Authors were found a few well-known oil-degrading bacterial groups to be enriched in the oil-contaminated planted soil samples compared with the untreated samples.

Distillery industry occupies a place of prominence in the Indian economy, but it contributes significantly toward the contamination of environment (Kumar and Chandra 2020; Chandra et al. 2018c,d). It is one of the highly polluting industries due to discharge of the huge amount of sludge as by-product during anaerobic digestion of spent wash (Chandra and Kumar 2017a,b; Kumar and Chandra 2019). The sludge is considered as a source of toxic HMs and various androgenic-mutagenic compounds, and its disposal in environment is problematic. In addition, distillery sludge also consists of a mixture of several recalcitrant organic compounds along with melanoidins (Chandra and Kumar 2017a,b). Distillery sludge is the most common habitat that harbors unique types of bacterial species, which are capable of running widespread in situ bioremediation activities (Chandra and Kumar 2017b). The growing autochthonous bacterial species act at the primary level to loosen the interaction of organometallic bond through their enzymatic action, which makes metal availability to plant. Generally, plants use varied mechanisms to uptake different organic and inorganic pollutants, which make the basis of phytoremediation technology (Cunningham et al. 1997; Alkorta et al. 2004). They employ numerous kinetic processes, including phytostabilization, phytovolatilization, and phytodegradation, for removal of organic compounds, whereas for inorganics, phytostabilization, phytoextraction, and phytovolatilization are involved (Figure 1.2a; Cunningham et al. 1997, 1995; Salt et al. 1998; Chandra and Kumar 2018).

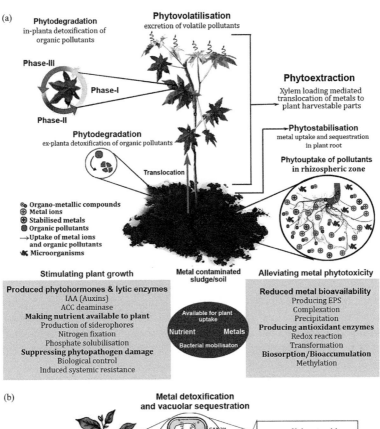

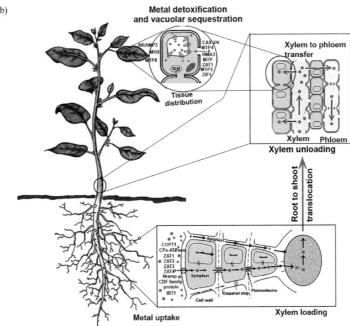

FIGURE 1.2 Phytoramediation strategies and heavy metals accumulation and detoxification in plant. (a) Illustration of various strategies of plant for phytoremediation of heavy metals and organic pollutants at contaminated sites. Plants-associated bacteria influence heavy metal accumulation in plant tissues and promote plant growth at heavy metals–contaminated site. (Modified from Chandra and Kumar 2018.) (b) Schematic mechanism of heavy metal transportation, detoxification, and accumulation in plant tissues. (Modified from Verbruggen et al. 2009.)

  i. Phytoextraction involves the use of pollutant-accumulating plants to remove HMs or organics from the soil, sludge, sediment, or water by concentrating them in harvestable parts of the plant. At the end of the growth period, plant biomass is harvested, dried, or incinerated, and the contaminant-enriched material is deposited in a special dump or added into a smelter. The energy gained from burning of the biomass could support the profitability of the technology, if the resultant fumes can be cleaned appropriately.

  ii. In phytodegradation, contaminants are taken up from soil or water, metabolized in plant tissues, and broken up to less toxic or nontoxic compounds within the plant by several metabolic processes via the action of compounds produced by the plant. Besides, the contaminants are also degraded in the rhizopshere by the proteins or enzymes produced by plants and their root-associated rhizospheric bacteria.

  iii. Phytostabilization: in this technique, metals are immobilized, and the mobility and bioavailability of metals to plant roots are reduced. Leachable constituents are adsorbed and form a stable complex with plant root from which the contaminants will not reenter the environment.

  iv. Phytovolatization: volatization of pollutants by plants from the soil, sludge, and/or sediment into the atmosphere.

Among them, the development of phytoextraction technique comes from the discovery of a variety of wild weeds, often endemic to naturally mineralized soils or even anthropogenic metal-enriched areas that concentrate high amount of essential and nonessential HMs in their harvestable parts (Fontem Lum et al. 2014; Hammami et al. 2016). Weed species appear to be a good choice for ecorestoration of metal-contaminated site since these hardy, tolerant species can quickly grow in most harsh conditions, including HMs stress, over large areas and give a good amount of biomass as a secondary product (Sahu et al. 2007; Chandra et al. 2018a). Weed plant can restrict the contaminant from being introduced into the food web and consequently reduce the risk to human health (Fontem Lum et al. 2014). Plants showing tolerance to toxic metals have a range of mechanisms at the cellular and molecular level that might be involved in the general homeostasis, detoxification, and tolerance to HMs stress (Chandra et al. 2018a,b). Some native hyperaccumulator plants have been reported for phytoextraction of HMs from disposed distillery sludge (Chandra et al. 2018b; Chandra and Kumar 2017a). These plants have strong potential and adaptation for survival and growth on organometallic containing distillery waste and capable of accumulating elevated concentrations of several HMs in their root and shoots during in situ phytoremediation. However, the rhizospheric bacterial communities of these growing plants are still unknown.

  HMs hyperaccumulation in plants is a multistep process that includes mobilization of soil at contaminated site, transport of the heavy metal across the plasma membrane of root cells, xylem loading and root-to-shoot translocation, detoxification, and sequestration of the HMs in the leaf tissue as shown in Figure 1.2b. Plant uptake of HMs from contaminated site either occurs passively with the mass flow of water into the roots or, through active transport, crosses the plasma membrane of root epidermal cells. For effective phytoextraction, metals not only must be taken up rapidly but should also be transported from the roots to aerial parts of the plant (Yang et al. 2005). Ultimately, the success of phytoextraction depends on several factors, including the extent of soil contamination, metal availability for uptake into roots (bioavailability), and plant ability to intercept, absorb, and accumulate metals in shoots (McGrath and Zhao 2003; Yoon et al. 2006). The concentration of bioavailable metals in the rhizosphere greatly influences the quantity of metal accumulation in plants because a large proportion of HMs are generally bound to various organic and inorganic constituents in polluted soil/sludge and their phytoavailability is closely related to their chemical speciation (Jing et al. 2007; Kumar and Chandra 2018). The bioavailability and bioaccumulation of metal in plant tissues are governed by several environmental factors such as pH, the presence of humic substances and other organic chelators, metal speciation, aging, nature of the metal, soil/sludge mineralogy, amount of metals cations exchange capacity, organic carbon content, and oxidation state of the system (Ghosh and Singh 2005; Alkorta et al. 2004). Plant root exudates contain a variety of compounds, such as organic acids and sugars, amino acids, fatty acids, vitamins, growth factors, hormones and antimicrobial compounds, nematicides, and flavonoids, are a major driving force in the regulation of microbial diversity and activity on plant roots and released into the soil by diffusion, ion channels, and vesicular transport (Wu et al. 2017; Li et al. 2019; Dennis et al. 2010). Root exudates are key determinants of rhizosphere microbiome structure that alter soil chemistry and provide nutrient sources for developing microbial communities in the rhizosphere. Moreover, the mobility and availability of HMs in soil are generally very low, especially when the soil is having high pH, clay, and organic matters, which have a metal binding tendency. It has been reported that melanoidins, major colorant of distillery waste, have net negative charges; hence, different heavy metals (i.e. Cu^{2+}, Cr^{3+}, Fe^{3+}, Zn^{2+}, Pb^{2+}, etc.) strongly bind with melanoidins to form an organometallic complex (Migo et al. 1997; Hatano et al. 2016). Hence, the metals in distillery waste occurs in complex form vary widely in their availability to the plant.

Plant-associated rhizospheric bacteria can potentially improve phytoextraction by altering the solubility, availability, and transport of heavy metal and nutrients by reducing soil pH, the release of chelators, surfactant, or redox changes (van Loon 2007; Chandra and Kumar 2015a; Rajkumar and Freitas 2008; Rajkumar et al. 2012). The metabolites released by plant associated bacteria can alter the uptake of HMs indirectly and directly: indirectly, through their effects on plant growth dynamics, and directly, through acidification, chelation, precipitation, immobilization, and oxidation-reduction reactions in the rhizosphere. Among the various metabolites produced by plant growth–promoting rhizobacteria (PGPR), the siderophores play a significant role in metal mobilization and accumulation (Sessitsch et al. 2013; Rajkumar et al. 2010); siderophores produced by rhizosphere bacteria bind HMs and thus enhance their bioavailability in the rhizosphere of plants. The resulting increase in trace metal uptake by the plants caused by microbial siderophores might enhance the effectiveness of phytoextraction processes of contaminated soil (Rajkumar et al. 2010). In addition to iron, siderophores can also form stable complexes with other metals that are of environmental concern, such as Al, Cd, Cu, Ga, In, Pb, and Zn, as well as with radionuclides, including U and Np. Additionally, bacteria colonizing in the rhizosphere of hyperaccumulators growing in metal-polluted soils improve plant growth through phytohormone production, secreting enzymes, and phosphate solubilization abilities (Sessitsch et al. 2013). Bacterial communities of rhizospheric soils play an important role in the tolerance and uptake of metal-tolerant/hyperaccumulating plants to metals. Bacterial communities degrade organic pollutants in plant rhizosphere and protect the plant to grow in the highly polluted environment, whereas plant accumulates the HMs in its various tissues. Both of them played a vital role in in situ phytoremediation of distillery sludge. Hence, the aim of this chapter is to provide a concise discussion of the role of rhizospheric bacteria in the phytoremediation of heavy metal from distillery sludge–polluted site. Further, we also discussed the challenges and future prospects for phytoremediation of HMs from distillery waste.

1.2 PLANT-ASSOCIATED BACTERIA

Plant-associated bacteria include endophytic, phyllospheric, and rhizospheric bacteria. These bacteria can promote plant growth and development and might even be able to degrade or detoxify the organic and inorganic pollutants (Weyens et al. 2009; Glick 2003, 2010, 2012). Endophytic bacteria are defined as bacteria colonizing the internal tissues of plants without causing symptoms of infection or negative effects on their host plants or environment (Compant et al. 2005; de Oliveira et al. 2012). Several studies revealed that endophytic bacteria mainly reside in the intercellular apoplast and in dead or dying cells. They also found in xylem vessels, within which they may be translocated from the root to the aerial parts of the plant (Turner et al. 2013). The highest densities of endophytic bacteria usually are observed in the roots and decrease progressively from the stem to the leaves (Weyens et al. 2009). Many endophytic bacteria, particularly those inhabiting plants growing in a polluted environment, produce degradative enzymes and contribute to the degradation of several types of organic compounds present in the rhizosphere and endosphere (Reinhold-Hurek and Hurek 2011; Mitter et al. 2013). The phyllosphere is the external region of plant parts that are above ground, including leaves, stems, blossoms, and fruits. Because the majority of the surface area available for colonization is located on the leaves, this is the dominant tissue of the phyllosphere. Bacteria residing in the phyllosphere are exposed to large and rapid fluctuations in temperature, solar radiation, and water availability. The rhizosphere, the narrow zone of soil that surrounds and is influenced by plant roots, is home to an overwhelming number of microorganisms and is considered to be one of the most dynamic interfaces on earth (Philippot et al. 2013). The term "rhizosphere" was originally introduced by Lorenz Hiltner in 1904 to illustrate the particular zone of soil surrounding plant roots in which microbe populations are stimulated by root exudates (Haldar and Sengupta 2015). The rhizosphere provides a peculiar environment where a huge variety of positive, negative, and neutral interactions between roots and microorganisms occur. Such interactions can significantly influence plant growth as well as the functioning, the abundance, and the diversity of rhizospheric microbial communities. However, plant roots exert strong effects on the rhizosphere through *rhizodeposition* (root exudation, production of mucilages, and release of sloughed-off root cells) and by providing suitable ecological niches for microbial growth (Mendes et al. 2013). All rhizospheric, phyllospheric, and endophytic bacteria can affect plant growth and development by fixing atmospheric nitrogen (diazotrophy) and/or synthesizing phytohormones and enzymes involved in plant growth hormone metabolism, as discussed in more detail here.

1.2.1 PLANT GROWTH–PROMOTING RHIZOBACTERIA

Bacteria inhabiting the rhizosphere are called *rhizobacteria*. The term *rhizobacteria* was coined by Kloepper and Schroth in 1978, based on their experiments with radishes. They defined these bacteria as a community that competitively colonizes plant roots and enhances their growth and also reduces plant diseases. Rhizobacteria can be classified into beneficial, deleterious, and neutral groups on the basis of their effects on plant growth. The rhizobacteria that stimulate the growth and health of the plant are referred to as *plant growth–promoting rhizobacteria* (Kloepper et al. 1989; Glick 2012). About 2%–5% of the rhizosphere bacteria are PGPR (Antoun and Prevost 2005). PGPR include a diverse group of free-living soil bacteria that can improve host plant growth and development in heavy metal–contaminated soils by mitigating toxic effects of heavy metals on the plants (Belimov et al. 2001). Predominant bacterial strains in the rhizosphere include gram-negative, rod-shaped, nonsporulating bacteria belonging to the group's *Proteobacteria* and *Actinobacteria* (Atlas and Bartha 1993; Teixeira et al. 2010) of which *Pseudomonas* is the most abundant. This may be attributed to the efficiency of gram-negative bacteria to utilize the root exudates, and hence they are stimulated by rhizodeposition, whereas the gram-positive bacteria are rather inhibited (Steer and Harris 2000). PGPR are further classified as extracellular plant growth–promoting rhizobacteria (ePGPR) or intracellular plant growth–promoting rhizobacteria (iPGPR) depending upon their intimacy in interaction with plants (Martinez-Viveros et al. 2010). PGPR have been shown to contribute to biodegradation of toxic organic compounds in polluted soil and could have the potential for improving phytoremediation. PGPR may facilitate plant growth either indirectly or directly as shown in Figure 1.3 (Glick 2012; Kong and Glick 2017).

1.2.1.1 Direct Plant Growth Promotion by Rhizobacteria

Direct plant growth promotion is based on either stipulation of the plants with favorable bacterial compounds or improving the nutrient uptake by the plant from the environment (Bashan and Holhuin 1998; Glick 2012, 2014; Kong and Glick 2017). Rhizobacteria promote plant growth and development through a variety of mechanisms, including the fixation of atmospheric nitrogen, and supply it to plants; synthesize iron chelators referred to as siderophores, which can solubilize and sequester iron from the soil and provide it to plant cells; synthesize and release different phytohormones, including auxins, cytokinins, and gibberellins, which can enhance various stages

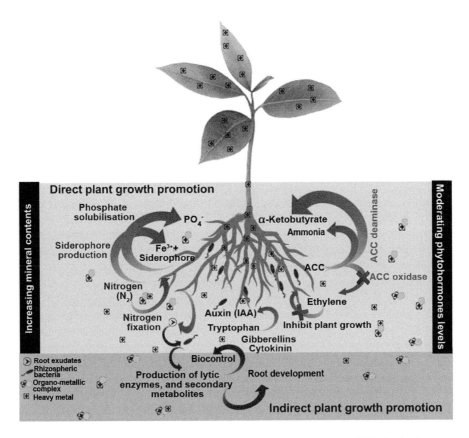

FIGURE 1.3 The mechanisms of plant growth–promoting rhizobacteria (PGPR) in the improvement of phytoremediation. (Modified from Kong and Glick 2017.)

of plant growth and symbiotic N_2 fixation; improve ammonia production; and have mechanisms for the solubilization of inorganic phosphate and mineralization of organic phosphate and/or other nutrients, which then become more readily available for plant growth (Glick et al. 2007; Glick 2012; Gutiérrez-Mañero et al. 2001; Kong and Glick 2017). In addition to this, PGPR enhance the tolerance capacity of the plant to a variety of environmental stresses through the production of 1-aminocyclopropane-1-carboxylate (ACC) deaminase that can modulate plant growth and development (Glick 2014).

1.2.1.1.1 Nitrogen Fixation

Even though 78% of the earth's atmosphere is nitrogen, this element is often alimiting factor for plant growth because atmospheric nitrogen exists as dinitrogen (N_2), a form that is inaccessible to all except a few specially adapted prokaryotes, including some eubacteria, cyanobacteria, and actinomycetes. Many free-living bacteria, including Rhizobia species, have the ability to fix nitrogen and then provide it to the plants. Generally, free-living bacteria can pass only a small amount of its fixed nitrogen to the associated plants (James and Olivares 1998; James 2000). These bacteria possess the enzyme nitrogenase, an oxygen (O_2)-sensitive enzyme, encoded by *nif* genes that catalyze the reduction of atmospheric nitrogen to ammonia (Brusamarello-Santos et al. 2012). Scientists believed that *nif* genes are isolated and properly characterized; then these genes could be genetically engineered to improve nitrogenfixation (Beringer and Hirsch 1984). Plants could be genetically engineered to fix their nitrogen.

1.2.1.1.2   Phosphate Solubilization

After nitrogen, phosphorus is the second essential macronutrient for plant growth and development. Although, the amount of phosphorus in the soil is generally quite high (often between 400 and 1,200 mg/kg of soil), most of this phosphorus is insoluble and therefore not available to support plant growth; however, only a very small fraction (~0.1%) is available to plants. Insoluble phosphorus is present as either an inorganic mineral such as apatite or as one of several organic forms, including inositol phosphate, phosphomonesters, and phosphotriesters. Plants can only take up phosphorous in monobasic ($H_2PO_4^-$) or dibasic (HPO_4^{2-}) soluble form. Under metal-stressed conditions, most metal-resistant plant growth–promoting bacteria (PGPB) can either convert these insoluble phosphates into available forms through acidification, chelation, exchange reactions, and release of organic acids (lactic, citric, 2-ketogluconic, malic, glycolic, oxalic, malonic, tartaric, valeric, piscidic, succinic, and formic acids) or mineralize organic phosphates by secreting extracellular phosphatases (Zaidi et al. 2006; Ahemad 2014a). Thus, solubilization and mineralization of phosphorus by phosphate-solubilizing bacteria (PSB) such as *Azotobaccter chrococcum, Bacillus* spp.*, Enterobacter agglomerans, Pseudomonas chlorraphis, Pseudomonas putida, Rhizobium* and *Bradyrhizobium* spp. are an important trait in PGPR as well as in plant growth–promoting fungi such as mychorrizae (Alori et al. 2017; Ahemad 2014b; Sharma et al. 2013). Additionally, PSB not only protect plants from phytopathogens through the production of antibiotics, HCN, phenazines, and antifungal metabolites, etc. (Upadhayay and Srivastava 2012; Singh et al. 2013) but also promote plant growth through N_2 fixation (He et al. 2010), siderophore production (Ahemad and Khan 2012a,b), phytohormone secretion (Oves et al. 2013), and lowering ethylene levels (Kumar et al. 2009). In addition, various plant growth–promoting traits of PSB, such as secretion of siderophores, IAA (indole-3-acetic acid) production, and ACC deaminase activity, contribute to enhancing the phytoremediation capability of plants (Glick 2012). PSB remediate metal-contaminated soils largely through facilitating either phytostabilization or. Examples of successful remediation of HMs by using plant-rhizosphere bacteria partnerships are listed in Table 1.1.

1.2.1.1.3   Production of Phytohormones

Phytohormones that are produced by plant-associated bacteria, such as auxins, cytokinins, and gibberellins, can frequently stimulate growth and indeed have been considered the causal agents for altered plant growth and reproduction, and protect plants against both biotic and abiotic stress (Egamberdieva et al. 2017). As the most studied phytohormones, auxin IAA produced by PGPR via the indole-3-pyruvate (IPyA) pathway can increase the number of root hairs, the number of lateral roots, and the total root surface, leading to an enhancement of root exudation and mineral uptake from the soil (Costacurta and Vanderleyden 1995). This latter effect may act to further enhance the colonization surface and the exudation of nutrients for bacterial growth; thus the IAA synthesized by PGPB is finely modulated in response to environmental stresses (such as salinity, HMs, acid pH) associated with the soil and plant. In addition, bacterial IAA loosens plant cell walls and, as a result, facilitates an increasing amount of root exudation that provides additional nutrients to support the growth of rhizospheric bacteria. IAA that was incorporated by the plant stimulated the activity of the enzyme 1-aminocyclopropane-1-carboxylic acid (ACC) synthase, resulting in increased synthesis of ACC, and a subsequent rise in ethylene that inhibited root elongation (Leveau and Lindow 2005; Glick 2012; Patel and Saraf 2017; Rajkumar et al. 2012). Overall, bacterial IAA increases root surface area and length and thereby provides the plant greater access to soil nutrients. In addition, rhizospheric IAA synthesized by both plants and bacteria may act as a signal for soil *Streptomyces* to increase their production of antibiotics, which are lethal to fungal and bacterial phytopathogens, thereby inhibiting the growth of competing microbes and simultaneously protecting plants from phytopathogens (Egamberdieva et al. 2017). However, the molecular mechanisms involved in PGPR-assisted phytoremediation of HMs–contaminated environments are still largely unknown.

TABLE 1.1
Examples of Successful Remediation of Heavy Metals by Using Plant-Rhizosphere Bacteria Partnership

Bacteria	Host plant	Pollutants	Isolation site/ contaminated site	Reference
Bacillus, Staphylococcus, and *Aerococcus.*	*Prosopis juliflora*	**Cr**	Tannery effluent–contaminated soil	Khan et al. 2014
Bacillus pumilus and *Micrococcus* spp.,	*Noccaea caerulescens*	Ni	Serpentine Ni-rich soil	Aboudrar et al. (2013)
Streptomyces AR17, *Agromyces* AR33	*Salix caprea*	Zn, Cd, Pb	Lead mining area	Kuffner et al. (2008)
Bacillus subtilis, Bacillus pumilus, Pseudomonas pseudoalcaligenes, and *Brevibacterium halotolerans*	*Zea mays* and *Sorghum bicolor*	Cr, Pb, Zn, Cu	Tannery effluent–contaminated site	Abou-Shanab et al. (2008)
Bacillus megaterium HKP-1	*Brassica juncea*	Pb, Zn	Lead-zinc mine	Wu et al. (2006)
Bacillus subtilis SJ-101	*Brassica juncea*	Ni	–	Zaidi et al. (2006)
Brevundimonas Kro13		Cd	–	Robinson et al. (2001)
Pseudomonas sp.	Soybean, mungbean, wheat	Ni, Cd, Cr	–	Gupta et al. (2002)
Ochrobactrum intermedium	Sunflower	Cr (VI)	Cr contaminated site	Faisal and Hasnain (2005)
Variovox paradoxus, Rhodococcus sp. *Flavobacterium*	*Brassica juncea*	Cd	Heavy metal polluted soil	Belimov et al. (2005)
Kluyvera ascorbate SUD165	Indian mustard, canola, tomato	Ni, Pb, Zn	–	Burd et al. (2000)
Ralstonia sp. TISTR 2219 and *Arthrobacter* sp. TISTR 2220	*Ocimum gratissimum*	Cd	Cd contaminated soil	Prapagdee and Khonsue (2015)
Burkholderia cepacia	*Sedum alfredii*	Cd, Zn	Pb/Zn mine site	Li et al. (2007)
Brevibacterium casei MH8a	*Sinapis alba* L.	Cd, Zn, Cu	Heavy metal-contaminated soil	Płociniczak et al. (2016)
Bacillus pumilus and *Micrococcus* sp.	*Noccaea caerulescens*	Ni	Serpentine Ni-rich soi	Aboudrar et al. (2013)
Microbacterium saperdae, Pseudomonas monteilii, and *Enterobacter cancerogenes,*	*Thlaspi caerulescens*	Zn	Heavy metal contaminated soil	Whiting et al. (2001)
Acinetobacter, Alcaligens, Listeria, Staphylococcus. Acinetobacter, Alcaligens, and *Listeria*	*Nymphaea pubescens*	Cu	–	Kabeer et al. (2014)
Methylobacterium oryzae, Berknolderia sp.	*Lycopersicon esculentom*	Ni, Cd	Metal contaminated soil	Madhaiyan et al. (2007)
Xanthomona sp. RJ3, *Azomonas* sp. RJ4, *Pseudomonas* sp. RJ10, *Bacillus* sp. RJ31	*Brassica napus*	Cd	Heavy metal contaminated soil	Sheng and Xia (2006)
Pseudomonas sp, *Bacillus* sp.	*Brassica*	Cr (VI)	Heavy metal contaminated soil	Rajkumar et al. (2006)
Brevibacillu	*Trifolium repens*	Zn	Zn contaminated soil	Vivas et al. (2006)

Cytokinins, as well as gibberellins, play an important role in regulating plant growth and development. Generally, cytokinins are involved in the regulation of metabolite transport, cell division, protein synthesis, chloroplast formation, activation of seed germination, stem morphogenesis, leaf senescence, and the functioning of the aerial plant organs Groskinsky et al. 2016. Furthermore, they can promote stomatal opening, stimulate shoot growth, and decrease root growth. When the soil is drying, natural cytokinin concentrations decrease in association with stomatal closure, and growth is redirected away from the shoots to the roots (Olanrewaju et al. 2017). Although decreased cytokinin levels induced by water shortage contribute to drought tolerance, Arkhipova et al. (2007) demonstrated that prevention of cytokinin loss in soil by inoculation with cytokinin-producing bacteria was beneficial for plant growth under moderate drought conditions. Some bacteria, particularly PGPB, can produce either cytokinins or gibberellins or both. For example, bacterial cytokinins have been detected in the cell-free medium of *Pseudomonas, Rhizobium, Azotobacter, Azospirillum, Bacillus*, and *Arthrobacter* (Ortíz-Castro et al. 2008; Glick 2010, 2012). Gibberellins (gibberellic acid) are involved in modifying plant morphology by plant tissue extension, in particular of the stem. Gibberellins, first and foremost, affect cell division and elongation in the intercalary meristem; they also are involved in activation of membranes and amylolytic enzyme synthesis and stimulation of florescence. Gibberellin functions in roots are less important than those of auxins, but they nevertheless play an indispensable role in normal root development (Tsavkelova et al. 2006; Egamberdieva et al. 2017). Gibberellins are synthesized by epiphytic and rhizospheric bacteria, including the genera *Azospirillum, Bacillus, Pseudomonas, Rhizobium, Azotobacter, Arthrobacter, Agrobacterium, Clostridium, Flavobacterium*, and *Xanthomonas* (Tsavkelova et al. 2006). Although gibberellin production by plant-associated bacteria seems to be less widespread, *Bacillus pumilus* and *Bacillus licheniformis* species producing this phytohormone have been isolated; Gutiérrez-Mañero et al. (2001) demonstrated that these bacteria promote plant growth and yield through the gibberellins they produce. A number of phytopathogenic bacteria have also been reported to produce cytokinins. It appears that the cytokinin levels produced by PGPB are lower than those from phytopathogens so that the effect of the cytokinins from PGPB on plant growth is stimulatory, whereas the effect of the pathogens is inhibitory (Glick 2012). However, in the past few years, it has been found that a number of plant growth–promoting bacteria contain the enzyme ACC deaminase (Glick et al. 2007; Glick 2012, 2014). This enzyme converts the ethylene precursor ACC to α-ketobutyrate and ammonia and promotes plant growth, especially during stress conditions, by reducing the level of stress ethylene to below the point where it is inhibitory to growth (Glick et al. 2007; Glick 2012). ACC is involved in the biosynthetic pathway of ethylene, as an intermediate in the conversion of methionine to ethylene following biosynthetic sequence: methionine–S-adenosylmethionine (SAM)–ACC–C_2H_4 (Grichko and Glick 2001). In general, ACC is exuded from plant roots or seeds and then taken up by the ACC-utilizing bacteria before its oxidation by the plant ACC oxidase and cleaved by ACC deaminase to α-ketobutyrate and ammonia. The bacteria utilize the ammonia evolved from ACC as a sole nitrogen source and thereby decrease ACC within the plant with the concomitant reduction of plant ethylene (Glick 2010; Kong and Glick 2017). The plant hormone ethylene plays an important role in root initiation and elongation, nodulation, senescence, abscission, and ripening as well as in stress signaling. The ethylene synthesized as a response to environmental stresses is called "stress ethylene." This increase in ethylene synthesis is typically associated with various environmental stresses, such as extreme temperatures, water stress, high salt, toxic metals, organic pollution, insect damage, radiation, wounding, and various pathogens, including viruses, bacteria, and fungi (Morgan and Drew 1997; Dubois et al. 2018). As part of a stress response, it inhibits root elongation, nodulation, and auxin transport, induces hypertrophies, speeds aging, and promotes senescence and abscission. During periods of environmental stress, plants produce high levels of "stress ethylene." However, ACC deaminase–containing bacteria can lower plant ethylene levels in plants and thereby provide some protection against the inhibitory effects of various stresses. Salinity stress has been shown to elevate ethylene levels (Mayak et al. 2004), which affects almost all aspects of plant growth and development, including the response to biotic and abiotic stresses (Cao et al. 2008). Bacteria that are associated with plants can affect ethylene levels via two main mechanisms. Some bacteria can balance the ethylene levels in plants through the auxins they synthesize, while the most commonly observed mechanism is via the activity of the bacterial enzyme ACC deaminase, which can lower the ethylene levels in plants and thereby protect plants against the inhibitory effects of various environmental stresses. ACC deaminase–containing PGPB can facilitate plant growth and development through conversion of the immediate ethylene precursor ACC into α-ketobutyrate and ammonia, thus reducing the levels of plant ethylene and providing some protection against growth inhibition caused by flooding, heavy metals, organic pollution, high salt, drought, and phytopathogens (Gamalero and Glick 2011; Glick 2012; Farwell et al. 2006; Rodriguez et al. 2008). Moreover, a number of studies have shown that ACC deaminase–containing PGPB can facilitate metal phytoremediation by reducing the level of stress ethylene that is induced by high levels of toxic HMs.

1.2.1.1.4 *Production of Siderophores*

Bacterial activities that improve mineral nutrient uptake by plants can facilitate plant growth under stressful conditions due to the production of siderophores, low-molecular-weight (200–2,000 Da) Fe-chelating secondary metabolites, which are produced by bacteria, fungi, and plants under Fe-limiting conditions (Chandra and Kumar 2015b). Siderophore-producing bacteria can stimulate plant growth either directly by improving plant Fe nutrition or indirectly by inhibiting the activity of plant pathogens in the rhizosphere through limiting their iron availability (Rajkumar et al. 2010; Ma et al. 2011).

Generally, Fe occurs mainly as Fe^{3+} and forms insoluble hydroxides and oxyhydroxides; thus, it is not easily available to both plants and microorganisms. All siderophores possess a higher affinity for Fe^{3+} than for Fe^{2+} or any other traces such as Cd, Cu, Ni, Pb, and Zn (Rajkumar et al. 2010). Supply of iron to growing plants under heavy metal stress becomes more important, as bacterial siderophores help to minimize the stress imposed by metal contaminants (Gamalero and Glick 2011). Interestingly, the binding affinity of phytosiderophores for iron is less than the affinity of bacterial siderophores, but plants require a lower iron concentration for normal growth than do bacteria (Meyer 2000). Plant members of the family Poaceae are capable of enhancing the availability of iron for uptake into roots (Neubauer et al. 2000). The efficiency of phytoextraction may be increased by growing siderophore-producing grass species in combination with accumulator plants. Although this approach holds promise, phytosiderophores obtain their specificity not by chelation specifically only of Fe in soils, but from their uptake of Fe-phytosiderophores by a membrane carrier (Neubauer et al. 2000). Improving plant nutrition and mobilizing metals enhance phytoextraction. Siderophore-producing bacteria have been shown to enhance chlorophyll content and growth of various crop plants in contaminated soil by selectively supporting iron uptake from the pool of trace element cations competing for import (Burd et al. 1998, 2000; Rajkumar et al. 2010). Moreover, complexation of trace elements by bacterial siderophores in the rhizosphere likely prevents generation of free radicals and oxidative stress. Siderophores are secreted and Fe^{3+}-siderophore complexes are recognized and scavenged from the environment by membrane receptor proteins. They are too large to pass membrane porins. In gram-negative bacteria, transport through the cytoplasmatic membrane is mediated by ABCtransporters, and the TonB protein transfers the necessary metabolic energy from the cytoplasmic to the outer membrane. In gram-positive bacteria, this process is less understood, and they lack TonB. Many bacteria produce more than one siderophore, and different molecules could be produced under different conditions.

1.2.1.2 Indirect Plant Growth Promotion by Rhizobacteria

Indirect plant growth promotion occurs when PGPR prevent the deleterious effects of phytopathogens, usually the fungi and the nematodes, through the production of antimicrobial compounds, thereby controlling the diseases, competition for iron and nutrients or for colonization sites, to mention a few mechanisms (Glick 2003, 2010, 2012). Pathogen suppression may be achieved through a variety of mechanisms such as production and release of cyanide, antibiotics, or extracellular lytic enzymes, including chitinases, proteases, β-1,3-glucanases, cellulases, and laminarinases, competition for nutrients and niches in the rhizosphere, parasitism, and predation. Moreover, some PGPB secrete antibiotics, which are particularly relevant for rhizosphere and rhizoplane colonization (Glick 2012). Well-known examples include 2,4-diacetylphloroglucinol (DAPG), hydrogen cyanide, oomycin A, phenazine, pyoluteorin, pyrrolnitrin, thiotropocin, tropolone, as well as many others such as cyclic lipopeptides, rhamnolipids, oligomycin A, kanosamine, zwittermicin A, and xanthobaccin. Many PGPR act antagonistically toward plant pathogens by producing antimicrobials or by interfering with virulence factors via effectors delivered by type-3-secretion systems (T3SSs).

1.3 MICROBIAL ECOLOGY OF THE RHIZOSPHERE

The rhizosphere is intriguingly complex and dynamic, and understanding its ecology and evolution is key to enhancing plant productivity and ecosystem functioning. The soil in the rhizosphere supports a typically diverse, and densely populated microbial community includes bacteria, fungi, oomycetes, nematodes, protozoa, algae, viruses, archaea, and arthropods and is subject to chemical transformations caused by the presence of root exudates and metabolites of microbial degradation (Mendes et al. 2013). The actual composition of the microbial community in the rhizosphere is dependent on root type, plant species, plant age, and soil type, as well as other factors such as exposure history of the plant roots to pollutants. Within the rhizomicrobiome, some microorganisms can promote plant growth and provide better plant health through several indirect or direct mechanisms. Rhizosphere organisms that have been well studied for their beneficial effects on plant growth and health are the nitrogen-fixing bacteria and archaea, arbuscular mycorrhizal fungi (AMF), ectomycorrhizal fungi, PGPR, biocontrol microorganisms, mycoparasitic fungi, and protozoa (Ahemad 2014b). Rhizosphere organisms that are deleterious to plant growth and health include the pathogenic bacteria, fungi, oomycetes, nematodes, and microarthropods. The third group of microorganisms that can be found in the rhizosphere is the human pathogens; particularly, the members of the family Enterobacteriaceae can invade the root tissue (Mendes et al. 2013). The microbial communities of the rhizosphere are important driving forces and the main participant in biogeochemical cycles of various life-providing elements in terrestrial ecosystems, and they also promote the absorption and utilization of plants by converting organic nutrients into inorganic nutrients. Microbes are not the only key to global biogeochemical cycles but also essential for the cleanup of polluted environments.

1.3.1 BACTERIA

The concentration of bacteria in the rhizosphere is 10–1,000 times higher than that in bulk soil; it is still 100-fold lower than that in the average laboratory medium (Lugtenberg and Kamilova 2009). The discovery of rhizosphere bacteria that are HMs resistant and able to promote plant growth in highly polluted environments has raised high hopes for reclamation of heavy metal–polluted soil (Sessitsch et al. 2013; Rajkumar and Freitas 2008). Plant growth–promoting effects by rhizosphere bacteria can greatly improve plant performance and also result in higher amounts of accumulated trace elements. Rhizosphere bacterial immobilization may lead to the reduction of HMs uptake in plants, whereas bacterial activity that enhances the mobility of HMs may cause the increase of heavy metal accumulation by plants (Rajkumar and Freitas 2008; Glick 2003).

Kunitoa et al. (2001) compared the characteristics of bacterial communities in the rhizosphere of *Phragmites* with those of nonrhizosphere soil in a highly Cu-contaminated area near a copper mine in Japan. Higher bacterial numbers were detected in the rhizosphere, which may be due to the lower Cu concentrations or due to the availability of root exudates that can serve as a carbon source for bacteria. Nevertheless, the percentage of highly resistant strains was higher in the rhizosphere than in nonrhizosphere soil. Turpeinen et al. (2004) studied the microbial community structure and activity in As-, Cr-, and Cu-contaminated soils. Dominant As-resistant isolates from contaminated soils were identified by their fatty acid methyl ester (FAME) profiles as *Acinetobacter, Edwardsiella, Enterobacter, Pseudomonas, Salmonella*, and *Serratia* species. Similarly, Bennisse et al. (2004) characterized rhizosphere bacterial populations of metallophyte plants growing in Pb-, Zn-, and Cu-contaminated soils from mining areas. There, gram-negative bacteria in the rhizosphere soil were dominant compared with the nonrhizosphere soils. Many reports have shown that gram-negative bacteria are more tolerant of heavy metal stress than gram-positive bacteria. The dominant gram-negative generic groups isolated were *Pseudomonas, Alcaligenes, Flavobacterium, Xanthomonas*, and *Acinetobacter*, whereas gram-positive bacteria were less abundant and were represented mainly in the *Bacillus* genera. Kim et al. (2006) explored the rhizosphere of diesel-contaminated soils planted with alfalfa (*Medicago sativa* L.) and showed that the total microbial activity was highest in diesel-contaminated rhizosphere soils. Furthermore, significantly more hydrocarbon degraders were found in diesel-contaminated rhizosphere soil compared with unplanted and uncontaminated soil. Palmroth et al. (2007) analyzed the functional diversity, extracellular enzymatic activities, and genetic diversity of bacteria microbial communities by denaturing gradient gel electrophoresis (DGGE) during phytoremediation of soil contaminated with weathered hydrocarbons and HMs. In addition, bulk soil and rhizosphere soil from pine and poplar plantations were analyzed separately to determine if the plant rhizosphere impacted hydrocarbon degradation. They found that prevailing microbial communities in the field site were both genetically and metabolically diverse and tree rhizosphere communities had greater hydrocarbon degradation potential than those of bulk soil. Navarro-Noya et al. (2010) studied the bacterial communities associated with the rhizospheres of pioneer plants *Bahia xylopoda* and *Viguiera linearis* were grown on silver mine tailings with a high concentration of HMs. The PCR-DGGE and 16S rRNA gene library analysis of metagenomic DNA showed a moderate bacterial diversity and 12 major phylogenetic groups, including *Proteobacteria, Acidobacteria, Bacteroidetes, Gemmatimonadetes, Chloroflexi, Firmicutes, Verrucomicrobia, Nitrospirae, and Actinobacteria* phyla, and divisions TM7, OP10, and OD1 were recognized in the rhizospheres. Only 25.5% from the phylotypes were common in the rhizosphere libraries, and the most abundant groups were members of the phyla *Acidobacteria* and *Betaproteobacteria* (*Thiobacillus* spp., *Nitrosomonadaceae*). The most abundant groups in bulk soil library were *Acidobacteria* and *Actinobacteria*, and no common phylotypes were shared with the rhizosphere libraries. Many of the clones detected were related with chemolithotrophic and sulfur-oxidizing bacteria, characteristic of an environment with a high concentration of HMs-sulfur complexes, and lacking carbon and organic energy sources.

Borymski et al. (2018) assess the biodiversity of the rhizosphere microbial communities of metal-tolerant plant species *Arabidopsis arenosa, A. halleri, Deschampsia caespitosa*, and *Silene vulgaris* when growing on various HMs–polluted sites. The result demonstrated that different rhizospheres showed distinctive profiles of microbial traits, which also differed significantly from bulk soil, indicating an influence from sampling site as well as plant species. However, total bacterial counts and PCR-DGGE profiles were most affected by the plants, whereas sampling site-connected variability was predominant for the phospholipid fatty acid (PLFA) profiles and interaction of both factors for BIOLOG-CLPP. Hemmat-Jou et al. (2018) were used metagenomics approach to investigate the microbial diversity of soils contaminated with different concentrations of Pb and Zn. The contaminated soils were collected from a Pb and Zn mine. The results indicated that the ten most abundant bacteria in all samples were *Solirubrobacter (Actinobacteria), Geobacter (Proteobacteria), Edaphobacter (Acidobacteria), Pseudomonas (Proteobacteria), Gemmatiomonas (Gemmatimonadetes), Nitrosomonas, Xanthobacter, Sphingomonas (Proteobacteria), Pedobacter (Bacterioidetes)*, and *Ktedonobacter (Chloroflexi)*, descendingly. Archaea were also numerous, and Nitrososphaerales, which are important in the nitrogen cycle, had the highest abundance in the samples. This study provided valuable insights into the microbial composition in heavy metal–contaminated soils. Pacwa-Płociniczak (2018) analyzed the impact of heavy metals and plant rhizodeposition on the structure of indigenous microbial communities in the rhizosphere and bulk soil that had been exposed to heavy metals for more than 150 years. Samples of the rhizosphere of *S. vulgaris* and nonrhizosphere soils 250 and 450 m from the source of emission that had different metal concentrations were collected for analyses. Unweighted pair group method with arithmetic mean cluster analysis of the DGGE profiles, as well as a cluster analysis that was generated on the phospholipid fatty acid profiles, showed that the bacterial community structure of rhizosphere soils depended more on the plant than on the distance and metal concentrations. The sequencing of the 16S rDNA genome revealed the representatives of the phyla *Bacteroidetes, Acidobacteria, Gemmatimonadetes, Actinobacteria,* and *Betaproteobacteria* in the analyzed soil with a predominance of the first three groups. The obtained results demonstrated that the presence of *S. vulgaris* did not affect the number of colony-forming units, except for those of Cd-resistant bacteria. However, the presence of *S. vulgaris* altered the soil bacterial community structure, regardless of the sampling site, which supported the thesis that plants have a higher impact on soil microbial community than metal contamination. A study conducted by Koshlaf et al. (2016) showed a shift in bacterial communities when pea straw was added to the diesel-contaminated soil. The metagenomic analysis indicated that the original soil contained hydrocarbon degraders (e.g., *Pseudoxanthomonas* spp.); however, treatment with the biostimulant (pea straw) made them active and accelerated the process of degradation. DGGE and Illumina 16S metagenomic analyses confirm shifts in bacterial communities compared with original soil after 12 weeks' incubation. In addition, the metagenomic analysis showed that original soil contained hydrocarbon degraders (e.g., *Pseudoxanthomonas* spp. and *Alcanivorax* spp.). Jung et al. (2016) studied the metagenomic and functional analyses of the consequences of reduction of bacterial diversity on soil functions and bioremediation in diesel-contaminated microcosms. A shift from *Proteobacteria-* to *Actinobacteria*-dominant communities was observed when species diversity was reduced. Metagenomic analysis showed that a large proportion of functional gene categories were significantly altered by the reduction in biodiversity. The abundance of genes related to the nitrogen cycle was significantly reduced in the low-diversity community, impairing denitrification. In contrast, the efficiency of diesel biodegradation was increased in the low-diversity community and was further enhanced by addition of red clay as a stimulating agent (Figure 1.4).

Abed et al. (2018) describe the bacterial diversity, using molecular (Illumina MiSeq sequencing) and cultivation techniques, in the rhizosphere soils of *Phragmites australis* from an oil-polluted wetland in Oman. Most sequences belonged to *Proteobacteria, Bacteriodetes, and Firmicutes*. The obtained isolates from the rhizosphere soils were planted growth-promoting properties and phylogenetically affiliated to *Serratia, Acinetobacter, Xenorhabdus, Escherichia*, and *Salmonella*. All strains were able to solubilize phosphate, and about half were capable of producing organic acids, deaminase, IAA, and siderophores. They conclude that the rhizosphere soils of *P. australis* in oil-polluted wetlands harbor diverse bacterial communities that could enhance the wetland performance through hydrocarbon degradation, nutrient cycling, and supporting plant growth. Kumar et al. (2018) used high-throughput sequencing to explore the rhizosphere microbial diversity in the alfalfa and barley-planted oil-contaminated soil samples. The analysis of 16s rRNA sequences showed *Proteobacteria* to be the most enriched (45.9%) followed by *Bacteriodetes* (21.4%) and *Actinobacteria* (10.4%) phyla. The results also indicated differences in the microbial diversity among the oil-contaminated planted soil samples. The oil-contaminated planted soil samples showed a higher richness in the microbial flora when compared with that of untreated samples.

1.3.2 Fungi

The structure of fungal communities in the rhizosphere is the result of complex interactions among selection factors that may favor beneficial or detrimental relationships. Rhizosphere fungi are closely linked to plant health and growth, owing to their roles in antagonizing pathogens, decomposing plant residues, and providing nutrients (Raaijmakers et al. 2009; Ehrmann and Ritz 2014). Antagonists are naturally occurring organisms with traits enabling them to interfere with a pathogen's growth, survival, infection, or plant attack (6). mechanisms responsible for antagonistic

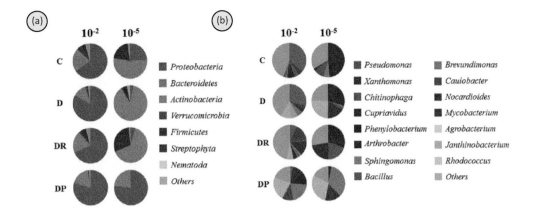

FIGURE 1.4   Analysis of microbial communities using rRNA gene sequences and MG-RAST (Metagenomics Rapid Annotation using Subsystem Technology) grown in diesel contaminated microcosm; taxa with abundance <5% are presented as "others." (a) Phyla, (b) genera. C, control; D, diesel-spiked soil; DR, diesel-spiked soil and red clay; DP, diesel-spiked soil and processed red clay. (Reprinted from Jung et al. 2016 with permission under a Creative Commons Attribution 4.0 (CC BY 4.0) International License http://creativecommons.org/licenses/by/4.0/.)

activity include (i) antibiosis via inhibition of the pathogen by antibiotics, toxins, and surface-active compounds called biosurfactants, (ii) competition for colonization sites, nutrients, and minerals, and (iii) parasitism, which may involve production of extracellular cell wall–degrading enzymes such as chitinase and β-1,3-glucanase.

AMF play major roles in ecosystem functioning such as carbon sequestration, nutrient cycling, and plant growth promotion. Many reports have shown that AMF may play an important role in soil phytoremediation processes (Yang et al. 2016; Cabral et al. 2015). Schneider et al. (2016) evaluate the occurrence and diversity of AMF and plant species as well as their interactions in soil contaminated with Pb from the recycling of automotive batteries. Thirty-nine AMF species from six families and 10 genera were identified. The *Acaulospora* and *Glomus* genera exhibited the highest occurrences both in the bulk (10 and 6) and in the rhizosphere soils (9 and 6). All of the herbaceous species presented mycorrhizal colonization. The highest Pb concentrations (mg/kg) in roots and shoots, respectively, were observed in *Vetiveria zizanoides* (15,433 and 934), *Pteris vitata* (9,343 and 865), *Pteridimaquilinun* (1,433 and 733), and *Ricinus communis* (1,106 and 625). Liang et al. (2015) examined the impact of transgenic high-methionine soybean ZD91 on the arbuscular mycorrhizal (AM) fungal community structure in rhizosphere soil. A total of 155 operational taxonomic units of AMF were identified based on the sequences of small subunit ribosomal RNA (SSU rRNA) genes. There were no significant differences found in AM fungal diversity in rhizosphere soil during the same growth stage between transgenic soybean ZD91 and its nontransgenic parental soybean ZD. In addition, plant growth stage and year had the strongest effect on the AM fungal community structure, whereas the genetically modified (GM) trait studied was the least explanatory factor.

1.3.3   ARCHAEA

The *Archaea* represent a significant component of the plant microbiome, whereas their function is still unclear. They have often been found in the rhizosphere and endosphere but rarely in the phyllosphere, which can be explained by the different abiotic conditions in these microenvironments. The fact that most of *Archaea* are difficult to cultivate and that plant-associated archaeal pathogens are currently not known may be attributed to the lack of knowledge. However, due to their ubiquitous occurrence on healthy plants, we assume that *Archaea* interact positively with plants. Within the rhizosphere, methane-producing archaea (methanogens) are involved with the production of the highly potent greenhouse gas methane. Methanogens are prevalent in rice paddies, wetlands, lake and ocean sediments, and other anoxic and flooded sediments. Archaea also play an important role in the nitrogen cycle as nitrogen fixers, denitrifiers, and ammonia oxidizers. Until the early 21st century, the ammonia-oxidizing capabilities of organisms such as *Nitrosomonas* were considered unique to bacteria. The extent of the role of archaea in the rhizosphere continues to be unknown, as new organisms and their functions are continually being discovered.

1.3.4 PROTOZOA

Protozoa are single-celled animals that feed primarily on bacteria and also eat other protozoa, soluble organic matter, and sometimes fungi. They are several times larger than bacteria, ranging from 1/5000 to 1/50 of an inch (5–500 μm) in diameter. Protozoa can affect plant health by mineralizing nutrients and altering the structure and activity of root-associated communities. Bacteria in the rhizosphere are strongly top-down regulated by grazing abilities of protozoa (Bonkowski et al. 2009). Grazing abilities of protozoa may, for instance, promote the production of plant growth hormones (Krome et al. 2010) or enhance the survival of beneficial microbes suppressing pathogens. The most important bacterial grazers in the soil are naked amoebae due to their high biomass and turnover and specialized feeding modes (Figure 1.2). Amoebae are grazing bacterial biofilms and colonies attached to the soil and root surfaces and thus have access to the majority of bacteria in soil. Protozoa are also an important food source for other soil organisms and help to suppress disease by competing with or feeding on pathogens. Protozoa feed selectively so they influence the bacteria composition and bacteria population in the rhizosphere.

1.3.5 HUMAN PATHOGENS

The occurrence of human pathogenic bacteria in the rhizosphere has been ascribed to several factors, including the high nutritional content, protection from UV radiation, and the availability of water for dispersal and for preventing desiccation (Berg et al. 2005; Tyler and Triplett 2008). Others have argued that the abundant and highly diverse indigenous rhizosphere microbial communities provide a strong barrier against the invasion of human pathogens. Nevertheless, many of the human pathogenic bacteria can be highly competitive for nutrients and produce various antimicrobial metabolites, allowing them to colonize and proliferate on plant surfaces in the presence of the indigenous microbial communities. Various studies have indicated that human pathogenic bacteria enter the root tissue at sites of lateral root emergence. This was shown for *Salmonella* and *Escherichia coli* O157:H7 on roots of *Arabidopsis* and lettuce and for *Klebsiella pneumoniae* on multiple plant species (Tyler and Triplett 2008). Fruits and vegetables infected with soft rot pathogens led to significant increases in populations of *Salmonella* and *E. coli* O157:H7 (Teplitski et al. 2011).

1.4 CATEGORIES OF A PLANT GROWN ON HEAVY METAL–CONTAMINATED SITES

Plants have been categorized according to three basic survival strategies for growth on HMs–contaminated sites: metal excluders, indicators, and accumulators (Figure 1.7; Baker 1981, Ghosh and Singh 2005).

1.4.1 METAL EXCLUDERS

Metal excluders are plants that effectively limit the levels of HMs translocation within them and maintain relatively low levels in their shoots over a wide range of soil- and sludge-contaminant levels, thus protecting the leaf tissues, particularly the metabolically active photosynthetic cells, from heavy metal damage.

1.4.2 METAL INDICATORS

Metal indicators are plants that accumulate HMs in their aboveground tissues, and the metal levels in the tissues of these plants generally reflect metal levels in the soil or sludge. Indicator plants are able to provide indirect or direct information on the impact of pollutants on the environment.

1.4.3 METAL ACCUMULATORS

Metal accumulation can occur in some plant species that grow mainly on metalliferous soils that concentrate metals in their aboveground tissues to levels far exceeding those present in the soil. If these plants continue to take up metals, they will eventually die.

1.5 HYPERACCUMULATOR PLANTS AND THEIR CHARACTERISTICS

Hyperaccumulators are a subgroup of accumulator species often endemic to naturally mineralized soils, which accumulate high concentrations of metals in aboveground tissues, without developing any toxicity symptoms (Baker and Brooks 1989; Raskin et al. 1997). Hyperaccumulator plants utilize different metabolic processes for the mobilization and uptake of metal ions from soils, based on the efficiency of metal translocation to the plant shoots via the symplast and apoplast (xylem), sequestration of metals within cells and tissues, and transformation of accumulated metals into metabolically less harmful forms (Maestri et al. 2010; Chandra et al. 2018a). Brooks et al. (1977) introduced the term "hyperaccumulators" to describe plants capable of accumulating more than 1,000 µg Ni g^{-1} on a dry leaf basis in their natural habitats. Since then, threshold values have also been established for other metals, such as Zn, Pb, Cd, Cu, Cr, Fe, and Mn (Brooks 1998). Usually, there are four main characteristics of hyperaccumulator: accumulation capacity, i.e., the minimum concentration in the shoots of a hyperaccumulator for As, Pb, Cu, Ni, and Co should be greater than 1,000 mg/kg dry mass, and Zn and Mn should be 10,000 mg/kg, Au is 1 mg/kg, and Cd is 100 mg/kg, respectively (Baker 1981); translocation capacity, i.e., elemental concentrations in the shoots of a plant should be higher than those in roots, i.e., TF>1 (translocation factor, concentration ratio of shoots to roots) (Ma et al. 2001; Yoon et al. 2006); bioconcentration (BCF) (concentration ratio of plant root to media), i.e., BCF value in shoots of plants should be higher than 1 (Wei et al. 2005); and tolerance capacity, i.e., a hyperaccumulator should have high tolerance to heavy metal (Baker and Brooks 1989; Maestri et al. 2010). The BCF and TF are calculated by the following equations:

$$BCF = \frac{\text{Metal concentration in plant root}}{\text{Metal concentration in soil / sludge}}$$

$$TF = \frac{\text{Metal concentration in plant shoot}}{\text{Metal concentration in plant root}}$$

Plants with both BCF and TF greater than one have the potential to be used in phytoextraction (Yoon et al. 2006). Besides, plants with BCF greater than one and TF less than one have the potential for phytostabilization (Yoon et al. 2006; Chandra et al. 2018b).

1.6 PHYTOEXTRACTION OF HEAVY METALS BY POTENTIAL NATIVE PLANTS FROM DISTILLERY WASTE AND OTHER COMPLEX ORGANOMETALLIC WASTE

Distillery industry occupies a place of prominence in the Indian economy, but it contributes significantly toward the contamination of the ecosystem. It is one of the highly polluting industries due to discharge of the huge amount of sludge as by-product during anaerobic digestion of spent wash

FIGURE 1.5   View of the in situ phytoremediation of distillery sludge dumping site. (a) Distillery sludge contained uncultured microbial communities dumped in open environment after methanogenesis of spent wash. (b) Initiation of plant growth at distillery sludge dumping site after microbial bioremediation at primary stage. (c) Some native weed plants grown luxuriantly at distillery sludge dumping site resulted in situ phytoremediation of toxic waste. (d) Roots of the plant growing on distillery sludge dumping site. (e) Ultramicroscopic view of the plant root showing the accumulation of heavy metals.

(Figure 1.5a; Chandra and Kumar 2017b; Chandra et al. 2018b). Sludge consists of a mixture of toxic organic compounds (organic matter), inorganic compounds (i.e. Na^+, Cl^-, SO_2^{-4}, PO_3^{-4}), and high content of toxic HMs, i.e., Cd^{2+}, Cu^{2+}, Mn^{2+}, Fe^{2+}, Pb^{2+}, Ni^{2+} and Zn^{2+}, and various androgenic and mutagenic compounds, viz., dodecanoic acid, octadecanoic acid, n-pentadecanoic acid, hexadecanoic acid, β-sitosterol, stigmasterol, β-sitosterol trimethyl ether, heptacosane, dotriacontane, lanosta-8, 24-dien-3-one,1-methylene-3-methyl butanol, and 1-phenyl-1-propanol as androgenic and mutagenic compounds (Chandra and Kumar 2017b; Chandra et al. 2018d). The organic pollutants bind with various HMs to make an organo-metallic complex that resulted in enhancing the vulnerability of organometallic complex toward its toxicity in the environment. The physicochemical characteristics and organic compounds detected and identified by by gas chromatography-mass spectrometry (GC-MS) analysis present in distillery sludge are presented in Tables 1.2–1.4. The dark color of distillery waste is generally attributed to the existence of naturally polymeric colorants such as melanoidins, plant polyphenols, ADPH, and caramels. Among them, melanoidins are major nitrogenous imparting organic compounds present in distillery waste. Melanoidins have antioxidant properties, which render them toxic to microorganisms and recalcitrance to biological wastewater treatments. Melanoidins with molecular weight distribution between 5 and 40 kDa consisting of polymeric, acidic, negatively charged, and highly dispersed colloids. It has been reported that due to their net negative charges various HMs ion such as Cu^{2+}, Cr^{3+}, Fe^{3+}, Zn^{2+}, Mn^{2+}, Co^{2+}, and Pb^{2+} bind with melanoidins to form an organometallic complex (Hatano et al. 2016; Migo et al. 1997). Consequently, the high HM ions binding tendency of melanoidins also enhances the vulnerability of organometallic complex toward its toxicity in the environment (Chandra et al. 2018a,b). Discharge of colored waste into open land or nearby water bodies results in a number of environmental water and soil pollution problem, including a threat to plant and animal lines (Figure 1.5a). Due to their highly toxic nature, no plants are able to grow this hazardous waste. But, the growth of microbial communities in nutrient-deficient anaerobically digested distillery sludge indicated the mineralizing capability of organometallic compounds

TABLE 1.2

Physicochemical Characteristics of Distillery Sludge Discharge after Biomethanation of Spent Wash (Chandra et al. 2018d)

S. No.	Parameter	Values
1.	pH	8.1 ± 0.00
2.	Electrical conductivity (µS/cm)	4.12 ± 0.01
3.	Sodium (mg/kg)	42.13 ± 1.00
4.	Chloride (mg/kg)	1272.74 ± 5.13
5.	Nitrate (mg/kg)	85.89 ± 0.76
6.	Phosphate (mg/kg)	2268.83 ± 1.70
7.	Sulfate (mg/kg)	145.07 ± 0.68
8.	Phenol (mg/kg)	501.34 ± 1.22
9.	Trace elements (mg/kg)	
a	Iron (Fe)	5264.49 ± 59.64
b	Zn (zinc)	43.47 ± 1.31
c	Cu (copper)	847.46 ± 1.00
e	Mn (manganese)	238.47 ± 0.83
f	Ni (nickle)	15.60 ± 0.54
g	Pb (lead)	31.22 ± 1.14

All values are mean three replicate ($n = 3$).

for their energy harvest. In a recent study, Chandra and Kumar (2017b) used restriction fragment length polymorphism approach to explore the microbial communities composition and function during in situ bioremediation of distillery sludge. The result indicated that *Bacillus* sp. and *Enterococcus* were found dominantly growing autochthonous bacterial communities during in situ bioremediation of distillery sludge due to the availability of diverse habitats and metabolization capabilities. Hence, this ability creates a specific niche of these grown bacterial communities that may lead to in situ phytoremediation of organic and inorganic compounds (Figure 1.5).

Phytoextraction is the most recognized, cost-effective, and applied phytoremediation technique for the removal of toxic metals from contaminated environments (Salt et al. 1995; Salt et al. 1998; Alkorta et al. 2004; Chandra and Kumar 2018). There are two main phytoextraction strategies proposed to clean up toxic HMs from soil or water. The first phytoextraction approach is the use of metal hyperaccumulator species, but their low annual biomass production tends to limit their phytoextraction ability (Baker and Brooks 1989; Baker et al. 1994; Raskin et al. 1997). The other possible alternative is the use of nonaccumulator plants, either high biomass crop plants such as Indian mustard (*Brassica juncea*), sunflower (*Helianthus annuus*), and maize (*Zea mays*) (Szabó and Fodor 2006; Cui et al. 2004; Turgut et al. 2004) or fast-growing high biomass trees such as willows (*Salix* spp.) and poplars (*Populus* spp.) that can be easily cultivated using established agronomic practices (Liphadzi et al. 2003; Ghosh and Singh 2005; Chandra et al. 2018a). In recent years, native plants involved in the uptake of elements can provide the most reliable method for the restoration of contaminated sites (Chandra and Kumar 2017a; Chandra et al. 2018b,c). Native plants can also become weeds when characteristics within their natural habitat change and enable them to better compete with other species and increase their population size and/or density. In comparison with other native plants, weed species are usually quick growers and have higher biomass production that show better tolerance to various adverse environments, including heavy metal stress. In most contaminated sites, hardy, tolerant weed species exist, and phytoremediation through these and other nonedible species can restrict the contaminant(s) from being introduced into the food chain. Weeds with a fibrous root system such as grasses are preferred for phytoremediation due to

TABLE 1.3

Identified EDCs and Other Organic Pollutants by GC-MS Analysis Present in (a) Ethyl Acetate and (b) n-Hexane Extract of Distillery Sludge (Chandra et al. 2018d)

(a)

S. No.	RT	Identified Compounds	S. No.	RT	Identified Compounds
1.	7.42	Silane, (4-ethylphenyl)trimethyl	16.	34.84	Docosanoic acid, TMS ester
2.	7.54	Benzene, 1-ethyl-2-methyl	17.	35.83	2-Monostearin TMS ether
3.	8.62	2,3-D-2-methylsuccinic acid 2TMS	18.	36.14	Octadecanoic acid, 2,3-bis[(TMS)oxy]propyl ester
4.	13.72	Ethanedioic acid, bis(TMS)ester	19.	37.11	Dotriacontane
5.	12.85	β-Eudesmol, TMS ether	20.	38.68	Hexacosanoic acid
6.	15.45	Benzoic acid, 2-methyl-, trimethylsilyl ester	21.	40.12	Silane, [[(3β, 22E)-ergosta-7,22-dien-3-yl]trimethyl
7.	18.48	Phenol, 2,4-bis(1,1-dimethylethyl)	22.	40.19	Octacosanol
8.	19.31	Phenol, 2, 6-bis(1,1-dimethylethyl)	23.	40.71	Stigmasterol TMS ether
9.	21.88	Tetradecanoic acid, TMS ester	24.	40.96	5α-Cholestane, 4-methylene
11.	24.74	9,12-Octadecadienoic acid (Z, Z)- TMS ester	25.	41.57	24-Ethyl-δ-(22)-coprostenol, TMS
10.	28.08	Benzoic acid, 3,4,5 tris(TMS oxy)-TMS ester	26.	41.98	Ergosten-3β-ol
12	30.90	11-trans-Octadecenoic acid, TMS ester	27.	42.57	Campesterol TMS
13.	31.87	cis-10-Nonadecenoic acid, TMS ester	28.	43.28	β-Sitosterol
14.	34.21	Hexadecanoic acid, 2-hydroxy-1-(hydroxymethyl)ethyl ester	29.	44.73	Lanosterol
15.	34.52	Hexadecanoic acid, 2,3-bis[(TMS)oxy]propyl ester	30.	45.37	9,19-Cyclocholestan-3-one, 4,14-dimethyl

(b)

S. No.	RT	Identified Compounds	S. No.	RT	Identified Compounds
1.	7.34	Acetic acid, [(TMS)oxy], TMS ester	11.	30.82	Octadecanoic acid, TMS ester
2.	9.07	Benzene, 1-ethyl-4-methyl	12.	31.34	1H-Purin-6-amine, [(2-fluorophenyl)methyl]
3.	13.71	2-Butenedioic acid, bis (TMS)ester	13.	32.56	Hexanedioic acid, dioctyl ester
4.	14.64	Docosane	14.	33.61	Tentradecanoic acid, TMS ester
5.	16.40	n-Pentadioic acid, bis(TMS) ester	15.	34.50	Hexadecanoic acid, 2,3-bis[(TMS)oxy]propyl ester
6.	16.85	Decanedioc acid	16.	35.80	2-Monostearin TMS ether
7.	19.11	Ethanol, 2(octadecyoxy)	17.	36.11	Octadecanoic acid, 2,3-bis[(TMS)oxy]propyl ester
8.	21.88	Benzoic acid, 3,4,5-tris(TMS oxy), TMS ester	18.	39.17	9,12-Octadecadienoic acid (Z, Z)-2,3-bis[(TMS)oxy]propyl ester
9.	27.97	Hexadecanoic acid, TMS ester	19.	44.37	Glycocholic acid methyl ester TMS
10.	29.34	Quercetin 7,3', 4'-trimethoxy			

TMS, trimethylsilyl; RT, retention time (min); EDC, endocrine-disrupting chemicals; GC-MS, gas chromatography-mass spectrometry.

TABLE 1.4

Identified Organic Compounds by GC-MS Analysis Present in n-hexane Extract of Distillery Sludge (Chandra and Kumar 2017b)

S.No.	RT	Name of Compound	S.No.	RT	Name of Compound
1.	8.15	2-Methyl-4-keto pentan-2-OL	14.	21.62	Heptacosane
2.	8.16	D-Lactic acid, TMS esther, TMS ester	15.	22.13	Dotriacontane
3.	11.13	1-Methylene-3-methyl-butanol	16.	23.01	Tert-hexadecanethiol
4.	12.46	Benzene, 1,3-bis(1,1-dimethylethyl)	17.	24.24	Tetradecanoic acid
5.	12.86	Phosphoric acid	18.	25.28	n-Pentadecanoic acid
6.	13.74	2-Isoropyl-5-methyl-1-heptanol	19.	27.15	Hexadecanoic acid, TMS ester
7.	14.84	1-Phenyl 1-propanol	20.	30.34	Octadecanoic acid
8.	15.70	Tetradecane	21.	35.91	2,6,10,14,18,22-Tetracosahexane 2,6,10,18,19,23 hexamethyl
9.	17.23	Decane, 2,3,5,8 tetramethyl	22.	41.33	Stigmasta-5,22-dien-3-ol(3β, 22E)
10.	17.50	Propanoic acid	23.	41.55	Stigmasterol
11.	19.12	1-Dodecanol	24.	42.14	Lanosta-8, 24 dien-3-one
12.	19.73	Docosane	25.	42.40	Spirostan-3-one (5α, 20β, 25R)
13.	20.73	Dodecanoic acid	26.	42.66	β-Sitosterol trimethyl ether

TMS, trimethylsilyl; RT, retention time (min); GC-MS: Gas Chromatography-Mass Spectrometry

their large root surface area, which can help establish active microbial activity and populations. Several plants growing in polluted soil and water host different types of rhizosphere bacteria able to degrade organic pollutants and have been isolated, and degradation pathways and genes involved in organic pollutants degradation have been identified (Fatima et al. 2015). Even though these bacteria showed high potential to degrade different persistent organic pollutants (POPs), these are unable to survive and proliferate in the contaminated soils. The luxuriant growth of native plants of stabilized distillery sludge indicates the capabilities of their rhizospheric bacterial communities, which might be detoxifying the complex organic compounds of distillery sludge during in situ phytoremediation. It is important to use the native plants of contaminated sites for phytoremediation because these plants are naturally adapted in terms of survival, growth, and reproduction under the environmental stresses compared with plants introduced from another environment. Several potential wetland plants that grow on distillery waste–contaminated sites under natural conditions have indicated the phytoremediation potential of the use of such plants at contaminated sites as natural hyperaccumulators of HMs from complex organic wastes (Figure 1.5d,e; Chandra and Kumar 2017a, Chandra et al. 2018d). Chandra and Kumar (2017a) and Chandra et al. (2018d) evaluated the phytoextraction potential of native plants growing on stabilized distillery sludge *Argemone mexicana, Saccharum munja, Cynodon dactylon, Pennisetum purpureum, Chenopodium album, Rumex dentatus, Tinospora cordifolia, Calotropis procera, Dhatura stramonium, Achyranthes* sp., *Kalanchoe pinnata, Trichosanthes dioica (parval), Parthenium hysterophorous, Cannabis sativa, Amaranthus spinosus* L., *Croton bonplandianum, Solanum nigrum, Sacchrum munja, Basella alba, Setaria viridis, Chenopodium, Ricinus communis,* and *Blumea lacera* revealed the high accumulation of Fe, Zn, Cu, Mn, Ni, and Pb. HMs accumulation pattern of native plants growing on organic pollutants–rich stabilized distillery sludge showed variable pattern of accumulation and distribution in different parts of plants as shown in Figure 1.6. Furthermore, the BCF and TF were

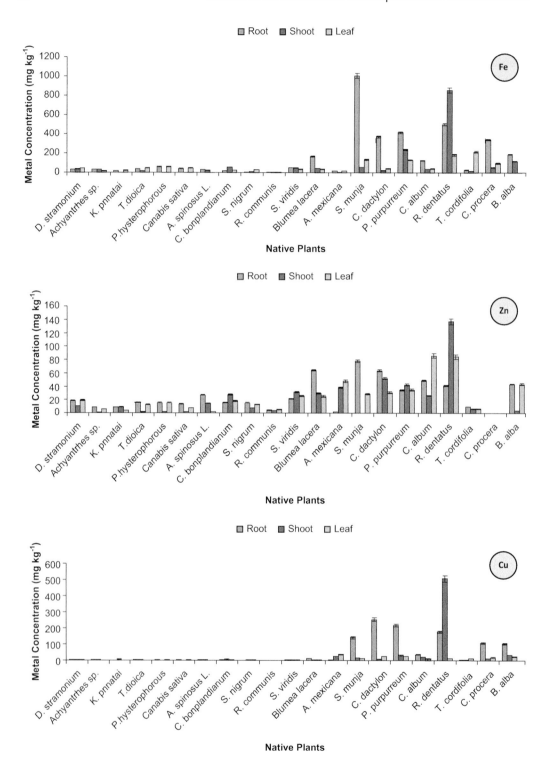

FIGURE 1.6 Accumulation of heavy metals in different parts of native plants collected from distillery sludge dumping site. (Reprinted from Chandra and Kumar 2017a and Chandra et al. 2018d with permission from Springer Nature and Elsevier V.B., respectively.)

(Continued)

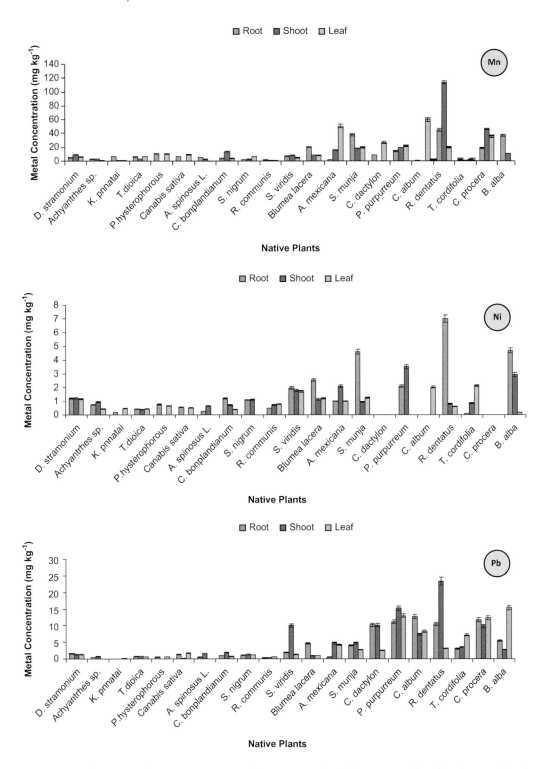

FIGURE 1.6 (CONTINUED)   Accumulation of heavy metals in different parts of native plants collected from distillery sludge dumping site. (Reprinted from Chandra and Kumar 2017a and Chandra et al. 2018d with permission from Springer Nature and Elsevier V.B., respectively.)

(Continued)

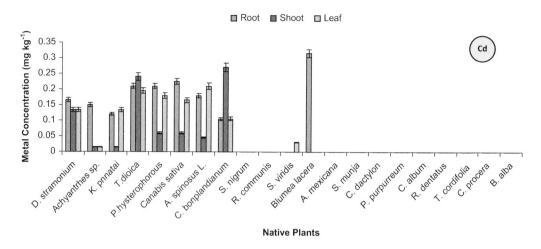

FIGURE 1.6 (CONTINUED)   Accumulation of heavy metals in different parts of native plants collected from distillery sludge dumping site. (Reprinted from Chandra and Kumar 2017a and Chandra et al. 2018d with permission from Springer Nature and Elsevier V.B., respectively.)

found >1 for the majority of plants for various metals. Thus, this gives strong evidence for hyperaccumulation tendency of these native weeds and grasses from complex polluted sites. The discharged sludge may contain very diverse organic pollutants that are potentially harmful to human health and wildlife animals. In addition, organic pollutants are toxic to plants and microorganisms; the presence of organic pollutants in soil and water decreases plant growth and its phytoremediation efficacy. However, some organic pollutants can be directly degraded and mineralized by plant enzymes through phytodegradation; many plants produce, and often secrete into the environment, enzymes that can be degraded to a wide range of organic compounds. Plants can accumulate organic pollutants from contaminated sites and detoxify them through their metabolic activities. From this point of view, green plants can be regarded as a "green liver" for the biosphere (Sandermann 1992). Phytoremediation of organic pollutants may also occur by phytostabilization (the use of certain plants to reduce the mobility and bioavailability of pollutants in the environment, thus preventing their migration to groundwater or their entry into the food chain) or by phytostimulation (the stimulation of microbial degradation of pollutants in the plant rhizosphere, sometimes called rhizodegradation). Moreover, plants have certain limits with respect to their capabilities to remove organic pollutants from the environment.

1.7 HEAVY METAL ACCUMULATION AND DETOXIFICATION MECHANISM OF THE PLANT DURING IN SITU PHYTOREMEDIATION OF DISTILLERY WASTE

HMs hyperaccumulation in plants is a multistep process that includes mobilization form soil into the soil solution, transport of the HMs across the plasma membrane of root cells, xylem loading and root-to-shoot translocation, detoxification, and sequestration of HMs in the leaf tissue (Figure 1.3). Most of the research on the mechanism of root and plant cell uptake has focused on the N, P, S, Fe, Ca, K, and Cl. However, little is known about the mechanisms of mobilization, uptake, and transport of HMs into plants in the presence of organic pollutants. It is clear that for a large proportion of these metals when soil bound, phytoextracting plants mobilize the metals from soil solutions and accumulate them in their harvestable parts (Raskin et al. 1997).

1.7.1 Metal Phytoavailability in the Rhizosphere

Distillery sludge generated after methanogenesis of spent wash is the most common spots that harbor unique types of microorganisms, which are capable of running extensive bioremediation activities (Chandra and Kumar 2017b). The huge range on carbon substrates present in sludge could also facilitate the development of complex bacterial communities. These microbes have been hardly analyzed mainly because the condition of this sludge is mixed with high physical and chemical heterogeneity (Chandra et al. 2018d). In another study, Chandra and Kumar (2017b) also assessed the microbial community composition and function during in situ bioremediation of distillery sludge. The stated that *Bacillus* sp. and *Enterococcus* sp. were found dominantly growing autochthonous bacterial communities during in situ bioremediation of distillery sludge due to the availability of diverse habitats and metabolization capabilities. However, there is no definite answer as to whether and how hyperaccumulator and nonhyperaccumulator plants, or their rhizospheric microbial communities, have different effects on metal availability in their rhizosphere. Several plants are known to possess highly specialized mechanisms to stimulate metal bioavailability in the rhizosphere for metal uptake. Any phytoremediation process starts at the soil-plant interface in the rhizosphere, and hence the process of heavy metal transformation occurs mainly in this region. A major proportion HMs in distillery sludge exist as the bound fraction with organic compounds at alkaline pH and need to be mobilized into the sludge solution to be made available for plant uptake. Phytoavailability of metals in soil/sludge is the first step for successful phytoextraction. Bioavailability refers to the fraction of a chemical that can be taken up or transformed by living organisms from the surrounding bioinfluenced zone where organism-mediated biochemical changes occur. pH plays an important role in the availability of metals to plantbecause of the alkaline pH of soil- or sludge-restricted metal mobility for accumulation in plants. The metal mobilization and bioavailability were higher reported in an acidic environment, which favors the phytoextraction process of the plant. With the exception of Fe, little is known about the active mobilization of trace elements by plant roots. Acidification of the rhizosphere, exudation of carboxylates, and mechanisms assisting in the acquisition of phosphorus contribute to increasing the bioavailability of certain micronutrients. HMs mobility in soils also depends on many factors, including pH, redox potential, cation exchange capacity, the presence of organic matter and the clay component, and biological activity (Yang et al. 2005). Therefore, plant roots and soil microbes and their interactions can improve metal phytoavailability in the rhizosphere through secretion of root exudates, siderophores, protons, amino acids, and enzymes (Yang et al. 2005). Plant root exudates contain not only organic acids (e.g., lactate, acetate, oxalate, glutamate, salicylic acid, succinate, fumarate, malate, and citrate), sugars, amino acids, and enzymes (e.g., proteins, lectins, proteases, acid phosphatases, peroxidases, hydrolases, and lipases) as the main components but also secondary metabolites (e.g., isoprenoids, alkaloids, and flavonoids), which are released to soil as rhizodeposits (Wenzel et al. 2003). Plant root activities that potentially increase metal/metalloid solubility and may change speciation include acidification/alkalinization, modification of the redox potential, and exudation of metal chelates and organic ligands (in particular low molecular organic acids and phytosiderophores) that compete with anionic species (e.g., arsenate) for binding sites. Moreover, microorganisms may either increase or decrease metal/metalloid solubility. Microorganisms can increase solubility and change speciation of metals/metalloids through the production of organic ligands via microbial decomposition of soil organic matter and exudation of metabolites (e.g., organic acids, comprising gluconate, 2-ketogluconate, oxalate, citrate, acetate, malate, succinate, etc.). An example is the release of 5-ketogluconic acid and microbial siderophores that can complex cationic metals or desorb anionic species (e.g., arsenate) by ligand exchange. Depending on the surface charge of soil minerals and below metal-specific pH values, siderophores produced by microbes (and plants) may also immobilizecationic metals such as cadmium, copper, or zinc. Microorganisms can produce and secrete an array of organic acids, such as gluconic acid, 2-ketogluconic acid, lactic acid, and acetic acid. The associated decrease in soil pH can also increase the solubility of some heavy metals. The biosurfactants produced by PGPB have also been demonstrated to enhance heavy metal mobilization in contaminated soils. The mechanisms enabling bacteria to promote the accumulation process are unclear.

The use of plant-associated bacteria in phytostabilization strategies not only may assist plant growth and tolerance to metals but can also reduce the metal uptake or translocation to aerial parts of plants by decreasing the metal bioavailability in the rooting medium. For survival under metal-stressed environment, plant-associated bacteria have evolved several mechanisms by which they can immobilize or transform metals rendering them inactive to tolerate the uptake of HMs ions. The mechanisms that are generally proposed for HMs resistance in bacteria are (i) exclusion of HMs by a permeability barrier or by active export of metal from the cell; (ii) intracellular physical sequestration of metal by binding extracellular polymers or extracellular sequestration; (iii) detoxification of heavy metals where metal is chemically modified to render it less active (1995). For instance, binding of metals to anionic functional groups (i.e., sulfhydryl, carboxyl, hydroxyl, sulfonate, amine, and amide groups) immobilizes the metal and prevents its entry into the plant root. Similarly, the metal-binding extracellular polymers comprising polysaccharides, proteins, humic substances, etc. may detoxify metals by chelating the HMs. The bacterial siderophores and organic acids can also reduce the metal bioavailability and toxicity by chelating the metal ions. Although the establishment of a successful vegetative cover on metal-contaminated soils is challenging, the beneficial bacteria immobilizing heavy metals and enhancing the plant tolerance to high metal concentrations and/or promoting plant growth could provide a practical tool for speeding up the phytostabilization process. In addition, certain PGPB have been shown to increase HMs mobilization by the secretion of low-molecular-mass organic acids comprising gluconate, 2-ketogluconate, oxalate, citrate, acetate, malate, and succinate. An example is the release of 5-ketogluconic acid by endophytic diazotroph *Gluconacetobacter diazotrophicus*, which dissolves various Zn sources, thus making Zn available for plant uptake (Saravanan et al. 2007). Although it is well accepted that organic acids produced by PGPB play an important role in the mobilization of HMs and mineral nutrients, the inoculation effects of organic acids producing bacteria on plant growth and metal accumulation in plants are still poorly understood. PGPB produce metal-chelating agents termed siderophores, which are able to bind metals and thus enhance their bioavailability in the rhizosphere through a complexation reaction. Biosurfactants are additional important metabolites produced by PGPB that have the potential to improve metal mobilization and phytoremediation. The biosurfactants produced by microbes form complexes with HMs at the soil interface, desorbing metals from the soil matrix, and thus increasing metal solubility and bioavailability (Rajkumar et al. 2012). The various mechanisms how microorganisms may interact with plants in relation to trace elements accumulation or resistance are summarized in Figure 1.7.

1.7.2 TRANSPORT OF HEAVY METALS ACROSS PLASMA MEMBRANES OF ROOT CELLS

Access of HMs to bare roots is confined to the first few millimeters of the root tip. Uptake and transport across root cellular membrane is an important process, which initiates metal absorption into plant tissues. Two different uptake routes have been reported: (i) passive uptake (apoplastic) driven only by the concentration gradient across the membrane and (ii) active uptake (symplastic) inducible substrate-specific and energy-dependent uptake mediated by membrane protein with transport functions. Either through passive or active uptake, root cells capture metals from soil that remain bound by their cell wall and then transported across the membrane. But the electrical charge on metal ions prevents their diffusion freely across the lipophilic cellular membranes into the cytosol. Therefore, metal transport into cells is also driven by ATP-dependent protein pumps that catalize H+ extrusion across the membrane. In the apoplastic pathway, metal ions or metal-chelate complexes enter roots through intercellular spaces. In contrast, the symplastic pathway is an energy-dependent process that is mediated by specific or generic metal ion carriers or channels. In the symplastic pathway, nonessential metal ions compete for the transmembrane carrier used by

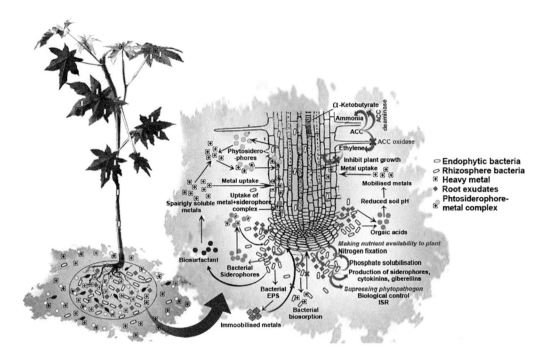

FIGURE 1.7   Schematic portrayal of the role of plant growth–promoting rhizobacteria bacteria in alleviation of heavy metal toxicity, phytoextraction, and phytostabilization.

essential trace elements. For example, Ni and Cd compete for the transmembrane carrier used by Cu and Zn. Even metal chelates, such as Fe-phytosiderophore complexes, can be transported by the symplastic pathway. Transporter proteins and an intracellular high-affinity binding site mediate the uptake of metals across the root cell plasma membrane. Several classes of protein families, including the cation diffusion facilitator (CDF), zinc-iron permease (ZIP), CPx-type ATPases, and Nramp familes, have been implicated in HMs transport in plants (Yang et al. 2005).

1.7.3   ROOT-TO-SHOOT TRANSLOCATION OF HEAVY METALS

Unlike nonaccumulator plants, which retain most HMs taken up from a contaminated site in their root cells, detoxifying the metals via chelation in the cytoplasm or storing them in vacuoles, hyperaccumulators efficiently translocate metal ions from the roots to the shoots via the xylem. Efficient translocation of metal ions to the shoot requires radial symplastic passage and active loading into the xylem, for which metal ions first have to cross the Casparian band on the endodermal layer, which is a water-impervious barrier that blocks the apoplastic flux of metal ions from the root cortex to the stele. To cross this barrier and to reach the xylem, metals must move symplastically; this is a rate-limiting step in metal translocation from roots to shoots. Xylem loading is a tightly regulated process mediated by membrane proteins such as the P-type ATPase-HMA (heavy metal–transporting ATPase), MATE, and oligopeptide transporter proteins. Once inside the plant, further movement of metal-containing sap from roots to the aerial parts is controlled by root pressure as well as the transpirational pull (Robinson et al. 2003). Metal transport to the shoot primarily takes place through the xylem. Efficient translocation of metal ions to the shoot requires radial symplastic passage and active loading into the xylem. Due to extreme toxicity of metals at high intracellular concentrations, plants catalyze redox reactions and alter the chemistry of these metal ions, thereby allowing their accumulation in nontoxic forms.

1.7.4 Detoxification and Sequestration of Heavy Metals in Aerial Parts of Plants

Great efficiency in detoxification and sequestration are key properties of hyperaccumulators, as these properties allow them to concentrate huge amounts of HMs in aboveground organs without suffering any phytotoxic effect. HMs detoxification and sequestration occurs in a location such as the epidermis, trichome, and even the cuticle (Rascio and Navari-Izzo 2011). A general mechanism for HMs detoxification in plants is the distribution of metals to the trichome and cell wall, with chelation of metals in complex with ligands, followed by the sequestration of the metal-ligand complexes into vacuoles. Small ligands such as organic acids, like malate and citrate, play major roles as detoxifying factors. Such ligands may be instrumental in preventing the persistence of HMs as free ions in the cytoplasm and even moreso in enabling their entrapment in vacuoles, where metal-organic acid chelates primarily become localized. The formation of metal-organic acid complexes is favored in the acidic environment of the vacuole. However, intracellular complexation involves peptide ligands, such as metallothioneins (MTs) and phytochelatins (PCs). MTs are low-molecular-weight proteins that are characterized by their high cysteine content and thus often give rise to metal-thiolate clusters. Most MTs have two metal clusters, one containing three and one containing four bivalent metal ions. MTs have been reported to play roles in other cellular processes, including regulation of cell growth and proliferation, DNA damage repair, and scavenging of reactive oxygen species. Although MTs are expressed throughout a plant, some have been found to be expressed in a tissue-specific manner. PCs form a family of peptides that consist of repetitions of a γ-Glu-Cys dipeptide followed by a terminal Gly residue. PCs are synthesized enzymatically from glutathione in the presence of certain heavy metals and metalloids, and they are ubiquitous in plants. In hyperaccumulators, PCs are mainly induced in the roots, in particular, by Cd, but not by Zn or Ni. Furthermore, the PC-metal complexes are pumped into vacuoles. Vacuoles are generally considered the main storage site for heavy metals in plant cells, and they often occupy between 60% and 95% of the cell volume in mature parenchymatous and epidermal cells. Several families of transporter proteins, such as the CDF, HMA, ZAT, NRAMP, CAX, MHX, and ABC families, are involved in sequestration of HMs in vacuoles.

The ultrastructural observation of root tissues of *P. purpureum* and *C. album* at low and high magnification showed deposition of metal granules inside the cell wall, cell membrane, cytoplasm, and nucleoplasm as shown in Figure 1.8. All the growing plants have been found for hyperaccumulation properties of HMs from organometallic-polluted site mixed with androgenic and mutagenic compounds. Native hyperaccumulator plants (weeds and grasses) such as *Dhatura stramonium, Achyranthes* sp., *Kalanchoe pinnata, Trichosanthes dioica, Parthenium hysterophorous, Cannabis sativa, Amaranthus spinosus* L., *Croton bonplandianum, Solanum nigrum, Ricinus communis, Setaria viridis, Blumea lacera, Argemone mexicana, Saccharum munja, Cynodon dactylon, Pennisetum purpureum, Chenopodium album, Rumex dentatus, Tinospora cordifolia, Calotropis procera*, and *Basella alba* have been reported for the potential growth on the organic pollutant–rich disposed distillery sludge during in situ phytoremediation (Chandra et al. 2018d; Chandra and Kumar 2017a). All the tested HMs have been accumulated in significant amount in their root, shoot, and leaves. The accumulation of HMs varied greatly among plant species, and uptake of heavy metals by any other plant species depends on the metal availability, pH of substrate, and chemical nature of other organic copollutants. The TEM observation in root tissues of all plant species showed an apparent formation of multinucleus, multinucleolus, multivacuoles, mitochondria, and dense deposition of metal granules in the cellular organelle of the plant, which supported the plant tolerance mechanism at higher concentration of HMs in the presence of other complex organic pollutants. HMs depositions near the cell wall play a significant role in HMs tolerance by preventing the circulation of free metals ions in the cytosol. The production of the greater number of nucleolus and vacuoles at an elevated level of HMs increases the production of ribosome and mRNA, which ultimately enhances the production of new proteins being involved in the HMs tolerance in the plant. All the plant species were luxuriantly growing

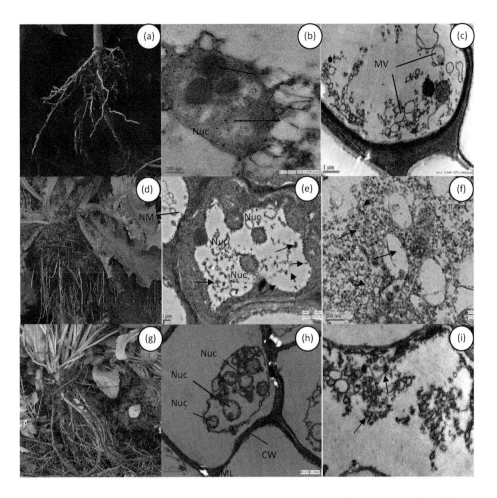

FIGURE 1.8 Electron micrographs of transverse section of plant roots after phytoextraction of heavy metals: (a–c) *Solanum nigrum*, (d–f) *Ricinus communis*, (h–i) *Saccharum munja*. V, vacuole; PM, plasma membrane; P, peroxisome; PL, plasmolysis; CW: cell wall; CM, cell membrane; N, nucleus; Nuc: nucleolus; IS: intercellular space; arrow (→) indicated heavy metal depositions. (Reprinted from Chandra and Kumar 2017a and Chandra et al. 2018d with permission from Springer Nature and Elsevier V.B. respectively.)

on disposed distillery sludge without showing any morphological deformities in their aerial parts. This indicated that these plant species have strong potential and adaptation for survival and growth on organometallic and EDCs containing distillery waste for in situ phytoremediation. These growing plants have also given strong evidence for HMs monitoring and phytoremediation of hazardous complex industrial waste for ecorestoration of polluted sites. Thus, the study recommended that these plants can be used for in situ monitoring and phytoextraction of HMs from organometallic waste–contaminated sites.

1.8 USE OF TRANSGENIC PLANTS FOR HEAVY METAL ACCUMULATION

An ideal plant for environmental cleanup can be envisioned as one with high biomass production, combined with superior capacity for pollutant tolerance, accumulation, and/or degradation, depending on the type of pollutant and the phytoremediation technology of choice. With the use of genetic engineering, it is feasible to manipulate a plant's capacity to tolerate, accumulate, and/or metabolize pollutants and thus to create the ideal plant for environmental cleanup. Many of the genes involved in metal uptake, translocation, and sequestration have been identified by using the model plant *Arabidopsis* or naturally hyperaccumulating plants. However, the phytoremediation capacities of these natural hyperaccumulators are limited by their small size, slow growth rates, and limited growth habitats. Therefore, if the genes were transferred to plant species such as poplar and willow, with their high biomass and extensive root systems, significant removal of the heavy metals could be achieved. Genetic engineering of plants for enhanced phytoremediation has obvious environmental benefits, yet some would see potential risks (Linacre et al. 2003). This is especially true when using genetically altered trees. Their long life cycle makes risk assessment more challenging, and so more specific research is needed in this realm. In a commentary on this topic, Linacre et al. (2003) described a risk assessment scenario for enhanced metal remediation. They stated that the risk of contamination of food with an engineered metal hyperaccumulator, for example, is low because plants used for phytoextraction would be in isolated, industrial-type areas, not in agricultural areas. Furthermore, crops used for phytoextraction would be harvested before the seed set, thus reducing the threat of crossing with other crops intended for food or entering the food supply. Plants engineered to hyperaccumulate toxic metals in foliage could be harmful to wildlife; however, studies have demonstrated that such foliage is not appealing in taste and is avoided. The best way to determine the ecological impact of transgenic plants for phytoremediation is by conducting field trials designed to assess risks (Linacre et al. 2003).

1.9 CONCLUSION

The ecorestoration of hazardous distillery waste–contaminated site presents not only a challenge for the country but also a threat to the scientific society. Phytoextraction as is a polishing green approach that whereby the action of plants and the microorganisms specifically associated with them is optimized to sequester, degrade, transform, assimilate, metabolize, or detoxify hazardous pollutants from the distillery sludge as well as contaminated site. Certain indirect attenuation mechanisms are involved in phytoremediation, such as the metabolism of contaminants by plant-associated microbes, and plant-induced changes in the contaminated environment. Microbes may reduce the toxicity of metals or increase their bioavailability and thus have some potential to improve phytoextraction efficiency. Different mechanisms have been proposed to explain the beneficial effect of root-associated bacteria, which include production of phytohormones, antioxidant enzymes, biosurfactants, metal-chelating compounds, siderophores, cell wall–degrading enzymes, ACC deaminase enzyme, solubilization of minerals, and induced systemic tolerance. The metal-chelating molecules synthesized by bacteria may also play as a key mechanism in HMs accumulation and phytoremediation technology. Their root systems are generally extensive and provide a conducive environment for microbial propagation and activity—both within and outside of plant tissues. This bestows plants with an exemplary capacity to extract and detoxify a wide range of harmful compounds from the distillery waste contaminated site. Several potential plants that grow on distillery waste–contaminated sites under natural conditions have indicated the phytoremediation potential of the use of such plants at contaminated sites as natural hyperaccumulators of HMs from complex organic wastes.

1.10 RECOMMENDATION AND FUTURE PROSPECTS

Discharge of distillery waste into open land or nearby water bodies results in a number of environmental, water, and soil pollution problems, including a threat to plant and animal lives. Thus, there is an urgent need for cheap and efficient, ecofriendly methods to clean up HMs contaminated industrial areas, and this could be achieved by using wild accumulator plants. From this, the following conclusions and recommendations can be made:

a. Weed plant appear to be a good choice for restoration of metal-contaminated site since these hardy, tolerant species can quickly grow in most harsh conditions, including HMs stress, over large areas and give a good amount of biomass as a secondary product.

b. Weed plant can restrict the contaminant from being introduced into the food web and consequently reduce the risk to human health.

c. To reduce the potential ecological risk on local ecosystems posed by nonnative weed species, more effective native weeds that are compatible with local habitats are preferred and need to be tested for use in phytoremediation.

d. Plants showing tolerance to HMs have a range of mechanisms at the cellular and molecular level that might be involved in the general homeostasis, detoxification, and tolerance to heavy metal stress. After harvesting biomass is burned, no additional carbon dioxide will be released into atmosphere beyond that which was originally assimilated by plants during growth.

e. As mentioned above, the biomass could be also used for bioenergy production and, in addition, the potential of phytoremediation with energy crops together with the production of biodiesel would be one of the economically feasible methods in the near future.

f. A synergistic interaction of bacteria in the plants may give more benefits to the plant, resulting better accumulation of HMs from the polluted soil by plants. Further research is needed to investigate the mechanisms involved in mobilization, degradation, various chemical aspects of metal accumulation, and transfer of HMs to develop future strategies and optimize phytoextraction by microbe-plant systems.

g. Although numerous studies on phytoremediation have been conducted at the laboratory scale under controlled environmental conditions, more research is still needed to gain a better understanding of the performance and potential for phytoremediation with weed plants over a longer term and in the field.

h. Successful phytoremediation is dependent on a high production of root biomass and high translocation of pollutants from the roots to aboveground plants tissues. Eco-friendly enhancement approaches in conjunction with phytoremediation are proposed to promote healthy plant growth by overcoming plant stress.

i. The acquisition of HMs by plants depends on the ability of a plant and its rhizospheric bacterial communities to solubilize and mobilize HMs in the rhizosphere. Research should thus be focused on increasing the population of PGPR in polluted soil and/or rhizosphere environments by inoculation of soils or plants.

j. An important question that currently remains unanswered is how the inoculated PGPR colonize plant roots and survive in co-contaminated sites.

k. Concentrated efforts are needed to explore the beneficial traits of pollutant-degrading PGPR to get the maximum outcome of the combined use of plants and PGPR in fields.

REFERENCES

Abed, R.M.M., Al-Kharusi, S., Gkorezis, P., Prigent, S., Headley, T. 2018. Bacterial communities in the rhizosphere of *Phragmites australis* from an oil-polluted wetland. *Archives of Agronomy and Soil Science* 64(3):360–370.

Aboudrar, W., Schwartz, C., Morel, J.L., Boularbah, A. 2013. Effect of nickel-resistant rhizosphere bacteria on the uptake of nickel by the hyperaccumulator Noccaea caerulescens under controlled conditions. *Journal of Soils and Sediments* 13(3):501–507.

Abou-Shanab, R.A., Ghanem, K., Ghanem, N., Al-Kolaibe, A. 2008. The role of bacteria on heavy metal extraction and uptake by plants growing on multi-metal-contaminated soils. *World Journal of Microbiology and Biotechnology* 24:253–262.

Abou-Shanab, R.A.I., Angle, J.S., van Berkum, P. 2007. Chromatetolerant bacteria for enhanced metal uptake by *Eichhornia crassipes* (Mart.). *International Journal of Phytoremediation* 9:91–105.

Ahemad, M. 2014a. Phosphate-solubilizing bacteria-assisted phytoremediation of metalliferous soils: a review. *3 Biotech.* DOI:10.1007/s13205-014-0206-0.

Ahemad, M. 2014b. Remediation of metalliferous soils through the heavy metal resistant plant growth promoting bacteria: paradigms and prospects. *Arabian Journal of Chemistry.* DOI:10.1016/j.arabjc.2014.11.020.

Ahemad, M., Khan, M.S. 2010a. Influence of selective herbicides on plant growth promoting traits of phosphate solubilizing *Enterobacter asburiae* strain PS2. *Research Journal of Microbiology* 5:849–857.

Ahemad, M., Khan, M.S. 2010b. Plant growth promoting activities of phosphate-solubilizing Enterobacter asburiae as influenced by fungicides. *Eur Asian J Biosci* 4:88–95.

Ahemad, M., Khan, M.S. 2012a. Biotoxic impact of fungicides on plant growth promoting activities of phosphate-solubilizing *Klebsiella* sp. isolated from mustard (*Brassica compestris*) rhizosphere. *Journal of Pest Science* 85:29–36.

Ahemad, M., Khan, M.S. 2012b. Effect of fungicides on plant growth promoting activities of phosphate solubilizing *Pseudomonas putida* isolated from mustard (*Brassica compestris*) rhizosphere. *Chemosphere* 86:945–950

Alkorta, I., Hernandez-Allica, J., Becerril, J., Amezaga, I., Albizu, I., Garbiscu, C. 2004. Recent finding on the phytoremediation of soil contaminated with environmentally toxic heavy metals and metalloids such as zinc, cadmium, lead, and arsenic, *Reviews in Environmental Science and Biotechnology* 3:71–90.

Alori, E.T., Glick, B.R., Babalola O.O. 2017. Microbial phosphorus solubilization and its potential for use in sustainable agriculture. *Frontiers in Microbiology* 8:971.

Antoun, H., Prevost, D. 2005. Ecology of plant growth promoting rhizobacteria. In Siddiqui, Z.A. (Ed.), *PGPR: Biocontrol and Biofertilization* (pp. 1–38). Dordrecht: Springer.

Arkhipova, T.N., Prinsen, E., Veselov, S.U., Martinenko, E.V., Melentiev, A.I., Kudoyarova G.R. 2007. Cytokinin producing bacteria enhance plant growth in drying soil. *Plant Soil* 292:305–315.

Atlas, R.M., Bartha, R. 1993. *Microbial Ecology: Fundamentals and Applications.* Redwood City, CA: Benjamin/Cummings.

Baker, A.J.M. 1981. Accumulators and excluders—strategies in the response of plants to heavy metals. *J Pant Nutr* 3:643–654.

Baker, A.J.M., Brooks, R.R. 1989. Terrestrial higher plants which hyperaccumulate metallic elements—a review of their distribution, ecology and phytochemistry. *Biorecovery* 1:81–126.

Baker, A.J.M., Reeves, R.D.A., Hajar, S.M. 1994. Heavy metal accumulation and tolerance in British populations of the metallophyte *Thlaspi caerulescens* J. &C. Presl (*Brassicaceae*). *New Phytologist* 127:61–68.

Bashan, Y., Holguin, G. 1998. Proposal for the division of plant growth- promoting rhizobacteria into two classifications: biocontrol-PGPB (plant growth-promoting bacteria) and PGPB. *Soil Biology and Biochemistry* 30:1225–1228.

Belimov, A.A., Hontzeas, N., Safronova, V.I., Demchinskaya, S.V., Piluzza, G., Bullitta, S., Glick, B.R. 2005. Cadmium-tolerant plant growth promoting rhizobacteria associated with the roots of Indian mustard (*Brassica juncea* L. Czern.). *Soil Biology and Biochemistry* 37:241–250.

Belimov, A.A., Safronova, V.I., Sergeyeva, T.A., Egorova, T.N., Mat-veyeva, V.A., Tsyganov, V.E., Borisov, A.Y., Tikhonovich, I.A., Kluge, C., Preisfeld, A., Dietz, K., Stepanok, V.V. 2001. Characterization of plant growth promoting rhizobacteria isolated from polluted soils and containing 1-aminocyclopropane-1-carboxylate deaminase. *Canadian Journal of Microbiology* 47:642–652.

Bennisse, R., Labat, M., ElAsli, A., Brhada, F., Chandad, F., Lorquin, J., Liegbott, P., Hibti, M., Qatibi A. 2004. Rhizosphere bacterial populations of metallophyte plants in heavy metal-contaminated soils from mining areas in semiarid climate. *World Journal of Microbiology and Biotechnology* 20:759–766

Berg, G., Eberl, L., Hartmann, A. 2005. The rhizosphere as a reservoir for opportunistic human pathogenic bacteria. *Environmental Microbiology* 7:1673–1685.

Beringer, E.J., Hirsch P.R. 1984. Genetic engineering and nitrogen fixation. *Biotechnology and Genetic Engineering Reviews* 1(1):65–88, DOI:10.1080/02648725.1984.10647781.

Bhargava, A., Carmona, F. F., Bhargava, M., Srivastava, S. 2012. Approaches for enhanced phytoextraction of heavy metals. *Journal of Environmental Management* 105:103–120.

Bonkowski, M., Villenave, C., Griffiths, B. 2009. Rhizosphere fauna: the functional and structural diversity of intimate interactions of soil fauna with plant roots. *Plant Soil* 321:213.

Borymski, S., Cycon, M., Beckmann, M., Mur, L.A.J., Piotrowska-Seget, Z. 2018. Plant species and heavy metals affect biodiversity of microbial communities associated with metal-tolerant plants in metalliferous soils. *Frontiers in Microbiology* 9:1425.

Brooks, R.R. (Ed.) 1998. *Plants that Hyperaccumulate Heavy Metals* (p. 384). Wallingford: CAB International.

Brooks, R.R., Lee, J., Reeves, R.D., Jaffrre, T. 1977. Detection of nickel ferrous rocks by analysis of her barium specimens of indicator plants. *Journal of Geochemical Exploration* 7:49–57.

Brusamarello-Santos, L.C.C., Pacheco, F., Aljanabi, S.M.M., et al. 2012. Differential gene expression of rice roots inoculated with the diazotroph *Herbaspirillum seropedicae*. *Plant Soil* 356:113.

Burd, G.I., Dixon, D.G., Glick, B.R. 1998. A plant growth promoting bacterium that decreases nickel toxicity in seedlings. *Applied and Environmental Microbiology* 64:3663–3668.

Burd, G.I., Dixon, D.G., Glick, B.R. 2000. Plant growth promoting bacteria that decrease heavy metal toxicity in plants. *Canadian Journal of Microbiology* 46:237–245.

Cabral, L., Soares, C.R.F.S., Giachini, A.J., Siqueira, J.O. 2015. Arbuscular mycorrhizal fungi in phytoremediation of contaminated areas by trace elements: mechanisms and major benefits of their applications. *World Journal of Microbiology and Biotechnology* 31:1655–1664.

Cao, Y.R., Chen, S.Y., Zhang, J.S., 2008. Ethylene signaling regulates salt stress response: An overview. *Plant Signal Behaviour* 3(10):761–763.

Chandra, R., Kumar, V. 2015a. Mechanism of wetland plant rhizosphere bacteria for bioremediation of pollutants in an aquatic ecosystem. In R. Chandra (Ed.), *Advances in Biodegradation and Bioremediation of Industrial Waste* (pp. 329–379). Boca Raton, FL: CRC Press.

Chandra R., Kumar V. 2015b. Biotransformation and biodegradation of organophosphates and organohalides. In R. Chandra (Ed.), *Environmental Waste Management* (pp. 475–524). Boca Raton, FL: CRC Press.

Chandra R., Kumar V., Singh K. 2018a. Hyperaccumulator versus nonhyperaccumulator plants for environmental waste management. In R. Chandra, N.K. Dubey, V. Kumar (Eds.), *Phytoremediation of Environmental Pollutants* (pp. 43–80). Boca Raton, FL: CRC Press.

Chandra R., Kumar V., Tripathi S., Sharma P. 2018b. Phytoremediation of industrial pollutants and life cycle assessment. In R. Chandra, N.K. Dubey, V. Kumar (Eds.), *Phytoremediation of Environmental Pollutants* (pp. 441–470). Boca Raton, FL: CRC Press.

Chandra, R., Kumar, V. 2017a. Phytoextraction of heavy metals by potential native plants and their microscopic observation of root growing on stabilized distillery sludge as a prospective tool for in-situ phytoremediation of industrial waste. *Environmental Science and Pollution Research* 24:2605–2619.

Chandra, R., Kumar, V. 2017b. Detection of androgenic-mutagenic compounds and potential autochthonous bacterial communities during in situ bioremediation of post methanated distillery sludge. *Frontiers in Microbiology* 8:887.

Chandra, R., Kumar, V. 2017c. Detection of *Bacillus* and *Stenotrophomonas* species growing in an organic acid and endocrine-disrupting chemicals rich environment of distillery spent wash and its phytotoxicity. *Environmental Monitoring and Assessment* 189:26.

Chandra, R., Kumar, V. 2018. Phytoremediation: A green sustainable technology for industrial waste management. In R. Chandra, N.K. Dubey, V. Kumar (Eds.), *Phytoremediation of Environmental Pollutants* (pp. 1–42). Boca Raton, FL: CRC Press.

Chandra, R., Kumar, V., Tripathi, S. 2018c. Evaluation of molasses-melanoidins decolourisation by potential bacterial consortium discharged in distillery effluent. *3 Biotech* 8:187. DOI.10.1007/s13205-018-1205-3.

Chandra, R., Kumar, V., Tripathi, S., Sharma, P. 2018d. Heavy metal phytoextraction potential of native weeds and grasses from endocrine-disrupting chemicals rich complex distillery sludge and their histological observations during in situ phytoremediation. *Ecological Engineering* 111:143–156.

Chandra, R., Saxena, G., Kumar, V. 2015. Phytoremediation of environmental pollutants: an eco-sustainable green technology to environmental management. In R. Chandra (Ed.), Advances in *Biodegradation and Bioremediation of Industrial Waste* (pp. 1–29). Boca Raton, FL: CRC Press.

Chaney, R.L. 1983. Plant uptake of inorganic waste. In J.E. Parr, P.B. Marsh, J.M. Kla (Eds.) *Land Treatment of Hazardous Waste* (pp. 50–76). Park Ridge, IL: Noyes Data Corp.

Chaney, R.L., Malik, M., Li, Y.M., Brown, S.L., Brewer, E.P., Angle, J.S., Baker, A.J.M. 1997. Phytoremediation of soil metals. *Current Opinion in Biotechnology* 8:279–284.

Compant, S., Duffy, B., Nowak, J., Clement, C., Barka, E.A. 2005. Use of plant growth-promoting bacteria for biocontrol of plant diseases: principles, mechanisms of action, and future prospects. *Applied and Environmental Microbiology* 71(9):4951–4959.

Costacurta, A., Vanderleyden, J. 1995. Synthesis of phytohormones by plant-associated bacteria. *Critical Review in Microbiology* 21(1):1–18.

Cui, Y.S., Wang, Q.R., Dong, Y.T., Li, H.F., Christie, P. 2004. Enhanced uptake of soil Pb and Zn by Indian mustard and winter wheat following combined soil application of elemental sulphur and EDTA. *Plant and Soil* 261:181–188.

Cunningham, S.D., Berti, W.R., Huang, J.W.W. 1995. Phytoremediation of contaminated soils. *Trends in Biotechnology* 13:393–397.

Cunningham, S.D., Shann, J.R., Crowley, D.E., Anderson, T.A. 1997. Phytoremediation of contaminated water and soil. In E.L. Kruger, T.A. Anderson, J.R. Coats (Eds.), *Phytoremediation of Soil and Water Contaminants*. ACS Symposium Series 664 (pp. 2–19). Washington, DC: American Chemical Society.

De Oliveira, N.C., Rodrigues, A.A., Alves, M.I.R., Filho, N.R.A., Sadoyama, G., Vieira, J.D.G. 2012. Endophytic bacteria with potential for bioremediation of petroleum hydrocarbons and derivatives. *African Journal of Biotechnology* 11:2977–2984.

Deng, H., Ye, Z.H., Wong, M.H. 2004. Accumulation of lead, zinc, copper and cadmium by 12 wetland plant species thriving in metal-contaminated sites in China. *Environmental Pollution* 132(1):29–40.

Dennis, P.G., Miller, A.J., Hirsch, P.R. 2010. Are root exudates more important than other sources of rhizodeposits in structuring rhizosphere bacterial communities? *FEMS Microbiology Ecology* 72:313–327.

Dubois, M., den Broeck, L.V., Inzé, D. 2018. The pivotal role of ethylene in plant growth. *Trends in Plant Science* 23(4). DOI:10.1016/j.tplants.2018.01.003.

Egamberdieva, D., Wirth, S.J., Alqarawi, A.A., Abd Allah, E.F., Hashem, A. 2017. Phytohormones and beneficial microbes: essential components for plants to balance stress and fitness. *Frontiers in Microbiology* 3(8):2104.

Ehrmann, J., Ritz, K. 2014. Plant: soil interactions in temperate multi-cropping production systems. *Plant Soil* 376:1–29.

Evangelou, M.W., Ebel, M., Schaeffer, A. 2007. Chelate assisted phytoextraction of heavy metals from soil. Effect, mechanism, toxicity, and fate of chelating agents. *Chemosphere* 68(6):989–1003.

Faisal, M., Hasnain, S. 2005. Bacterial Cr (VI) reduction concurrently improves sunflower (*Helianthus annuus* L.) growth. *Biotechnology Letters* 27:943–947.

Farwell, A., Vesely, S., Nero, V., Rodriguez, H., Shah, S., Dixon, D., et al. 2006. The use of transgenic canola (*Brassica napus*) and plant growth-promoting bacteria to enhance plant biomass at a nickel-contaminated field site. *Plant and Soil* 288(1):309–318.

Fatima K., Afzal M., Imran A., Khan Q.M. 2015. Bacterial rhizosphere and endosphere populations associated with grasses and trees to be used for phytoremediation of crude oil contaminated soil. *Bulletin of Environmental Contamination and Toxicology* 94:314–20.

Fontem Lum A., Ngwa E.S.A., Chikoye D., Suh C.E. 2014. Phytoremediation potential of weeds in heavy metal contaminated soils of the bassa industrial zone of Douala, Cameroon. *International Journal of Phytoremediation* 16(3):302–319.

Gamalero, E., Glick, B.R. 2011. Mechanisms used by plant growth-promoting bacteria. In D.K. Maheshwari (Ed.), *Bacteria in Agrobiology: Plant Nutrient Management* (pp. 17–46). Berlin and Heidelberg: Springer.

Garbisu, C., Alkorta, I. 2001. Phytoextraction: a cost-effective plant-based technology for the removal of metals from the environment. *Bioresource Technology* 77(3):229–236.

Ghosh, M., Singh, S.P. 2005. A review on phytoremediation of heavy metals and utilization of its byproducts. *Applied Ecology and Environmental Research* 3(1):1–18.

Glick B.R., Cheng Z., Czarny J., Duan J. 2007. Promotion of plant growth by ACC deaminase-containing soil bacteria. *European Journal of Plant Pathology* 119:329–339.

Glick, B.R. 2003. Phytoremediation: synergistic use of plants and bacteria to clean up the environment. *Biotechnology Advances* 21:383–393.

Glick, B.R. 2010. Using soil bacteria to facilitate phytoremediation. *Biotechnology Advances* 28:367–374.

Glick, B.R. 2012. Plant growth-promoting bacteria: mechanisms and applications. *Scientifica* 2012, Article ID 963401.

Glick, B.R. 2014. Bacteria with ACC deaminase can promote plant growth and help to feed the world. *Microbiology Research* 69(1):30–39.

Grichko V.P., Glick B.R. 2001. Amelioration of flooding stress by ACC deaminase-containing plant growth-promoting bacteria. *Plant Physiology and Biochemistry* 39:11–17.

Großkinsky, D.K., Tafner, R., Moreno, M.V., Stenglein, S.A., de Salamone, I.E.G., NelsonL.M., Novák, O., Strnad, M., van der Graaff, E., Roitsch, T. 2016. Cytokinin production by pseudomonas fluorescens G20-18 determines biocontrol activity against. *Pseudomonas syringae in Arabidopsis. Scientific Reports* 6:23310. DOI:10.1038/srep23310.

Gupta, A., Meyer, J.M., Goel, R. 2002. Development of heavy metal resistant mutants of phosphate solubilizing *Pseudomonas* sp. NBRI4014 and their characterization. *Current Microbiology* 45:323–332.

Gutiérrez-Mañero, F.J., Ramo-Solano, B., Probanza, A., Mehouachi, J., Tadeo, F.R., Talon, M. 2001. The plant-growth-promoting rhizobacteria *Bacillus pumilus* and *Bacillus licheniformis* produce high amounts of physiologically active gibberellins. *Physiologia Plantarum* 111:206–211.

Haldar, S., Sengupta, S. 2015. Plant-microbe cross-talk in the rhizosphere: insight and biotechnological potential. *Open Microbiology Journal* 9:1–7.

Hammami, H., Parsa, M., Mohassel, M.H.R., Rahimi S., Mijani, S. 2016. Weeds ability to phytoremediate cadmium-contaminated soil. *International Journal of Phytoremediation* 18(1):48–53.

Hatano, K., Kanazawa, K., Tomura, H., Yamatsu, T., Kubota, K.T.K. 2016. Molasses melanoidin promotes copper uptake for radish sprouts: the potential for an accelerator of phytoextraction. *Environmental Sciences and Pollution Research* 23:17656.

He, L.Y., Zhang, Y.F., Ma, H.Y., Su, L.N., Chen, Z.J., Wang, Q.Y., Meng, Q., Fang, S.X. 2010. Characterization of copper resistant bacteria and assessment of bacterial communities in rhizosphere soils of copper-tolerant plants. *Applied Soil Ecology* 44:49–55.

Hemmat-Jou, M.H., Safari-Sinegani, A.A., Mirzaie-Asl, A., Tahmourespour, A. 2018. Analysis of microbial communities in heavy metals-contaminated soils using the metagenomic approach. *Ecotoxicology* 27(9):1281–1291.

Hiltner, L. (1904). Uber neuere Erfahrungen und Probleme auf dem Gebiete der Bodenbakteriologie unter besonderden berucksichtigung und Brache. Arb. Dtsch. Landwirtsch. Gesellschaft 98, 59–70.

James, E.K. 2000. Nitrogen fixation in endophytic and associative symbiosis. *Field Crops Research* 65:197–209.

James, E.K., Olivares, F.L. 1998. Infection and colonization of sugarcane and other graminaceous plants by endophytic diazotrophs. *Critical Reviews in Plant Sciences* 17:77–119.

Jing, Y.-de, He, Z., Yang, X.-e. 2007. Role of soil rhizobacteria in phytoremediation of heavy metal contaminated soils. *Journal of Zhejiang University Science B* 8(3):192–207.

Jung, J., Philippot, L., Park, W. 2016. Metagenomic and functional analyses of the consequences of reduction of bacterial diversity on soil functions and bioremediation in diesel-contaminated microcosms. *Scientific Reports* 6:23012.

Kabeer, R., Varghese R., Kannan, V.M., Thomas, J.R., Poulose, S.V. 2014. Rhizosphere bacterial diversity and heavy metal accumulation in *Nymphaea pubescens* in aid of phytoremediation potential. *Journal of Bioscience and Biotechnology* 3(1):89–95.

Khan, M.U., Sessitsch, A., Harris, M., Fatima, K., Imran, A., Arslan, M., Shabir, G., Khan, Q.M., Afzal, M. 2014. Cr-resistant rhizo- and endophytic bacteria associated with *Prosopis juliflora* and their potential as phytoremediation enhancing agents in metal-degraded soils. *Frontier in Plant Science* 1(755):1–10.

Kim, J., Kang, S., Min, K., Cho, K., Lee, I. 2006. Rhizosphere microbial activity during phytoremediation of diesel-contaminated soil. *Journal of Environmental Science and Health Part A* 41:1–14.

Kloepper, J.W., Lifshitz, R., Zablotowicz, R.M. 1989. Free-living bacterial inocula for enhancing crop productity. *Trends in Biotechnology* 7:39–43.

Kloepper J. W. and Schroth M. N. (1978) Plant growth promoting rhizobacteria on radishes. In Proceedings of the 4th International Conference on Plant Pathogenic Bacteria. ed. Station de Pathologic Vegetal et Phytobacteriologic. Vol. 2, pp. 879–882. Angers, France.

Kong, Z., Glick, B.R. 2017. The role of plant growth-promoting bacteria in metal phytoremediation. *Advances in Microbial Physiology* 71:97–132.

Koshlaf, E., Shahsavari, E., Aburto-Medina, A., Haleyur, N., Makadiaa, P.D., Morrison T.H., Ball, A.S. 2016. Bioremediation potential of diesel-contaminated Libyan soil. *Ecotoxicology and Environmental Safety* 133:297–305.

Krome, K., Rosenberg, K., Dickler, C., Kreuzer, K., Ludwig-Müller, J., UllrichEberius, C., et al. 2010. Soil bacteria and protozoa affect root branching via effects on the auxin and cytokinin balance in plants. *Plant Soil* 328:191–201.

Kuffner, M., Puschenreiter, M., Wieshammer, G., Gorfer, M., Sessitsch, A. 2008. Rhizosphere bacteria affect growth and metal uptake of heavy metal accumulating willows. *Plant and Soil* 304:35–44.

Kumar, K.V., Srivastava, S., Singh, N., Behl, H.M. 2009. Role of metal resistant plant growth promoting bacteria in ameliorating fly ash to the growth of *Brassica juncea*. *Journal of Hazardous Materials* 170:51–57.

Kumar, V., AlMomin, S., Al-Aqeel, H., Al-Salameen, F., Nair, S., Shajan, A. 2018. Metagenomic analysis of rhizosphere microflora of oil contaminated soil planted with barley and alfalfa. *PLoS ONE* 13(8):e0202127.

Kumar, V., Chandra, R. 2018. Bacteria-assisted phytoremediation of industrial waste pollutants and ecorestoration. *In* R. Chandra, N.K. Dubey, V. Kumar (Eds.), *Phytoremediation of Environmental Pollutants* (pp. 159–200). Boca Raton, FL: CRC Press.

Kumar, V., Chandra, R. 2018.Characterisation of manganese peroxidase and laccase producing bacteria capable for degradation of sucrose glutamic acid-Maillard products at different nutritional and environmental conditions. *World Journal of Microbiology & Biotechnology* 34:82.

Kumar, V., Chandra, R. 2020. Bioremediation of melanoidins containing distillery waste for environmental safety. In R.N. Bharagava, G. Saxena (Eds.), *Bioremediation of Industrial Waste for Environmental Safety. Vol II—Microbes and Methods for Industrial Waste Management*. Singapore: Springer.

Kunitoa, T., Saekib, K., Nagaokac, K., Oyaizud, H., Matsumoto, S. 2001. Characterization of copper-resistant bacterial community in rhizosphere of highly copper-contaminated soil. *European Journal of Soil Biology* 37:95–102.

Leitenmaier, B., Küpper, H. 2013. Compartmentation and complexation of metals in hyperaccumulator plants. *Frontiers in Plant Sciences* 4:374.

Leveau, J.H.J., Lindow, S.E. 2005. Utilization of the plant hormone Indole-3-acetic acid for growth by *Pseudomonas putida* strain. *Applied and Environmental Microbiology* 71(5):2365–2371.

Li, H., Qu Y., Tian, Y., Feng, Y. 2019. The plant-enhanced bio-cathode: root exudates and microbial community for nitrogen removal. *Journal of Environmental Sciences* 77:97–103.

Li, W.C., Ye, Z.H., Wong M.H. 2007. Effects of bacteria on enhanced metal uptake of the Cd/Zn-hyperaccumulating plant, Sedum alfredii. *Journal of Experimental Botany* 58:4173–4182.

Liang, J., Meng, F., Sun, S., Wu, C., Wu, H., Zhang, M., et al. 2015. Community structure of arbuscular mycorrhizal fungi in rhizospheric soil of a transgenic high-methionine soybean and a near isogenic variety. *PLoS ONE* 10(12): e0145001.

Linacre, N.A., Whiting, S.N., Baker, A.J.M., Angle, S., Ades, P.K. 2003. Transgenics and phytoremediation: the need for an integrated risk assessment, management, and communication strategy. *International Journal of Phytoremediation* 5:181–185.

Liphadzi, M.S., Kirkham, M.B., Mankin, K.R., Paulsen, G.M. 2003. EDTA-assisted heavy-metal uptake by poplar and sunflower grown at a long-term sewage-sludge farm. *Plant and Soil* 257:171–182.

Lugtenberg, B., Kamilova, F. 2009. Plant-growth-promoting rhizobacteria. *Annual Reviews in Microbiology* 63:541–556.

Ma, L.Q., Komar, K.M., Tu, C., Zhang, W., Cai, Y., Kenelly, E.D. 2001. A fern that hyperaccumulates arsenic. *Nature* 409:579–582.

Ma, Y., Prasad, M.N.V., Rajkumar, M., Freitas, H. 2011. Plant growth promoting rhizobacteria and endophytes accelerate phytoremediation of metalliferous soils. *Biotechnology Advances* 29:248–258.

Madhaiyan, M., Poonguzhali, S., Sa, T. 2007. Metal tolerating methylotrophic bacteria reduces nickel and cadmium toxicity and promotes plant growth of tomato (*Lycopersicon esculentum* L.). *Chemosphere* 69:220–228.

Maestri, E., Marmiroli, M., Visioli, G., Marmiroli, N. 2010. Metal tolerance and hyperaccumulation: costs and trade-offs between traits and environment. *Environmental and Experimental Botany* 68:1–13.

Martinez-Viveros, O., Jorquera, M.A., Crowley, D.E., Gajardo, G., Mora, M.L. 2010. Mechanisms and practical considerations involved in plant growth promotion by rhizobacteria. *Journal of Soil Science and Plant Nutrition* 10:293–319.

Mayak, S., Tirosh, T., Blick, B.R., 2004. Plant growth-promoting bacteria confer resistance in tomato plants to salt stress. *Plant Physiology and Biochemistry* 42(6):565–572.

McGrath, S.P., Zhao, F.J. 2003. Phytoextraction of metals and metalloids from contaminated soils. *Current Opinion in Biotechnology* 14:1–6.

Mendes R., Garbeva P., Raaijmakers J.M. 2013. The rhizosphere microbiome: significance of plant beneficial, plant pathogenic, and human pathogenic microorganisms. *FEMS Microbiology Reviews* 37:634–663.

Meyer, J.M. 2000. Pyoverdines: pigments siderophores and potential taxonomic markers of fluorescent Pseudomonas species. *Archives of Microbiology* 174(3):135–142.

Migo, V.P., Del Rosario, E.J., Matsumura, M. 1997. Flocculation of melanoidins induced by inorganic ions. *Journal of Fermentation Bioengineering* 83:287–291.

Mitter, B., Petric, A., Shin, M.W., Chain, P.S., Hauberg-Lotte, L., Reinhold-Hurek, B., Nowak, J., Sessitsch, A. 2013. Comparative genome analysis of *Burkholderia phytofirmans* PsJN reveals a wide spectrum of endophytic lifestyles based on interaction strategies with host plants. *Frontiers in Plant Science* 4:1–15.

Morgan, P.W., Drew, M.C. 1997. Ethylene and plant response to stress. *Physiological and Plantarum* 100:620–630.

Muehe, E.M., Weigold, P., Adaktylou, I.J., Planer-Friedrich, B., Kraemer, U., Kappler, A., Behrens, S. 2015. Rhizosphere microbial community composition affects cadmium and zinc uptake by the metal-hyperaccumulating plant *Arabidopsis halleri*. *Applied and Environmental Microbiology* 81:2173–2181.

Navarro-Noya, Y.E., Jan-Roblero, J., del Carmen González-Chávez, M., Hernandez-Gama, R., Hernandez-Rodrıguez, C. 2010. Bacterial communities associated with the rhizosphere of pioneer plants (*Bahia xylopoda* and *Viguiera linearis*) growing on heavy metals-contaminated soils. *Antonie van Leeuwenhoek* 97:335–349.

Neubauer, U., Furrer, G., Kayser, A., Schulin, R., 2000. Siderophores, NTA, and citrate: potential soil amendments to enhance heavy metal mobility in phytoremediation. *International Journal of Phytoremediation* 2: 353–368.

Olanrewaju, O.S., Glick, B.R., Babalola, O.O. 2017. Mechanisms of action of plant growth promoting bacteria. *World Journal of Microbiology and Biotechnology* 33(11):197.

Ortíz-Castro, R., Valencia-Cantero, E., López-Bucio, J. 2008. Plant growth promotion by *Bacillus megaterium* involves cytokinin signalling. *Plant Signaling & Behavior* 3(4):263–265.

Oves, M., Khan, M.S., Zaidi, A. 2013. Chromium reducing and plant growth promoting novel strain *Pseudomonas aeruginosa* OSG41 enhance chickpea growth in chromium amended soils. *European Journal of Soil and Biology* 56:72–83.

Pacwa-Płociniczak, M., Płociniczak, T., Yu, D., Kurola, J.M., Sinkkonen, A., Piotrowska-Seget, Z., Romantschuk, M. 2018. Effect of *Silene vulgaris* and heavy metal pollution on soil microbial diversity in long-term contaminated soil. *Water Air Soil Pollution* 229(1):13.

Palmroth, M.R., Koskinen, P.E., Kaksonen, A.H., Münster, U., Pichtel, J., Puhakka, J.A. 2007. Metabolic and phylogenetic analysis of microbial communities during phytoremediation of soil contaminated with weathered hydrocarbons and heavy metals. *Biodegradation* 18(6):769–782.

Patel, T., Saraf, M. 2017. Biosynthesis of phytohormones from novel rhizobacterial isolates and their in vitro plant growth-promoting efficacy. *Journal of Plant Interactions* 12(1):480–487.

Philippot, L., Raaijmakers, J.M., Lemanceau, P., van der Putten, W.H. 2013. Going back to the roots: the microbial ecology of the rhizosphere. *Nature Reviews Microbiology* 11:789–799.

Płociniczak, T., Sinkkonen, A., Romantschuk, M., Sułowicz, S., Piotrowska-Seget, Z. 2016. Rhizospheric bacterial strain *Brevibacterium casei* MH8a colonizes plant tissues and enhances Cd, Zn, Cu phytoextraction by white mustard. *Frontiers in Plant Science* 16(7):101.

Prapagdee, B., Khonsue, N. 2015. Bacterial-assisted cadmium phytoremediation by *Ocimum gratissimum* L. in polluted agricultural soil: a field trial experiment. *International Journal of Environmental Science and Technology* 12:3843–3852.

Raaijmakers, J.M., Paulitz, T.C., Steinberg, C., Alabouvette, C., Moënne-Loccoz, Y. 2009. The rhizosphere: a playground and battlefield for soil borne pathogens and beneficial microorganisms. *Plant Soil* 321:341–361.

Rajkumar, M., Ae, N., Prasad, M.N., Freitas, H. 2010. Potential of siderophore-producing bacteria for improving heavy metal phytoextraction. *Trends in Biotechnology* 28(3):142–149.

Rajkumar, M., Freitas, H. 2008. Influence of metal resistant-plant growth-promoting bacteria on the growth of *Ricinus communis* in soil contaminated with heavy metals. *Chemosphere* 71:834–842.

Rajkumar, M., Nagendran, R., Kui, J.L., Wang, H.L., Sung, Z.K. 2006. Influence of plant growth promoting bacteria and Cr (VI) on the growth of Indian mustard. *Chemosphere* 62:741–748.

Rajkumar, M., Sandhya, S., Prasad, M.N., Freitas, H. 2012. Perspectives of plant-associated microbes in heavy metal phytoremediation. *Biotechnology Advances* 30(6):1562–1574.

Rascio, N., Navari-Izzo, F. 2011. Heavy metal hyperaccumulating plants: how and why do they do it? And what makes them so interesting. *Plant Science* 180(2):169–181.

Raskin, I., Smith, R.D., Salt, D.E. 1997. Phytoremediation of metals: using plants to remove pollutants from the environment. *Current Opinion in Biotechnology* 8(2):221–226.

Reinhold-Hurek, B., Hurek, T. 2011. Living inside plants: bacterial endophytes. *Current Opinion in Plant Biology* 14:435–443.

Robinson, B., Russell, C., Hedley, M.J., Clothier, B. 2001. Cadmium adsorption by rhizobacteria: implications for New Zealand Pastureland. *Agriculture, Ecosystem & Environment* 87:315–321.

Robinson, B.H., Fernández, J.E., Madejón, P., Marañón, T., Murillo, J.M., Green, S.R., Clothier, B.E. 2003. Phytoextraction: an assessment of biogeochemical and economic viability. *Plant Soil* 249(1):117–125

Rodriguez, H., Vessely, S., Shah, S., Glick, B.R. 2008. Effect of a nickel-tolerant ACC deaminase-producing *Pseudomonas* strain on growth of non transformed and transgenic canola plants. *Current Microbiology* 57(2):170–174.

Sahu, R.K., Naraian, R., Chandra, V. 2007. Accumulation of metals in naturally grown weeds (aquatic macrophytes) grown on an industrial effluent channel. *Clean* 35(3):261–265.

Salt, D.E., Blaylock, M., Kumar, N.P.B.A., Dushenkov, V., Ensley, B.D., Chat I., Raskin, I. 1995. Phytoremediation: A novel strategy for the removal of toxic metals from the environment using plants. *Nature Biotechnology* 12(5):468–474.

Salt, D.E., Smith, R.D., Raskin, I. 1998. Phytoremediation. *Plant and Molecular Biology* 49:643–668.

Sandermann, H.J. 1992. Plant metabolism of xenobiotics. *Trends in Biochemical Sciences* 17(2):82–84.

Saravanan, V.S., Kalaiarasan, P., Madhaiyan, M., Thangaraju, M. 2007. Solubilization of insoluble zinc compounds by *Gluconacetobacter diazotrophicus* and the detrimental action of zinc ion (Zn^{2+}) and zinc chelates on root knot nematode *Meloidogyne incognita*. *Letters in Applied Microbiology* 44:235–241.

Schneider, J., Bundschuh, J., Nascimento, C.W.A. 2016. Arbuscular mycorrhizal fungi-assisted phytoremediation of a lead-contaminated site. *Science of the Total Environment* 572:86–97.

Sessitsch, A., Kuffner, M., Kidd, P., Vangronsveld, J., Wenzel, W.W., Fallmann, K., Puschenreiter, M. 2013. The role of plant-associated bacteria in the mobilization and phytoextraction of trace elements in contaminated soils. *Soil Biology Biochemistry* 60:182–194.

Sharma, S.B., Sayyed, R.Z., Trivedi, M.H., Gobi, T.A. 2013. Phosphate solubilizing microbes: sustainable approach for managing phosphorus deficiency in agricultural soils. *Springer Plus* 2:587.

Sheng, X.F., Xia, J.J. 2006. Improvement of rape (*Brassica napus*) plant growth and cadmium uptake by cadmium-resistant bacteria. *Chemosphere* 64:1036–1042.

Singh, Y., Ramteke, P.W., Shukla, P.K. 2013. Isolation and characterization of heavy metal resistant *Pseudomonas* spp. and their plant growth promoting activities. *Advances in Applied Science Research* 4:269–272.

Steer, J., Harris, J.A. 2000. Shifts in the microbial community in rhizosphere and non-rhizosphere soils during the growth of *Agrostis stolonifera*. *Soil Biology and Biochemistry* 32:869–878.

Szabó, L., Fodor, L. 2006. Uptake of microelements by crops grown on heavy metal-amended soil. *Communications in Soil Science and Plant Analysis* 37:2679–2689.

Tchounwou, P.B., Yedjou, C.G., Patlolla, A.K., Sutton, D.J. 2012. Heavy metals toxicity and the environment. *EXS* 101:133–164.

Teixeira, L.C.R.S., Peixoto, R.S., Cury, J.C., Sul, W.J., Pellizari, V.H., Tiedje, J., Rosado, A.S. 2010. Bacterial diversity in rhizosphere soil from Antarctic vascular plants of Admiralty Bay, maritime Antarctica. *ISME Journal* 4:989–1001.

Teplitski, M., Warriner, K., Bartz, J., Schneider, K.R., 2011. Untangling metabolic and communication networks: interactions of enterics with phytobacteria and their implications in produce safety. *Trends in Microbiology* 19:121–127.

Tsavkelova, E.A., Klimova, S.Y., Cherdyntseva, T.A., Netrusov, A.I. 2006. Microbial producers of plant growth stimulators and their practical use: a review. *Applied Biochemistry and Microbiology* 42:117–126.

Turgut, C., Pepe, M.K., Cutright, T.J. 2004. The effect of EDTA and citric acid on phytoremediation of Cd, Cr, and Ni from soil using *Helianthus annuus*. *Environmental Pollution* 131:147–154.

Turner, T.R., James, E.K., Poole, P.S. 2013. The plant microbiome. *Genome Biology* 14(6):209.

Turpeinen, R., Kairesalo, T., Haggblom, M.M. 2004. Microbial community structure and activity in arsenic-, chromium- and copper-contaminated soils. *FEMS Microbiology Ecology* 47:39–50.

Tyler, H.L., Triplett, E.W. 2008. Plants as a habitat for beneficial and/or human pathogenic bacteria. *Annual Review in Phytopathology* 46:53–73.

Upadhayay, A., Srivastava, S. 2012. Evaluation of multiple plant growth promoting traits of an isolate of *Pseudomonas fluorescens* strain Psd. *Indian Journal of Experimental Biology* 48:601–609.

van Loon, L.C. 2007. Plant responses to plant growth-promoting rhizobacteria. *European Journal of Plant Pathology* 105:513–517.

Verbruggen, N., Hermans, C., Schat, H. 2009. Molecular mechanisms of metal hyperaccumulation in plants. *New Phytologist* 181:759–776.

Vivas, A., Biro, B., Ruiz-Lozano, J.M., Barea, J.M., Azcon R. 2006. Two bacterial strains isolated from a Zn-polluted soil enhance plant growth and mycorrhizal efficiency under Zn toxicity. *Chemosphere* 52:1523–1533.

Wei, S.H., Zhou, Q.X., Wang, X. 2005. A newly discovered Cd-hyperaccumulator *Solanum nigrum* L. *Science Bulletin* 50:33–38.

Wenzel, W.W., Bunkowski, M., Puschenreiter, M., Horak, O. 2003. Rhizosphere characteristics of indigenous growing nickel hyperaccumulator and excluder plants on serpentine soil. *Environmental Pollution* 123:131–138.

Weyens, N., van Der Lelie, D., Taghavi, S., Newman, L., Vangronsveld, J. 2009. Exploiting plantmicrobe partnerships to improve biomass production and remediation. *Trends in Biotechnology* 27:591–598.

Whiting, S.N., Desouza, M.P., Terry, N. 2001. Rhizosphere Bacteria Mobilize Zn for Hyperaccumulation by *Thlaspi caerulescens*. *Environmental Science and Technology* 35:3144–3150.

Wu, S.C., Cheung K.C., Luo Y.M., Wong M.H. 2006. Effects of inoculation of plant growth-promoting rhizobacteria on metal uptake by *Brassica juncea*. *Environmental Pollution* 140(1):124–135.

Wu, C.H., Wood, T.K., Mulchandani, A., Chen, W. 2006. Engineering plant-microbe symbiosis for rhizoremediation of heavy metals. *Applied and Environmental Microbiology* 72:1129–1134.

Wu, H., Wang, X., Hea, X., Zhang, S., Liang, R., Shen, J. 2017. Effects of root exudates on denitrifier gene abundance, community structure and activity in a micro-polluted constructed wetland. *Science of the Total Environment* 598:697–703.

Wu, J., Hsu, F.C., Cunningham, S.D. 1999. Chelate-assisted pb phytoextraction: pb availability, uptake, and translocation constraints. *Environmental Science and Technology* 33(11):1898–1904.

Yang, X., Feng, Y., He, Z., Stoffella, P.J. 2005. Molecular mechanisms of heavy metal hyperaccumulation and phytoremediation. *Journal of Trace Elements in Medicine and Biology* 18(4):339–353.

Yang, Y., Liang, Y., Han, X., Chiu, T., Ghosh, A., Chen, H., Tang, M. 2016. The roles of arbuscular mycorrhizal fungi (AMF) in phytoremediation and tree-herb interactions in Pb contaminated soil. *Scientific Reports* 6:20469.

Yoon, J., Cao, X., Zhou, Q., Ma, L.Q. 2006. Accumulation of Pb, Cu, and Zn in native plants growing on a contaminated Florida site. *Science of the Total Environment* 368(2):456–464.

Zaidi, S., Usmani, S., Singh, B.R., Musarrat, J. 2006. Significance of *Bacillus subtilis* strain SJ 101 as a bioinoculant for concurrent plant growth promotion and nickel accumulation in *Brassica juncea*. *Chemosphere* 64:991–997.

2 Microbial Enzymes for Eco-friendly Recycling of Waste Paper by Deinking

Sanjeev Balda, Aarjoo Sharma,
Neena Capalash, and Prince Sharma
Panjab University

CONTENTS

2.1   INTRODUCTION

Despite the general belief that advancement in information technology and computerization would result in a paperless global society, the demand for paper continues to grow due to increasing literacy rate and rise in per capita consumption of paper. Worldwide, over 40% of the industrial wood harvest goes into paper products, and by 2050, it is expected that pulp and paper production will account for over half of the world's industrial wood demand. The world's total paper production amounted to about 395 million tons in 2010 and is expected to reach 490 million tons by 2020. The global average is about 57 kg of paper per person, with the extremes being the United States' 300 kg and Africa's about 7 kg. Use of fresh wood-based pulp is not economical or environmentally sound, especially in developing countries where industrialization and urbanization have drastically affected the forest cover. So, there have been continuous efforts to replace wood-based raw material with agro-based and industry-based fibers such as wheat straw or waste paper. The major problems industries using agro-based fibers as raw material face are these are used as feed for domestic animals and not available throughout the year, competition with other industries, and the high amount of silica (3%~8%) present in nonwood material. High silica content makes the recovery of cooking chemicals difficult, thus making the use of nonwood fibers an environmental hazard. Alternatively, recycling of used waste paper is an important resource-saving technology for the pulp and paper industry (Bajpai 2014).

There are many advantages of using recycled paper, e.g., 1 ton of recycled paper can save 3 cubic yards of landfill space, requires about 28%–70% less energy, and emits about 38% less greenhouse gas than paper making from virgin wood pulp. Deinking, "the most crucial step in paper recycling," can also be called "pulp laundering," as it is an industrial operation of removing ink and stickles from fibers of the waste paper. Deinking, which is traditionally handled by using large quantities of non–environmentally friendly chemical agents (such as sodium hydroxide, sodium carbonate, diethylenetriaminepentacetic acid, sodium silicate, hydrogen peroxide, and surfactants), helps in forest conservation. Many xenobiotic compounds, such as resin acids, chlorophenols, dioxins, and furans, are formed, thereby turning paper mill effluents into "a Pandora's box of waste chemicals" (Martin-Sampedro et al. 2015). Paper recycling is preferred over land filling, as land-filled paper decomposes and produces methane, a greenhouse gas with much higher heat-trapping power than carbon dioxide. So, enzymes provide an eco-friendly alternative approach to provide novel means to convert recycled fiber into the quality product, as use of enzymatic deinking will reduce the consumption of chemicals, which will further decrease a load of pollutants in the environment. Many enzymes such as cellulase, hemicellulase (viz., xylanase), pectinase, lipase, laccase, cutinase, and esterase have been employed in the deinking of waste paper (Soni et al. 2008).

2.2 PAPER RECYCLING: STATUS AND USES

A major source of waste paper includes newsprint paper and mixed office paper (i.e., photocopied paper, laser-printed and inkjet-printed paper). Over the past decade, paper recycling in an increasingly environmentally conscious world is gaining importance. Worldwide, recycled paper demand has grown by around 60 million tons (40%) in 10 years (2005–2015). The recycling rate in Europe reached 70.4% in 2011; 13 European countries exceeded the 70% recycling rate (Bajpai 2014). Recovery rate in the United States rose to 66.8% in 2015 from 51.5% in 2005 (Figure 2.1). Also, land filling of waste paper decreased from 35.4 million tons in 2005 to 21.1 million tons in 2015.

The Indian Paper Industry accounts for about 2.6% of the world's production of paper and paperboard and about 3.3% of the world's recovered paper market (Annual Statistics 2015). Paper production is expected to rise to 44 million tons in 2025 from 15.3 in 2015. In spite of the continued focus on digitization, India's demand for paper is expected to increase 53% in the next 6 years, basically because of a sustained increase in the number of school-going children in rural areas. Also, waste paper–based raw material consumption has increased from 47% in 2011 to 70% in 2015 (CPPRI Annual Report 2015–2016).

According to the data for the year 2015 (United States), 33.7% of the recovered paper was used for containerboard production, while 40% was exported to China and other countries (Annual Survey of Paper, Paperboard and Pulp Capacity, AF & PA 2016) as shown in Figure 2.2. The USEPA (United States Environmental Protection Agency) has recommended significantly different levels of total recovered fibers in certain paper grades ranging from 10% for printing and writing grades to 100% for newsprint and packaging (Bajpai 2014).

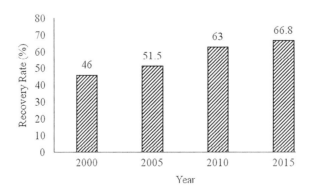

FIGURE 2.1 Changing recovery rate pattern in the US pulp and paper industry.

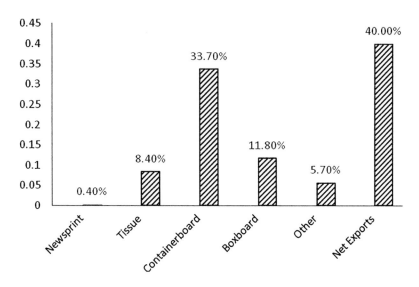

FIGURE 2.2 Utilization of recovered paper in different paper grades in the United States in 2015. (Data Source: Annual Survey of Paper, Paperboard and Pulp Capacity, AF & PA 2016.)

2.3 DEINKING OF WASTE PAPER: CHALLENGES AND PROCESSES

The ease or difficulty with which secondary fiber can be deinked depends on ink type, printing process, and fiber type. Most difficult to deink are nonimpact printed papers, e.g., mixed office waste (MOW) and cross-linking inks, due to which quantity of these types of papers continues growing as a proportion of total recovered paper. Photocopiers and laser printers have been used to print large portion of MOW paper, which fuse the ink to fibers, thus making it difficult to remove by conventional methods. Thus, removal of ink is the major step that requires continuous technical and economical improvements for efficient recycling of waste paper (Bajpai 2014; Bajpai et al. 1999).

The principal process steps in deinking plants are pulping, deflaking, deinking (flotation or washing), and final product. Pulping being the first step aims at defibring the paper for detachment of ink from fibers to produce dispersed ink particles of the size and geometry that can be efficiently removed in subsequent ink removal steps. Smaller ink particles cause problem, as these can be redeposited on fibers. Devices used for pulping are low-consistency pulpers, medium-consistency pulpers, and drum pulpers. Due to the nature of chemicals involved in this step, it is analogous to sulfite "cooking." For deflaking, machines are used which defiber flakes by fiber-to-fiber rubbing or hydraulic shear to form individual fibers. Most deflakers can handle consistencies of 3%–6%. For removal of staples and heavy materials, high-density cleaning is performed. Deinking is the most crucial step in waste paper recycling, which helps in detachment and removal of ink from fibers. It is done by two methods: flotation and washing. Flotation is preferred as in the case of washing fiber loss is more (up to 40%) as compared to flotation (up to 10%). Flotation works on the principle of difference of surface physicochemical properties between the ink and fiber for removal of hydrophobic particles (e.g., ink particles), where the ink particles conjoin with air bubbles and float up to the surface where these are removed. Flotation aids such as surfactants can be added to improve the attachment of ink particles with air bubbles. Washing is based on the principle of filtration and separation, which causes higher fiber losses. Some modern plants use a combination of flotation and washing for effective ink removal. Flotation or washing is followed by chemical deinking followed by thickening, kneading, dispersion, and final processing. Steps involved in waste paper recycling are shown in Figure 2.3.

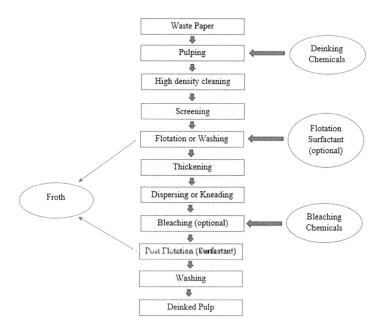

FIGURE 2.3 Steps in waste paper recycling process.

2.4 METHODS FOR DEINKING OF WASTE PAPER

Traditionally, deinking is performed by large quantities of chemicals. But physical and enzymatic methods of deinking are gaining more importance due to their environmentally friendly nature.

2.4.1 CHEMICAL DEINKING

Flotation and washing use different physicochemical properties and require different types of chemicals, such as sodium hydroxide (NaOH), sodium carbonate (Na_2CO_3), diethylenetriaminepentacetic acid (DTPA), sodium silicate (Na_2SiO_3), hydrogen peroxide (H_2O_2), and surfactants. These chemicals are used at different stages of deinking. The costs for pulping chemicals are often around €10–20 per ton and almost the same or even double for bleaching sequence.

The right combination of chemicals is required for maximal performance and minimal cost. Sodium hydroxide changes the pH toward alkaline causing ink detachment because of fiber swelling and chemical hydrolysis of bonds between substrate and ink particles. Alkaline conditions are also responsible for ionization of carboxylic group of cellulose fiber, saponification of fatty acids, and hydrolysis of ink resins. Deinking efficiency increases with increase in pH; in some cases, higher pH causes yellowing, which can be countered by the addition of hydrogen peroxide (Ferguson 1992; Bajpai et al. 1999). Sodium silicate performs various functions: stabilizes hydrogen peroxide, buffering, saponification, acts as a chelating agent, helps in ink dispersion, and acts as ink collector (Mahagaonkar et al. 1997). It stabilizes hydrogen peroxide by deactivating metal ions which can decompose peroxide. Hydrogen peroxide, which is mostly used in flotation deinking, is added to pulper or added to bleach towers. It reduces the yellowing effect of NaOH by forming per hydroxyl ions which attack chromophores (Ferguson 1992).

Chelating agents such as diethylenetriaminepentaacetic acid (DTPA) and ethylenediaminetetraacetic acid (EDTA) are also known to form soluble complexes with metal ions, thus preventing decomposition of hydrogen peroxide. Surfactants are composed of hydrophobic as well as hydrophilic parts and act on fiber surface for the release of ink particles and avoid their redeposition by dispersion in water. Some nonionic surfactants are also used in flotation as dispersion agents. These mainly consist of a polymer of ethylene oxide and propylene oxide in varying ratio. Fatty acids act as collecting chemicals by aggregating smaller ink particles.

2.4.2 Physical Deinking

The chemicals used in deinking have a negative effect on paper properties as well as the environment. So, many researchers tried physical methods such as microwaving, ultrasonication, and UV treatment in combination with chemical or enzymatic deinking with the aim of reducing chemical consumption without compromising the paper quality. The earliest known attempt to use ultrasound to deink paper was by using a mechanically vibrating whistle jet device (Turai and Teng 1978). Ultrasonic treatment for 1 min following flotation deinking increases brightness by 20% at the cost of 1.4 times more energy consumption as compared to conventional flotation deinking process (Tatsumi et al. 2000).

Ultrasonication in combination with flotation for 2 min of disintegration resulted in 58% whiteness, which was similar to 30 min of disintegration time without sonication and increased to 59.32% after 20 min of ultrasonication disintegration and froth flotation under alkaline conditions (Lim et al. 2012). The ultrasonic treatment was effective in reducing particle size even under the condition where no caustic soda or surfactant was used at all. Ultrasonication for 20 min during deinking increased the brightness and decreased the dirt area as compared to control pulp for laser-printed paper. Also combined deinking using UV radiation and enzyme cocktail enhances the brightness and decreases dirt area drastically (Zhenying et al. 2009). A combination of physical (sonication and microwaving) and enzymatic treatments resulted in improved brightness by 28.8% and decreased effective residual ink concentration (ERIC) value by 73.9% as compared to chemical deinking (Virk et al. 2013). Effects of physical and combined methods of deinking on different paper properties are shown in Table 2.1.

2.4.3 Enzymatic Deinking

Paper recycling is gaining more importance due to increasing environmental and economic concerns. Paper recovery rates are increasing continuously in the United States and European countries. Due to rapid developments and increasing efficiency of technology, the quality of the recycled paper is

TABLE 2.1
Effects of Physical and Combined Methods of Deinking on Paper Properties

Treatment Type	Enzymes	Type of Paper	Brightness		ERIC		References
			Before Treatment	After Treatment	Before Treatment	After Treatment	
UV radiation	Cellulase and amylase	Laser print	68.50	70.56	–	–	Zhenying et al. 2009
Microwaving	No	Old newsprint	49.0	52.8	–	–	Virk et al. 2013
Microwaving	Xylanase and laccase	Old newsprint	49.0	61.73	535.41	148.51	Virk et al. 2013
Ultrasonication	No	Old newsprint	61.6	66.7	–	–	Tatsumi et al. 2000
Ultrasonication	No	Laser print	53.3	78.0	–	–	Tatsumi et al. 2000
Ultrasonication	No	Old newsprint	49.0	52.1	–	–	Virk et al. 2013
Ultrasonication	Xylanase and laccase	Old newsprint	49.0	61.45	535.41	145.20	Virk et al. 2013
Microwaving and ultrasonication	Xylanase and laccase	Old newsprint	49.0	63.12	535.41	139.04	Virk et al. 2013

approaching that of virgin paper. However, there are environmental and ecological concerns related to paper recycling. The biggest problem is the solid waste rejects and sludge recovered from paper processing mills which are in the range between 5% and 40%, depending on different paper grades.

Chemical deinking is a messy process that uses a large quantity of chemicals and produces harmful by-products and emissions. So, enzymes provide an eco-friendly approach for waste paper deinking. Various enzymes from both fungal and bacterial origins have been reported for deinking of waste paper, for example, lipases, esterases, pectinases, hemicellulases, cellulases, cutinases, and ligninolytic enzymes such as laccases. These enzymes have been used separately and in combination to study their effect on ink removal and paper properties. Depending upon the paper grade and pulp type (chemical or mechanical), the paper contains varying amount of lignin in addition to cellulose and hemicellulose. The enzymatic deinking mechanism involves attacking the fiber surfaces or the ink for removal of ink particles. The probable mechanisms involved in enzymatic deinking are as follows (Figure 2.4):

a. Lipase and esterases attack directly on ink and degrade vegetable oil–based inks.
b. The cellulases depolymerize surface fibers to small fibrils (a phenomenon known as "peeling-off"), which leads to ink detachment from the surface. It also increases pulp freeness.
c. Hemicellulases release lignin by breaking the lignin–carbohydrate complex.
d. Cellulases, hemicellulases, pectinases, and laccases may alter the fiber surface or bonds in the vicinity of the ink particles. The free ink can be removed by washing or flotation.
e. Removal of small fibrils by lignocellulolytic enzymes changes hydrophobicity of ink particles, which also facilitates ink removal by flotation or washing. This change in hydrophobicity is due to the removal of small fibrils.
f. Amylases attack the starch layer on the paper surface in different waste papers, which helps in the removal of toner particles adhered on the paper surface and their separation from pulp suspension by flotation/washing.
g. Laccase has been used alone and in combination with other enzymes. It attacks lignin present in secondary fiber, thus loosening fiber bonds, and facilitates ink removal. It may also be involved in direct decolorization of inks.

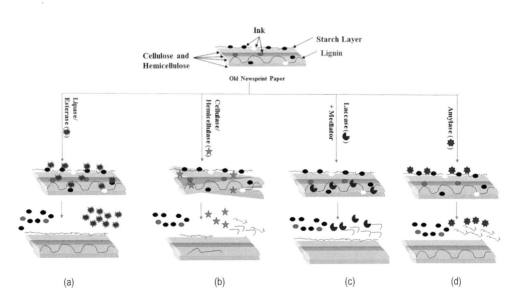

FIGURE 2.4 Probable mechanisms of deinking of old newsprint paper pulp by (a) lipase and esterase, (b) cellulase and hemicellulase, (c) laccase, and (d) Amylase.

Enzymatic deinking may involve a combination of these mechanisms depending upon the type of paper, ink composition, and enzyme cocktail. Various factors affect enzymatic deinking such as point of enzyme addition, a dose of enzyme, pulp consistency, and retention time.

2.5 RECENT DEVELOPMENTS IN ENZYMATIC DEINKING

Cellulolytic enzymes consist of three classes of enzymes based on their substrate specificities: endoglucanases (EC 3.2.1.4), exoglucanases (EC 3.2.1.74 and EC 3.2.1.91), and β-glucosidases (EC 3.2.1.21). Synergistic effect of these enzymes leads to the release of glucose. So, optimization of enzyme dosage and retention time is crucial to enzyme deinking using cellulase. Deinking of composite waste paper using cellulases from *Aspergillus* sp. AMA, *Aspergillus terreus* AN1, and *Myceliophthora fergusii* T4I increased brightness by 4.32%, 3.56%, and 3.01% ISO. Also, these cellulases showed deinking efficiency of 53%, 52.7%, and 40.32% for *Aspergillus* sp. AMA, *A. terreus* AN1, and *M. fergusii* T4I, respectively, which was better than conventional chemical deinking. Better deinking efficiency was attributed to multiple endoglucanases lacking a cellulose-binding module (CBD) (Soni et al. 2008). Monocomponent endoglucanases were found more efficient toward the deinking of MOW papers compared to crude cellulase (Elegir et al. 2000). Immobilization of cellulase reduced ink content for photocopied papers to 50 ppm and laser-printed papers to 110 ppm, which was much lower than free cellulase, i.e., 350 ppm for both photocopied papers and laser-printed papers (Zuo and Saville 2005). Addition of nonionic surfactant along with commercial cellulolytic enzyme at neutral pH resulted in lower residual ink particles and enhanced brightness and mechanical properties compared to a chemically deinked pulp. β-Glucosidase from *Penicillium purpurogenum* lowered the kappa number of newspaper pulp, resulting in enhanced brightness. Deinking of composite waste paper by using commercial cellulase (endo-1,4-cellulase) from *Trichoderma viride* reduced pulp drainage degree and increased strength properties.

Hemicellulases such as xylanases attack lignin–carbohydrate complexes to release lignin fragments along with ink particles from fiber surfaces assisting in deinking of waste paper. Xylanase from *Bacillus* sp. CKBx1D increased the brightness of laser-printed paper pulp by 6.3% ISO compared to control (Maity et al. 2012). The brightness of old newsprint (ONP) paper enhanced by 22% as compared to control pulp using cellulase-free xylanase from *Aspergillus niger* DX-23 (Desai and Iyer 2016). Coconut oleic acid, palmitic acid, rosin, and xylanase were used to prepare xylanase-loaded, biomass-based deinking agent. After flotation, brightness, ERIC, burst index, and tensile index of handsheet were 66.2% ISO, 223 ppm, 2.82 kPa m^2/g, and 28.4 Nm/g, respectively, which were higher than the control pulp (Zhang et al. 2017).

Laccases remove the surface lignin in the presence of chemical mediators and seem promising for the deinking of waste paper which mainly forms lignin-rich mechanical pulp. The ONP pulp deinked by laccase–violuric acid system (LVS) was 13% and 20% higher in tear and tensile strength, respectively, and the brightness was 4.2% ISO higher than that of the control. Deinking of ONP with other laccase-mediator systems (LMS) lowered the ERIC value and increased brightness and strength properties (Xu et al. 2011). *Myceliophthora thermophile* laccase improved the brightness of the inkjet-printed paper pulp by five units in the presence of acetosyringone as a mediator (Nyman and Hakala 2011). A novel *Rheinheimera* sp. laccase has been reported to deink ONP pulp without any mediator supplementation and was able to reduce ERIC by 62.2% (Virk et al. 2013).

A lipase preparation from *Pseudomonas aeruginosa* in combination with a neutral surfactant was used for effective deinking of soybean oil–based ink–printed paper. This may be due to partial degradation of the binder of the soybean oil–based inks by lipases, finally removing ink particles from paper (Morkbak et al. 1999). Mill trials with esterase on ONP and old magazine waste paper resulted in reduced sticky particle size and hence brighter paper (Jones and Fitzhenry 2003). The action of bacterial α-amylase on coated, colored, printed magazine resulted in larger ink particle reduction (Elegir et al. 2000). Use of xylanase and pectinase for deinking in combination with chemical deinking reduced the chemical consumption by 50% and a gain of 10.71%, 7.49%,

10.52%, and 6.25% in viscosity, breaking length, burst factor, and tear factor, respectively. Also, there was a decrease of 20.15% in biochemical oxygen demand (BOD) and 22.64% in chemical oxygen demand (COD) of effluents (Singh et al. 2012).

Effects of cellulase-xylanase combination, cellulase, and lipase were studied on 7-month-old newsprint paper followed by conventional flotation deinking. Cellulase-xylanase combination and lipase had constant tensile and burst indices and improved drainage rate, brightness, and ERIC, while a decrease in tear indices was observed (Spiridon and de Andrade 2005). 96% dirt removal efficiency was achieved by using a combination of cellulase and amylase and surfactant for the deinking process, which was higher as compared to cellulase (93%) and amylase (89%) (Elegir et al. 2000). A mixture of cellulase and xylanase (in equal concentration) with lipase in ratio 60:40 improved the deinking performance in comparison to cellulase-xylanase mixture alone. Along with improved pulp drainage and improved pulp yield, there was an increase of 3.2%, 7.4%, and 7.1% in breaking length, burst index, and tear index, respectively (Gu et al. 2004). Ibarra et al. (2012) compared deinking efficiency of commercial cellulase/hemicellulase with commercial laccase-mediator system for newspaper and magazine pulp. Reduction of kappa number and improvement in brightness of waste paper were evaluated using lignin peroxidase (LiP), manganese-dependent peroxidase (MnP), laccase from white rot fungi, and mixture of these enzymes (Selvam et al. 2005).

Enzyme cocktails produced pulp with superior quality in terms of brightness and strength properties in comparison to single enzymes due to the synergistic effect of enzyme combinations. Combination of xylanase/laccase for deinking of ONP pulp reduced chemical consumption by 50% along with an increase of 21.6%, 16.5%, 4.2%, 6.9%, 13%, and 10.3% in brightness, breaking length, burst factor, tear factor, viscosity, and cellulose crystallinity, respectively. And decrease in ERIC and kappa number was 65.8% and 22%, respectively. The surface appeared more fibrillar along with changes in surface functional groups. Furthermore, combining physical methods (sonication and microwaving) with xylanase/laccase system improved brightness by 28.8% and decreased ERIC by 73.9%. Morphological changes were observed after enzymatic treatment by scanning electron microscopy (SEM) analysis which revealed that fiber surface became rough and fibrillation appeared on the surface indicating delignification (Virk et al. 2013). Treatment of newspaper pulp with xylanase and cellulase from *Trichoderma longibrachiatum* MDU-6 increased brightness by 9.6% compared to control pulp. Fibrillation and perforation were revealed when deinked newspaper stained with safranin and malachite green was observed under scanning and transmission electron microscopes (Chutani and Shrama 2016). Combination of cutinase and amylase was able to deink MOW paper with an increase in brightness (8.99%), tensile index (64.11%), and tear index (78.2%) (Wang et al. 2018).

2.6 ENZYMATIC DEINKING: BENEFITS AND LIMITATION

Enzymatic deinking has been proven the economical and ecofriendly method of waste paper recycling. Enzyme-deinked pulp is superior to a chemical-deinked pulp in terms of higher brightness, enhanced strength properties, and lowered effective residual ink concentration. Enzymes also improve drainage property of pulp, hence better runnability on paper machines. There are some reports on loss in pulp yield due to enzymatic treatment which can be controlled by controlling enzyme dose and retention time. Deinking using some cellulase/hemicellulase preparations was reported to lower strength properties, which was compensated by treatment with laccase enzyme (Virk et al. 2013). Various reports suggest a decrease in chemical consumption after enzymatic deinking which in turn lowers COD and BOD, thus decreasing the load of pollutants in the final discharge; hence, minimal treatment of wastewater is required. As shown in Figure 2.3, traditional chemical deinking also involves bleaching stage for better paper brightness. But the use of enzymes also reduces the requirement of bleaching chemicals (Table 2.2).

TABLE 2.2

Effect of Enzymatic Deinking on Pulp and Paper Properties

Enzymes	Source	Type of Paper	% Change (+/–)							References
			Brightness % ISO	Breaking Length	Burst Index	Tensile Index	Tear Index	ERIC	Chemical Consumption	
Cutinase	*Thermobifida fusca*	Mixed office waste	(+)6.67	–	–	(+)40.9	(+)41.5	–	–	Wang et al. 2018
Cutinases	*Thermobifida fusca*	Old newsprint	(+)5.13	–	–	(+)4.30	–	–	–	Hong et al. 2017
	Fusarium solani		(+)4.38	–	(+)1.24	(+)2.39	(+)6.67	–	–	
Xylanase-loaded biomass	Novozymes	–	(+)4.1	–	(+)48.4	(+)31.6	–	(–)18.6	–	Zhang et al. 2017
Xylanase	*Aspergillus niger*	Old news print	(+)21.9	–	–	–	–	–	–	Desai and Iyer 2016
Xylanase	*Bacillus halodurans*	Old news print	(+)8.41	(–)8.01	(+)3.59	–	–	(–)47.97	–	Virk et al. 2013
Cellulase (OPTIMASE™ CX 40L)	Genencor International	Photocopier	(–)2.1	–	(+)15.3	(+)2.7	(–)21.9	(–)24.6	–	Pathak et al. 2011
Cellulases	*Aspergillus* sp.	Composite paper	(+)5.33	–	(+)70.5	(+)46.23	(+)2.15	–	–	Soni et al. 2008
	Aspergillus terreus		(+)4.40	–	(+)74.78	(+)52.72	(+)5.80	–	–	
	Myceliphthora sp.		(+)1.0	–	(+)70.5	(+)52.13	(+)9.78	–	–	
Laccase	*Rheinheimera* sp.	Old news print	(+)12.04	(+)7.23	(+)2.45	–	–	(–)62.22	–	Virk et al. 2013

(Continued)

TABLE 2.2 (Continued)
Effect of Enzymatic Deinking on Pulp and Paper Properties

Enzymes	Source	Type of Paper	Brightness % ISO	Breaking Length	Burst Index	Tensile Index	Tear Index	ERIC	Chemical Consumption	References
			% Change (+/−)							
Cellulase and xylanase	*Penicillium rolfsii* c3-2(1) IBRL	Laser-printed paper	–	–	(–)52	(–)32	(–)60	(–)35	–	Lee et al. 2017
Cellulase and xylanase	*Trichoderma longibrachiatum* MDU-6	Old news print	(+)9.6	–	–	–	–	–	–	Chutani and Shrama 2016
Xylanase and pectinase	*Bacillus pumilus* AJK	School waste paper	(+)10.21	–	(+)10.52	–	(+)6.25	(–)25.03	(–)50	Singh et al. 2012
Laccase and xylanase	*Rheinheimera* sp. and *Bacillus halodurans*	Old news print	(+)21.63	(+)16.57	(+)4.24	–	–	(–)65.8	(–)50	Virk et al. 2013
Cellulase, xylanase, and lipase	Commercial	Old newsprint	–	(+)3.2	(+)7.4	–	(+)71	–	–	Gu et al. 2004
Cutinase and amylase	*Thermobifida fusca* and Novozymes	Mixed office waste	(+)8.99	–	–	(+)64.11	(+)78.2	–	–	Wang et al. 2018
Xylanase, lipase, and amylase	–	Mixed office waste	–	–	(+)22.9	(+)3.2	(+)12.8	(–)19.9	–	Gao et al. 2018

2.7 CONCLUSION AND FUTURE PROSPECTS

Enzymatic deinking is energy-saving, cost-effective, and environmentally friendly alternate to conventional chemical deinking process. It is also able to deink composite waste paper which is not easily deinkable by chemical methods. Enzyme conditions can be optimized according to paper type, ink type, and printing method. Enzymes directly affect the paper properties and ink particle size. Synergistic effect of two or more enzymes has more potential for future applications. Recent advances in fermentation processes have lowered the cost of enzyme production. Also, the heterologous expression of deinking enzymes holds the key for cost-effective production of these enzymes. Nanotechnology can be utilized to improve enzymatic deinking performance, allowing larger application in the paper recycling industry.

GLOSSARY

Basis weight or grammage (GSM): is defined as the weight per unit area and expressed as gram per square meter (g/m² or GSM).

Breaking length (km): is the length of a paper strip of uniform width such that when the strip is suspended by one end, it would break of its own weight.

Brightness (%ISO): is a measure of numerical reflectance of paper with respect to the blue spectrum of light at wavelength 457 nm. Brightness instrument employs 45° illumination and 0° viewing geometry.

Burst index (kPa*m²/g): is hydrostatic pressure required to rupture the material when pressure is increased at a constant and controlled rate.

Effective residual ink concentration (ERIC): is quantification of the amount (ppm) of residual ink left in recycled paper.

Tear index (mN*m²/g): is defined as tearing strength divided by basis grammage weight.

Tensile index (Nm/g): is a measure of the force required to tear the paper when pulled at opposite ends and in opposite directions.

REFERENCES

AF & PA (American Forest & Paper Association) (2016) 56th Annual Survey of Paper, Paperboard and Pulp Capacity, AF & PA, Washington, DC.

Annual Report (2015–2016) Central Pulp and Paper Research Institute (CPPRI). Saharanpur (U.P.), India.

Annual Statistics (2015) Confederation of European Paper Industries (CEPI). Brussels, Belgium.

Bajpai P (2014) Deinking with enzymes. In: *Recycling and Deinking of Recovered Paper*, ed. Bajpai P Elsevier, 139–153.

Bajpai P, Bajpai PK, Kondo R (1999) Enzyme deinking. In: *Biotechnology for Environment Protection in the Pulp and Paper Industry*, ed. Bajpai P, Bajpai PK, Kondo R Springer, Berlin, Heidelberg, and New York, 91–108.

Chutani P, Shrama KK (2016) Concomitant production of xylanases and cellulases from *Trichoderma longibrachiatum* MDU-6 selected for the deinking of paper waste. *Bioprocess Biosyst Eng* 39:747–758.

Desai DI, Iyer BD (2016) Biodeinking of old newspaper pulp using a cellulase-free xylanase preparation of *Aspergillus niger* DX-23. *Biocatal Agric Biotechnol* 5:78–85.

Elegir G, Panizza E, Canetti M (2000) Neutral enzyme deinking of office waste with amylase/cellulase xerographic assisted mixture. *TAPPI J* 83:40–44.

Ferguson LD (1992) Deinking chemistry: Part 2. *TAPPI J* 75(8):J49–J58.

Gao S, Li L, Lin Y, Han S (2018) Deinkability of different secondary fibers by enzymes. *Nord Pulp Pap Res J* 33(1):12–20.

Gu QP, You JX, Yong Q, Yu SY (2004) Enzymatic deinking of ONP with lipase/cellulase/xylanase. *China Pulp Paper* 23(2):7.

Hong R, Su L, Chen S, Long Z, Wu J (2017) Comparison of cutinases in enzymic deinking of old newsprint. *Cellulose* 24(11):5089–5099.

Ibarra D, Monte MC, Blanco A, Martinez AT, Martinez MJ (2012) Enzymatic deinking of secondary fibers: Cellulases/hemicellulases versus laccase-mediator system. *J Ind Microbiol Biotechnol* 39:1–9.

Jones DR, Fitzhenry JW (2003) Esterase type enzymes offer recycled mills an alternative approach to stickies control. *Pulp Pap Canada* 77(2):28–31.

Lee KC, Tong WY, Ibrahim D, Arai T, Murata Y, Mori Y, Kosugi A (2017) Evaluation of enzymatic deinking of non-impact ink laser-printed paper using crude enzyme from *Penicillium rolfsii* c3-2(1) IBRL. *Appl Biochem Biotechnol* 181:451–463.

Lim GI, Hwang IS, Kim JW, You KS, Ahn JW, Han C (2012) Sonication effects on froth flotation for deinking from old newspaper. *J Korean Inst Resour Recycl* 21:44–49.

Mahagaonkar M, Banham P, Stack K (1997) The effects of different furnishes and flotation conditions on the deinking of newsprint. *Prog Pap Recycl* 6(2):50–57.

Maity C, Ghosh K, Halder SK, Jana A, Adak A, Das Mohapatra PK, Pati BR, Mondal KC (2012) Xylanase isozymes from the newly isolated *Bacillus* sp. CKBx1D and optimization of its deinking potentiality. *Appl Biochem Biotechnol* 167:1208–1219.

Martin-Sampedro R, Miranda J, García-Fuentevilla LL, Hernández M, Arias ME, Diaz MJ, Eugenio ME (2015) Influence of process variables on the properties of laccase biobleached pulp. *Bioprocess Biosyst Eng* 38:113–123.

Morkbak AL, Degn P, Zimmermann W (1999) Deinking of soybean oil based ink-printed paper with lipases and a neutral surfactant. *J Biotechnol* 67:229–236.

Nyman K, Hakala T (2011) Decolorization of inkjet ink and deinking of inkjet-printed paper with laccase–mediator system *Bioresources* 6:1336–1350.

Pathak P, Bharadwaj NK, Singh AK (2011) Optimization of chemical and enzymatic deinking of photocopier waste paper. *Bioresources* 6:447–463.

Selvam K, Swaminathan K, Rasappan K, Rajendran R, Michael A, Pattabi S (2005) Deinking of waste papers by white rot fungi *Fomes lividus*, *Thelephora* sp. and *Trametes versicolor*. *Nat Env Poll Tech* 4:399–404.

Singh A, Yadav DR, Kaur A, Mahajan R (2012) An ecofriendly cost effective enzymatic methodology for deinking of school waste paper. *Bioresour Technol* 120:322–327.

Spiridon I, de Andrade AM (2005) Enzymatic deinking of old newspaper (ONP). *Prog Pap Recycl* 14(3):14.

Soni R, Nazir A, Chadha BS, Saini HS (2008) Novel sources of fungal cellulases for efficient deinking of composite paper waste. *Bioresources* 3:234–246.

Tatsumi D, Higashihara T, Kawamura S, Matsumoto T (2000) Ultrasonic treatment to improve the quality of recycled pulp fibre. *Wood Sci* 46:405–409.

Turai LL, Teng CH (1978) Ultrasonic deinking of wastepaper. *TAPPI J* 61(2):31–34.

Virk AP, Puri M, Gupta V, Capalash N, Sharma P (2013) Combined enzymatic and physical deinking methodology for efficient eco-friendly recycling of old newsprint. *PLoS ONE* 8(8):e72346.

Wang F, Zhang X, Zhang G, Chen J, Sang M, Long Z, Wang B (2018) Studies on the environmentally friendly deinking process employing biological enzymes and composite surfactant. *Cellulose* 25(5):3079–3089.

Xu QH, Wang YP, Qin MH, Fu YJ, Li Z, Zhang FS, Li JH (2011) Fiber surface characterization of old newsprint pulp deinked by combining hemicellulase with laccase-mediator system. *Bioresour Technol* 102:6536–6540.

Zhang M, Li Z, Yang R (2017) Preparation of xylanase loaded biomass-based deinking agents and their application in secondary fiber recycling. *Bioresorces* 12(2):2818–2829.

Zhenying S, Shijin D, Xuejun C, Yan G, Junfeng L, Hongyan W, Zhang SX (2009) Combined de-inking technology applied on laser printed paper. *Chem Eng Process Process Int* 48:587–591.

Zuo Y, Saville BA (2005) Efficacy of immobilized cellulase for deinking of mixed office waste. *J Pulp Pap Sci* 31:3.

3 Advances in Industrial Wastewater Treatment

Maulin P. Shah
Enviro Technology Limited

CONTENTS

3.1   INTRODUCTION

Population growth leads to rapid industrialization and urbanization. Therefore, the world has been facing two major problems. The first is the decline in fossil fuels contamination, and second is the natural environment (Shah et al. 2013a-c). Human activities in the water for domestic use, agriculture, aquaculture, industry, transport, and radioactive wastes; other than any industrial pollution, mobile combustion, burning of fuel, ionizing radiation, and cosmic radiation particles in the air; and household wastes, industrial wastes, agricultural chemicals and fertilizers, acid rain, and animal wastes are negative impacts on biotic and abiotic components of several natural ecosystems (Shah et al. 2013b-d). Although the water, air, and land are equally important, drinking water is especially of higher significance. Two-thirds of the earth's surface consists of water. This is also undoubtedly the most precious natural resource that exists on our planet (Shah et al. 2013d). The fact that industries are polluting water resources is a common occurrence emanating from effluent. Especially, drinking water has become largely contaminated because in many cases, the water lost its originality. The highly colored wastewater discharge in the sources of drinking water will soon make the planet into a desert. So, it is a serious environmental matter. Indiscriminate and uncontrolled unloading of industrial and municipal waste in the environmental sink has become a topic of major global concerns in the development and densely populated countries such as India (Chandra and Kumar 2017a,b; Chandra et al. 2018). They are a main source of direct and often continuous supply of pollutants to aquatic ecosystems with long-term implications for ecosystem function, including changes in food availability and an extreme threat to self-regulation of the biosphere (Shah et al. 2013a). Based on the nature of different amounts of pollutants, they can be disposed of directly into the environment or indirectly through public sewers. Wastewater from the industry sanitary waste from workers, process waste from production, wash water, and relatively uncontaminated water from heating and cooling operation of high levels of pollutants in river water systems lead to an increase in the biological oxygen demand, chemical oxygen demand (COD), total dissolved solids, total suspended solids, toxic metals such as Pb, Cn, Cd, Cr, and Ni to make such water unsuitable for drinking and irrigation (Shah et al. 2013e). The Indian textile industry is one of the leading industries in the world, and it depends largely on textile manufacturing, and export section plays an important or a crucial role in India's economical growth (Shah 2014). India earns almost 25%–27% of its total foreign exchange through the export of textiles. In addition, the textile industry of India also contributes 14% of total industrial production of the country, whereas it sums 4% of the total GDP of India (Robinson et al. 2001). The textile industry in India is the largest provider of employment after agriculture. This industry is one of the first industries of India that came into existence, and nowadays, it is currently the second largest industry in the world after China. Over the years, this industry has proven to be the supplier of the basic needs of the population (Savin and Butnaru 2008). In this chapter, we have tried to attempt to explain the current state of the problem that is mainly created by the presence of a large number of azo dyes in textile wastewaters. In order to understand the problems of the textile industry wastewater, a variety of processes within the textile sector have been highlighted in this chapter, which provides to increase wastewater requiring treatment. Various organic and inorganic chemical compounds that are used in the textile industry and classifications are discussed in a systematic way. Since many of these dyes have toxic effects on human health and the ecosystems, all these things are also discussed briefly.

3.2 BRIEF HISTORY OF DYES

Since the beginning of mankind, people have been using dyes for dyeing and painting their environment, their skin, and their clothes. The first evidence of the use of coloring materials by the man can be found as far as 15,000–9,000 BC, in the walls of the cave of Altamira in Spain. The drawings were made with inorganic pigments such as soot, manganese oxide, ochre, and hematite (Akan et al. 2009). Historically, there is a dye derived from animal sources. This is very important, although time does not matter and is not commercially available. The shoot is purple (Figure 3.1), and the pigment itself is not the molluscs, however, when the extracted precursor can be converted to dyeing by air or light. The presence of this dye goes as far as 1,400 BC in the Late Bronze Age as recently discovered in Lebanon. It has always been rare and expensive to be used by Roman emperors and high ranking ecclesiastics (Kuberan et al. 2011).

Another former dye, which is still in use, but is not naturally occurring today, is Indigo. It is extracted from *Indigofera tinctoria* fermentation and had a unique blue color (see Figure 3.2). It was used by the Romans as a pigment because it was chemically reduced to become water soluble. First produced synthetically by Adolf von Baeyer in 1880 and, in fact, is used to color the denim

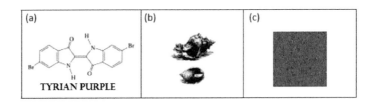

FIGURE 3.1   (a) Chemical structure of tyrian purple. (b) Sea shells from which tyrian purple was extracted. (c) A purple-dyed fabric.

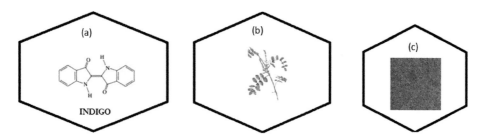

FIGURE 3.2   (a) Chemical structure of indigo. (b) *Indigofera tinctorium*. (c) Denim.

(Faryal and Hameed 2005). Until the late 19th century, natural colors, obtained mainly from plants, have been the most important dyes used in the textile dyeing procedures. The main disadvantages of using natural dyes are the need for a number of stages in the process of dyeing, the diversity of sources and procedures relating to the application, rapidly changing trends, and the demand for good properties of different substrates, the strength of which require extensive database describing potential applications (Yusuff and Sonibare 2004).

The groundbreaking synthesis of mauveine by W.H. Perkins started the era of synthetic dyes with chemical and physical properties of more modern standards, better quality, and more reproducible use of the techniques. It has also allowed the development and expansion of the use of certain products (Babu et al. 2007). For example, the development of synthetic fibers such as polyester and cellulose tri-acetate would be difficult without the design and synthesis of appropriate characteristics of dyes. Since then, thousands of dyes have been synthesized, and dye production has become an important part of the chemical industry. Today, when the care of the environment is an important issue, it is tempting to assume that the use of natural colors is an alternative that respects the environment in the current prac-tice. Several researchers study the use of natural dyes in the modern dyeing industry. Some advantages of using these compounds are the absence of toxicity on humans, the use of sustainable sources, and adjusting the tracks in natural biodegradation of dyebaths released (Jin et al. 2007).

3.3   CLASSIFICATION

All the molecules absorb electromagnetic radiation, but there are differences in the specific absorbed wavelengths. Some molecules of the light absorption are in the visible spectrum (400–800 nm), and as a result, they are themselves colored. The dyes are molecules having delocalized electron systems with conjugated double bonds, which contain two groups: the chromophore and the auxochromic. The chromophore is a group of atoms, which controls the color of the dye, and it is usually an electron-withdrawing group. The main chromophores are $-C = C-$, $-C = N-$, $-C = O$, $-N = N-$, $-NO_2$, and $-NO$ groups. It is an auxochromic electron-donating substituent, which can enhance the color of the chromophore by changing the total energy of the electron system and provides solubility and compliance with the dye on the fiber. The main auxochromes are $-NH_2$, $-NR_2$, NHR, $-COOH$, $-SO_3H$, $-OH$, and $-OCH_3$ groups (15). On the basis of the chemical structure or chromophore, 20–30 different dye groups can be identified. Azo, anthraquinone, phthalocyanine, and triarylmethane dyes are quantitatively the most important chromophores (Figure 3.3).

The majority of the commercial dyes are classified in terms of color, structure, or method of application in the color index, which is adapted to 3 months since 1924 by the "Society of Dyers and Colourists" and "American Association of Textile Chemists and Colorists." The latest edition of the Color Index contains about 13,000 different dyes. Each dye is assigned to a color index. Generic names are defined by the application and color.

Application and characteristics of dyes are listed in Table 3.1.

Some common classes of dyes, which are based on their chromophore groups, are shown in Figure 3.4.

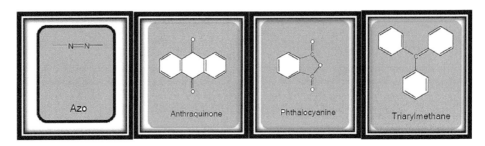

FIGURE 3.3 Chormophore groups.

TABLE 3.1
Dyes Classes and Their Characteristics

Class of Dyes	Characteristics
Direct dyes	Their flat shape and length enables them to bind alongside cellulose fiber and maximizing the Van der Waals and dipole hydrogen bonds. Only 30% of the 1,600 structures still produced because of their lack of color fastness during washing. The most common structures are almost always sulfonated azo dyes.
Reactive dyes	To form covalent bonds with –OH, –NH, or –SH groups in cotton, wool, silk, and nylon. The problem of effluents associated with the use of color these dyes is due to the hydrolysis of reactive groups that occurs during the dyeing process. The most common structures are azo, metal complex azo, anthraquinone, and phthalocyanine.
Acid dyes	Highly soluble in water due to the presence of sulfonic acid groups. Forming ionic interactions between the features of the protonated fibers ($-NH3^+$) and negative charge of the dyes. Also, Van der Waals, dipolar links and form hydrogen. The most common structures are azo, anthraquinone, and triarilmetano.
Mordant dyes	Mordants are generally metal salts, such as sodium or potassium dichromate. They act as "fixing agent" to improve color fastness. They used wool, leather, silk, and cellulose fibers modified. The most common structures are azo or oxazina triarilmetano.
Basic dyes	The basic dyes work well in acrylics because of the strong ionic interaction between functional groups of dyes such as $-NR3+$ or $=NR2+$ and the negative charges in the copolymer. The most common structures are azo, diarilmetano, triarilmetano, and anthraquinone.
Disperse dyes	These are nonionic structures with a polar functionality that $-NO_2$ and –CN improve water solubility, Van der Waals forces, dipole forces, and color. They are generally used in polyester. The most common structures are azo, nitro, anthraquinones, or azo metal complex.
Vat dyes	Vat dyes are insoluble in water but may be solubilized using an alkali reduction (sodium dithionite in the presence of sodium hydroxide). The leuco form produced is absorbed by the cellulose (Van der Waals forces) and can oxidize again, usually hydrogen peroxide, to its insoluble form. The most common structures are anthraquinones or indigoides.
Pigment dyes	These compounds or salts are insoluble nonionic, representing 25% of total trade names of dyes, and retain their crystalline structure or particles throughout its implementation. The most common structures are azo or metal phthalocyanine complex.
Sulfur dyes	Sulfur dyes are aromatic heterocyclic polymer complexes containing rings, which represent around 15% of global production of contrast medium. Dyeing with sulfur dyes (mainly cellulose fibers) means reduction and oxidation processes, comparable with the dye vat.
Solvent dyes	Nonionic dyes are used for dyeing substrates, which can dissolve as plastics, varnishes, inks, and waxes. These are often used for textile processing. The most common structures are diazo molecular compounds that undergo some change, triarilmetano, anthraquinone, and phthalocyanine.
Other classes	Food dyes are not used as textile dyes. The natural dyes used in textile processing operations are very limited. They mask the fluorescent brighteners yellowish tint of natural fibers by absorbing ultraviolet light and weakly emitting blue light. They are not registered in a separate class in the color index; many metal complex dyes can be found (usually chromium, copper, cobalt, or nickel). The metal complex dyes are generally azocompounds.

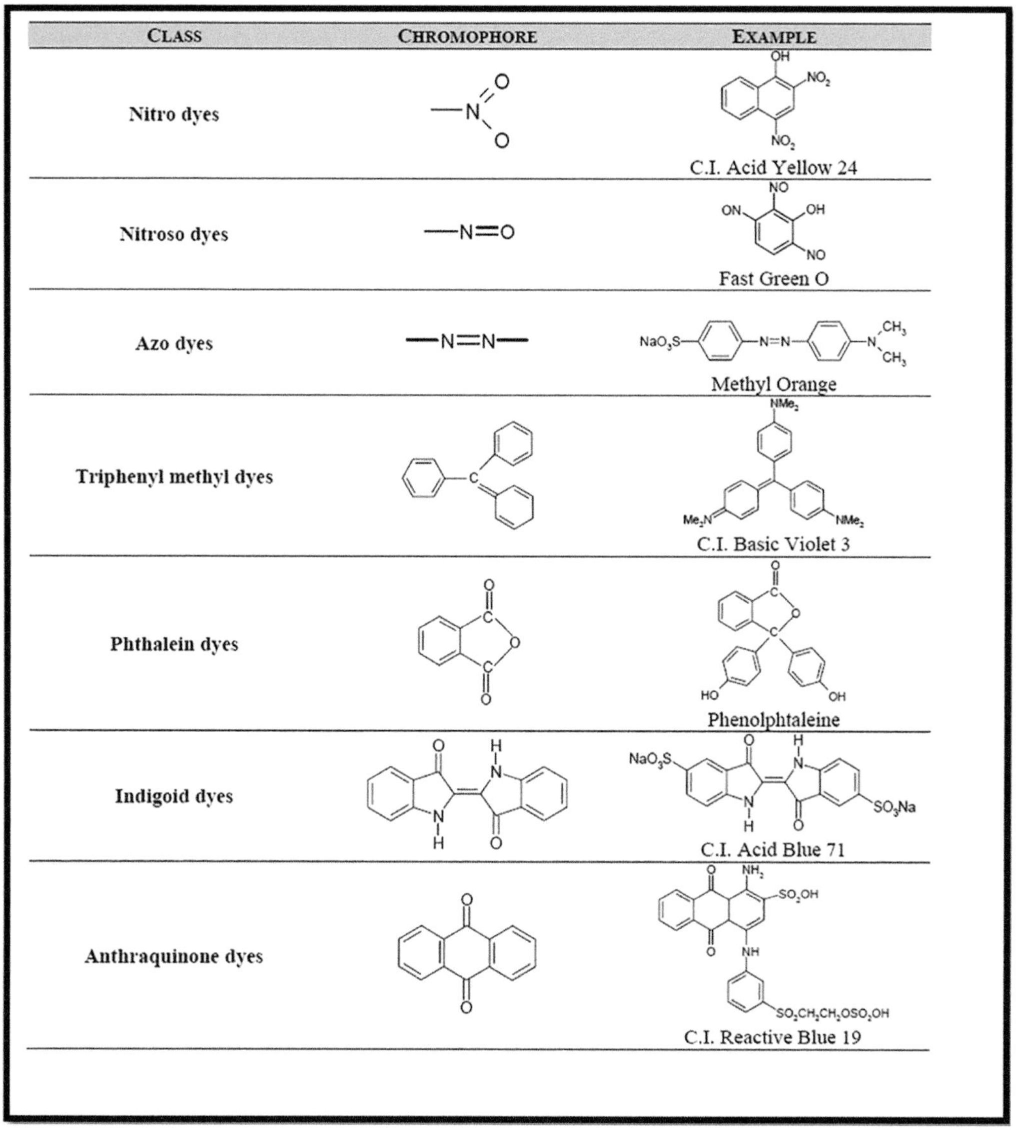

FIGURE 3.4 Some common classes of dyes, which are based on their chromophore groups.

3.4 AZO DYES AND THEIR ECOTOXICITY

The azo dyes are the largest group of synthetic dyes and pigments of industrial application due to the relatively simple synthesis and virtually unlimited number and type of substituents. In world production, these organic dyes are currently estimated at 460,000 tons/year, with almost 60,000 tons/year lost in the effluent during application and production. Azo dyes contain at least one N = N double bond, and many different structures are possible. Monoazoico dyes have a single N = N double bond, whereas diaztriazol poliazo dyes contain two, three, or more N = N double bonds, respectively. Azo groups are commonly associated with benzene and naphthalene rings, but heterocyclic aromatic enolizable aliphatic groups can also be joined. The general structure of the azo dye molecule can be seen in Figure 3.5.

Ar-N=N-R

FIGURE 3.5 General structure of azo dyes (where R can be an aryl, heteroaryl or – CH = C(OH) – alkyl derivative).

These side groups are necessary to obtain colors with different shades and intensities. Azo dyes go to the shade of greenish yellow to orange, red, purple, and brown. The colors depend largely on the chemical structure, whereas the different shades depend more on physical properties. However, the disadvantage, which limits their commercial application, is that most of them are reds and none are green. The synthesis of most of the azo dyes involves diazotizing a primary aromatic amine to give a diazonium salt. The diazonium compound is then coupled with one or more nucleophilic agents. The amino and hydroxyl groups are the only coupling agents used. The coupling reaction is generally in para position with respect to the amino group or to the hydroxyl groups (Fu and Viraraghavan 2001). The general regime of dye azo synthesis is illustrated in Figure 3.6.

The azo compound is considered to be the most unstable part of an azo dye. The enzyme-linked, but the thermal and photochemical properties, can also occur. Degradation of azo dyes may be obtained by reduction or oxidation. Reducing releases colorless composite parts (Figure 3.7).

A variety of azo dyes can be reduced by many different bacteria, all of which suggests nonspecific nature of this reaction. The potential to reduce azo dyes therefore be considered a universal property of anaerobic bacteria. A distinction accepted the mechanisms for reducing azo dyes that can be made between reducing enzyme directly or indirectly by enzymatic reduction and chemical reduction (Figure 3.8).

The mechanism of oxidation, in general, is more difficult to establish because of the high reactivity of free radicals normally involved in the degradation process. The chemical oxidation of azo dye for chlorine in an acidic medium is shown in Figure 3.9 (Puvaneswari et al. 2006). A similar route was observed in the enzymatic oxidation (Ghoreishi and Haghighi 2003). The electron withdrawal of azo-character groups generates electron deficiency, thus, making the compounds less susceptible to oxidative catabolism; consequently, many of these chemicals tend to persist under aerobic environmental conditions.

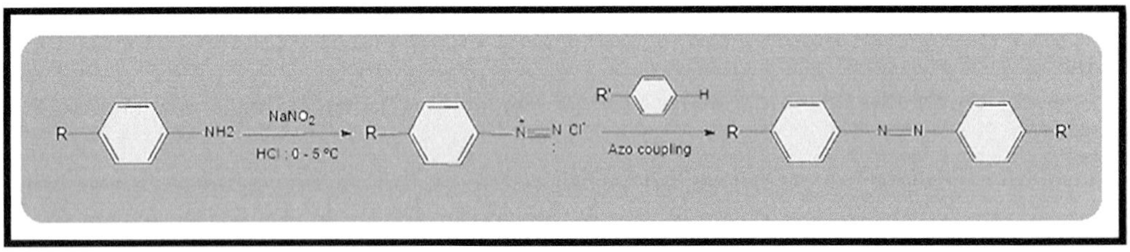

FIGURE 3.6 Synthesis of azo dye.

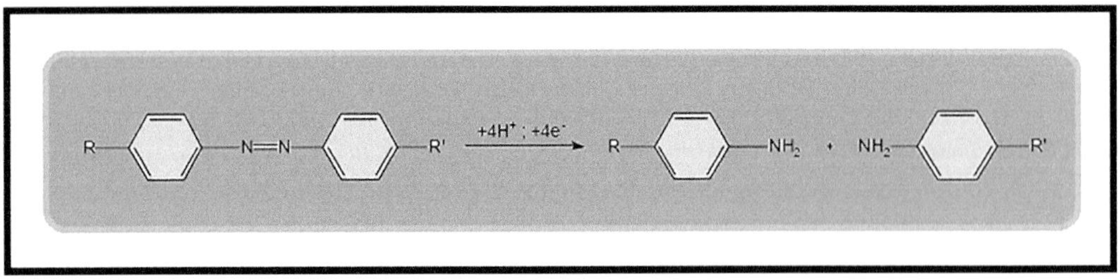

FIGURE 3.7 Reduction of azo dye.

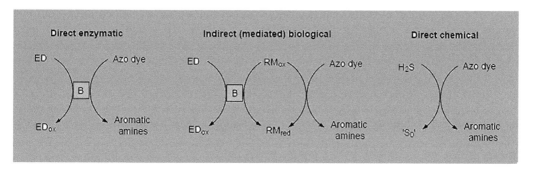

FIGURE 3.8 Schematic representation of the different mechanisms of the azo dye reduction (ED = electron donor; B = bacteria (enzyme system); RM = redox mediator).

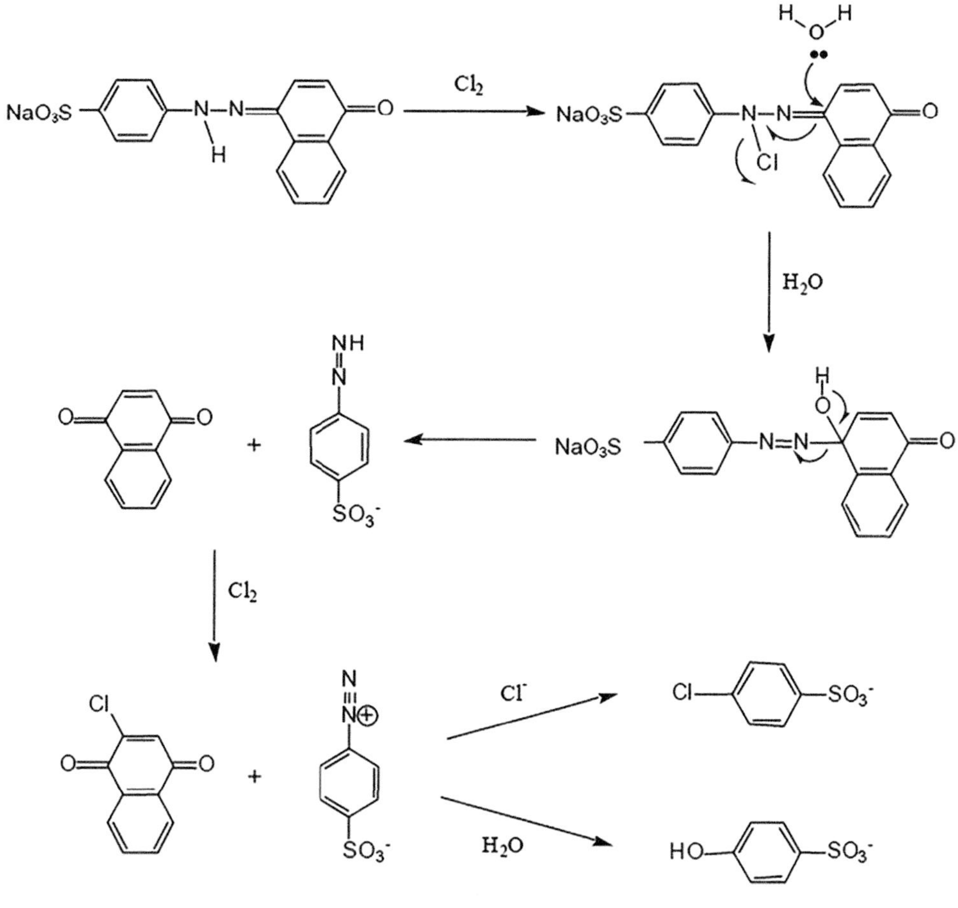

FIGURE 3.9 Acidic media containing orange azo dye oxidation.

Most of the dyes are perceptible in water at concentrations as low as 1 mg/L. In textile-processing wastewater, typically, the dye concentration is between 10 and 200 mg/L (Forgacs et al. 2004) and is, therefore, usually highly colored and discharged in open waters, which presents an esthetic problem. Dyes are designed and chemically photolytically stable; they are very persistent in natural environments. The release of dyes can therefore present a risk of ecotoxic and presents a threat of bioaccumulation, which may ultimately affect male traffic through the food chain. The trend of dyestuffs bioaccumulation in fish has been extensively investigated in research promoted by the association of ecological and toxicological organic pigments and dyes manufacturers. The bioconcentration factor of almost 70–75 different kinds of dyes application was determined and compared with the partition coefficient of n-octanol/water with each other. Soluble dyes in water had low Kow. In addition, research has been conducted on the effects of dyestuffs and dye-containing effluents on the activity of aerobic and anaerobic bacteria in the wastewater treatment systems. The acute toxicity of dyestuffs is generally low. Algal growth tested in, respectively, 60 and 50 commercial dyes, in general, was not inhibited at concentrations below 1 mg/L (Chacko and Subramaniam 2011). Test fish mortality showed that 2% of 3,000 sales dyes tested had LC50 values below 1 mg/L. The most acutely toxic to fish are basic dyes, especially those that have a structure of triphenylmethane. Many fish also appear to be relatively sensitive to acidic dyes. Therefore, likelihood of human mortality due to acute toxicity colorant is probably very low (Saratale et al. 2011). However, acute sensitization reactions by humans often occur in dyes. Especially, disperse dyes have found to cause allergic reactions, i.e., eczema or contact dermatitis. In general, all of genotoxicity associated with traces of aromatic amines benzidine and some aromatic amines, toluene, naphthalone, and aniline remains. Toxicity of aromatic amines greatly depends on the spatial structure of the molecule or, in other words, the location of amino group (Gogate and Pandit 2004a). In an extensive review of literature data on genotoxicity and carcinogenicity of sulfonated aromatic amines, it was concluded that sulfonated aromatic amines, in contrast to some of their unsulfonated analogs, have generally no or very low genotoxic and tumorigenic potential.

3.5 DIFFERENT AZO DYE REMOVAL TECHNOLOGIES

A variety of physical, chemical, and biological pretreatments and the main techniques of treatment and posttreatment can be used to eliminate the color of wastewater containing dyes. Physicochemical techniques comprise membrane ion exchange, ion pair extraction filtration, coagulation/flocculation, precipitation, flotation, adsorption, mineralization, ultrasound, electrolysis, advanced oxidation, and chemical reduction (Gogate and Pandit 2004b). Biological techniques comprise fungal and bacterial biodegradation in aerobic, anaerobic, anoxic, or anaerobic-aerobic treatment processes/combined. There are several factors that determine the technical and economic viability of each individual dye removal technique:

- Type of dye
- Composition of wastewater
- Dosage and costs of chemicals required
- Operating costs (energy and materials)
- Destination management and environmental costs of waste products generated

In general, each technique has its limitations. Using an individual process often cannot be sufficient to achieve complete bleaching. Strategies for eliminating dyes consist, therefore, largely a combination of different techniques. The most important techniques of removing dye are discussed briefly in the following sections (Papic et al. 2004).

3.5.1 PHYSICAL METHODS

3.5.1.1 Membrane Filtration

Reverse osmosis and nanofiltration can be applied as primary or posttreatment process for the separation, purification, and reuse of salts and larger molecules, including dyes and dye bath effluent of major wastewater processing textiles. In reverse osmosis, effluent is forced under moderate pressure through a semipermeable membrane to form a purified and concentrated permeate. Membrane filtration process can remove up approximately 98% of the impurities in the water with a molecular relationship then greater too (Arslan-Alaton 2003). Although membrane filtration is an effective cleanup technology to water and wastewater contaminants, it has a relatively high cost capital and high operation costs. Although the degree of efficiency, reverse osmosis, and nanofiltration have some disadvantages. The membranes must be cleaned regularly and can be attacked by dye materials or other constituents of the effluent changing surface characteristics (Muruganandham and Swaminathan 2004). Moreover, these techniques have a relatively high capital and high operating costs.

3.5.1.2 Flocculation and Coagulation

The inorganic coagulants—lime, aluminum, magnesium, and iron salts—have been used for coagulation in the treatment of textile-processing wastewater to partly remove total suspended solids, biochemical oxygen demand (BOD), COD, and color over many years (Neamtu et al. 2002). The principle of the process is the addition of a coagulant followed by a generally rapid association between the coagulant and the pollutants. Thus formed coagulates or flocks subsequently precipitated are then removed by flotation, settling, filtration, or other physical techniques to generate a sludge that is normally further treated to reduce its water content and toxicity (Shu and Huang 1995). Organic anionic, cationic, or nonionic coagulant polymers have been developed in the past years for color removal treatments, and in general, they offer advantages over inorganic: lower sludge production, lower toxicity, and improved color removal ability (Shu and Chang 2005). The use of any adsorbent, whether it is an ion exchange, activated carbon, or a high-surface inorganic material, to eliminate species of liquid fluid depends on the balance between adsorbed and free species (Shu and Chang 2005). The coloring effluents are multicomponent mixtures with different absorption degrees and concentrations. In the same cases, lower limits are formed with the adsorbent, and certain materials can be released into the stream. The range of adsorbents described in the literature for application covers the range of activated carbons, inorganic surface product materials, synthetic ion exchange resins, and adsorbents based on chitin, synthetic cellulose, and other bioadsorbents (Sarria et al. 2003). Standard ion exchange systems have not been widely used for the treatment of coloring effluents because of the high cost of organic solvents to regenerate the ion exchanger and because of the very large inorganic filler from the effluent. Carbon is reasonably effective in removing many different dyes from streams. However, the actual cost of the high-temperature regeneration process, including the replacement cost and the yield of the waste sludge, makes their regeneration unattractive for small business (Gogate and Pandit 2004b). As a result, several low-cost adsorbents have been studied as an alternative to activated carbon. The use of inorganic adsorbents such as silica, clay, and ash on the top surface was judged by a number of dyes (Chen et al. 2002; Shu and Huang 1995; Cisneros et al. 2002; Aleboyeh et al. 2003). Its effectiveness depends on the type of dye in the effluent stream or, more particularly, in the charge on the dye molecule. Bioadsorbents are biodegradable natural polymers at low prices, which have a high binding capacity of the dye and can act as ion exchangers. Various biomaterials can be used as bioadsorbents: corn, wheat, rice husk, wood chips, sawdust, bark, cotton bone bagasse waste, cellulose, microbial biomass, fungal biomass, yeast biomass, etc.

3.5.1.3 Ion Exchange

Standard ion exchange is not widely used for the treatment of wastewater containing dyes, mainly of the opinion that the ion exchangers cannot adapt to a wide variety of dyes and coloring agents and performance strongly influenced by the presence of additives (Neamtu et al. 2002). In this technique, the wastewater enters into the ion exchange resin, until all the available exchange sites are saturated. Anionic and cationic dyes are effectively removed in this process. A disadvantage of this method is the high cost of organic solvents to regenerate the ion exchanger.

3.5.2 CHEMICAL TREATMENTS

3.5.2.1 Photochemical Treatment

Processes such as UV/H_2O_2, UV/TiO_2, reactive UV/Fenton, and UV/O_3 and other photochemical methods are based on a result of UV radiation and formation of free radicals. Degradation is due to the production of high concentrations of hydroxyl radicals, and the dye molecule decomposes CO_2 and H_2O. Speed of flotation affects the intensity of UV-radiation, pH, color structure, and composition of the dye bath. When the oxidizing agent H_2O_2 is used, UV light activates the decomposition of H_2O_2 to two hydroxyl radicals. This method does not produce sludge and reduces odors (Muruganandham and Swaminathan 2004). It is also possible to efficiently use sunlight or UV radiation, which is significant cost saving, particularly for large-scale operations. Faster, cheaper, and more efficient photocatalytic processes are based on the solid semiconductor materials for catalysis, mainly TiO_2 particles. TiO_2 catalyzed by UV treatment of a wide range of dyes may be mineralized (Arslan and Balcioglu 2001). Light degradation dyes of this method depend greatly on the chemical structure of the dye.

3.5.2.2 Fenton Treatment

Salts of H_2O_2-Fe (II) are used when the wastewater is resistant to biological treatment or is toxic to microbial biomass. Fenton's reaction occurs in an acidic pH to form a strong, oxidizes the hydroxyl radical and ferric iron. Both forms of iron are coagulants, thus conferring the dual function of oxidation and coagulation to this process. Therefore, the mechanism of removal of color also means union or sorption of dyes dissolved in floccules formed (Azbar et al. 2004). The formation of large amounts of sludge-concentrated dyes and iron is the main disadvantage of this procedure. Due to the removal of color, this procedure is suitable for different classes of dyes.

3.5.2.3 Ozone

Ozone is an excellent oxidizing agent for its immense unsteadiness with respect to chlorine and H_2O_2. Ozone is used to treat, a large number of contaminants, such as phenols, pesticides, and aromatic hydrocarbons were decomposed, and this method was used in the early 1970s for wastewater treatment (Arslan 2001). The main disadvantage of this method is the use of ozone in a short half-life—decomposes in 20 min—requiring continuous ozonation, so this method is expensive to implement. Although oxidation caused by the thermodynamics of ozone may be favorable, kinetic factors tend to dictate whether ozone is polluting to oxidize within a reasonable time. It is useful for posttreatment due to the presence of a reducing agent, and the blowing agents include reduction in color removal (Neamtu et al. 2004). Ozone can also be used as a hydrogen peroxide activator. Advantages of using most of the above are as follows: no waste or sludge and no formation of toxic metabolites, and the application of gaseous does not increase the amount of wastewater.

3.5.2.4 Electrolysis Process

The method of electrolysis is based on an application of an electric current to the wastewater through the electrodes. Organic compounds such as dyes react via a combination of electrochemical oxidation, electrochemical reduction, electrocoagulation, and electroflotation reactions. For example, when iron is the sacrificial anode, Fe (II) ions are released into solution, and bulk and acid dyes are sorbed on the precipitated $Fe(OH)_2$(Xu 2001). In addition, Fe(II) can reduce the azo on arylamines. In addition, water can also be oxidized, resulting in the formation of O_2 and O_3, and if chloride is present, there is also the formation of Cl_2 and oxychloride anions. In the cathode, water is reduced to H_2 and OH^-. Thus, to improve the performance, various materials were tested in the electrodes such as carbon fiber, Ti/Pt, and aluminum. The main drawback of these types of methods is cost, the initial capital costs, energy, and electrode replacement. The formation of undesirable decomposition products and the foam are disadvantages of this method. The main advantages are the compact dimensions of the device, simplicity of operation, high speed removal of pollutants, and reduced quantity of sludge produced. The method is effective for color, BOD, COD, total organic carbon, total dissolved solids, and total suspended solids and removes heavy metals (Bandara et al. 1996).

3.5.2.5 Advanced Oxidation Processes

Advanced oxidation can be classified by the oxidation compounds with higher oxidation potential of the oxygen, namely, hydrogen peroxide, ozone, and hydroxyl radical. Hydrogen peroxide alone, however, is generally not powerful enough. Advanced oxidation processes are based on mostly the generation of highly reactive radical species that can react with a wide range of compounds, also compounds that are otherwise difficult to degrade, for example, dye molecules. The four advanced oxidation process have been studied: ozonation, H_2O_2, UV reagent Fenton, and UVs/TiO_2. In the process of ozonation, hydroxyl radicals are formed when water breaks down O_3:

$$H_2O + O_3 \rightarrow HO_3^+ + OH^- \rightarrow 2\, HO_2$$

$$\xrightarrow{+\,2\, O_3} HO\bullet + 2\, O_2$$

Although ozone itself is a powerful oxidizer, the hydroxyl radicals are even more reactive. Decomposition of ozone requires a high pH (>10). The ozone treatment of organic molecules is alkaline at a neutral pH or acid where ozone is the main oxidant. Ozone quickly decolorizes water-soluble dyes, but insoluble dyes react much slower (Torrades et al. 2004). Textile-processing wastewater usually also contains many refractory materials. Constituents other than dyes will react with ozone, thus increasing ozone request. It is therefore advisable to pretreat the wastewater before applying ozonation. Instead, ozone converts compounds in small molecules such as dicarboxylic acids and aldehydes (Verma et al. 2004). The reduction in COD is therefore low, whereas some of the ozonation products are very toxic. It is therefore preferable to treat the effluent from the ozonation stage, logically using inexpensive biological methods. The oxidation of Fenton is based on the generation of hydroxyl radicals from the Fenton reagent when ferrous iron is oxidized by hydrogen peroxide.

$$Fe^{2+} + H_2O_2 \rightarrow Fe(OH)^{2+} + HO^{\cdot}$$

In comparison with ozonation, the process is relatively inexpensive and usually has a greater COD reduction, although treatments are still needed. A disadvantage for the application of Fenton oxidation or Fenton-like oxidation for the treatment of the most common alkaline-textile processing wastewaters is that the process requires low pH. At higher pH, large quantities of waste sludge are produced by precipitation of iron and iron salts, and the method loses effectiveness as H_2O_2 catalytically decomposes oxygen or Fenton oxidation. Furthermore, it will be negatively affected by the presence of radicals and strong chelating agents in the wastewater (Moraes et al. 2002).

3.5.3 Biological Treatment

Environmental bioremediation is becoming more imperative as it is cost-effective and environmentally friendly and produces less sludge (Poulios et al. 2003; Kumar et al. 2018; Chandra and Kumar 2015a) and it is functional for genetic diversity and metabolic versatility of microorganisms. Chen et al. (2002) pointed out that microorganisms are known to play a principal role in the mineralization of biopolymers and foreign substances, such as azo dyes. Wastewater treatment of concentrated dye mostly relies on biological technique/s, which includes any mineralization or color in harmless inorganic compounds such as carbon dioxide and water, creating small amounts of relatively harmless sludge and, when used correctly, operating costs than other remedies. Among the various technologies, discoloration of bioremediation using microbial cells has been widely used. A more recent approach to the use of a microbial enzyme is promising for effective bleaching or industrial wastewater from the dyeing industry, and the degradation of ecosystems contaminated materials to color (Chandra et al. 2015). The fate of environmental pollutants is largely determined by abiotic processes, such as photooxidation, and by the metabolic activities of microorganisms. Since the catabolic enzymes are more or less certain, they can act more than their natural substrates. This explains why most of xenobiotics are subject to arbitrarily metabolism (Sugiarto et al. 2002) and several groups explore the capacities for microbial bioremediation of dyes.

The limits of biological processes are mainly caused by limited biodegradability of particular xenobiotic compounds such as dyes, by toxic or inhibitory effect of pollutants to the microbial population and by the slow rate of biodegradation of pollutants in particular (An et al. 2002). Dyestuffs and polymers are generally grueling to biodegrade, and many members are absolutely in congruous for conventional biological treatment. For textiles, in particular, the emphasis is on physical, chemical, and biological treatment systems. Biodegradation methods such as bleaching by fungi, microbial degradation, adsorption, and living or dead systems are usually applied in the treatment of industrial effluents due to many microorganisms such as bacteria, yeasts, algae, and fungi, which are able to accumulate and degrade various pollutants (Horikoshi et al. 2002; Chandra and Kumar 2015b), and all biological systems require a continuous input of effluent. Therefore, where there is water discharge process, relatively small or likely to be discontinuous, then physical or chemical treatments are more appropriate (Zhang et al. 2003). Biological treatment requires a large land area and is constrained by sensitivity to day variation as well as the toxicity of some chemicals and less flexibility in design and operation. Biological treatment is incapable of obtaining satisfactory color displacement by current conventional biodegradation process. Moreover, however, many organic molecules are degraded much more because of their complex chemical structure and synthetic organic origin. Specifically, because of their xenobiotic nature, the biodegradability of azo dyes is very limited.

A wide range of structurally different dyes are consumed within a very short time during fabric processing. Therefore, cloth leaves vary in composition (Kusvuran et al. 2004). It provides the need for a general nonspecific fabric treatment process of waste. Alternative methods of using microbial biocatalysts to remove dyes for washed fabrics offer potential advantages over conventional process for little environmental impact and cost effective. This cloth treatment process is essentially self-sufficient and does not require rigorous monitoring. Options for biological treatment of washed cloth can be single-phase aerobic or anaerobic or a combination of two. The limitations of the biological process are mainly because of limited biodegradability of xenobiotic compounds such as dyes, toxic or inhibitory effects of dirt for the microbial population, and the slow rate of biodegradation of specific pollutants.

3.5.3.1 Bacterial Biodegradation

3.5.3.1.1 Aerobic Biodegradation

Bacterial dye biotransformation focuses on the most abundant class of dyes, i.e., azo dyes. The electron-withdrawing nature of the azobonds interferes with the sensitivity of azo molecules of oxidative reactions (Ge and Qu 2003). It was reported that aerobic treatment plants reduce the level of colorants by adsorption at the sludge biomass in a nonenzymatic process (Joseph et al. 2000). Therefore, conventional aerobic wastewater treatment plants are not useful for discoloration of liquid azo dyes and related colored compounds (Chen 2004). Peroxidase-producing bacterial strains of the *Streptomyces* genus *Sphingomonas* was shown to decolorize azo dyes (Destaillats et al. 2000; Tezcanli-Güyer and Ince 2004). Later research revealed a number of limitations of aerobic bacterial azo dye decolourisation. Leitner et al. (1997) reported that these bacterial strains of *Streptomyces* genus *Sphingomonasgenus* was only decolorize azo dyes in the presence of additional carbon and energy sources where they are used, *Bacillus subtilis*, and p-amino-azobenzene decolorization supply in the presence of glucose into aniline. This is economically not viable in a commercial scale because the carbon source is expensive.

3.5.3.1.2 Anaerobic Biodegradation

Under anaerobic conditions, many bacteria reduce the highly electrophilic azo linkage in the dye molecule by a nonspecific enzymatic action (Harbel et al. 1991; Solpan and Güven 2002). The nonspecific action of anaerobic bacteria allows biodegradation of a wide range of textile dyes, making this process more suitable for application in a commercial scale. The anaerobic reduction of azo dyes by bacteria appears to be better suitable for fading in sewage treatment systems (Matthews and McEvoy 1992). This process offers the following advantages: (1) reactions take place at a neutral pH and are extremely nonspecific when low redox molecular weight mediators are available, (2) in static cultures, oxygen depletion is easily accomplished, allowing mandatory and optional anaerobic bacteria to reduce azo dyes, and (3) sewage systems often provide additional sources of carbon which, in general, increase reduction rates. These sources of carbon also facilitate training and regeneration of reducing equivalents by their oxidation.

All processes, however, tend to have disadvantages. The major restriction of anaerobic azo dye reduction is that the aromatic amines formed from reductive cleavage cannot be further mineralized (Kopf et al. 2000). Their accumulation is a serious cause of concern since they are presumed carcinogenic and these amines may be formed in the lower intestine after ingestion of dye-containing foods (Chen et al. 2002). Fortunately, the potential carcinogenic producing dyes have been banned from the market (Guivarch et al. 2003).

3.5.3.1.3 Anaerobic-Aerobic Biodegradation

An absolute mineralization of aromatic amines and sulfonated amino aromatics by aerobic bacteria (Neyens and Baeyens 2003; Mrowetz et al. 2003; Stock et al. 2000) has led to the insinuation of combining the anaerobic cleavage of azo dyes with aerobic treatment to mineralize the aromatic amines generated in the anaerobic step. This can be done sequentially or simultaneously. In the former processes, the anaerobic phase and aerobic phases may be combined alternatively in the same vessel or in a continuous system with separate vessels (Kim et al. 2004; Naim and Abd 2002; Hu et al. 2005). Simultaneous treatment systems utilize anaerobic zones within aerobic bulky phases as demonstrated in biofilms, granular sludge, or biomass immobilized in other matrices (Muthuraman and Palanivelu 2005; Koyuncu et al. 2004; Bes-Piá et al. 2002). Both treatments require auxiliary substrates to feed the bacteria in the anaerobic zones with carbon and energy sources as well as reducing equivalents for azo bond cleavage. For the most part, it can be inferred that, in persistent anaerobic—high-impact frameworks, there is finish decolorization of colors and noteworthy lessening in BOD and COD levels. In the ensuing oxygen-consuming stage, the rest of the BOD from the assistant substrates might be totally mineralized (Bes-Piá et al. 2003).

The initial step of bacterial metabolism of azo dyes under anaerobic conditions involves reductive cleavage of azo dyes. This process is catalyzed by a type of soluble cytoplasmic enzymes with low substrate specificity, known as "azoreductases" (Dulkadiroglu et al. 2002). Under anoxic conditions, these enzymes transfer electrons through the melt flavin azo dyes, which are then reduced. The role of the cytoplasmic enzyme in vivo is, however, uncertain. The work of Zaoyab and coworkers showed, however, that the cytoplasmic "azoreductases" such as flavin reductases and others have a vital importance in in vivo reduction of sulfonated azo compounds (Zaoyan et al. 1992). Another possibility for the extracellular reduction of azo compounds under anaerobic conditions is the action of reduced organic compounds formed as end products in certain strictly anaerobic bacterial metabolic reactions (Panswad et al. 2001). Furthermore, mineralization of formed amines is not possible under anaerobic conditions. This is why there are few studies that propose a combined anaerobic-aerobic system for the removal of dyes from wastewaters with a

consortium/sludge. The use of consortia offers great advantages over the use of pure culture prevailing artificial dyes. The individual forms may attack pigment molecules in various positions
or may use the decomposition products made by another strain for further decay. However, the
composition of the mixed culture can be changed during the process of decomposition interfering
with the control system. The consortium used activated sludge system, which consists mainly of
bacteria, but also with the usual presence of fungi and protozoa. Possible models for the anaerobic
azo dye reduction are shown in Figure 3.9. The first step in bacterial azo dye metabolism under
anaerobic conditions involves the reductive cleavage of the azo-coupling. This process is catalyzed by several soluble cytoplasmic enzymes with a low substrate specificity, which is known
as "azoreductases" (Koyuncu et al. 2004; Bes-Piá et al. 2002, 2003; Dulkadiroglu et al. 2002).
Under anoxic conditions, these enzymes facilitate the transfer of electrons to the azo dye, which
is then reduced via soluble flavins. The possibilities of noncytoplasmic azoreductases are amplified because it is highly unlikely that highly charged sulfonated azo dyes or azo polymeric dyes
pass bacterial cell wall (Muthuraman and Palanivelu 2005). A membrane-bound "azoreductase"
was found by Guivarch and coworkers in the cell wall of *Sphingomonas* sp. This strain possesses both cytoplasmic and membrane-bound activities of azoreductase (Guivarch et al. 2003).
Another model for unspecific reduction of azo dyes by bacteria was proposed based on studies on
Sphingomonas xenophaga (Neyens and Baeyens 2003). They found an increase in the degree of
reduction when quinones, such as anthraquinone-2-sulfonate or 2-hydroxy-1,4-naphthoquinone,
are added to the culture medium. Authors suggested then that quinones added to the medium, or
a decomposition products released by the cells occurred to the medium as redox mediators that
are reduced enzymatically by the bacterium cells and the hydroquinones formed reduced to the
azo dye in a purely chemical redox reaction. Swaminathan suggests an extracellular azoreductase
activity done in studies of bacteria isolated from human intestine, mainly *Eubacterium* sp. and
Clostridium sp. (Swaminathan et al. 2003). Another possibility for the extracellular reduction of
azo compounds, under anaerobic conditions, is the effect of reduced inorganic compounds (for
example, Fe^{2+}, H_2S), which are formed as end products of certain strictly anaerobic bacterial
metabolic reactions (Figure 3.10).

The additional mineralization formed amines, not possible under anaerobic conditions (Panswad
et al. 2001; Lin and Lin 1993). Therefore, there are several studies that provide anaerobic-aerobic
combined system for the removal of dyes in wastewater with a consortium/sludge. The use of consortia offers significant advantages over the use of pure cultures in the degradation of synthetic
dyes. Individual logs can fall to the dye molecule at various positions or the decomposition products
produced by other strain for subsequent use degradation, however, by interfering with the control
system to modify the composition of the mixed cultures during the degradation process. The most
commonly used consortium is the activated sludge system, consisting mainly of bacteria, but also
with the usual presence of fungi and protozoa.

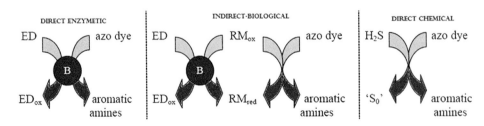

FIGURE 3.10   Mechanisms of anaerobic azo dye reduction by bacteria [RM_{ox} = oxidized redox mediator;
RM_{red} = reduced redox mediator; ED = electron donor; ED_{ox} = oxidized electron donor; B = bacteria (enzyme
system)].

3.6 CONCLUSION

This chapter recapitulates a momentous analysis of current technologies available for discoloration, as well as effluent treatment, and suggests effective and textile economic alternatives. Textiles for wastewater treatment before discharging are of great importance to reduce the production costs and the burden of pollution. Conventional technologies for wastewater treatment textiles include various combinations of biological factors, and chemical and physical methods, but these methods need capital and exploitation costs. So far, there are no simple and economical treatment attractive cans effectively coloring the bleaching effluent. In recent years, remarkable achievements have been made in the use of biotechnological applications in textile effluents not only for the color elimination but also for the complete degradation of the dyes. The different microorganisms, such as aerobic and anaerobic bacteria, fungi, and physicochemical methods, have been initiated to catalyze bleaching of the dye. Physicochemical versus biological treatment systems that can efficiently remove dyes from large volumes of wastewater at low cost alternatives are preferable. Colors are very important emission classes and have influence on the general human and aquatic life drastically. To minimize the negative impact of color-contaminated water to people and the environment, wastewater must be treated carefully before discharge. Several color removal strategies, including conventional techniques such as adsorption, oxidation, coagulation, and biological treatment, and the relatively new techniques of reverse micellar extraction have been developed to remove color from the wastewater. Indeed, new techniques bring about several improvements. Most of the techniques are able to achieve more than 80% removal of paint and slightly exceed 90%. Really, great progress in the removal of paint from wastewater has been reported over the past few years, and it is very encouraging that several reported methods are very fast and have low costs with exciting color removal efficiency. So, it is advised that more research be conducted in this direction because water is the second most precious abiotic factors of our ecosystem, and the safe treatment and preservation is the duty of every person living on this planet. We hope to have more sophisticated technology developed so that water can be treated easily, at low cost, on industrial and pilot scales.

3.7 MY FEELINGS

Finally, I am inspired to write something about environmental microbiology whenever I look upon my son's shining smiling face. I guess it is the innocence and helplessness that I really see. I look at him and wonder if he understands the place in time that we now occupy on earth in relation to our environment and if he and his friends will have a place to live where the water is clean-free of disease and pollution free. Will he have a chance to grow to old age and gain the wisdom to understand that he was born of water and needs to be sustained by it?

REFERENCES

Akan, J.C., Abdulrahman, F.I., Ayodele, J.T., Ogugbuaja, V.O. Impact of tannery and textile effluent on the chemical characteristics of Challawa River, Kano State Nigeria. *Aust. J. Basic Appl. Sci.* 3, 1933–1947, 2009.

Aleboyeh, A., Aleboyeh, H., Moussa, Y. "Critical" effect of hydrogen peroxide in photochemical oxidative decolorization of dyes: Acid Orange 8, Acid Blue 74 and Methyl Orange. *Dyes Pigments* 57, 67, 2003.

An, T-C., Zhu, X-H., Xiong, Y. Feasibility study of photoelectrochemical degradation of methylene blue with threedimensional electrode-photocatalytic reactor. *Chemosphere* 46, 897, 2002.

Arslan, I. Treatability of a simulated disperse dye-bath by ferrous iron coagulation, ozonation, and ferrous iron-catalyzed ozonation. *J. Hazard. Mater.* B 85, 229, 2001.

Arslan, I., Balcioglu, I.A. Advanced oxidation of raw and biotreated textile industry wastewater with O_3, H_2O_2 /UV-C and their sequential application. *J. Chem. Technol. Biotechnol.* 76, 53, 2001.

Arslan-Alaton, I. A review of the effects of dye-assisting chemicals on advanced oxidation of reactive dyes in wastewater. *Color. Technol.* 119, 345, 2003.

Azbar, N., Yonar, T., Kestioglu, K. Comparison of various advanced oxidation processes and chemical treatment methods for COD and color removal from a polyester and acetate fiber dyeing effluent. *Chemosphere* 55, 35, 2004.

Babu, B.R., Parande, A.K., Raghu, S., Kumar, T.P. Cotton textile processing: Waste generation and effluent treatment. *J. Cotton Sci.* 11, 141–153, 2007.

Bandara, J., Morrison, C., Kiwi, J., Pulgarin, C., Peringer, P. Degradation/decoloration of concentrated solutions of Orange II. Kinetics and quantum yield for sunlight induced reactions via Fenton type reagents. *J. Photochem. Photobio. A Chem.* 99, 57, 1996.

Bes-Piá, A., Mendoza-Roca, J.A., Alcaina-Miranda, M.I., Iborra-Clar, A., Iborra-Clar, M.I. Reuse of wastewater of the textile industry after its treatment with a combination of physicochemical treatment and membrane technologies. *Desalination* 149, 169, 2002.

Bes-Piá, A., Mendoza-Roca, J.A., Alcaina-Miranda, M.I., Iborra-Clar, A., Iborra-Clar, M.I. Combination of physico-chemical treatment and nanofiltration to reuse wastewater of a printing, dyeing and finishing textile industry. *Desalination* 157, 73, 2003.

Chacko, J.T., Subramaniam, K. Enzymatic degradation of azo dyes: A review. *Int. J. Environ. Sci.* 1, 1250–1260, 2011.

Chandra, R., Kumar, V. Mechanism of wetland plant rhizosphere bacteria for bioremediation of pollutants in an aquatic ecosystem. In R. Chandra (Ed.), *Advances in Biodegradation and Bioremediation of Industrial Waste.* CRC Press, Boca Raton, FL, 2015a, pp. 329–379.

Chandra, R., Kumar, V. Biotransformation and biodegradation of organophosphates and organohalides. In R. Chandra (Ed.), *Environmental Waste Management.* CRC Press, Boca Raton, FL, 2015b, pp. 475–524.

Chandra, R., Kumar, V. Detection of androgenic-mutagenic compounds and potential autochthonous bacterial communities during in situ bioremediation of post methanated distillery sludge. *Front. Microbiol.* 8, 887, 2017a.

Chandra, R., Kumar, V. Detection of *Bacillus* and *Stenotrophomonas* species growing in an organic acid and endocrine-disrupting chemicals rich environment of distillery spent wash and its phytotoxicity. *Environ. Monit. Assess.* 189, 26, 2017b.

Chandra, R., Kumar, V., Tripathi, S. Evaluation of molasses-melanoidins decolourisation by potential bacterial consortium discharged in distillery effluent. *3 Biotech* 8, 187, 2018. DOI:10.1007/s13205-018-1205-3.

Chandra, R., Kumar, V., Yadav, S. Extremophilic ligninolytic enzymes. In R. Sani, R.N. Krishnaraj (Eds.), *Extremophilic Enzymatic Processing of Lignocellulosic Feedstocks to Bioenergy.* Springer International Publishing AG, Cham, 2015.

Chen, G. Electrochemical technologies in wastewater treatment. *Sep. Purif. Technol.* 38(1), 11, 2004.

Chen, H., Jin, X., Zhu, K., Yang, R. Photocatalytic oxidative degradation of acridine orange in aqueous solution with polymeric metalloporphyrins. *Water Res.* 36, 4106, 2002.

Chen, Y.H., Chang, C.Y., Huang, S.F., Chiu, C.Y., Ji, D., Shang, N.C., Yu, Y.H., Chiang, P.C., Ku, Y., Chen, J.N. Decomposition of 2-naphthalenesulfonate in aqueous solution by ozonation with UV radiation. *Water Res.* 36, 4144, 2002.

Cisneros, R.L., Espinoza, A. G., Litter, M.I. Photodegradation of an azo dye of the textile industry. *Chemosphere* 48, 393, 2002.

Destaillats, H., Colussi, A. J., Joseph, J.M., Hoffmann, M.R. Synergistic effects of sonolysis combined with ozonolysis for the oxidation of azobenzene and methyl orange. *J. Phys. Chem. A* 104, 8930, 2000.

Dulkadiroglu, H., Dogruel, S., Okutman, D., Kabdasli, I., Sozen, S. and Orhon, D. Effect of chemical treatment on soluble residual COD in textile wastewaters. *Wat. Sci. Tech.* 45(12), 251–259, 2002.

Faryal, R., Hameed, A. Isolation and characterization of various fungal strains from textile effluent for their use in bioremediation. *Pak. J. Bot.* 37, 1003–1008, 2005.

Forgacs, E., Cserhati, T., Oros, G. Removal of synthetic dyes from wastewaters: A review. *Environ. Int.* 30, 953–971, 2004.

Fu, F., Viraraghavan, T. Fungal decolorization of dye wastewaters: A review. *Bioresour. Technol.* 79, 251–262, 2001.

Ge, J., Qu, J. Degradation of azo dye acid red B on manganese dioxide in the absence and presence of ultrasonic irradiation. *J. Hazard. Mater.* 100, 197, 2003.

Ghoreishi, M., Haghighi, R. Chemical catalytic reaction and biological oxidation for treatment of non-biodegradable textile effluent. *Chem. Eng. J.* 95, 163–169, 2003.

Gogate, P.R., Pandit, A.B. A review of imperative technologies for wastewater treatment I: Oxidation technologies at ambient conditions. *Adv. Environ. Res.* 8, 501, 2004a.

Gogate, P.R., Pandit, A.B. A review of imperative technologies for wastewater treatment II: Hybrid methods. *Adv. Environ. Res.* 8, 553, 2004b.

Guivarch, E., Trevin, S., Lahitte, C., Oturan, M.A. Degradation of azo dyes in water by Electro-Fenton process. *Environ. Chem. Lett.* 1, 38, 2003.

Harbel, R., Urban, W., Gehringer, P., Szinovatz, W. Treatment of pulp-bleaching effluents by activated sludge, precipitation, ozonation and irradiation. *Wat. Sci. Tech.* 24, 229, 1991.

Horikoshi, S., Hidaka, H. and Serpone, N. Environmental remediation by an integrated microwave/UV-illumination method. 1. Microwave-assisted degradation of rhodamine-B dye in aqueous TiO2 dispersions. *Environ. Sci. Technol.* 36, 1357, 2002.

Hu, H., Yang, M., Dang, J. Treatment of strong acid dye wastewater by solvent extraction. *Sep. Purif. Technol.* 42, 129, 2005.

Jin, X.C., Liu, G.Q., Xu, Z.H., Tao, W.Y. Decolorization of a dye industry effluent by Aspergillus fumigatus XC6. *Appl. Microbiol. Biotechnol.* 74, 239–243, 2007.

Joseph, J.M., Destaillats, H., Hung, H-M., Hoffmann, M.R. The sonochemical degradation of azobenzene and related azo dyes: Rate enhancements via Fenton's reactions. *J. Phys. Chem. A* 104, 301, 2000.

Kim, T-H., Park, C., Yang, J., Kim, S. Comparison of disperse and reactive dye removals by chemical coagulation and Fenton oxidation. *J. Hazard. Mater. B* 112, 95, 2004.

Kopf, P., Gilbert, E., Eberle, S.G. TiO_2 photocatalytic oxidation of monochloroacetic acid and pyridine: Influence of ozone. *J. Photochem. Photobio. A Chem.* 136, 163, 2000.

Koyuncu, I., Topacik, D., Yuksel, E. Reuse of reactive dyehouse wastewater by nanofiltration: Process water quality and economical implications. *Sep. Purif. Technol.* 36, 77, 2004.

Kuberan, T., Anburaj, J., Sundaravadivelan, C., Kumar, P. Biodegradation of azo dye by Listeria sp. *Int. J. Environ. Sci.* 1, 1760–1770, 2011.

Kumar V, Shahi SK, Singh S (2018) Bioremediation: an eco-sustainable approach for restoration of contaminated sites. In: Singh J, Sharma D, Kumar G, Sharma NR (eds) Microbial bioprospecting for sustainable development. Springer, Sharma.

Kusvuran, E., Gulnaz, O., Irmak, S., Atanur, O.M., Yavuz, H.I., Erbatur, O. Comparison of several advanced oxidation processes for the decolorization of Reactive Red 120 azo dye in aqueous solution. *J. Hazard. Mater. B* 109, 85, 2004.

Leitner, N.K.V., Bras, E.L., Foucault, E., Bousgarbiès, J.-L. A new photochemical reactor design for the treatment of absorbing solutions. *Wat. Sci. Tech.* 35(4), 215, 1997.

Lin, S.H. and Lin, C.M. Treatment of Textile waste effluents by ozonation and chemical coagulation. *Wat. Res.* 27, 1743, 1993.

Matthews, R.W., McEvoy, S.R. A comparison of 254 nm and 350 nm excitation of TiO_2 in simple photocatalytic reactors. *J. Photochem. Photobiol. A Chem.* 66, 355, 1992.

Moraes, S.G., Freire, R.S., Durán, N. Degradation and toxicity reduction of textile effluent by combined photocatalytic and ozonation processes. *Chemosphere* 40, 369, 2002.

Mrowetz, M., Pirola, C., Selli, E. Degradation of organic water pollutants through sonophotocatalysis in the presence of TiO_2. *Ultrason. Sonochem.* 10, 247, 2003.

Muruganandham, M., Swaminathan, M. Photochemical oxidation of reactive azo dye with UV–H2O2 process. *Dyes Pigments* 62, 269, 2004.

Muthuraman, G., Palanivelu, K. Selective extraction and separation of textile anionic dyes from aqueous solution by tetrabutyl ammonium bromide. *Dyes Pigments* 64, 251, 2005.

Naim, M.M., Abd, Y.M.E. Removal and recovery of dyestuffs from dyeing wastewaters. *Sep. Purif. Methods* 31(1), 171, 2002.

Neamtu, M., Siminiceanu, I., Yediler, A., Kettrup, A. Kinetics of decolorization and mineralization of reactive azo dyes in aqueous solution by the UV/H2O2 oxidation. *Dyes Pigments* 53, 93, 2002.

Neamtu, M., Yediler, A., Siminiceanu, I., Macoveanu, M., Antonius Kettrup, A. Decolorization of disperse red 354 azo dye in water by several oxidation processes—A comparative study. *Dyes Pigments* 60, 61, 2004.

Neyens, E., Baeyens, J. A review of classic Fenton's peroxidation as an advanced oxidation technique. *J. Hazard. Mater. B* 98, 33, 2003.

Panswad, T., Techovanich, A., Anotai, J. Comparison of dye wastewater treatment by normal and anoxic+anaerobic/aerobic SBR activated sludge processes. *Wat. Sci. Tech.* 43(2), 355, 2001.

Papic, S., Koprivanac, N., Bozic, A.L., Metes, A. Removal of some reactive dyes from synthetic wastewater by combined Al(III) coagulation/carbon adsorption process. *Dyes Pigments* 62, 291, 2004.

Poulios, I., Micropoulou, E., Panou, R., Kostopoulou, E. Photooxidation of eosin Y in the presence of semiconducting oxides. *Appl. Catal. B* 41, 345, 2003.

Puvaneswari, N., Muthukrishnan, J., Gunasekaran, P. Toxicity assessment and microbial degradation of azo dyes. *Indian J. Exp. Biol.* 44, 618–626, 2006.

Robinson, T., McMullan, G., Marchant, R., Nigam, P. Remediation of dyes in textile effluent: A critical review on current treatment technologies with a proposed alternative. *Bioresour. Technol.* 77, 247–255, 2001.

Saratale, R.G., Saratale, G.D., Chang, J.S., Govindwar, S.P. Bacterial decolorization and degradation of azo dyes: A review. *J. Taiwan Inst. Chem. Eng.* 42, 138–157, 2011.

Sarria, V., Deront, M., Péringer, P., Pulgarin, C. Degradation of a biorecalcitrant dye precursor present in industrial wastewaters by a new integrated iron (III) photoassisted–biological treatment. *Appl. Catal. B Environ.* 40, 231, 2003.

Savin, I.I., Butnaru, R. Wastewater characteristics in textile finishing mills. *Environ. Eng. Manage. J.* 7, 859–864, 2008.

Shah, M.P. Microbiological removal of phenol by an application of pseudomonas spp. ETL: An innovative biotechnological approach providing answers to the problems of FETP. *J. Appl. Environ. Microbiol.* 1(2), 6–11, 2014.

Shah, M.P. Patel, K.A., Nair, S.S., Darji, A.M. Microbial decolorization of methyl orange dye by pseudomonas spp. ETL-M. *Int. J. Environ. Biorem. Biodegrad.* 2(1), 54–59, 2013a.

Shah, M.P., Patel, K.A., Nair, S.S., Darji, A.M., Maharaul, S. Microbial degradation of azo dye by pseudomonas spp. MPS-2 by an application of sequential microaerophilic and aerobic process. *Am. J. Microbiol. Res.* 43(1), 105–112, 2013b.

Shah, M.P., Patel, K.A., Nair, S.S., Darji, A.M., Maharaul, S. Optimization of environmental parameters on decolorization of Remazol Black B using mixed culture. *Am. J. Microbiol. Res.* 3(1), 53–56, 2013c.

Shah, M.P., Patel, K.A., Nair, S.S., Darji, A.M. An innovative approach to biodegradation of textile dye (Remazol Black) by Bacillus spp. *Int. J. Environ. Biorem. Biodegrad.* 2(1), 43–48, 2013d.

Shah, M.P., Patel, K.A., Nair, S.S., Darji, A.M. Microbial degradation and decolorization of reactive orange dye by strain of pseudomonas Spp. *Int. J. Environ. Biorem. Biodegrad.* 1(1), 1–5, 2013e.

Shah, V., Verma, P., Stopka, P., Gabriel, J., Baldrian, P., Nerud, F. Decolorization of dyes with copper(II)/organic acid/hydrogen peroxide systems. *Appl. Catal. B: Environ.* 46, 287, 2003.

Shu, H.Y., Chang, M.C. Decolorization effects of six azo dyes by O_3, UV/O_3 and UV/H_2O_2 processes. *Dyes Pigments* 65, 25, 2005.

Shu, H.Y., Huang, C.R. Degradation of commercial azo dyes in water using ozonation and UV enhanced ozonation process. *Chemosphere* 31, 3813, 1995.

Solpan, D., Güven, O. Decoloration and degradation of some textile dyes by gamma irradiation. *Rad. Phy. Chem.* 65, 549, 2002.

Stock, N.L., Peller, J., Vinodgopal, K., Kamat, P.V. Combinative sonolysis and photocatalysis for textile dye degradation. *Environ. Sci. Technol.* 34, 1747, 2000.

Sugiarto, A.T., Ohshima, T., Sato, M. Advanced oxidation processes using pulsed streamer corona discharge in water. *Thin Solid Films* 407, 174, 2002.

Swaminathan, K., Sandhya, S., Sophia, A.S., Pachhade, K., Subrahmanyam, Y.V. Decolorization and degradation of H-acid and other dyes using ferrous–hydrogen peroxide system. *Chemosphere* 50, 619, 2003.

Tezcanli-Güyer, G., Ince, N.H. Individual and combined effects of ultrasound, ozone and UV irradiation: A case study with textile dyes. *Ultrasonics* 42, 603, 2004.

Torrades, F., García-Montaño, J., García-Hortal, J.A., Domènech, X., Peral, J. Decolorization and mineralization of commercial reactive dyes under solar light assisted photo-Fenton conditions. *Solar Energy* 77, 573, 2004.

Verma, P., Baldrian, P., Nerud, F. Decolorization of structurally different synthetic dyes using cobalt(II)/ascorbic acid/hydrogen peroxide system. *Chemosphere* 50, 975, 2003.

Verma, P., Shah, V., Baldrian, P., Gabriel, J., Stopka, P., Trnka, T., Nerud, F. Decolorization of synthetic dyes using a copper complex with glucaric acid. *Chemosphere* 54, 291, 2004.

Xu, Y. Comparative studies of the $Fe^{3+/2+}$ -UV, H_2O_2-UV, TiO_2-UV/vis systems for the decolorization of a textile dye X-3B in water. *Chemosphere* 43, 1103, 2001.

Yusuff, R.O., Sonibare, J.A. Characterization of textile industries' effluents in Kaduna Nigeria and pollution implications. *Global Nest Int. J.* 6, 212–221, 2004.

Zaoyan, Y., Ke, S., Guangliang, S., Fan, Y., Jinshan, D., Huanian, M. Anaeroicaerobic treatment of a dye wastewater by combination of RBC with activated sludge. *Wat. Sci. Tech.* 26(9–11), 2093, 1992.

Zhang, W., An, T., Xiao, X., Fu, J., Sheng, G., Cui, M., Li, G. Photoelectrocatalytic degradation of reactive brilliant orange K-R in a new continuous flow photoelectrocatalytic reactor. *Catal. A Gen.* 255, 221, 2003.

4 Microbial Degradation of Pesticides in the Environment

Prashant S. Phale, Kamini Gautam, and Amrita Sharma
Indian Institute of Technology—Bombay

CONTENTS

4.1 INTRODUCTION

Today, the world population is more than 7.6 billion, and by 2050, it is expected to reach 9.8 billion (Thatcher et al. 2018). Hence, production of food and useful products (e.g., medicines, biofuels, etc.) will need to increase with the same frequency or more to serve the needs of the population. In the food chain, plants are considered as the first link that captures energy from sunlight and converts it into compounds that directly or indirectly provide the food essential for other living organisms. Conventionally, plants have been improved through a selection that involves evaluation of a breeding population for one or more traits in the agricultural field. Today, this has contributed to high yielding varieties such as hybrids (hybrid seeds), which constitute the basis for green revolution. The green revolution in the 20th century brought developments in the agricultural sector and related industries. To enhance the quality and quantity of crops, agricultural industries started applying pesticides in the field. A pesticide can be defined as any substance or mixture of substances intended to prevent, destroy, repel, or mitigate any pest (rats, insects, nematodes, mites, etc.). Pesticides such as herbicides, insecticides, fungicides, and various other substances are used to control pests. Some of the commonly used pesticides structures are shown in Figure 4.1. The extensive use of pesticides causes serious environmental concerns, as only 5% or less from the applied pesticides reach the target organisms, which resulted in contamination of soil and water bodies. The continuous and excessive use of pesticides has led to their accumulation in an environment that endangered the entire population by their multifaceted toxicity. In addition to causing toxic effects to humans, there is a high risk of contamination in the ecosystem as well as hitting directly nontarget vegetation. There are chronic threats to human life, caused by long-term, low-dose exposure to pesticides. It can cause diminished intelligence, reproductive abnormalities, and hormonal disruption (Table 4.1). The constant mobility of applied pesticides through sorption, leaching, and volatilization led to contamination of different levels in the environment.

Due to adverse effects, the removal of pesticides from the environment is a major concern. Pesticides are known to be highly hydrophobic in nature. Their persistence in the environment

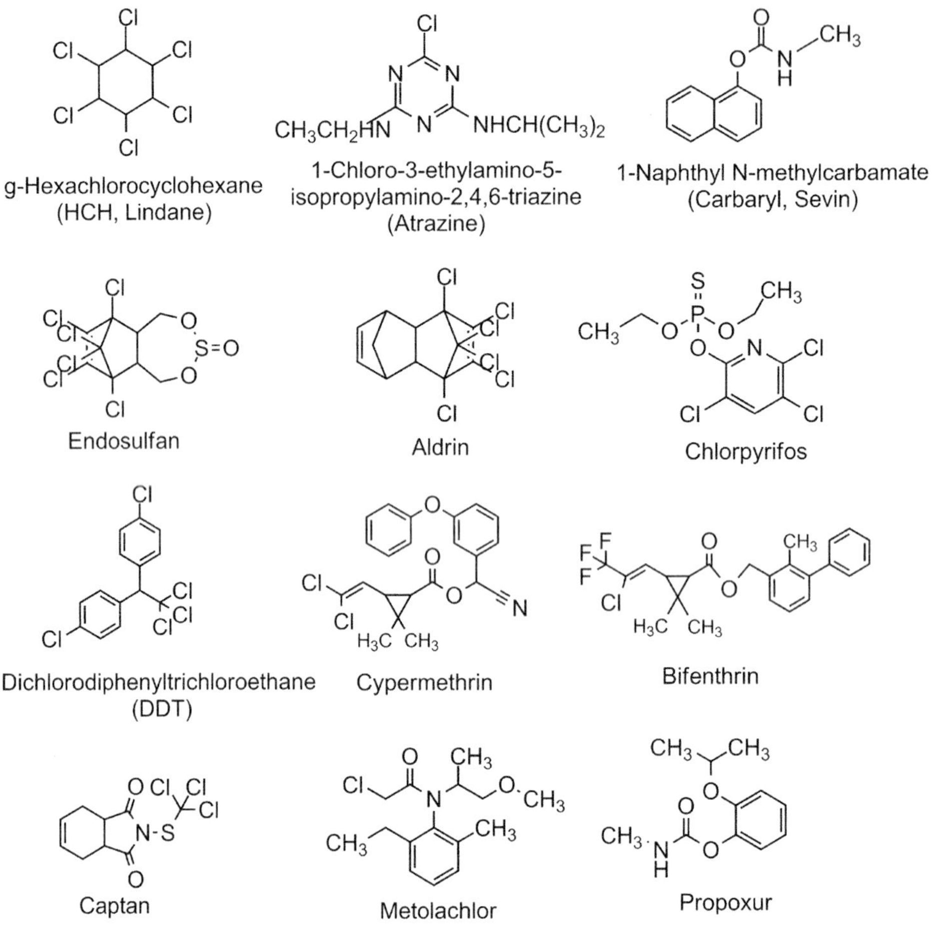

g-Hexachlorocyclohexane
(HCH, Lindane)

1-Chloro-3-ethylamino-5-
isopropylamino-2,4,6-triazine
(Atrazine)

1-Naphthyl N-methylcarbamate
(Carbaryl, Sevin)

Endosulfan

Aldrin

Chlorpyrifos

Dichlorodiphenyltrichloroethane
(DDT)

Cypermethrin

Bifenthrin

Captan

Metolachlor

Propoxur

FIGURE 4.1   Structure of various pesticides.

can be explained by low water solubility and good association with the soil sediments, where they become buried and persist until degraded. Hydrophobicity, environmental persistence, and genotoxicity increase with more number of fused ring structures in the pesticides. Several methods have been reported to reduce xenobiotics/pesticides level from the environment. These are divided into abiotic and biotic methods. The abiotic method involves chemical oxidation, incineration, immobilized enzyme, photocatalysis, etc., whereas the biotic method involves the use of plants and microorganisms (Bertilsson and Widenfalk 2002; Rivas 2006). Abiotic methods are mostly expensive and may give rise to intermediates, which are more toxic and recalcitrant than the parent compound (Elespuru et al. 1974; Shea and Berry 1983). Compared with a biotic methods, biotic methods are more efficient, cost-effective, eco-friendly, and economical. Therefore, microbial degradation is very useful/beneficial for (i) the detoxification of these compounds into nontoxic products or (ii) their ability to utilize them as carbon, nitrogen, or energy source, thus removing them completely. In literature, various organisms are reported degrading the pesticides (Table 4.1).

4.2   MICROBIAL DEGRADATION OF PESTICIDES

Microbial degradation is defined as the breakdown of a compound into simple metabolic intermediates. The microorganism involved in the degradation process utilizes these compounds and their subsequent metabolites as the sole source of carbon, nitrogen, and energy (Kumar et al. 2018;

TABLE 4.1
List of Pesticides and Their Degradation by Microbes

Pesticides	Hazardous Effects	Microbial Degraders	References
γ-HCH	Changes in the level of sex hormones, liver and kidney damage, blood disorders, possibly carcinogenic	*Sphingobium paucimobilis* UT26A, *Sphingobium indicum* B90A, *Sphingobiumfrancense, Anabeana* sp., *Nostoc ellipsosporum, Devosia lucknowensis, Sphingobium chinhatense* strain IP26T, *Pandoraea* sp. strain SD6-2, *Pontibacter indicus* sp.	Sahu et al. 1990; Kuritz and Wolk 1995; Cérémonie et al. 2006; Nagata et al. 2010; Dua et al. 2013; Niharika et al. 2013; Pushiri et al. 2013; Singh et al. 2014
Atrazine	Affects the function of immune, central nervous, and cardiovascular system endocrine disruptor, delayed puberty, intrauterine growth retardation, carcinogenic	*Pseudomonas* sp. ADP, *Rhodococcus* sp. strain MB-P1, *Ralstonia* sp. M91–3., *Clavibacter* sp., *Agrobacterium* sp. J14, *Alcaligenes* sp. SG1, *Arthrobacter aurescens* TC1, *Rhodococcus* sp. NI86/21, *Streptomyces* sp. PS1/5, *Nocardia* sp., *Arthrobacter* sp. AK_YN10, *Pseudomnonas* sp. AK_AAN5, AK_CAN1	Nagy et al. 1995; Boundy-Mills et al. 1997; de Souza et al. 1998; Sadowsky et al. 1998; Fazlurrahman et al. 2009; Sagarkar et al. 2014
Carbaryl	Skin and eye irritation, bronchoconstriction, cholinesterase inhibition, muscle weakness, memory loss, and headache	*Nocardia* sp., *Xanthomonas* sp., *Achromobacter* sp., *Pseudomonas* sp., *Rhodococcus* sp. NCIB 12038, *Blastobacter* sp., *Arthrobacter* sp., *Micrococcus* sp., *Pseudomonas* sp. C4, C5pp, C6, C7, *Burkholderia* sp. C3	Sud et al. 1972; Larkin and Day 1986; Chapalamadugu and Chaudhry 1991; Hayatsu and Nagata 1993; Doddamani and Ninnekar 2001; Swetha and Phale 2005; Seo et al. 2013, Trivedi et al. 2017
DDT	Accumulate in the fatty tissues of ingesting organisms along the food chain, have estrogenic activity, cause male infertility, miscarriage, developmental delay, nervous disruptor	*Pseudomonas aeruginosa* 640X, *Alcaligenes eutropha* A5, *Alcaligenes* sp. JB1, *Aspergillus flavus, Phanerochaete chrysosporium*	Boul et al. 1994; Foght et al. 2001; Nadeau et al. 1994; Subba-Rao and Alexander 1985; Bumpus et al. 1993
Endosulfan	Hyperexcitability, tremors, and convulsions, gonadal toxicity, genotoxicity, neurotoxicity	*Klebsiella oxytoca* KE8, *Bacillus* spp., *Pandoraea* sp., *Micrococcus* sp.	Singh and Pandey 1990; Chaudhuri et al. 1999; Kwon et al. 2005; Awasthi et al. 2003; Siddique et al. 2003; Guha et al. 2000
Aldrin	Headaches, giddiness, muscle twitching, convulsions, loss of consciousness, gastrointestinal disturbances, carcinogenicity	*Clostridium* spp., *Pseudomonas* sp., *Phanerochaete chrysosporium, Aerobacter aerogenes, Trichoderma koningi*	Hayes 1957; Matsumoto et al. 2009; Maule et al. 1987; Kennedy et al. 1990; Anderson et al. 1970
Chlorpyrifos	CNS neural cell loss, abnormalities of synaptic function and behavior, neuroteratogen, disruption of motor activity	*Alcaligenes faecalis, Bacillus pumilus* C2A1, *Bacillus cereus, Enterobacter* sp., *Pseudomonas nitroreducens* PS-2, *Pseudomonas aeruginosa, Synechocystis* sp. strain PUPCCC 64	Aldridge et al. 2005; Levin et al. 2002; Chishti et al. 2013

(Continued)

TABLE 4.1 (*Continued*)
List of Pesticides and Their Degradation by Microbes

Pesticides	Hazardous Effects	Microbial Degraders	References
Propoxur	Blurred vision, nausea, vomiting, sweating, tachycardia, potent inhibitor of cholinesterase	*Pseudomonas* sp.	Kamanavalli and Ninnekar 2000
Captan	Hypothermia, depression, diarrhea, weight loss, carcinogenic, genotoxic, adenocarcinomas, mutagen	*Bacillus circulans*	Hines et al. 2008; Megadi et al. 2010

Chandra and Kumar 2015). Microbes can degrade a variety of pesticides/aromatic pollutants *via* aerobic or anaerobic pathways (Bondarenko and Gan 2004; DiDonato et al. 2010). In aerobic degradation, it is necessary to convert compound in the highly oxidative state making them water soluble and thus susceptible to further microbial degradation (Malmstrom 1982). For this, the first enzyme involved in the degradation pathway is very important. Some of the reported enzymes belong to hydrolases, carboxylases, oxygenases, or dehydrogenases. Hydrolases and oxygenases are the most studied enzymes. Hydrolases catalyze the hydrolysis of certain chemical bonds in the toxic compound that will lead to the reduction in its toxicity. Depending on the chemical bond that is hydrolyzed, they are grouped as an esterase (act on ester bond), protease/peptidase (act on peptide bond), acid anhydride hydrolases (act on acid anhydride), halide hydrolases (act on halide bond), and much more. For the biodegradation of organophosphates and carbamates insecticides, hydrolases play a very important role. Some of the examples are carbaryl hydrolase (esterase) and atrazine chlorohydrolase (halide hydrolase), which is the first enzyme involved in carbaryl and atrazine degradation, respectively, as shown in Figure 4.2.

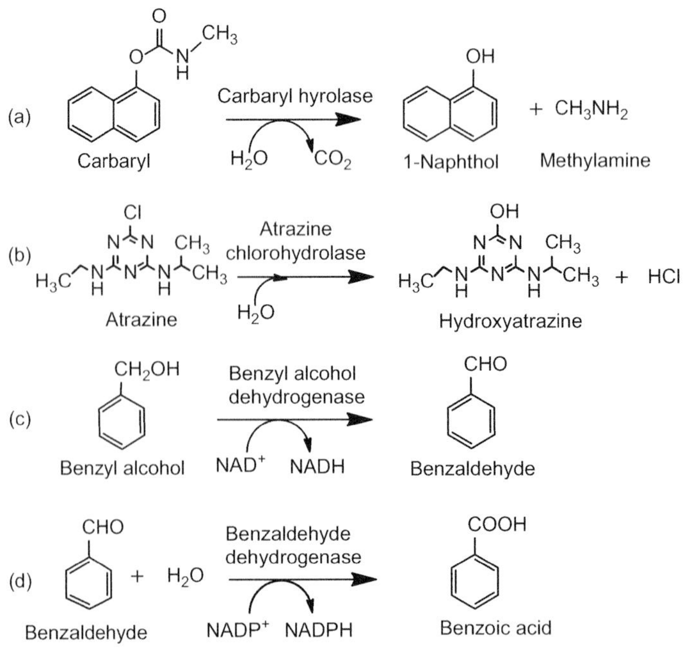

FIGURE 4.2 Examples of hydrolases and dehydrogenases involved in pesticides degradation: (a) ester hydrolase; (b) halide hydrolase; (c) alcohol dehydrogenase; and (d) aldehyde dehydrogenase.

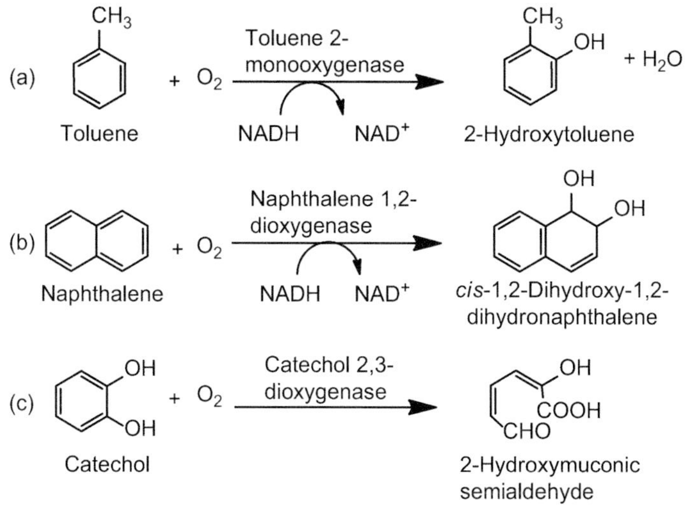

FIGURE 4.3 Oxygenases reactions by various enzymes: (a) monooxygenase, (b) ring-hydroxylating dioxygenase, and (c) ring-cleaving dioxygenase.

Similarly, dehydrogenases are also responsible for the oxidation of a wide variety of compounds, e.g., benzylalcohol and benzaldehyde dehydrogenases (Figure 4.2). They display wide substrate specificity. Another class of enzymes are called "oxygenases," which participate in oxidation of reduced substrates by incorporating oxygen atom into them. Oxygenases are sub-classified into monooxygenase and dioxygenase depending on the number and position of an oxygen atom(s) incorporated into the aromatic ring as shown in Figure 4.3. Monooxygenases convert the aromatic compound into the monohydroxylated compound with the addition of a single oxygen atom from molecular oxygen (e.g., toluene-2-monooxygenase). Dioxygenases incorporate both atoms of oxygen into the substrate giving either dihydroxylated compound (e.g., naphthalene dioxygenase) or ring cleavage of the aromatic diol (e.g., catechol-2,3-dioxygenase) (Figure 4.3).

Underanaerobic condition, microorganisms use inorganic compounds such as sulfate, nitrate, or metal ions as electron acceptors to generate cell mass, hydrogen sulfide (H_2S), nitrogen gas (N_2), methane (CH_4), and reduced forms of metals as products (Evans and Fuchs 1988; DiDonato et al. 2010). In this chapter, the aerobic mode of pesticide degradation pathways and enzymes from several bacterial strains are discussed.

4.2.1 LINDANE

The organochloride insecticide γ-hexachlorohexane (γ-HCH, lindane) is an extensively used pesticide by farmers throughout the world. It is synthesized by photochemical chlorination of benzene, giving the following stable isomers α-, β-, γ-, δ-, ε-, η-, and θ-HCH. All isomers are persistent in the environment and have a tendency of bioaccumulation. Lindane residues are found in soil, water, air, meat, and food grains. Extensive use of lindane has led to environmental contamination due to its persistence and slow degradation. Exposure of lindane causes many health hazards, affecting the central nervous system (Abalis et al. 1985), there productive system in rats (Simic et al. 2012), and much more. Therefore, it has been banned in many countries. Microorganisms are known to degrade lindane under aerobic and anaerobic conditions. The major studies have been done on the aerobic degradation by *Sphingomonas japonicum* UT26, *Sphingomonas indicium* B90A, and *Sphingobium francense* (Nagata et al. 2010; Pal et al. 2005; Cérémonie et al. 2006). The degradation pathway involves two sequential dechlorination reactions to give the putative product 1,3,4,6-tetrachloro-1,4-cyclohexadiene (1,3,4,6-TCDN) *via* the observed intermediate γ-pentachlorocyclohexene

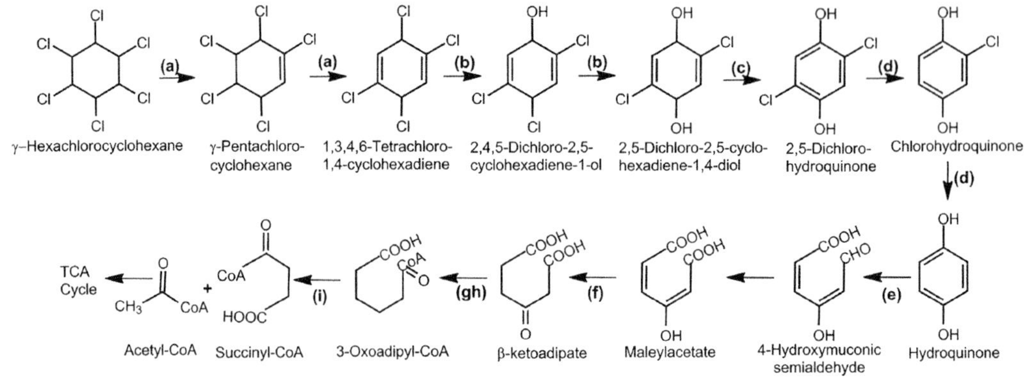

FIGURE 4.4 γ-HCH degradation pathway. Enzymes involved in the pathway are (a) γ-HCH dehydrochlorinase, (b) 1,3,4,6-tetrachloro-1,4-cyclohexadiene hydrolase, (c) 2,5-dichloro-2,5-cyclohexadiene-1,4-diol dehydrogenase, (d) 2,5-dichlorohydroquinone reductive dechlorinase, (e) chlorohydroquinone/hydroquinone 1,2-dioxygenase, (f) maleylacetate reductase, (g, h) 3-oxoadipate CoA-transferase α-subunit and β-subunit, (i) acetyl-CoA-acetyltransferase.

(γ-PCCH) (Figure 4.4). 1,3,4,6-TCDN undergo double hydrolytic dechlorination reaction forming 2,5-dichloro-2,5-cyclohexadiene-1,4-diol (2,5-DDOL) with an intermediate product of 2,4,5-trichloro-2,5-cyclohexadiene-1-ol (2,4,5-DNOL). Finally, 2,5-DDOL is converted to 2,5-dichlorohydroquinone (2,5-DCHQ) by dehydrogenation reaction. This completes the upper pathway (Nagata et al. 1994) (Figure 4.4). In the lower pathway, 2,5DCHQ is metabolized to chlorohydroquinone and to hydroquinone with two dehydrochlorination reactions. With the help of hydroquinone-1,2-dioxygenase, hydroquinone is ring-cleaved forming hydroxymuconic semialdehyde, which is further metabolized *via* succinyl-CoA and acetyl-CoA to TCA cycle intermediates (Endo et al. 2005; Nagata et al. 2007) (Figure 4.4).

4.2.2 ATRAZINE

Atrazine (2-chloro-4-ethylamino-6-isopropylamino-1,3,5-triazine) is one of the most widely used herbicides that belongs to the *s*-triazine family (Abigail 2012). It is sold under various trade names such as Aatrex, Atratol, Gesaprim, Fogard, and more. It is used against broadleaf and some grassy weeds in corn, sorghum, sugarcane fields, etc. (Buchholtz 1963). It also used in nonagricultural sites such asroadsides, airfields, parks, playground, and athletic fields to control the growth of unwanted grasses. Due to repeated applications, the concentration has reached beyond the permissible limit on the surface water (0.1 µg/mL). The increased levels are affecting animals as well as humans. It has toxic, endocrine-disrupting, and teratogenic effects (Batra et al. 2009). Several bacterial systems have been reported to metabolize atrazine by various routes into ammonia and carbon dioxide (Govantes et al. 2009; Nagy et al. 1995; Sadowsky et al. 1998; Batra et al. 2009). Studies reported three different pathways to degrade atrazine, one is hydrolytic and others are anoxidative-hydrolyticmix, which has common metabolic intermediate, i.e., cyanuric acid (Figure 4.5). In oxidative-hydrolytic route, mostly studied from *Rhodococcus* species, first atrazine undergoes oxidative removal of ethyl or isopropyl group catalyzed by *N*-dealkylase or atrazine monooxygenase to give deethylatrazine or deisopopylatrazine as products, respectively (Nagy et al. 1995). Deethylatrazine undergoes second oxidative removal of isopropyl group giving deisopropyldeethylatrazine. Subsequently, removal of ring substituents from deisopropylatrazine and deisopropyldeethylatrazine yields cyanuric acid (Figure 4.5). The hydrolytic degradation pathway is mostly studied in *Ralstonia* species and *Pseudomonas* species like *Pseudomonas* sp. strain ADP (Mandelbaum et al. 1995;

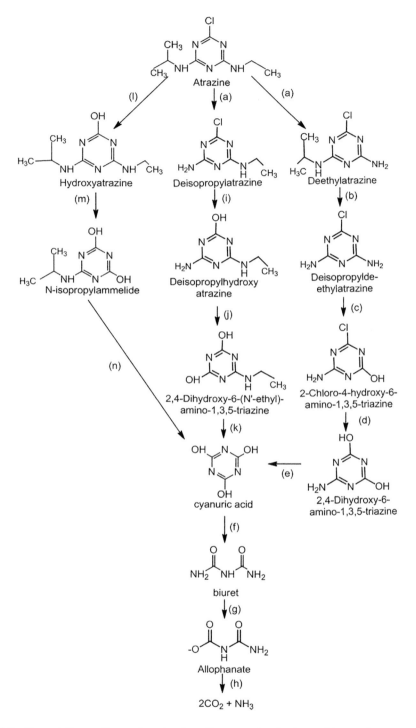

FIGURE 4.5 Degradation pathway for atrazine. Enzymes involved are (a) atrazine monooxygenase (atrazine *N*-dealkylase), (b) deethylatrazine monooxygenase, (c) *N*-ethylammeline chlorohydrolase, (d) hydrochloro atrazine ethylamino hydrolase, (e) *N*-isopropyl ammelide isopropyl aminohydrolase, (f) cyanuric acid amidohydrolase, (g) biuret hydrolase, (h) allophanate hydrolase, (i) *N*-ethylammeline chlorohydrolase, (j) deisopropyl hydroxyatrazine aminohydrolase, (k) 2,4-dihydroxy-6-(*N*-ethyl) amino-1,3,5-triazine aminohydrolase, (l) atrazine chlorohydrolase (AtzA), (m) hydroxyatrazine *N*-ethylaminohydrolase, (n) *N*-isopropyl ammelideisopropyl aminohydrolase.

Martinez et al. 2001), *Pseudomonas* sp. strain AKN5 (Sagarkar et al. 2014), and others. In this route, atrazine is dechlorinated by atrazine chlorohydrolase to give hydroxyatrazine. Furthermore, with the help of sequential hydrolases, hydroxyatrazine is converted to cyanuric acid (Boundy-Mills et al. 1997; Sadowsky et al. 1998). Then, cyanuric acid is converted to biuret with the help of cyanuric acid amidohydrolase. Then, biuret hydrolase converts biuret to allophanate, which is finally converted to ammonia and carbon dioxide by allophanate hydrolase (García-González et al. 2005) (Figure 4.5), thus enabling the removal of atrazine from the environment.

4.2.3 CARBARYL

Carbaryl (1-naphthyl *N*-methyl carbamate) is a third most widely used carbamate pesticide in the agriculture. Carbaryl itself does not persist in soil for a long time. However, due to repeated applications, the carbaryl level in the soil had increased (Felsot et al. 1981). Carbaryl is absorbed easily through inhalation or *via* the oral route and less readily absorbed *via* dermal routes. Early symptoms of carbaryl exposure may include muscle weakness, headache, stomach cramps, nausea, restlessness, and sweating. These symptoms of carbaryl poisoning develop quickly after absorption and disappear rapidly after exposure ends (Hayes 1982; Gossel and Bricker 1994). High exposure to carbaryl may lead to tearing, excessive salivation, pinpoint pupils, vomiting, diarrhea, nasal discharge, problems in coordination, and muscle twitching. However, acute carbaryl poisoning can result in coma, convulsions, and death. Under different conditions, in soil, carbaryl half-life varies from 4 to 72 days (US EPA 2003). It is highly toxic to marine and aquatic invertebrates, such as stoneflies and shrimps (US EPA 2003). Acute toxicity of carbaryl is a major concern, and it is necessary to study all the possible pathways through which it degrades. Chemical, physical, and biological processes can degrade it. Chemical methods include pH-dependent hydrolysis. It is stable at an acidic pH, while under alkaline conditions the ester bond (responsible for carbaryl toxicity) between *N*-methyl carbamic acid and 1-naphthol hydrolyzed into 1-naphthol, methylamine, and CO_2 (Larkin and Day 1986). Physical methods include ultraviolet irradiations and ultrasound wave exposure. Combination of high-frequency ultrasound wave and UV irradiation is also used for carbaryl degradation, which is considerably more effective than ultrasound or ultraviolet light alone (Khoobdel et al. 2010). Chemical and physical degradation method yield toxic and recalcitrant compound, 1-naphthol. Various organisms have been reported to degrade carbaryl (Singh et al. 2013). The metabolic steps and enzymes involved in carbaryl degradation in different bacteria are shown in Figure 4.6. The degradation pathway is initiated by conversion of carbaryl to 1-naphthol by carbaryl hydrolase; 1-naphthol is further metabolized *via* 1,2-dihydroxynaphthalene, 2-hydroxybenzalpyruvic acid, salicylaldehyde, salicylate, andgentisate to TCA cycle intermediates (Swetha and Phale 2005; Singh et al. 2013). However, *Pseudomonas* sp. such as 50552 and 50581 degrades carbaryl to salicylate, which is further metabolized *via* catechol to TCA cycle (Chapalamadugu and Chaudhry 1991) (Figure 4.6).

4.3 THE RESIDUAL PESTICIDES IN THE ENVIRONMENT AND THEIR CHALLENGES FOR BIODEGRADATION

The fate of pesticides in the environment depends not only on the presence of the microbial population but also on various abiotic and biotic parameters. Therefore, understanding the factors and mechanisms that affect biodegradation is of great significance from the perspective of efficient and complete removal of pesticides. Abiotic factors include soil type, temperature, pH, oxygen availability, moisture content, etc. The effect of soil pH on pesticide degradation depends on whether a pesticide is susceptible to acid or alkaline-catalyzed hydrolysis (Aislabie and Lloyd-Jones 1995). Temperature is considered as one of the key factors that affect the adsorption of pesticide by altering its solubility and hydrolysis (Racke et al. 1997). The biotic factor includes the substrate bioavailability, metabolic

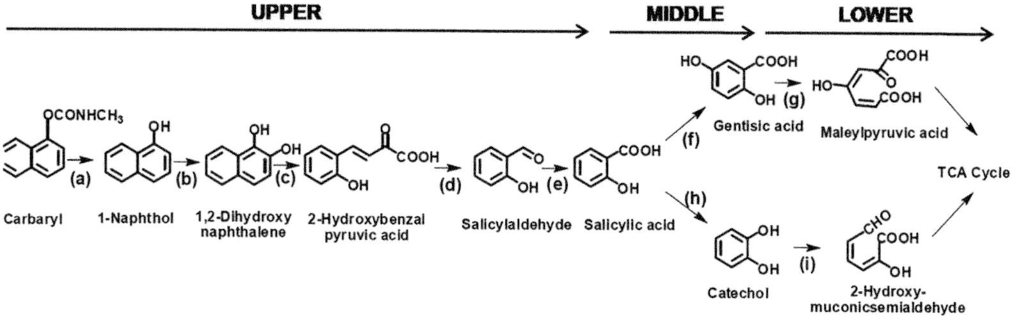

FIGURE 4.6 Carbaryl degradation pathway. Enzymes involved in the pathway are (a) carbaryl hydrolase, (b) 1-naphthol 2-hydroxylase, (c) 1,2-dihydroxynaphthalene dioxygenase, (d) *trans* o-hydroxy benzylidene pyruvate hydratase-aldolase, (e) salicylaldehyde dehydrogenase, (f) salicylate 5-hydroxylase, (g) gentisate 1,2-dioxygenase, (h) salicylate 1-hydroxylase, (i) catechol 1,2-dioxygenase. (Adapted from Singh et al. 2013.)

capacity, diversity, metabolic efficiency, etc. In soil, pesticides are reported to bind tightly to soil organic matter that reduces its bioavailability for degradation. Also in nature, microbes have inherent characters like the specificity for utilization of substrate that affects the substrate degradation. Moreover, the optimum growth conditions required for microbes are unpredictable, as the vitality of them is important for efficient function (Srivastava et al. 2014). The availability of substrate for the microbes is limited when the pesticides are less/not soluble in water. This can be overcome by pretreatment of contaminated soil with surfactants that enhance the pesticide solubility (Zhu and Aitken 2010). The concentration of pesticides at the toxicity level is one of the factors that also affect the bioremediation process as it can kill the microbes (Abdel-Shafy and Mansour 2016).

4.4 EVOLUTION OF PESTICIDES DEGRADATION PATHWAYS

Metabolic degradation pathways are evolved from the existing pathways or enzymes because of selective pressure. In another word, to degrade various pesticides/pollutants, microorganisms adapt themselves in the contaminated environment by acquiring new genes/evolving new enzymes (Feldgarden et al. 2003; Nojiri et al. 2004). These adaptations can be acquired by several ways such as (i) by recombination and rearrangement of genomic DNA, (ii) changes in intragenomic DNA sequences like small deletions or insertions of one or a few nucleotides, and/or (iii) lateral or horizontal gene transfer (LGT or HGT) of gene(s) involved in the biodegradation from phylogenetically distinct or related organisms (Ochman and Moran 2001). In such adaptations, besides HGT, the mobile genetic elements (MGEs) also play a major role in transferring the catabolic genes involved in pesticide degradation (Burrus et al. 2002). The major catabolic MGEs include acatabolic plasmid, elements using phage-like integrases, transposable elements, and catabolic and conjugative transposons. Interestingly, both on the chromosome and on the plasmid, catabolic genes may be flanked by insertion sequences (IS elements), which play a very important role in the mobilization of catabolic genes. In various bacteria, the genes encoding pesticide-/aromatic pollutant–degrading enzymes (hydrolase, oxygenase, transferase, hydroxylase, isomerase, and decarboxylase) are found to be located on either plasmid or chromosomal DNA. For example, in the 1990s, atrazine-degrading *Pseudomonas* sp. strain ADP was isolated from an atrazine spill site (Mandelbaum et al. 1995). In strain ADP, the *atz* genes (involved in atrazine degradation) are found to be present on a 108-kb single self-transmissible plasmid, pADP-1. The whole pADP-1 plasmid has been sequenced, and it has been demonstrated that the catabolic genes *atz*ABCDEF are responsible for the complete degradation of atrazine (Martinez et al. 2001). In brief, atrazine is converted into cyanuric acid by three amidohydrolases encoded by genes *atzA, atzB,* and *atzC.* Furthermore, cyanuric acid is completely

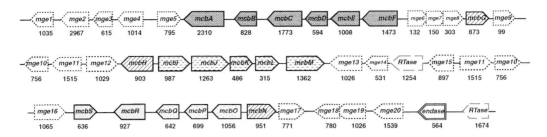

FIGURE 4.7 The arrangement of genes involved in the carbaryl metabolism. The direction of gene transcription is indicated by arrow; various mobile genetic elements indicated by *mge,* the genes encoding the "upper" pathway, are marked with dots, "middle" pathway with broken lines, and "lower" pathway with circles. Striped lines depict the probable functional regulators. Gray color represents the regulator genes probably not related to the carbaryl metabolism. Reverse transcriptase and endonuclease are represented by "RTase" and "endase," respectively. (Adapted from Trivedi et al. 2016.)

degraded into NH_3 and CO_2 by three other hydrolases encoded by gene *atzD, atzE,* and *atzF,* which are found to be organized in an operon-like structure (García-González et al. 2005; Govantes et al. 2010). The *atzA, atzB,* and *atzC* genes are found to be highly conserved and widely dispersed across the several atrazine-degrading bacterial strains (de Souza et al. 1998). Genes encoding γ-HCH-degrading pathway enzymes are also identified and characterized by several γ-HCH-degrading organisms (Lal et al. 2010). In Sphingomonads, the γ-HCH degradation pathway or genes encoding pathway enzymes are found to be recruited and established by HGT, which is mediated by mobile genetic elements. In addition, lindane genes were found to be located on the chromosome as well as on the plasmids (Lal et al. 2010).

In *Pseudomonas* sp. strain C5pp, carbaryl degradation pathway was hypothesized to be acquired *via* HGT events and probably evolved under the positive selection pressure of carbaryl (Trivedi et al. 2016). The draft genome analysis revealed the probable origin of genes and evolutionary events for the evolution of carbaryl degradation pathway. The carbaryl degradation property in this strain was found to be stable, noncurable, and nonconjugable. The carbaryl-degrading pathway is divided into "upper" (carbaryl to salicylate), "middle" (salicylate to gentisate), and "lower" (gentisate to TCA cycle) segments (Figure 4.6). Genes encoding the pathway enzymes are present on the chromosome and organized into three distinct putative operons (Singh et al. 2013; Trivedi et al. 2016). Genes encoding "upper" pathway were found to be present as a part of integron, whereas "middle" and "lower" pathway genes were found to be present in two distinct classI composite transposons (Figure 4.7). The "upper" pathway genes were found to harbor classI integron with features like transposases. Additionally, it was found that strain C5pp possesses a catabolic transposon, which harbors "middle" pathway (*mcbIJKL*) genes and exhibits classI composite transposon features. Similarly, the "lower" pathway genes were found to be a part of the classI composite transposon. Overall, it indicates that the acquisition of carbaryl degradation pathway in strain C5pp is a result of multiple transposition events (Trivedi et al. 2016). This seems to be an essential adaptation process for the organism to survive. Overall, this study helps in understanding the molecular mechanisms involved in the construction and evolution of new metabolic pathway/enzymes. Studies suggest that the new pathways for any pesticide/xenobiotic degradation are assembled either as promiscuity of enzymes or as "patchwork" of enzymes from the preexisting pathways.

4.5 APPLICATIONS AND FUTURE DIRECTIONS

A large number of microorganisms that carry out bioremediation reactions are reported. The isolation and characterization of pure cultures have been and will continue to be crucial for bioremediation itself and analysis of molecular ecology. Investigations of their biodegradation capacity will also help to understand the metabolic products formed and their fate in the environment. Under laboratory conditions, the degradation of contaminants has been well established. However, in the field, survival of these organisms and the degradation of pesticide remain challenging. Many physical, chemical, and biological factors, e.g., substrate availability, soil properties, moisture, pH, temperature, suboptimal growth conditions, suppression of certain catabolic genes, and competition with indigenous microorganisms, determine the activity and fate of microorganisms into the environment (van Veen et al. 1997). However, there are several methods used to enhance the degradation capability of microorganisms in contaminated sites. One such method involves the use of immobilized cells for environmental application. There are many advantages of using immobilized cells such as longer retention of the bioremediation agents (microbial cells), stability and survival of the cells, reuse of the cells, cost-effective, increased tolerance to unfavorable conditions, and long-term storage. Also, soil can be supplemented with inorganic nutrients, cyclodextrins (increases the solubility of the pollutants and its desorption in the soil), surface active agents, and surfactants (due to their low toxicity and biodegradability) (van Veen et al. 1997). Several atrazine-degrading microorganisms are used in the field study. Immobilized *Pseudomonas* sp. strain ADP is considered a stable inoculant for the bioremediation of atrazine. It was originally isolated from a site heavily contaminated with atrazine and uses atrazine as a sole nitrogen source. Strain ADP is the model organism for the full mineralization of this *s*-triazine herbicide (Mandelbaum et al. 1995). Strain ADP immobilized onto a zeolite carrier was found to remain viable and retained its ability to degradeatrazine for the complete test period of 10 weeks at 25°C (Stelting et al. 2012). Degradation of atrazine has been studied at the field level also by using strain ADP. In this study, recombinant cells expressing atrazine chlorohydrolase were used for bioremediation and applied to a site (Wackett et al. 2002).

Another approach could be the use of genetic engineering techniques to construct a single chassis strain, which will be effective against an array of pesticides and aromatic pollutants. The genetic arrangement and acquisition of numerous pesticides'/aromatic pollutants' metabolic pathways are known as, viz., atrazine, HCH, phenanthrene, naphthalene, carbaryl, etc. In this perspective, the information becomes particularly important giving an avenue for genetic engineering a suitable candidate to bioremediate the array of pesticides/aromatic pollutants from the contaminated sites, which catabolizes to central carbon pathway *via* common intermediates.

4.6 CONCLUSION

Bioremediation is one of the environmentally friendly and cost-effective methods to reduce the recalcitrant and toxic pesticides from the environment with the help of microorganisms. There is need for further understanding the myriad of factors such as certain environmental factors (temperature, pH, growth conditions, soil properties, etc.) and genetic potential that influence the growth of the microorganism and rate of degradation for efficient removal of environmental pollutants in the contaminated sites. Understanding of metabolic pathway evolution and its regulation is also necessary to develop genetically engineered candidate for the bioremediation of the array of pesticides. With this, the future prospect is to investigate the degrading capability by recombinant and indigenous bacteria in order to restore the soil quality. This may help farmers to move back to natural organic farming than chemical farming for better productivity.

REFERENCES

Abalis IM, Eldefrawi ME, Eldefrawi AT. 1985. High-affinity stereospecific binding of cyclodiene insecticides and γ-hexachlorocyclohexane to γ-aminobutyric acid receptors of rat-brain. *Pesticide Biochemistry Physiol* 24:95–102.
Abdel-Shafy HI, Mansour MS. 2016. A review on polycyclic aromatic hydrocarbons: source, environmental impact, effect on human health and remediation. *Egypt J Pet* 25:107–123.

Abigail EA. 2012. Microbial degradation of atrazine, commonly used herbicide. *Intel Adv Biological Res* 2:16–23.

Aislabie J, Lloyd-Jones G. 1995. A review of bacterial-degradation of pesticides. *Aust J Soil Res* 33:925–942.

Aldridge JE, Meyer A, Seidler FJ, Slotkin TA. 2005. Alterations in central nervous system serotonergic and dopaminergic synaptic activity in adulthood after prenatal or neonatal chlorpyrifos exposure. *Environ Health Perspect* 113:1027–1031.

Anderson JPE, Lichtenstein EP, Whittingham WF. 1970. Effect on Mucor alternans on the persistence of DDT and dieldrin in culture and in soil. *J Econ Entomol* 63:1595–1599.

Awasthi N, Singh AK, Jain RK, Khangarot BS, Kumar A. 2003. Degradation and detoxification of endosulfan isomers by a defined co-culture of two *Bacillus* strains. *Appl Microbiol Biotechnol* 62:279–283.

Batra M, Pandey J, Suri CR, Jain RK. 2009. Isolation and characterization of an atrazine-degrading *Rhodococcus* sp. strain MB-P1 from contaminated soil. *Lett Appl Microbiol* 49:721–729.

Bertilsson S, Widenfalk A. 2002. Photochemical degradation of PAHs in freshwaters and their impact on bacterial growth–influence of water chemistry. *Hydrobiolog* 469:23–32.

Bondarenko S, Gan J. 2004. Degradation and sorption of selected organophosphate and carbamate insecticides in urban stream sediments. *Environ Toxicol Chem* 23:1809–1814.

Boul HL, Garnham ML, Hucker D, Baird D, Aislabie J. 1994. Influence of agricultural practices on the levels of DDT and its residues in soil. *Environ Sci Technol* 28:1397–1402.

Boundy-Mills KL, De Souza ML, Mandelbaum RT, Wackett LP, Sadowsky MJ. 1997. The *atzB* gene of *Pseudomonas* sp. strain ADP encodes the second enzyme of a novel atrazine degradation pathway. *Appl Environ Microbiol* 63:916–923.

Buchholtz KP. 1963. Use of atrazine and other triazine herbicides in control of quackgrass in corn fields. *Weeds* 11:202–205.

Bumpus JA, Powers RH, Sun T. 1993. Biodegradation of DDE (1,1-dichloro-2,2-bis(4-chlorophenyl)ethene) by *Phanerochaete chrysosporium*. *Mycol Res* 97:95–98.

Burrus V, Pavlovic G, Decaris B, Guedon G. 2002. Conjugative transposons: the tip of the iceberg. *Mol Microbiol* 46:601–610.

Cérémonie H, Boubakri H, Mavingui P, Simonet P, Vogel TM. 2006. Plasmid-encoded γ-hexachlorocyclohexane degradation genes and insertion sequences in *Sphingobium francense* (ex-*Sphingomonas paucimobilis* Sp+). *FEMS Microbiol Lett* 257:243–252.

Chandra R., Kumar V. 2015. Biotransformation and biodegradation of organophosphates and organohalides. In R. Chandra (ed.), *Environmental Waste Management*. Boca Raton, FL: CRC Press, pp. 475–524.

Chapalamadugu S, Chaudhry GR. 1991. Hydrolysis of carbaryl by a *Pseudomonas* sp. and construction of a microbial consortium that completely metabolizes carbaryl. *Appl Environ Microbiol* 57:744–750.

Chaudhuri K, Selvaraj S, Pal AK. 1999. Studies on the genotoxicity of endosulfan in bacterial systems. *Mutat Res* 439:63–67.

Chishti Z, Hussain S, Arshad KR, Khalid A, Arshad M. 2013. Microbial degradation of chlorpyrifos in liquid media and soil. *J Environ Manage* 114:372–380.

de Souza ML, Wackett LP, Sadowsky, MJ. 1998. The atzABC genes encoding atrazine catabolism are located on a self-transmissible plasmid in *Pseudomonas* sp. strain ADP. *Appl Environ Microbiol* 64:2323–2326.

DiDonato RJJr, Young ND, Butler JE, Chin KJ, Hixson KK, Mouser P, Lipton MS, DeBoy R, Methe BA. 2010. Genome sequence of the deltaproteobacterial strain NaphS2 and analysis of differential gene expression during anaerobic growth on naphthalene. *PLoS One* 5:e14072. doi:10.1371/journal.pone.0014072.

Doddamani HP, Ninnekar HZ. 2001. Biodegradation of carbaryl by a *Micrococcus* species. *Curr Microbiol* 43:69–73.

Dua A, Malhotra J, Saxena A, Khan F, Lal R. 2013. *Devosia lucknowensis* sp. nov., a bacterium isolated from hexachlorocyclohexane (HCH) contaminated pond soil. *J Microbial* 51:689–694.

Elespuru R, Lijinsky W, Setlow JK. 1974. Nitrosocarbaryl as a potent mutagen of environmental significance. *Nature* 247:386–387.

Endo R, Kamakura M, Miyauchi K, Fukuda M, Ohtsubo Y, Tsuda M, Nagata Y. 2005. Identification and characterization of genes involved in the downstream degradation pathway of γ-hexachlorocyclohexane in *Sphingomonas paucimobilis* UT26. *J Bacteriol* 187:847–853.

Evans WC, Fuchs G. 1988. Anaerobic degradation of aromatic compounds. *Annu Rev Microbiol* 42:289–317.

Fazlurrahman, Batra M, Pandey J, Suri CR, Jain RK. 2009. Isolation and characterization of an atrazine-degrading *Rhodococcus* sp. strain MB-P1 from contaminated soil. *Lett Appl Microbiol* 49:721–729.

Feldgarden M, Byrd N, Cohan FM. 2003. Gradual evolution in bacteria: evidence from *Bacillus systematics*. *Microbiology* 149:3565–3573.

Felsot A, Maddox JV, Bruce W. 1981. Enhanced microbial degradation of carbofuran in soils with histories of Furadan use. *Bull Environ Contam Toxicol* 26:781–788.

Foght J, April T, Biggar K, Aislabie J. 2001. Bioremediation of DDT-contaminated soils: a review. *Bioremediat J* 5:225–246.

García-González V, Govantes F, Porrúa O, Santero E. 2005. Regulation of the *Pseudomonas* sp. strain ADP cyanuric acid degradation operon. *J Bacteriol* 187:155–167.

Gossel T, Bricker JD. 1994. *Principles of Clinical Toxicology*. 3rd ed. New York: Raven Press.

Govantes F, García-González V, Porrúa O, Platero AI, Jiménez-Fernández A, Santero E. 2010. Regulation of the atrazine-degradative genes in *Pseudomonas* sp. strain ADP. *FEMS Microbiol Lett* 310:1–8.

Govantes F, Porrúa O, García-González V, Santero E. 2009. Atrazine biodegradation in the lab and in the field: enzymatic activities and gene regulation. *Microb Biotechnol* 2:178–185.

Guha A, Kumari B, Bora TC, Deka PC, Roy MK. 2000. Bioremediation of endosulfan by *Micrococcus* sp. *Indian J Environ Health* 42:9–12.

Hayatsu M, Nagata T. 1993. Purification and characterization of carbaryl hydrolase from *Blastobacter* sp. strain M501. *Appl Environ Microbiol* 59:2121–2125.

Hayes, WJ, Jr. 1957. Dieldrin poisoning in man. *Publ Health Rep* 72:1087–1091.

Hayes WJ. 1982. *Pesticides Studied in Man*. San Francisco, CA: Williams & Wilkins.

Hines CJ, Deddens JA, Jaycox LB, Andrews RN, Striley CA, Alavanja MC. 2008. Captan exposure and evaluation of a pesticide exposure algorithm among orchard pesticide applicators in the Agricultural Health Study. *Ann Occupational Hyg* 52:153–166.

Kamanavalli CM, Ninnekar HZ. 2000. Biodegradation of propoxur by *Pseudomonas* species. *World J Microbiol Biotechnol* 16:329–331.

Kennedy DW, Aust SD, Bumpus JA. 1990. Comparative biodegradation of alkyl halide insecticides by the white rot fungus, *Phanerochaete chrysosporium* (BKM-F-1767). *Appl Environ Microbiol* 56:2347–2353.

Khoobdel M, Shayeghi M, Golsorkhi S, Abtahi M, Vatandoost H, Zeraatii H, Bazrafkan S. 2010. Effectiveness of ultrasound and ultraviolet irradiation on degradation of carbaryl from aqueous solutions. *Iran J Arthropod Borne Dis* 4:47–53.

Kumar V., Shahi SK, Singh S. 2018. Bioremediation: an eco-sustainable approach for restoration of contaminated sites. In J. Singh, D. Sharma, G. Kumar, N.R. Sharma (eds.), *Microbial Bioprospecting for Sustainable Development*. Singapore: Springer, pp. 115–136.

Kuritz T, Wolk CP. 1995. Use of filamentous cyanobacteria for biodegradation of organic pollutants. *Appl Environ Microbiol* 61:234–238.

Kwon GS, Sohn HY, Shin KS, Kim E, Seo BI. 2005. Biodegradation of the organochlorine insecticide, endosulfan, and the toxic metabolite, endosulfan sulfate, by *Klebsiella oxytoca* KE-8. *Appl Microbiol Biotechnol* 67:845–850.

Lal R, Pandey G, Sharma P, Kumari K, Malhotra S, Pandey R, Raina V, Kohler HP, Holliger C, Jackson C, Oakeshott JG. 2010. Biochemistry of microbial degradation of hexachlorocyclohexane and prospects for bioremediation. *Microbiol Mol Biol Rev* 74:58–80.

Larkin MJ, Day MJ. 1986. The metabolism of carbaryl by three bacterial isolates, *Pseudomonas* spp. (NCIB 12042 & 12043) and *Rhodococcus* sp. (NCIB 12038) from garden soil. *J Appl Bacteriol* 60:233–242.

Levin ED, Addy N, Baruah A, Elias A, Christopher NC, Seidler FJ, Slotkin TA. 2002. Prenatal chlorpyrifos exposure in rats causes persistent behavioral alterations. *Neurotoxicol Teratol* 24:733–741.

Malmstrom BG. 1982. Enzymology of oxygen. *Annu Rev Biochem* 51:21–59.

Mandelbaum RT, Allan DL, Wackett LP. 1995. Isolation and characterization of a *Pseudomonas* sp. that mineralizes the s-triazine herbicide atrazine. *Appl Environ Microbiol* 61:1451–1457.

Martinez B, Tomkins J, Wackett LP, Wing R, Sadowsky MJ. 2001. Complete nucleotide sequence and organization of the atrazine catabolic plasmid pADP-1 from *Pseudomonas* sp. strain ADP. *J Bacteriol* 183:5684–5697.

Matsumoto E, Kawanaka Y, Yun SJ, Oyaizu H. 2009. Bioremediation of the organochlorine pesticides, dieldrin and endrin, and their occurrence in the environment. *Appl Microbiol Biotechnol* 84:205–216.

Maule A, Plyte S, Quirk AV. 1987. Dehalogenation of organochlorine insecticides by mixed anaerobic microbial populations. *Pestic Biochem Physiol* 27:229–236.

Megadi VB, Tallur PN, Hoskeri RS, Mulla SI, Ninnekar HZ. 2010. Biodegradation of pendimethalin by *Bacillus circulans*. *Indian J Biotechnol* 9:173–177.

Nadeau LJ, Menn FM, Breen A, Sayler GS. 1994. Aerobic degradation of 1,1,1-trichloro-2,2-bis(4-chlorophenyl)ethane (DDT) by *Alcaligenes eutrophus* A5. *Appl Environ Microbiol* 60:51–55.

Nagata Y, Endo R, Ito M, Ohtsubo Y, Tsuda M. 2007. Aerobic degradation of lindane (gamma-hexachlorocyclohexane) in bacteria and its biochemical and molecular basis. *Appl Microbiol Biotechnol* 76:741–752.

Nagata Y, Ohtomo R, Miyauchi K, Fukuda M, Yano K, Takagi M. 1994. Cloning and sequencing of a 2,5-dichloro-2,5-cyclohexadiene-1,4-diol dehydrogenase gene involved in the degradation of gamma-hexachlorocyclohexane in *Pseudomonas paucimobilis*. *J Bacteriol* 176:3117–3125.

Nagata Y, Ohtsubo Y, Endo R, Ichikawa N, Ankai A, Oguchi A, Fukui S, Fujita N, Tsuda M. 2010. Complete genome sequence of the representative γ-hexachlorocyclohexane-degrading bacterium *Sphingobium japonicum* UT26. *J Bacteriol* 192:5852–5853.

Nagy I, Compernolle F, Ghys K, Vanderleyden J, DeMot R. 1995. A single cytochrome P-450 system is involved in degradation of the herbicides EPTC (S-ethyl dipropylthiocarbamate) and atrazine by Rhodococcus sp. strain NI86/21. *Appl Environ Microbiol* 61:2056–2060.

Niharika N, Sangwan N, Ahmad S, Singh P, Khurana JP, Lal R. 2013. Draft genome sequence of *Sphingobium chinhatense* strain IP26T, isolated from a hexachlorocyclohexane dumpsite. *Genome Announc* 1:e00680-13.

Nojiri H, Shintani M, Omori T. 2004. Divergence of mobile genetic elements involved in the distribution of xenobiotic-catabolic capacity. *Appl Microbiol Biotechnol* 64:154–174.

Ochman H, Moran NA. 2001. Genes lost and genes found: evolution of bacterial pathogenesis and symbiosis. *Science* 292:1096–1099.

Pal R, Bala S, Dadhwal M, Kumar M, Dhingra G, Prakash O, Prabagaran SR, Shivaji S, Cullum J, Holliger C, Lal R. 2005. Hexachlorocyclohexane-degrading bacterial strains *Sphingomonas paucimobilis* B90A, UT26 and Sp+, having similar lin genes, represent three distinct species, *Sphingobium indicum* sp. nov., *Sphingobium japonicum* sp. nov. and *Sphingobium francense* sp. nov., and reclassification of [*Sphingomonas*] *chungbukensis* as *Sphingobium chungbukense* comb. nov. *Int J Syst Evol Microbiol* 55:1965–1972.

Pushiri H, Pearce SL, Oakeshott JG, Russell RJ, Pandey G. 2013. Draft genome sequence of *Pandoraea* sp. strain SD6-2, isolated from lindane-contaminated Australian soil. *Genome Announc* 1:e00415-13.

Racke KD, Skidmore MW, Hamilton DJ, Unsworth JB, Miyamoto J, Cohen SZ. 1997. Pesticide fate in tropical soils. *Pure and Appl Chem* 69:1349–1371.

Rivas FJ. 2006. Polycyclic aromatic hydrocarbons sorbed on soils: a short review of chemical oxidation based treatments. *J Hazard Mater* 138:234–251.

Sadowsky MJ, Tong Z, de Souza M, Wackett LP. 1998. AtzC is a new member of the amidohydrolase protein superfamily and is homologous to other atrazine-metabolizing enzymes. *J Bacteriol* 180:152–158.

Sagarkar S, Nousiainen A, Shaligram S, Bjorklof K, Lindstrom K, Jorgensen KS, Kapley A. 2014. Soil mesocosm studies on atrazine bioremediation. *J Environ Manage* 139:208–216.

Sahu SK, Patnaik KK, Sharmila M, Sethunathan N. 1990. Degradation of alpha-, beta-, and gamma-hexachlorocyclohexane by a soil bacterium under aerobic conditions. *Appl Environ Microbiol* 56:3620–3622.

Seo JS, Keum YS, Li QX. 2013. Metabolomic and proteomic insights into carbaryl catabolism by *Burkholderia* sp. C3 and degradation of ten N-methylcarbamates. *Biodegradation* 24:795–811.

Shea TB, Berry ES.1983. Toxicity and intracellular localization of carbaryl and 1-naphthol in cell cultures derived from goldfish. *Bull Environ Contam Toxicol* 30:99–104.

Siddique T, Okeke BC, Arshad M, Frankenberger WT. 2003. Biodegradation kinetics of endosulfan by *Fusarium ventricosum* and a *Pandoraea* species. *J Agric Food Chem* 51:8015–8019.

Simic B, Kmetic I, Murati T, Kniewald J. 2012. Effects of lindane on reproductive parameters in male rats. *VetArhiv* 82:211–220.

Singh AK, Garg N, Lata P, Kumar R, Negi V, Vikram S, Lal R. 2014. *Pontibacter indicus* sp. nov., isolated from hexachlorocyclohexane-contaminated soil. *Int J Syst Evol Microbiol* 64:254–259.

Singh SK, Pandey RS. 1990. Effect of sub-chronic endosulfan exposures on plasma gonadotrophins, testosterone, testicular testosterone and enzymes of androgen biosynthesis in rat. *Indian J Exp Biol* 28:953–956.

Singh R, Trivedi VD, Phale PS. 2013. Metabolic regulation and chromosomal localization of carbaryl degradation pathway in *Pseudomonas* sp. strains C4, C5 and C6. *Arch Microbiol* 195:521–535.

Srivastava J, Naraian R, Kalra SJS, Chandra H. 2014. Advances in microbial bioremediation and the factors influencing the process. *Int J Environ Sci Technol* 11:1787–1800.

Stelting S, Burns RG, Sunna A, Visnovsky G, Bunt CR. 2012. Immobilization of *Pseudomonas* sp. strain ADP: a stable inoculant for the bioremediation of atrazine. *Appl Clay Sci* 64:90–93.

Subba-Rao RV, Alexander M. 1985. Bacterial and fungal cometabolism of 1,1,1-trichloro-2,2-bis(pchlorophenyl) ethane (DDT) and its breakdown products. *Appl Environ Microbiol* 49:509–516.

Sud RK, Sud AK, Gupta KG. 1972. Degradation of sevin (1-naphthyl-N-methyl carbamate) by *Achromobacter* sp. *Arch Mikrobiol* 87:353–358.

Swetha VP, Phale PS. 2005. Metabolism of carbaryl via 1,2-dihydroxynaphthalene by soil isolates *Pseudomonas* sp. strains C4, C5, and C6. *Appl Environ Microbiol* 71:5951–5956.

Thatcher A, Waterson P, Todd A, Moray N. 2018. State of Science: ergonomics and global issues. *Ergonomics* 61:197–213.

Trivedi VD, Bharadwaj A, Varunjikar MS, Singha AK, Upadhyay P, Gautam K, Phale PS. 2017. Insights into metabolism and sodium chloride adaptability of carbaryl degrading halotolerant *Pseudomonas* sp. strain C7. *Arch Microbiol* 199:907–916.

Trivedi VD, Jangir PK, Sharma R, Phale PS. 2016. Insights into functional and evolutionary analysis of carbaryl metabolic pathway from *Pseudomonas* sp. strain C5pp. *Sci Rep* 6:38430. doi:10.1038/srep38430.

US EPA. 2003. Revised EFED Risk Assessment of Carbaryl in Support of the Reregistration Eligibility Decision (RED). https://www.epa.gov/espp/effects/carb-riskass.pdf.

van Veen JA, van Overbeek LS, van Elsas JD. 1997. Fate and activity of microorganisms introduced into soil. *Microbiol Mol Biol Rev* 61:121–135.

Wackett LP, Sadowsky MJ, Martinez B, Shapir N. 2002. Biodegradation of atrazine and related s-triazine compounds: from enzymes to field studies. *Appl Microbiol Biotechnol* 58:39–45.

Zhu H, Aitken MD. 2010. Surfactant-enhanced desorption and biodegradation of polycyclic aromatic hydrocarbons in contaminated soil. *Environ Sci Technol* 44:7260–7265.

5 Dissimilatory Iron-Reducing Bacteria

A Template for Iron Mineralization and Nanomaterial Synthesis

Abhilash
CSIR-National Metallurgical Laboratory

CONTENTS

5.1 INTRODUCTION

Dissimilatory iron-reducing bacteria (DIRB) was first studied on the *Geobacter metallireducens* in 1988 (Lovley and Phillips 1988), and since then, much work has been done on it, leading to isolation of novel species. DIRB, as the word suggests, is the reduction of iron Fe(III), depicting it as the terminal electron acceptor using organic sources or H_2 or aromatic substances as the electron donor, thereby generating energy through this metabolism. Much interest has been generated in the scientific community toward these microorganisms not only because of their significant capability of releasing metals and nutrients in the aquatic environment (Lovley 2004), thus helping in the biogeochemical nutrient cycling, but also because of their uniqueness to form extracellular iron particles, contrary to magnetotactic bacteria, which synthesizes intracellular particles giving each cell a permanent magnetic dipole to align themselves parallel to earth's geomagnetic axis (Table 5.1). Though the Fe(III) reducers are diverse, maximum attention has been given to

TABLE 5.1
Salient Features Governing Biologically Controlled and Induced Mineralization

Features	Biologically Controlled Mineralization	Biologically Induced Mineralization
Site of particle formation	Intracellular	Extracellular
Mechanism	• Ions are taken inside an isolated compartment formed inside the cell, up to a supersaturation level • Nucleation process takes place depending on steric hindrance	• Dissimilatory iron-reducing bacteria respire with Fe(III) oxyhydroxides reducing and secreting Fe(II) extracellularly • Fe(II) reacts with excess oxyhydroxides forming magnetite
Examples of microorganisms	*Magnetotactic magnetospirillium, Magnetotacticum bavaricum*	*Geobacter metallireducens, Shewanella putrefaciens*
Habitat	Aquatic, present at the oxic-anoxic transition zone	Mostly aquatic, present in mesophilic or thermophilic conditions
Characteristics of magnetite formed	• Size depends on the intracellular compartment where nucleation occurs • Uniform particles • Properly characterized • Size 5–100 nm	• Size depends on environmental factors • Nonuniform particles • Poorly characterized • Size 10–50 nm
Controlling factors	Genetic control	Environmental factors (pH, temperature)

mesophilic iron reducers, *Geobacter* and *Shewanella* species. The DIRB have been spoken together with Mn(IV) reduction. Here, we give a summarized review emphasizing on DIRB to understand it better for it to aid in further research in the field of nanoparticle biosynthesis. Until recently, it was believed that the iron reduction in the environment was nonenzymatic, led by the belief that anoxic environment was created by the microbial consumption of oxygen and by the production of reduced metabolites in the sediments, finally ensuing Fe(III) reduction (Lovley 1997). This belief has now been rejected on the basis that a mere redox environment is not enough for Fe(III) reduction to Fe(II) (Vargas et al. 1998). Proofs have been found where it is observed that the extent of abiotic reduction of Fe(III) is far less than in the presence of iron-reducing microorganisms. The nanoparticles of magnetite formed on the reduction of Fe(III) have been observed not only on earth but also in martian meteorite, thus enhancing the importance of Fe(III) reduction by microbes.

Iron biomineralization process has been developed by a wide diversity of organisms belonging to a large number of phyla. Even after the immense energy expenditure in such a biomineralization process, microbes perform this process because of the increased number of profits. Formation of iron particles takes care of the future metabolic needs of the organism without increasing the metal toxicity inside the cell. Not only that these iron particles provide certain necessary attributes to the microorganism, which is highly needed by them such as hardness and magnetism, but iron biomineralization process also increases the sustainability of the environment. There have been various examples of microorganism where aromatic and organic compounds have been degraded along with Fe(III) reduction and biomineralization such as in the case of *G. metallireducens* (Lovley et al. 1993). *G. metallireducens* has been known to oxidize organic compounds along with the reduction of Fe(III). Magnetic iron nanoparticle formation has been given the credit for magnetization of sediment. Finally, this iron biomineralization process has led to the development of a biomimetic process where synthesis of iron biominerals by microorganism has been used for biomedical purposes of magnetic labeling of tissues, molecular imaging, and drug delivery techniques, thereby increasing the potential application of these minerals.

The interaction between iron minerals and their reduction by bacteria is an important phenomenon that has become known in the last century though much evidence has been provided of Fe(III) reduction as the earliest form of respiration on earth (Vargas et al. 1998). Bacterial Fe(III) reduction plays a crucial role in the biogeochemical cycling of iron where Fe(III) minerals act as the terminal electron acceptor. Not only the redox cycling of iron but dissimilatory iron reducers also have been studied for their ability to remediate toxic metals, remediation by utilization of aromatic compounds, and metal corrosion (Lovley 1997). The iron reducers range from archaea to bacteria. Most of them, though being facultative aerobes, reduce iron under anaerobic conditions. The iron reducers mainly reduce Fe(III) by completely oxidizing multicarbon compounds to CO_2 or by incompletely oxidizing the multicarbon compounds.

Until recently, it was believed that the iron reduction in the environment was nonenzymatic, led by the belief that anoxic environment was created by the microbial consumption of oxygen and by the production of reduced metabolites in the sediments, which finally ensued Fe(III) reduction. This belief has now been rejected on the basis that a mere redox environment is not enough for Fe(III) reduction to Fe(II) (Vargas et al. 1998). Proofs have been found where it is observed that the extent of abiotic reduction of Fe(III) is far less than in the presence of iron-reducing microorganisms.

Though DIRB have been known for long until the last decade, it had been assumed that earliest form of microbial respiration has been by the sulfur-reducing bacteria. A recent observation has shown that, instead of sulfur-reducing bacteria, it might have been by Fe(III) reduction (Vargas et al. 1998). Vargas et al. (1998) give evidence that the bacteria and archaea most related to the LCA (last common ancestor) respire by reducing Fe(III) to Fe(II) (Vargas et al. 1998). Also, hyperthermophiles such as *Thermatoga maritima* thought to use sulfate as the terminal electron acceptor can grow by reduction of Fe(III). DIRB are known to be present under extreme conditions too. Generally, its habitat can be from psychrophilic to the hyperthermophilic environment. Table 5.2 gives a view of the example of a few DIRB present under various temperature profiles.

TABLE 5.2
Examples of Dissimilatory Iron-Reducing Bacteria, Growth Optimum Temperature, and Electron Donors

Condition	Optimum Temperature	Microorganisms	Electron Donor
Hyperthermophilic	75°C–90°C	*Thermoterrabacterium ferrireducens* (isolated from hot spring)	Not known
Thermophilic	60°C	*Bacillus infernus*	H_2/CO_2
	65°C	*Deferribacter thermophilus*	Organic compounds
Thermophilic acidophiles	55°C–60°C	*Sulfobacillus thermosulfidooxidans*, *Sulfobacillus acidophiles*, and *Acidimicrobium ferrooxidans*	Glycerol, tetrathionate
Mesophilic	30°C	*Desulfuromonas acetoxidans*	Organic compounds
	30°C	*Geobacter metallireducens*	Organic compounds
	30°C	*Geothrix fermentans*	Acetate
	25°C	*Ferrimonas futtsuensis* sp. Nov. and *Ferrimonas kyonanensis*	Selenate

5.2 TYPES OF DIRB

DIRB are classified as fermentative iron-reducing bacteria (IRB), sulfur-oxidizing IRB, organic acid–oxidizing IRB, hydrogen-oxidizing IRB, and aromatic compound–oxidizing IRB.

5.2.1 Fermentative Fe(III)-Reducing Bacteria

The earliest form of Fe(III) reducers has been the aerobic and anaerobically cultivated fermentative microorganisms. Among the aerobically fermentative Fe(III) reducers were *Escherichia coli*, *Lactobacillus lactis*, and *Clostridium pasteurianum* (Runov 1926), *Clostridium sporogenes* and *E. coli* reduced Fe(III) while growing on glucose or peptone media anaerobically (Starkey and Halvorson 1997). All these microorganisms did not gain enough energy to grow from Fe(III) reduction, as they transferred a very small amount of the reducing equivalents obtained from fermentation of glucose or others to Fe(III). It was then believed that if Fe(III) reducers completely oxidize glucose to CO_2 while reducing Fe(III), much more energy yield could be attained. Later, this hypothesis was contradicted (Lovley 1998). An assumption was made by Mcinerney and Beaty (1988) that the competition for organic substrates is not for the amount of energy available per mole of substrate metabolized but for the amount of energy released per mole of electrons transferred (Mcinerney and Beaty 1988). Thermodynamically, it is also known that the energy yield for fermenting glucose is more than the energy yield on oxidizing glucose completely to CO_2. Also, the energy is greater in the case where Fe(III) acts as an electron acceptor while fermenting than only fermentation. Therefore, it is not surprising that a wide range of Fe(III) fermentative reducers are present.

5.2.2 Sulfur-oxidizing Fe(III)-Reducing Bacteria

As previously mentioned in Table 5.2, thermophilic acidophiles such as *Sulfobacillus thermosulfidooxidans*, *Sulfobacillus acidophiles*, and *Acidimicrobium ferrooxidans* can reduce Fe(III) while oxidizing sulfur. In the case of *Acidithiobacillus ferrooxidans*, a hydrogen sulfide ferric iron oxidoreductase (Sugio et al. 1987, 1989) has been found, which is assumed to perform this function.

5.2.3 Organic Acid–Oxidizing Fe(III)-Reducing Bacteria

When first fermentative Fe(III) reducers were observed and it was believed that organisms completely oxidizing organic sources to CO_2 would provide far greater yield, many trials had been conducted to isolate a pure culture of such a bacterium. The earliest evidence of such a microbe was observed in paddy fields where, under anaerobic conditions, it was found that the added acetate was removed in the form of CO_2 and Fe(III) was reduced to Fe(II) (Kamura et al. 1963). The first pure culture of an organic acid–oxidizing Fe(III) reducer is *G. metallireducens* GS (Lovley and Phillips 1986). It was able to oxidize acetate while reducing Fe(III) obtaining enough energy to support growth (Eq. 5.1).

$$\text{Acetate}^- + 8\text{Fe(III)} + 4\,H_2O \rightarrow 2\,HCO_3^- + 8\text{Fe(II)} + 9\,H^+ \tag{5.1}$$

Shewanella putrefaciens is an organic acid Fe(III) reducer. It has the capability to oxidize formate and incompletely oxidize pyruvate and lactate, coupling them with Fe(II) or Mn(IV) reduction (Eqs. 5.2–5.3).

$$\text{Formate}^- + 2\text{Fe(III)} + H_2O \rightarrow HCO_3^- \_ 2\text{Fe(II)} + 2\,H^+ \tag{5.2}$$

$$\text{Lactate}^- + 4\text{Fe(III)} + 2H_2O \rightarrow \text{Acetate}^- + HCO_3^- + 4\text{Fe(II)} + 5\,H^+ \tag{5.3}$$

5.2.4 Hydrogen-Oxidizing Fe(III)-Reducing Bacteria

Proof for hydrogen-oxidizing Fe(III)-reducing bacteria was initially obtained in sediments where there was a drop in the steady-state hydrogen in sediments where Fe(III) was added in comparison with those that have no reducible Fe(III) (Eq. 5.4). Among the hydrogen-oxidizing Fe(III) reducer are *S. putrefaciens* and *Pseudomonas* species. Both cultures do not show autotrophic growth. Though both the cultures have the capability to metabolize a large number of compounds, only a restricted number of these compounds can be coupled with Fe(III) reduction.

$$H_2 + 2Fe(III) \rightarrow 2H^+ + 2Fe(II) \tag{5.4}$$

5.2.5 Aromatic Compound–Oxidizing Fe(III)-Reducing Bacteria

Aromatic compounds that present as a source of contaminant in groundwater have been remediated by Fe(III) reducers such as GS-15. This particular aspect of Fe(III) reducers has much applicability. Given below are the Eqs. (5.5–5.8) of the aromatic metabolites, which can be oxidized by reducing Fe(III).

$$benzoate^- + 3OFe(III) + 19H_2O \rightarrow 7HCO_3^- + 30Fe(II) + 36H^+ \tag{5.5}$$

$$toluene + 36Fe(III) + 21H_2O \rightarrow 7HCO_3^- + 36Fe(II) + 43H^+ \tag{5.6}$$

$$phenol + 28Fe(III) + 17H_2O \rightarrow 6HCO_3^- + 28Fe(II) + 34H^+ \tag{5.7}$$

$$p\text{-}cresol + 34Fe(III) + 20H_2O \rightarrow 7HCO_3^- + 34Fe(II) + 41H^+ \tag{5.8}$$

Biomineralization, as the term suggests, is the mineralization by the biological system, which has a long history. As indicated by fossil records, biomineralization was initiated 525 million years ago. Kirschvink and Hagadorn (2000) refer to the Cambrian evolutionary explosion when biomineralization was observed in major animal phyla, further suggesting that to understand the complexity of the biochemical pathway mechanism for crystallization, a simpler ancient bio-system needs to be first scrutinized (Kirschvink and Hagadorn 2000). This is so as much of the modern biomineralization processes are seemed to be developed by the expectation of a previous ancient system. A second requirement for mineralization of magnetite is to use the deposited ferric hydride in a controlled fashion to form uniform crystallites, which can then harden the tooth caps of polyphora, this being performed by certain proteins. These similar requirements form a basis of crystallite formation from magnetite deposition in human brain cells to magnetite.

Iron mineralization in nature, solitarily in prokaryotes, has been under much discussion and controversies in the past, a process quite as old as the early Cambrian period (Kirschvink and Hagadorn 2000). It is only in recent years that biomineralization is classified into two processes:

- Biologically induced mineralization (BIM)
- Biologically controlled mineralization (BCM)

BIM is a process observed in iron-reducing bacteria (and also in sulfate-reducing bacteria) where extracellular mineralization occurs by reduction of surrounding Fe(III) minerals to Fe(II) and further complex formation of this Fe(II) with excess of Fe(III) or phosphorus silica sulfur or carbonates to form secondary mineral crystallites of iron phosphates, silicates, sulfides, carbonates, and oxides of iron. The crystals formed are usually poorly crystalline impure and environmentally controlled. BCM, on the other hand, is an intercellular crystallite formation method seen in biomineralizing bacteria where Fe(III) ions are imported into the cell converted to Fe(II) ions and rated with other anions to form complexes. Biologically controlled, this mechanism leads to uniform crystals (Kirschvink and Hagadorn 2000).

The characteristics of crystallite formed by the BCM process, which distinguishes it from the crystals formed by the BIM process, are as follows:

- The perfect crystalline structure of the crystals
- The magnetite crystals formed by the BIM process are small octahedral crystal. BCM magnetite crystals are cubic in shape.
- Lattice dislocation, chemical impurity, and crystal defects are often seen in BIM crystal but not in BCM crystals.
- BIM crystals are usually large and uncontrolled, whereas BCM crystals are within a single domain region and uniform in size.

The mechanism for both the biomineralization is further discussed later.

In both mineralization processes, Fe(III) reduction is an important criterion. It is believed by some that Fe(III) reduction is the earliest form of respiration that gives evidence of last common ancestors that reduce Fe(III) to Fe(II) to conserve energy contradicting the old hypothesis of elemental sulfur, thus being the sole electron acceptor for the hyperthermophiles (Vargas et al. 1998).

The hypothesis of Fe(III) respiration on early earth is noted to be a distinct possibility cause of the high amount of Fe(III) present in early earth. Among the various models for the evolution of life on earth, one of them is the ferrous wheel. Earlier Fe(III) reduction and complexation with sulfur had initially led to the formation of membrane-like complexes of iron cellular minerals, with UV radiation being a perpetual source of energy. Iron cycle carried on in earlier earth. The inorganic membranes were replaced by organic membranes. Later on, early-day microorganisms came into being. Further on, as hypothesis developed, so developed the Fe(II)-based hypothesis generating Fe(III) and organic hypothesis directly converting Fe(II) to Fe(III). The condition, therefore, that was present at that time was more suitable for Fe(III)-reducing bacteria, leading scientists to believe in evolution through ferrous wheel. The schematic diagram (Figure 5.1) explains this process.

To better understand the iron reduction and mineralization by these microorganisms, one needs to understand the metal-microbe interaction. The next section attempts to do so by describing the various cell surfaces and how they play a role in microbe-metal interaction.

5.3   MICROBE-METAL INTERACTION

The outer cell structure of a microorganism determines the interaction between the organism and metal. The more the anionic charge, the more the binding will be. The gram-positive cell contains a polymer such as teichoic acid and teichuronic acid. This polymer along with the peptidoglycan layer provides the site of metal-binding *Bacillus subtilis* as studies by Beveridge and Murray (1980) considered gram-positive organism as a model to study the metal sorption process through X-ray diffraction and electron scattering technique (Beveridge and Murray 1980). The important metal chelating sites such as the carboxyl group of glutamic acid, technical phosphodiester group, and the amino group were modified by carbodiimide linkage of glycine ethyl ester, mild alkaline treatment to remove teichoic acid, and S-acetyl mercapto succinic anhydride reaction of the amino group, respectively. Final observations made showed that changes in the carbonyl group had the most effective in the reduction of metal-binding capacity, showing these are the main sites of metal interaction. Removal of teichoic acid also reduced the metal sorption limits though it was not as drastic as that of the blocking of the carboxyl groups. Amine linkage blocking showed the least result as expected. This and further electron microscope images prove that the thick peptidoglycan layer of gram-positive cell wall has comparatively more anionic charge than that of the gram-negative cell wall (Beveridge 1999).

Gram-negative cell wall, differing from the gram-positive one, has a very thin peptidoglycan layer (3 nm). An example of metal binding to a gram-negative cell (*Escherichia coli* K-12) has been given by Beveridge and Koval (1981). In this study (Beveridge and Koval 1981), the bacterial

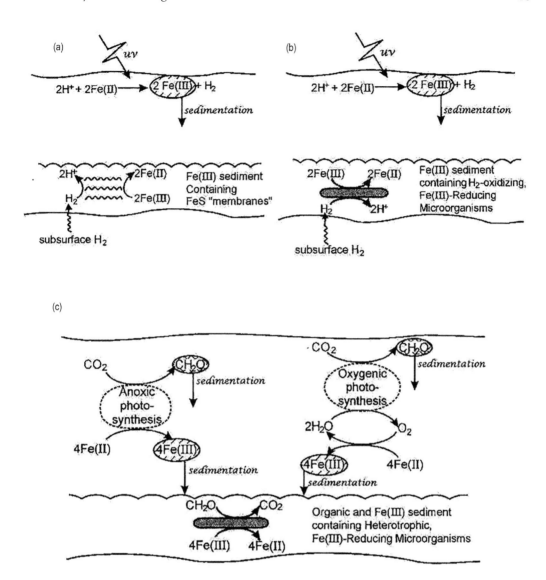

FIGURE 5.1 Hypothetical evolution of ferrous wheel on early earth. (a) Fe(III) and hydrogen generated from UV-mediated oxidation of dissolved Fe(II); (b) hydrogen oxidation coupled to Fe(III) reduction by early microorganisms; (c) enhanced precipitation of Fe(III) with evolution of Fe(II)-based and oxygen-based photosynthesis, resulting in coprecipitation of organic matter and Fe(III) oxide, which supports heterotrophic Fe(III) reduction. (Reprinted with permission from Lovley 2000.)

envelope separated from the cell was used rather than the whole cell. The cell envelope of *E. coli* K-12 is similar to that of another gram-negative bacterium though lacking an O side chain. It too has more of lipopolysaccharide (LPS) on its outer surface and more phospholipid in its inner surface making the inner surface more hydrophilic. It is this hydrophilic phospholipid layer where the metals are thought to bind to the lipid polar heads. The transport of the metals across the envelope is again through the many transmembrane proteins present. Some amount of metal sorption also is reported to take place on anionic sites of the outer LPS layer. Atomic absorption spectroscopy and X-ray fluorescence spectroscopy have revealed the intermediate sorption ability of Fe(III) along with other metals, such as Pb, Zn, and Mn, to *E. coli* K-12 cell surface.

The gram-negative bacterium most studied for iron reduction is *G. metallireducens* (Lovley et al. 1993). The cell wall is similar to that of other gram-negative bacterium. However, *Geobacter* has certain periplasmic cytochrome that acts as sensors for the Fe(III) source present in the environment. Thereby sensing, the microorganism has the ability to produce flagella, which increases the availability of iron source (Childers et al. 2002) (Figure 5.2).

Also because of their anionic charge, the amorphous layer, capsule, forms another cation-binding site through electrostatic interaction. The bacterial capsule is a unique structure providing the bacteria with the ability to adhere, form biofilm, and act as a protective guard against toxic substances. Capsules also are known to provide pathogenicity to a microbe. The capsule-metal interaction has been studied deeply in case of *Bacillus licheniformis* in Figure 5.3 (Mclean et al. 1992). In this case, a γ-glutamyl capsular polymer was used. Different metal cations showed that metals such as Cu^{2+}, Al^{3+}, Cr^{3+}, and Fe^{3+} caused flocculation on the binding. The Fe^{3+} also formed ferrihydrite precipitates. Metal-binding capacity varied with different metals. The reason behind this can be the difference in the conformational change on the binding. Carboxylate groups are present being the main reactive site as seen in other gram-positive organism mentioned earlier.

Not only the metal and capsule characteristics are the decisive characters for the amount of metal binding, but it is also the oxidation state of the metal, which determines the capsule's binding ability. Fe(III) binding is more favorable to Fe(II) binding in *B. licheniformis* cells (Mclean et al. 1992). The reason for this being as postulated an increased charge density of Fe(III) ion and its ability in the hydrated bound state to further form metal complexes. Sheaths are another surface layer that interacts with metal layers and help in mineralization in iron-oxidizing bacteria.

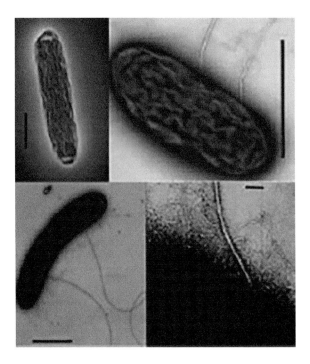

FIGURE 5.2   Above: *Geobacter metallireducens* cell showing an absence of flagella while growing in Fe(III) citrate. Below: Cells of *Geobacter metallireducens* with flagella while growing in Fe(III) oxides. (Reprinted from Childers et al. 2002 with permission from Springer Nature.)

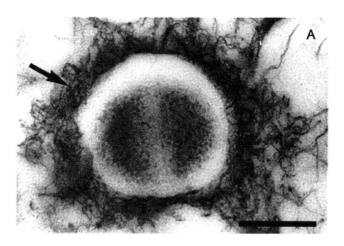

FIGURE 5.3 Bacterial cell with iron-rich capsule. (Reprinted with permission from Abhilash et al. 2011.)

5.4 THE MECHANISM FOR CRYSTAL FORMATION

5.4.1 MAGNETOTACTIC BACTERIA

A mechanism for magnetite synthesis has been proposed by Noguchi et al. (1999) (in magnetotactic bacteria). The process for magnetite to be formed occurs by the following steps:

- Intake of Fe(III) into the cell
- Reduction of Fe(III) to Fe(II) during transport
- Reoxidation of Fe(II) to Fe(III), forming ferrihydrite
- Ferrihydrite reacting with Fe(II) to form magnetite

Magnetotactic bacteria have an array of enzymes performing this process. Magnetotactic bacteria *Magnetospirillum magnetotacticum* produces a compound known as siderophore, which helps in solubilizing and binding Fe(III). This bound Fe(III) is reduced to Fe(II) by iron reductase, which separates the Fe(III) from the siderophore inside the cellular compartment. Iron reductase is a 36 kDa, single subunit enzyme, which performs the function of Fe(III) reduction to Fe(II) using NADH as an electron donor. Iron reductase is present in the cytoplasm and on the inner surface of the membrane and is inhibited by Zn^{2+} in the solution. The reduced Fe(II) now needs to be partially converted to Fe(III) since magnetite is a compound having both Fe(II) and Fe(III). This reoxidation is catalyzed by Fe(II)-nitrite oxidoreductase. This nitrite as has been proved acts as an oxidizing agent for Fe(II) conversion to ferrihydrite conversion. The Fe(II)-nitrite oxidoreductase in *M. magnetotacticum* has been denoted as cytochrome cd1. The cytochrome cd1 provides denitrifying conditions for magnetite synthesis. Nucleation is the final step for the crystal formation process. The nucleation occurs at a specific site in a specific magnetotactic bacteria (Figure 5.4). For this, a certain protein has been discovered, which is believed to decide and initiate the nucleation process.

5.4.2 IRON-REDUCING BACTERIA

Contradicting to magnetotactic bacteria when it comes to iron-reducing bacteria, the problem arises from the bacteria to transport electrons outside the cell. To resolve these, dissimilatory iron-reducing bacteria such as *Shewanella* and *Geobacter*, were preferred owing to developed membrane cytochrome proteins. Innermost layer present in the membrane is the 89 kDa cytochrome followed by 9 kDa cytochrome existing in the periplasmic space and finally a 94 kDa cytochrome, which then transfers it to Fe(III) reducing to Fe(II). The crystallization process that occurs thereof is similar to

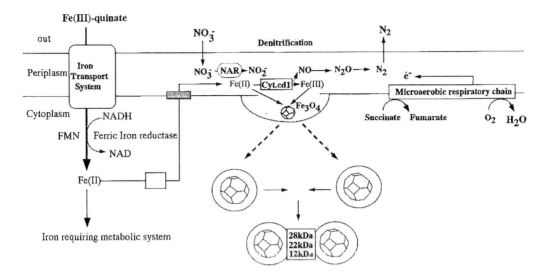

FIGURE 5.4 Schematic diagram for the mechanism of crystallization in magnetotactic bacteria *Magnetospirillum magnetotacticum*. (Reprinted with permission from Schröder and Johnson 2003.)

that, which occurs inorganically (Abhilash et al. 2011). The Fe(II) is accumulated on the cell wall of the microorganism, due to its high binding capacity complexes with the surrounding Fe(III) ion present forming ferrihydrite-like compound. This ferrihydrite-like compound then finally gets converted to compounds such as magnetite and hematite, based on the environmental factors present.

5.5   THE DIVERSITY OF BIOMINERALS FORMED

The discussion until now might have given the readers an idea on the diverse group of iron minerals, which these microbes can synthesize. A single microorganism can produce different salts of iron crystals at a different environment, for example, under a high concentration of carbonate buffer. *G. metallireducens* have been observed to form metal carbonates, and when the concentration of phosphate buffer increases, it is the vivianite structure that is observed. The various biominerals formed are described as follows.

5.5.1   OXIDES

As has been previously mentioned, magnetite formation has been seen to occur in both prokaryotes and eukaryotes. In prokaryotes, magnetite production has occurred through both BIM and BCM processes (Bazylinski and Frankel 2000). The crystals of magnetite formed through the BIM process are similar to those being formed through inorganic processes, being poorly crystallized and in the size distribution of superparamagnetic range (<35 nm). Fe(III) reducers under specific conditions of Eh and pH produce magnetite while reducing insoluble Fe(III) oxyhydroxides present in the environment. The excess insoluble Fe(III) oxyhydroxide reacts with Fe(II) reduced by the microbes to form green rust which age to form magnetite. The irregularity of the shape in the crystals formed by the BIM process is because of the rapid uncontrolled growth of magnetite formation. In comparison with the BIM process, the BCM process leads to regular shaped, morphologically specific crystals of magnetite with high structural perfection. The crystal formation is under the genetic control of the magnetotactic bacterium and is often surrounded by organic vesicles formed by the bacteria known as magnetosome. Magnetite formed by magnetotactic bacteria has a size range of 35–100 nm. The crystals are usually octahedral, prismatic, or bullet-shaped differing from the irregular crystals of the BIM process.

Magnetite formed by the magnetotactic bacterium provides the bacterium with the specific ability to move along the magnetic axis of the earth and thereby obtain a microaerophilic to anoxic habitat required by the bacteria to survive. The magnetotactic bacterium in the northern hemisphere moves along the axis to the north, and the ones present in the southern hemisphere move toward the south. The bacterium present at the equator moves toward both sides. In comparison, the extracellular magnetite formed by the DIRB does not have any specific function in the bacteria but has its significance in the magnetization of the sediments. Also, the magnetotactic bacteria seem to generate energy for growth by the mineralization process in contrast to the dissimilatory iron reducers where biomineralization does not provide energy for utilization and is just a passive process where the negative cell wall and the high surface area to volume ratio act as a site for anion binding. Nucleation occurs with the sorbed metal ion joining the initial ions. This clustering of metal ions reduces the activation energy barrier for mineral formation (Frankel and Bazylinski 2003) than in magnetotactic bacterium. Evidence has been provided, which suggests that as biomineralization is organism controlled in a magnetotactic bacterium, in DIRB the mineralization process is environmentally controlled, and as has been mentioned, the Eh and pH play a main role. The range of redox potential and pH has to fall in the magnetite stability field for magnetite formation. Also, temperature stability is required during the formation period. Though magnetite formation has been observed in both mesophilic and few thermophilic bacteria, still temperature dependency is not clearly understood.

The thermophilic bacterium where magnetite formation has been studied in detail is the Fe(III)-reducing bacterium strain TOR-39 (*Thermoanaerobacter ethanolicus*) (Frankel and Bazylinski 2003). Other few magnetite mineralizing thermophilic bacteria (Vargas et al. 1998) are the Archaeon *Pyrobaclum islandicum* and the bacteria *Thermotoga species*. There are ample amount of mesophilic Fe(III) reducers that crystallize magnetite, but the most studied are *G. metallireducens* and *S. putrefaciens*. Among the magnetotactic bacteria, much is known about *M. magnetotacticum* [MV-1].

5.5.2 Sulfides

Some reducing bacteria and magnetotactic bacteria present below the oxic-anoxic transition zone (OATZ), in high sulfidic conditions, produce sulfide crystals in the form of monosulfides such as greigite, mackinawite, or pyrrhotite. The magnetic properties of some of the iron sulfides such as greigite perform similar magnetotactic functions as iron oxide crystals. Also, the arrangements in the cell- and species-specific characters are similar to those of intracellular magnetite crystals.

Magnetotactic bacteria forming iron sulfide crystals have been seen in two cases:

- Many-celled magnetotactic prokaryotes (MMPs)
- Large rod-shaped bacterium

The many-celled magnetic prokaryotes or MMPs are a bunch of 10–30 cells that are arranged orderly with greigite crystals present intracellularly (Figure 5.5). The crystals are present in each cell parallel to each other. The MMPs are placed just below the OATZ, whereas the large magnetotactic bacterial rods are present further deeper down (Figure 5.6).

It is believed that the greigite formed in MMPs has initial stages of inorganic poor crystalline FeS. This FeS is initially converted to nonmagnetic mackinawite and finally to magnetic greigite by losing one-fourth of its iron (Pósfai et al. 2006). TEM images show tetrahedral crystals, whereas the stored sample of the cells had greigite crystals in them. In the large magnetotactic rods which belong to the group of gammabacteria (Figure 5.7), both chains of magnetite and greigite have been seen in the cell. The two chains have different magnetosome membrane and were thought to have been produced by two different genes. The greigite crystals in such cases are rectangular prismatic in shape. The formation of crystals is dependent on the iron content of the habitat.

Similarly, biologically induced mineralization leads to the production of iron sulfide crystals though usually much more formation of pyrite is seen in the sediments than that of greigite.

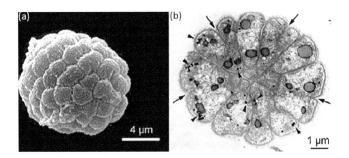

FIGURE 5.5 Magnetotactic cells producing sulfidic magnetosome. (a) SEM image of multicellular magnetic prokaryote. (b) Ultrathin section of a multicellular magnetic prokaryote with the arrow showing the sites of cell division. (Reprinted with permission from Pósfai et al. 2006.)

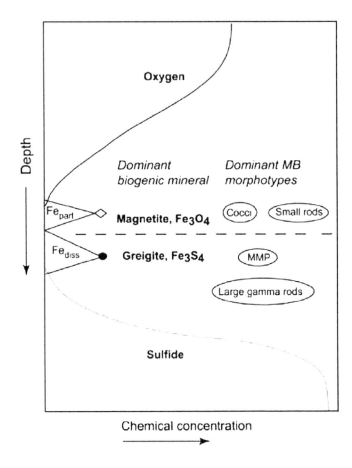

FIGURE 5.6 The position of the type of magnetotactic bacteria in the water column with respect to the depth and concentration of oxygen and sulfides. (Reprinted with permission from Pósfai et al. 2006.)

It is the sulfate-reducing bacteria (or as later known sulfate-reducing prokaryotes (SRPs)) that synthesizes the formation of pyrite. The SRPs reduce sulfate ions to form the sulfide. Some portion of this sulfide reacts with iron, which is initially bound to the anionic surface of the cell to form iron sulfide crystals.

$$2H_2O + SO_4^{2-} \rightarrow 2HCO_3^- + H_2S \tag{5.9}$$

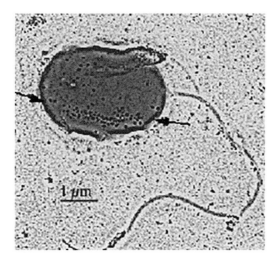

FIGURE 5.7 Single rod-shaped magnetotactic cell with arrows pointing iron sulfide magnetosomes. (Reprinted with permission from Pósfai et al. 2006.)

The production of the extracellular mineral is dependent on iron and organic matter content, which the SRPs oxidize to reduce the sulfate. There are some SRP species that reduce organic matter incompletely and some species that oxidize organic matter completely to carbon dioxide to reduce sulfate. The availability of iron for pyritization can be best understood (Raiswell and Canfield 1998) by the dissolution mechanism of SRP, leading to greigite formation such as lepidocrocite, goethite, ferrihydrite, and hematite. It is believed that under aerobic condition, these iron oxides give the sediment a high degree of pyritization (DOP), whereas the DOP is minimal in cases of anaerobic sediments.

Nucleation of iron sulfide as has been discussed earlier is on the anionic surface of the cell. Completely encrusted SRPs have been observed. The crystals formed initially are of small mackinawite crystals of 100 nm size. This size increases up to 1–2 µm. The mackinawite crystals transforming into greigite or pyrite further depend on the presence or absence of aldehydic carbonyls in the solution, respectively (Rickard et al. 2001). Also, the pyrite formation is further restricted by iron content entering the cell. The pyrite formed has many morphologies being euhedral, irregular, oxidic pyrite, or framboidal.

SRPs, a diverse group of microorganism, have been classified into spore-forming and non–spore-forming SRPs. This classification has been done mainly on the basis that spore-forming bacteria *Desulfovibrio orientis* and Coleman's organism are obligate mesophiles having cell pigments cytochrome C3 and desulfoviridin. Mesophile bacteria growing under thermophilic conditions do produce spores but lack pigments. While the spore-forming cells have oscillatory motility, the nonsporulating cells have motility with a wobble (Postgate 1959). Among the non–spore-forming, SRP is the *Desulfovibrio* species.

5.5.3 SILICATES

Microbial activity usually determines the chemistry of water (Konhuser et al. 1994). The bacterial population in nature, just as it mineralizes iron sulfides and iron oxides, can form iron silicates. The formation process is quite similar to that of iron sulfide crystal formation, i.e., iron present in minerals, usually in its oxide form in nature, is solubilized and absorbed onto the anionic surface of the cell. The silica thereby reacts with the iron to form iron silicates by hydrogen bonding of hydroxyl groups mobilizing Fe from solution to sediment. In the order of the above theory, it can be believed that silicate crystal formation undergoes the transformation from the initial stage of

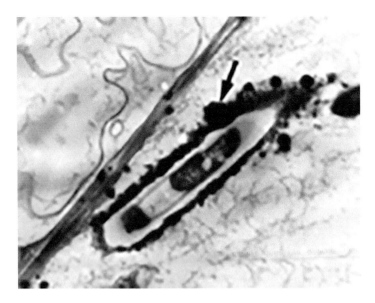

FIGURE 5.8 TEM image of a siliceous sphere embedded with a dense iron-rich capsule. Arrow points amorphous iron silicate grain. (Reprinted with permission from Konhauser 1997.)

amorphous solid to rigid crystals of iron silicate. This occurs by removal of water by the hydrogen bonding of the hydroxyl groups.

Contradicting it, reports have been cited on electrostatic interaction prevailing between silicate (SiO_3^{2-}) and ammonium (NH_4^+) group on the cell wall (Konhuser et al. 1994). This hypothesis has been rejected on the basis that silica ions present in the natural environment are mostly neutral or negatively charged. Thus, it suffices to say that iron-silica-microbe interaction is not through electrostatic forces. Recently, it was concluded that bacterial surfaces give no significant amount of silica immobilization than blanks having no bacteria (Phoenix et al. 2003). The statement states that in regions of high iron concentrations, the iron oxy-hydroxides formed bind onto the bacterial surface due to their affinity; the reactive silica further getting immobilized on these iron oxides and oxy-hydroxides has been proved void such that now it is assumed that, at circumneutral pH, maximum of the iron present reacts with silica to form iron silicates decreasing the ability of iron oxides and hydroxides to bind onto the bacterial surface (Figure 5.8). An experimental study by Phoenix et al. (2003) using *B. subtilis* demonstrated that bacterial population certainly do not enhance the silica immobilization, i.e., the iron sorbed on the bacterial surface does not act as a salt bridge for silica binding (Phoenix et al. 2003). Nevertheless, iron silicate biominerals have been observed in hot spring and freshwater bodies. Minerals precipitation of [Fe, Al] silicates by epilithic biofilms has been noted where the formation of gel-like to crystalline phases has been studied. The gel chamosite clay was observed to be reactive to silicic acid (H_4SiO_4).

5.5.4 PHOSPHATES

Phosphate mineralization events have been thought to form since the Precambrian period. Evidence has been observed of a bacterial population, which is an integral part of phosphate mineralization by the presence of a bacterial filament. O'Brien et al. (1990) speaks of a phosphorite formation on the east Australian continental margin and gives similar evidence of a bacterial role in crystallization (O'Brien et al. 1981). Microbial activity seemed to a maximum at modern phosphorite zones. Phosphate mineralization mechanism has been noticed in nature through the dissolution of the

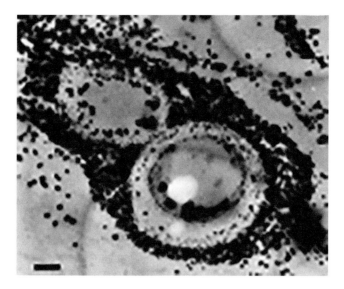

FIGURE 5.9 TEM image of two bacterial cells encrusted with iron phosphate. (Reprinted with permission from Konhauser and Ferris, 1997.)

phosphorus present and its solubilization. These phosphorus ions are transported concentrated and immobilized within a biofilm. The biofilms by the closed recycling methods inhibited the loss of phosphorus iron present in the same environment in the form of oxyhydroxides that are made available by the solubilization mechanism of DIRB. The iron and phosphorus present in close proximity react to form phosphate minerals. These phosphate granules (Figure 5.9) can be of a varied size from 200 to 50 nm in diameter and can be formed by the microbes in a variety of secondary structures such as vesicles or can be associated with periplasmic space or cytoplasm.

5.6 MOLECULAR MECHANISM OF MICROBIAL Fe(III) REDUCTION

Though much data are present on the mechanism of microbial Fe(III) reduction, still it is not understood fully. The Fe(III) present in nature is mostly in the insoluble form such as Fe(III) oxides. The Fe(III) reduction is determined by the bioavailability of the iron. This increase in the surface area contact between the bacteria and the Fe(III) source can be bought about by addition of a chelator such as nitriloacetic acid and/or an electron shuttling compound, which shuttles electron from the Fe(III) reducers (3). Other examples of chelator are EDTA, ethanoldiglycine (EDG), and polyphosphates. Humic substances act as possible electron shuttles. The schematic figure (Figure 5.10) given below shows the electron shuttling between 2,6-anthraquinone disulfonate and its reduced form.

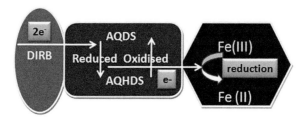

FIGURE 5.10 Pathway for electron transport of anthraquinone disulfonate (AQDS) in iron reduction. (Reprinted with permission from Abhilash et al. 2011.)

Aromatic hydrocarbons do act as electron shuttlers in the natural environment. DIRB have the ability to form particles of siderite vivianite magnetite and green rust (Abhilash et al. 2011) along with more stable particles, such as goethite, hematite, and lepidocrocite (Abhilash et al. 2011). The DIRB transfer the reduced Fe(II) intracellularly via cytochrome present between the outer and inner membrane oxidizing NADH to NAD. This reduced Fe(II) along with the excess Fe(III) present forms magnetite particles extracellularly. Nanoparticle synthesis occurs at high pH. Usually, the mesophilic bacterium produces superparamagnetic magnetite of the size ≤35 nm, whereas psychrotolerant bacterium produces single-domain magnetite >35 nm (Kirschvink and Hagadorn 2000). Many control factors have been observed in the Fe(III) reduction and nanoparticle synthesis by the DIRB. It has been demonstrated that the accumulation of dissolved Fe(II) acts as a negative feedback control for Fe(III) reduction (Mclean et al. 1992). Though this problem can be done away with the removal of Fe(II), an increase in Fe(III) oxide thermodynamic stability in case further reduces Fe(III) reduction. The reduction rate of the bacterium determines the transformation of hydrous ferric oxide to goethite magnetite or green rust. High reduction rates form goethite, intermediate rates form magnetite, and low rates form green rust. The size of the particle formed extracellularly can also be determined by many factors such as organic or inorganic compounds being released by bacteria, which reduce the size of the crystal formed. The particle size may be determined by the template effect as in the case of iron-oxidizing bacteria (Banefield et al. 2000; Teremova et al. 2016).

5.7 CONCLUSIONS

Though much is known about the DIRB and their manner of producing extracellular crystals, still much of it is a mystery, for example, the role of DIRB in extracellular crystal formation still remains as a fact shrouded in mystery. Applicability of these extracellular biogenic particles such as magnetite while being used for various biomedical applications such as molecular imaging has brought many commercial benefits, but still the exact conditions and control factor for extracellular crystal formation, which have not been outlined properly, remain as a hurdle for the scientists.

REFERENCES

Abhilash, Revathi K, Pandey BD. Microbial synthesis of iron based nanomaterials—A review. *Bulletin of Material Science*, 2011, 34(2), 1–8.

Banefield JF, Welch SA, Zhang H, Ebert TT, Penn RL. Aggregation based crystal growth and microstructure development in natural iron oxydroxide biomineralisation products. *Science*, 2000, 289, 751–754.

Bazylinski DA, Frankel RB. Magnetic iron oxide and iron sulfide minerals within microorganisms. In Baeuerlein E (ed.), *Biomineralisation from Biology to Biotechnology and Medical Application*, Wiley-VCH, Weinheim, Cambridge, 2000, pp. 25–47.

Beveridge TJ, Koval SF. Binding of metals to cell envelopes of Escherichia coli K-12. *Applied and Environmental Microbiology*, 1981, 42, 325–335.

Beveridge TJ, Murray RGE. Site of metal deposition in the cell wall of Bacillus subtilis. *Journal of Bacteriology*, 1980, 141, 876–887.

Beveridge TJ. Structures of gram-negative cell walls and their derived membrane vesicles. *Journal of Bacteriology*, 1999, 181, 4725–4733.

Childers SE, Ciufo S, Lovley DR. Geobacter metallireducens accesses insoluble Fe(III) oxide by chemotaxis. *Nature*, 2002, 416, 767–769.

Frankel RB, Bazylinski DA. Biologically induced mineralization by bacteria. *Reviews in Mineralogy and Geochemistry*, 2003, 54, 95–114.

Fukumori Y. Enzymes for magnetite synthesis in Magnetosprillium magnetotacticum. In Baeuerlein E (ed.), *Biomineralisation from Biology to Biotechnology and Medical Application*, Wiley-VCH, Weinheim, 2000, pp. 93–106.

Kamura T, Takai Y, Ishikawa K. Microbial reduction mechanism of ferric iron in paddy soils (Part 1). *Soil Science and Plant Nutrition*, 1963, 9, 171–175.

Kirschvink JL, Hagadorn JW. *A Grand Unified Theory of Biomineralisation, The Biomineralisation of Nano- and Macro-Structures*, Wiley-VCH Verlag GmbH, Weinheim, 2000, pp. 139–150.

Konhauser KO. Bacterial iron biomineralisation in nature. *FEMS Microbiology Reviews*, 1997, 20, 315–326.

Konhuser KO, Schultze-Lam S, Ferris FG, Fyfe WS, Longstaffe FJ, Beveridge TJ, Mineral Precipitation by Epilithic Biofilms in the Speed River, Ontario Canada. *Applied and Environmental Microbiology*, 1994, 60, 549–553.

Lovley DR, Phillips EJ. Novel mode of microbial energy metabolism: Organic carbon oxidation coupled to dissimilatory reduction of iron or manganese. *Applied and Environmental Microbiology*, 1988, 54, 1472–1480.

Lovley DR, Giovanonni SJ, White DC, Champine JE, Philips EJP, Gorby YA, Godwin S. Geobacter metallireducens gen. nov. sp. nov., a microorganism capable of coupling the complete oxidation of organic compounds to the reduction of iron and other metals. *Archives of Microbiology*, 1993, 159, 336–344.

Lovley DR, Phillips EJ. Organic matter mineralization with reduction of ferric iron in anaerobic sediments. *Applied and Environmental Microbiology*, 1986, 51, 683–689.

Lovley DR. Dissimilatory Fe(III) and Mn(IV) reduction. *Advances in Microbial Physiology*, 2004, 49, 219–286.

Lovley DR. *Fe(III) and Mn(IV) Reduction, Environmental Microbe- Metal Interaction*, ASM Press, Washington D.C., 2000, pp. 3–30.

Lovley DR. Microbial Fe(III) reduction in subsurface environments. *FEMS Microbiology Reviews*, 1997, 20, 305–313.

Lovley DR. Organic matter mineralization with the reduction of ferric iron: A review. *Geomicrobiology Journal*, 1998, 5, 375–399.

McInerney MJ, Beaty PS. Anaerobic community structure from a nonequilibrium thermodynamic perspective. *Canadian Journal of Microbiology*, 1988, 34, 487–493.

Mclean RJC, Beauchemin D, Beveridge TJ. Influence of oxidation state on iron binding by Bacillus lincheniformis capsule. *Applied and Environmental Microbiology*, 1992, 58, 405–408.

Nakagawa T, Lino T, Suzuki K, Harayama S. Ferrimonas futtsuensis sp. nov. and Ferrimonas kyonamensis sp. nov. selenate reducing bacteria belonging to the gammaproteobacteria isolated from Tokyo bay. *International Journal of Systematic and Evolutionary Microbiology*, 1996, 56, 2639–2645.

Noguchi Y, Fujiwara T, Yoshimatsu K, Fukumori Y. Iron reductase for magnetite synthesis in the magnetotactic bacterium Magnetospirillum magnetotacticum. *Journal of Bacteriology*, 1999, 181(7), 2142–2147.

O'Brien GW, Harris JR, Milnes AR, Veeh HH. Bacterial origin of east Australian continental margin phsophorites. *Nature*, 1981, 294(3), 442–444.

O'Brien GW, Milnes, AR, Veeh, HH, Heggie, DT, Riggs, SR, Cullen, DJ, Marshall, J F, Cook, PJ. Sedimentation dynamics and redox iron-cycling: controlling factors for the apatite-glauconite association on the East Australian continental margin. In Notholt AJG. & Jarvis I. (eds), *Phosphorite Research and Development Geological Society Special Publication No. 52*, London, 1990, pp. 61–86.

Phoenix VR, Konhauser KO, Ferris FG. Experimental study of Iron and Silica immobilization by bacteria in mixed Fe-Si systems: Implications for microbial silicification in hot springs. *Canadian Journal of Earth Science*, 2003, 40, 1669–1678.

Pósfai M, Kasama T, Dunin-Borkowski RE. Characterization of bacterial magnetic nanostructures using high-resolution transmission electron microscopy and off-axis electron holography. In Schüler D. (eds.), *Magnetoreception and Magnetosomes in Bacteria*. Microbiology Monographs, Vol. 3. Springer, Berlin and Heidelberg, 2006.

Postagate JR. Sulphate reduction by bacteria. *Annual Review of Microbiology*, 1959, 13, 505–520.

Raiswell R, Canfield DE. Sources of iron for pyrite formation in marine sediments. *American Journal of Sciences*, 1998, 298, 219–245.

Rickard D, Butler IB, Oldroyd A. A novel iron sulphide mineral switch and its implication for earth and planetary sciences. *Earth and Planetary Science Letters*, 2001, 189, 85–91.

Runov EV. Die Reduktion der Eisenoxyde auf microbiologichem Wege. *Vestn. Bakter-Agronomich. Stantsii*, 1926, 24, 75–82.

Starkey RL, Halvorson HO. Studies on the transformations of iron in nature. II. Concerning the importance of microorganisms in the solution and precipitation of iron. *Soil Science*, 1997, 24, 381–402.

Sugio T, Katagiri T, Inagaki K, Tano T. Actual substrate for elemental sulfur oxidation by sulfur: Ferric ion oxidoreductase purified from Thiobacillus ferrooxidans. *Biochimica et Biophysica Acta (BBA) – Bioenergetics*, 1989, 973, 250–256.

Sugio T, Mizunashi W, Inagaki K, Tano T. Purification and some properties of sulfur: Ferric ion oxidore-
    ductase from Thiobacillus ferrooxidans. *Journal of Bacteriology*, 1987, 169, 4916–4922.
Teremova MI, Petrakovskaya EA, Romanchenko AS, Tuzikov FV, Gurevich, YL, Tsibina OV, Yakubailik
    EK, Abhilash. Ferritisation of industrial waste water and microbial synthesis of iron-based magnetic
    nanomaterials from sediments. *Environmental Progress and Sustainable Energy (AiChE)*, 2016, 35(5),
    1407–1414.
Vargas M, Kashefi K, Blunt-Harris EL, Lovley DR. Microbiological evidence for Fe(III) reduction on early
    earth. *Nature*, 1998, 395, 65–67.

6 Heat and Chemical Pretreatment of Bacterial Cells to Enhance Metal Removal

Ling Sze Yap and Adeline Su Yien Ting
Monash University Malaysia

CONTENTS

6.1 INTRODUCTION

The industrial revolution, characterized by rapid urbanization and civilization, has unfortunately led to an increase of pollution in the environment. Soil, water, and atmosphere are laden with pollutants, presenting global hazards that are detrimental to man and the environment (Chandra and Kumar 2015). Water environments are of the greatest concern, as ~70% of earth is composed of water bodies. There are two main pollutants contaminating the water environment; the organic pollutants (solvents, organic compounds, pesticides, and insecticides) and inorganic pollutants (metals and fertilizers) (Vijayaraghavan and Yun 2008). This review will focus on toxic metals, as they are used extensively in the many industries; from manufacturing of paints and batteries, synthesis of pesticides and fertilizers, to the mining and smelting of metals (Duruibe et al. 2007). Metals, in minute amounts, are, however, beneficial to most living organisms, including humans, as they are the coenzyme factors for physiological processes (Fu and Wang 2011). Hence, high metal concentrations in the environment demand attention and immediate management to reduce the impact of metal toxicity to living organisms.

The high metal concentrations in the environment are the result of bioaccumulation through time, as they are often nondegradable and resistant to photolytic or biological activity (Mohammed et al. 2011). As a result, metals accumulate in soils and water, and when consumed or ingested by living organisms, these metals react with the biomolecules in the organisms to form extremely stable biotoxic compounds where dissociation of this compound is almost impossible (Duruibe et al. 2007). Mercury (Hg^{2+}), lead (Pb^{2+}), cadmium (Cd^{2+}), and chromium (Cr^{6+}) are metals with extreme toxicity, causing severe damage to the kidney, gastrointestinal system, nervous system, and have

TABLE 6.1
Examples of Toxic Metals, Their Sources, and Effect on Human Health

Toxic Metals	Major Sources	Effect on Human Health	Permissible Level (ppm)
Arsenic (As^{3+})	Pesticides, fungicides, metal smelting	Bronchitis, dermatitis	0.02
Cadmium (Cd^{2+})	Welding, electroplating, pesticides, fertilizers, nuclear fission plant	Kidney damage, bronchitis, gastrointestinal disorder, bone marrow cancer	0.06
Lead (Pb^{2+})	Paint, pesticide, cigarrettes, automobile emission, mining, burning of coal	Liver and kidney damage, gastrointestinal disorder, mental retardation in children	0.10
Manganese (Mn^{2+})	Welding, fuel addition, ferromanganese production	Inhalation or contact causes damage to central nervous system	0.26
Mercury (Hg^{2+})	Pesticides, batteries, paper industry	Damage to nervous system, protoplasm poisoning	0.01
Zinc (Zn^{2+})	Refineries, brass manufacturing, Metal plating, plumbing	Zinc fumes have corrosive effect on skin and causes damage to nervous system	15.00

Source: Modified from Alluri et al. (2007).

been linked to bronchitis, cancer, and mental retardation in children (Table 6.1). On the contrary, cobalt (Co^{2+}), copper (Cu^{2+}), nickel (Ni^{2+}), and zinc (Zn^{2+}) are regarded as less toxic, but their unusually high levels and presence in the environment often contribute to metal toxicity. Such high levels of metal could initiate oxidative damage of soft tissues as in the liver tissues, thus disrupting the functionality of the liver (Gaetke and Chow 2003).

6.2 STRATEGIES IN METAL REMOVAL

Various treatments have been explored and implemented to eliminate or reduce toxic metal pollutants from the environment. These approaches are classified into two: the nonbiological approaches and the biological approaches. The former constitutes of techniques that are more conventional and energy demanding, whereas the latter comprises approaches that are more environmentally friendly and sustainable.

6.2.1 NONBIOLOGICAL APPROACHES

The removal of metal ions in the early days was achieved via the application of nonbiological approaches, which includes several of the following physicochemical techniques: chemical precipitation, reverse osmosis, electrodialysis, and ion exchange. To date, chemical precipitation remains the most widely used process in many industries to treat industrial wastewaters. Chemical precipitation works by incorporating coagulants, such as alum, lime, and iron salts, into the wastewaters to precipitate metals (Barakat 2011). For reverse osmosis, metals are separated from wastewater by a semipermeable membrane due to selection pressure that is greater than the osmotic pressure (Shahalam et al. 2002). For electrodialysis, metals are removed through separation via a semipermeable ion-selective membrane with the aid of electricity (Sadrzadeh et al. 2009). In this process, the electrical potential between two different electrodes elicits the migration of cations and anions toward the respective electrodes. For metal removal via ion exchange, ions held by electrostatic forces by resin are exchanged for metal ions in the aqueous solutions (e.g., wastewaters) (Ku and Jung 2001). Although these conventional methods are established methods and have been used extensively, the techniques implemented have several limitations. Some of the major issues arising from these techniques are incomplete metal

TABLE 6.2

Advantages and Limitations Arising from the Various Conventional Approaches to Metal Removal

Techniques	Advantages	Limitations	References
Chemical precipitation	Low cost, simple procedure, applicable to most metals	Excessive sludge production, slow process, requires large amount of chemical	Barakat (2011)
Reverse osmosis	High removal efficiency	High cost	Shahalam et al. (2002)
Electrodialysis	High separation selectivity	Formation of metal hydrides (clogs the membrane), corrosion	Sadrzadeh et al. (2009)
Ion Exchange	Able to exchange most cations and anions	High cost, incomplete removal of metal, highly sensitive to pH	Ku and Jung (2001)

removal, high-energy requirements, generation of toxic sludge, the high cost incurred, and low removal efficiencies for solutions with low metal concentrations. The advantages and limitations of conventional approaches are further summarized in Table 6.2.

6.2.2 BIOLOGICAL APPROACHES

Several new technologies have been discovered as alternatives to address the limitations of the conventional approaches. One such alternative is the bioremediation approach. Bioremediation adopts the use of biological organisms to remove or break down pollutants from the contaminated site (Kumar et al. 2018). As such, the level of pollutants in the environment can be reduced, or the pollutants can be biodegraded into less toxic or nontoxic simpler compounds (Mejáre and Bülow 2001). The bioremediation approach can be achieved via phytoremediation, bioaccumulation, biodegradation, or biosorption. Phytoremediation uses living plants to remove pollutants in soil and water (Chandra et al. 2015, 2018; Chandra and Kumar 2018). Although this approach is sustainable and effective, it requires a long period of time to effectively remove metals. It is also dependent on plant regeneration/growth (Ahalya et al. 2003) and has higher demands for space use. Bioaccumulation, on the other hand, uses living organisms to absorb and accumulate the toxic pollutants into the cells. With bioaccumulation, biodegradation may also occur as the pollutant molecules are either compartmentalized within the cell (bioaccumulation) or degraded (biodegradation) via catabolism or excretion (Chandra and Kumar 2015; Gadd 1990). While bioaccumulation and biodegradation are ideal, these methods have been discovered to cause chronic poisoning in the organisms. Irrespective of whether phytoremediation, bioaccumulation, or biodegradation is implemented for metal removal, all three approaches rely on the biosorption process to a certain degree.

Biosorption is an adsorption process in which biological materials are used to bind and concentrate pollutants. The difference between biosorption and the other techniques (phytoremediation, bioaccumulation, biodegradation) is that biosorption is a physicochemical and metabolically independent process. As such, biosorption can occur when biomass of either living or nonliving organisms is used. On the contrary, bioaccumulation, biodegradation, and phytoremediation require that the organisms remain viable to permit bioaccumulation and biodegradation of the pollutants (Fomina and Gadd 2014). Comparatively, the biosorption approach in bioremediation is deemed a safer approach, as dead organisms (i.e., nonviable microbial cells) can be used as sorbents, with minimum risks to the environment. Other beneficial attributes of capitalizing the biosorption approach include high metal retention capacity, require low cost and minimum nutrients (less cost to upkeep dead biomass), lesser production of chemical or biological sludge (compared with conventional practices), and high amenability to metal recovery and reuse of biosorbents (metal binding on the biosorbents can be washed off by desorption) (Veglio and Beolchini 1997). Hence, biosorption has good potential and feasibility for metal removal and recovery.

For metal removal via biosorption, the same concept applies. Biosorption of metals occurs in which pollutant molecules (i.e., metals) bind to the biosorbents (from biological material). The binding of metal ions onto the biosorbents is attributed to several mechanisms (in no particular order): physical adsorption, chemical adsorption, or ion exchange (Javanbakht et al. 2014). These mechanisms depend solely on the interaction of the affinity of biosorbents with the biosorbates (pollutant molecules, i.e., metal cations). The biosorbents typically bind the biosorbates through the role of certain chemical functional groups present on the surface of the biosorbent. The interaction between functional groups and metal ions is, however, dependent on the pH of the metal solution. Solutions with pH value that is higher than the dissociation constant (pK_a) would result in the deprotonation of the functional groups, subsequently leading to the binding of metal cations (positively charged cations) by the negatively charged ligands. On the contrary, solutions with pH value lower than pK_a would result in protonated functional groups (positively charged), which would attract negatively charged metal anion (Kayalvizhi et al. 2015). The metal cations or anions bind to the biosorbent and are concentrated on the biosorbent, hence separating them from the aqueous solutions (Veglio and Beolchini 1997). The biosorption process can occur in almost any biosorbent as it involves the interaction and affinity differences between biosorbent and biosorbate. Dead or live biomass of plant and microbial origin are often the favorite choices for biosorbents. In this chapter, the focus will be on bacterial biosorbents.

6.2.3 Metal Removal by Bacteria

Bacteria are prokaryotic microorganisms, which have simple structures and morphologies. A typical bacterial cell comprises the cell wall, cell membrane, ribosomes, inclusion bodies, and nucleus, and in some, flagella and surface layer (S-layer) are found (Vijayaraghavan and Yun 2008). There are two types of cell walls found in bacteria, giving rise to the two main groups of bacteria: the gram-positive and gram-negative bacteria. The main difference between the two groups of bacteria is the peptidoglycan layer present in the cell wall. The peptidoglycan layer constitutes of linear polymers of alternating units of two sugar derivatives: N-acetylglucosamine (NAG) and N-acetylmuramic acid (NAM) (Wang and Chen 2009), which forms the basis of binding sites for metal cations. As such, bacterial biosorbents are useful for metal removal.

Gram-positive bacteria have a thicker peptidoglycan layer, which is cross-linked heavily in the cell wall as compared with gram-negative bacteria; therefore, they exhibit lower levels of complexation using functional groups present on the cell surface. On the contrary, the outer membrane of gram-negative cell consists of a single monolayer peptidoglycan layer, phospholipids, lipopolysaccharides, enzymes, and other proteins (Schleifer and Kandler 1972). This complex layer is responsible for the binding of metal cations in gram-negative bacteria, as most of these structures are exposed on the cell walls.

The bacteria species commonly used for metal removal is summarized in Table 6.3. Both gram-positive and gram-negative bacteria are used extensively in biosorption of metal. Gram-positive bacteria that are widely studied are *Bacillus* sp., *Mycobacterium* sp., *Staphylococcus* sp., and *Streptomyces* sp., with *Bacillus* sp. as the most common gram-positive bacteria studied. As for the gram-negative bacteria, *Escherichia coli*, *Pseudomonas* sp., and *Stenotrophomonas* sp. are examples of commonly studied bacteria for biosorption studies. Each different species demonstrates biosorption efficacy that is highly influenced by conditions, such as temperature, pH, contact time, and concentrations of biomass and metal.

Several advantages arise from the use of bacteria for metal removal. They include reduced operational costs, as microorganisms are inexpensive, easily available, and culturable (Aryal and Liakopoulou-Kyriakides 2015). They are also amenable to upscaling, mass regeneration of biomass,

TABLE 6.3
Examples of Bacteria Species and Their Biosorption Efficacy in Removing Metals

Gram	Bacteria	Metal Ion(s) Adsorbed	Biosorption Capacity (mg/g)	References
Positive	*Bacillus subtilis*	Cr^{3+}	23.9	Aravindhan et al. (2012)
	Bacillus sp. FM1	Cr^{6+}	64.102	Masood and Malik (2011)
	Bacillus sp.	Mn^{2+}	43.5	Hasan et al. (2012)
	Bacillus thuringiensis strain OSM29	Ni^{2+}, Cu^{2+}, Cd^{2+}	43.13	
	Bacillus cereus RC-1	Cd^{2+}	31.95	Huang et al. (2013), Oves et al. (2013)
	Bacillus cereus	Pb^{2+}	23.25	Çolak et al. (2011)
	Mycobacterium sp.	Cr^{3+}	87.09	Aryal and Liakopoulou-Kyriakides (2014)
		Cr^{6+}	61.51	
	Staphylococcus xylosus	Mn^{2+}	59	Gialamouidis et al. (2010)
		As^{3+}, As^{5+}	54.35, 61.34	Aryal et al. (2010)
	Streptomyces lunalinharesii	Cu^{2+}	26	Veneu et al. (2013)
Negative	*Escherichia coli* HD701	Cd^{2+}, Zn^{2+}	162.1	Morsy (2011)
	Pseudomonas sp.	Mn^{2+}	109	Gialamouidis et al. (2010)
		Fe^{3+}	86.20	Aryal and Liakopoulou-Kyriakides (2013)
	Pseudomonas fluorescens	Ni^{2+}	65.1	Wierzba and Latała (2010)
	Pseudomonas aeruginosa ASU 6a	Zn^{2+}	83.33	Joo et al. (2010)
	Pseudomonas sp. LKS06	Cd^{2+}	27.5	Huang and Liu (2013)
	Stenotrophomonas maltophilia	Cu^{2+}	26	Ye et al. (2013)

and genetic manipulation to enhance metal uptake (Kapoor and Viraraghavan 1997). In some instances, bacteria from industrial fermentation process can also be used to remove metals (Wei et al. 2016; Vimalnath and Subramanian 2018). In recent years, improvements have been introduced to enhance metal uptake in bacterial cells. This includes chemical modification (physical and chemical pretreatments, grafting) and biological modification (genetic and protein engineering). This chapter will focus on the use of pretreatment as a means to improve the metal removal process by bacteria biosorbents.

6.3 INNOVATIONS TO ENHANCE METAL REMOVAL

The common approach in enhancing or improving biosorption activities in biosorbents is through the use of pretreatments. Pretreatment is the modification performed onto the biosorbents during the preparation of biomass. It can be achieved through the implementation of physical- or chemical-based methods, such as heating and autoclaving, or using chemical reagents such as acids, alkalis, and organic chemicals, respectively (Gabr et al. 2008; Kumar and Gaur 2011; Guo et al. 2012; Huang et al. 2013; Boeris et al. 2016; Kinoshita et al. 2016; Kiran et al. 2016; Haris et al. 2018). The pretreatment process modifies the surface functional group either to remove or eliminate inhibiting functional groups or to destruct the cell membranes to expose more intracellular components, hence revealing more metal-binding sites (Wang and Chen 2009). As such, the pretreatment processes render inevitable changes to the availability of functional groups of the biosorbents, impacting the biosorption efficacy.

6.3.1 Heat Pretreatment Approach

Heat pretreatment, or also known as thermal pretreatment, is one of the most common approaches used in physical pretreatment. This is performed using different types of heat: moist heat (autoclaving and boiling) and dry heat (oven-drying) (Selatnia et al. (2004). Autoclaving is performed by subjecting the bacterial biomass to standard autoclaving condition (121°C, 20–30 min) while boiling is achieved by heating the bacterial biomass up to 100°C for at least 15 min. For oven-drying, the biomass is dried in an oven (~90°C) prior to use to remove metals.

The superiority of pretreated cells in removing metals compared with nonpretreated cells is evident. Boeris et al. (2016) discovered that autoclaved cells of *Pseudomonas putida* A were more efficient in removing cationic Al^{3+} from aqueous solution compared with viable cells without pretreatment. This was attributed to the increased availability of phosphatidylcholine, exposing more metal-binding sites. Similarly, higher biosorption efficacy was also demonstrated by autoclaved cells of *Bacillus cereus* RC-1 (Huang et al. 2013), *Streptomyces ciscaucasicus* (Li et al. 2010), and *Bacillus subtilis* (Al-Gheethi et al. 2017). All autoclaved bacterial biomass demonstrated higher metal adsorption capacity than the untreated biomass. The heat from the autoclaving process disrupted the cellular structure of the bacterial cells, causing more surface metal-binding sites to be exposed, increasing the available surface area to bind metals, leading to higher metal uptake capacity (Table 6.4).

The other moist-heat approach, boiling, has also been discovered as an effective pretreatment to enhance metal removal. Boiling is performed by heating the bacterial biosorbents at high temperature, and this has been tested at temperatures 60°C–100°C with varying results. Nevertheless, the concept is similar to autoclaving, whereby the high temperature disrupts the cell membranes, exposing more functional groups for metal binding. This has been observed in *Pseudomonas cepacia* 120S and *Bacillus subtilis* 117S, where boiled cells significantly improve the uptake of Ni^{2+} than the nontreated biomass (Abdel-Monem et al. 2010). In another study, heat-pretreated lactic acid bacterium, *Weisella viridescens* MYU205, showed a higher rate of mercury (Hg^{2+}) biosorption (90%) compared with nontreated cells (80%) (Kinoshita et al. 2016). The

TABLE 6.4
Typical Heat Pretreatments Performed on Various Bacterial Cells and Their Metal Removal Potential

Pretreatments	Condition	Bacteria	Metal Cations adsorbed	References
Autoclaving	1°C, 30 min	*Pseudomonas putida*	Al^{3+}	Boeris et al. (2016)
	121°C, 20 min	*Bacillus cereus*	Cd^{2+}	Huang et al. (2013)
	121°C, 30 min	*Streptomyces ciscaucasicus*	Zn^{2+}	Li et al. (2010)
	110°C, 10 min	*Bacillus subtilis*	Ni^{2+}, Pb^{2+}, Zn^{2+}, Cu^{2+}, Cd^{2+}	Al-Gheethi et al. (2017)
Boiling	100°C, 15 min	*Pseudomonas cepacia* *Bacillus subtilis*	Ni^{2+}	Abdel-Monem et al. (2010)
	100°C, 15 min	*Weisella viridescens*	Hg^{+}	Kinoshita et al. (2016)
	Kept in boiled water for 20 min	*Pseudomonas plecoglossicida*	Cd^{2+}	Guo et al. (2012)
	60°C–100°C, 60 min	*Escherichia coli*	Tungsten anions (W^{-})	Ogi et al. (2016)
Drying	90°C, 24 h	*Pseudomonas aeruginosa*	Ni^{2+}, Pb^{2+}	Gabr et al. (2008)

researchers further hypothesize that the oral ingestion of *W. viridescens* MYU205 may help in Cd^{2+} and Hg^{2+} absorption in the body, thus preventing their accumulation and toxicity in the body. Ogi et al. (2016) illustrated the effects of different heating temperatures on *E. coli* cells. *E. coli* cells were heated at 60°C, 70°C, 80°C, 90°C, and 100°C for 60 min. Cells heated at 100°C and 90°C showed higher removal of tungsten as residual tungsten ions (W^-) in the solutions were low (0.04 and 0.02 mmol/L, respectively after 60 min of incubation period). The decrease in heating temperature has an inverse relationship to tungsten removal, with biomass heated at 80°C, 70°C, and 60°C revealing 0.23, 0.38, and 0.40 mmol/L of residual tungsten ions in the solutions, respectively. And, in general, the pretreated cells were all more effective than the nontreated cells (0.52 mmol/L residual tungsten ions in solution).

In addition to moist heat, dry heat was also found to enhance metal removal in the treated bacterial cells effectively. Heat-dried *Pseudomonas aeruginosa* ASU 6a had a higher maximum biosorption capacity for Pb^{2+} and Ni^{2+} ions (123 and 113.6 mg/g, respectively) as compared with the untreated cells (79 and 70 mg/g, respectively) (Gabr et al. 2008). The passive uptake of metal ions by heat-treated cells is enhanced by the increase in both surface area and exposure of intracellular binding sites after the heat treatment.

Comparatively, moist heating methods (autoclaving and boiling) are more commonly studied than the dry heating (oven-drying) approach. Nevertheless, Yap (2016) evaluated these three approaches (autoclaving, boiling, and oven-drying) on *Stenotrophomonas maltophilia* and results revealed that all appeared to be equally effective in enhancing the removal of Pb^{2+} and Cu^{2+}. In some studies, metal removal efficacy between pretreated cells and live cells was discovered to be insignificantly different. Guo et al. (2012) reported this observation when boiled cells of *Pseudomonas plecoglossicida* (in deionized water, 20 min) adsorbed similar amount of Cd^{2+} ions as the live cells. Several other studies have also reported the higher biosorption capacity of the live cells compared with the pretreated cells (Bai et al. 2014; Chen et al. 2005; Paul et al. 2012). In these examples, it is presumed that the live cells may have removed metal ions via bioaccumulation, which is another useful mechanism other than biosorption.

6.3.2 Chemical Pretreatment Approach

Chemical pretreatment of bacterial cells using reagents of inorganic and organic origin has also been found to enhance metal removal efficacy. The exposure to chemicals ruptures the cell wall, exposing more functional groups for metal binding. In some instances, the pretreatment with chemicals, such as acids, alkalis, and organic chemicals, can modify the existing functional groups, or they can provide additional binding sites for the metal ions. Chemical reagents, which are commonly used, are acids or alkali, and the reagents are typically applied by soaking the biomass with the chemical reagents or by coincubating them (biomass, chemical reagents) with heat.

Hydrochloric acid (HCl) is the most commonly used acid for pretreatments. Colica et al. (2012) pretreat bacterial cells with HCl and discovered that the maximum adsorption capacity of HCl-pretreated biomass was two to three times higher than those of the nontreated cells. The HCl pretreatment also increased the metal binding specificity to ruthenium ions (Ru^+), as the adsorption rate of Ru^+ was the highest when tested together with Ni^{2+}, Cu^{2+}, and Zn^{2+} ions. Kumar and Gaur (2011) used HCl on cyanobacterial mats, and this significantly improved the metal binding capacity of Pb^{2+}, Cu^{2+}, and Cd^{2+} ions compared with other chemical reagents, such as calcium chloride ($CaCl_2$), nitric acid (HNO_3), sodium hydroxide (NaOH), and sodium dodecyl sulfate (SDS). Another weak acid, nitric acid (HNO_3), has also been found to enhance metal removal. Nitric acid (0.1 M) was used to pretreat the gram-negative bacterium *Yersinia* sp. strain SOM-12D3 (for 10 min at 150 rpm), which resulted in biosorption efficacy that is 4.7 times higher than the nontreated cells (Haris et al. 2018). Acids are therefore effective chemical reagents, as they cause the functional groups on the bacterial cell surface to be protonated; hence the pretreated cells are more effective in removing metal anions and transition ions (Table 6.5).

TABLE 6.5
Various Chemical Pretreatments Applied on Bacterial Cells to Enhance Uptake of Different Metal Species

Pre-treatments	Methods	Bacteria	Metal Adsorped	References
Nitric acid (HNO_3)	Soaking	*Yersinia* sp.	As^{3+}	Haris et al. (2018)
	Shaking	*Lyngbya putealis*	Cr^{6+}	Kiran et al. (2016)
Hydrochloric acid (HCl)	Dialysis	*Rhodopseudomonas palustris*	Ru^+	Colica et al. (2012)
	Shaking	Cyanobacterial mats	$Pb^{2+}, Cu^{2+}, Cd^{2+}$	Kumar and Gaur (2011)
	Shaking	*Lyngbya putealis*	Cr^{6+}	Kiran et al. (2016)
Sodium hydroxide (NaOH)	Soaking	*Oscillatoria* sp.	Cd^{2+}	Azizi et al. (2012)
	Soaking	*Streptomyces rimosus*	Cd^{2+}	Selatnia et al. (2004a)
			Fe^{3+}	Selatnia et al. (2004b)
			Pb^{2+}	Selatnia et al. (2004c), Selatnia et al. (2004b)
			Ni^{2+}	Selatnia et al. (2004d)
	Coincubation	*Stenotrophomonas maltophilia*	Cu^{2+}, Pb^{2+}	Yap (2016)
Detergent	Coincubation	*Stenotrophomonas maltophilia*	Cu^{2+}, Pb^{2+}	Yap (2016)

On the contrary, alkali reagents improve metal binding of treated cells by deprotonating the functional groups. Sodium hydroxide (NaOH) is the most widely used alkali reagent. When used to pretreat cells, such as the case for *Streptomyces rimosus*, biosorption of Cd^{2+}, Pb^{2+}, Ni^{2+}, Fe^{3+} were significantly enhanced with 63.3, 135, 32.6, and 122 mg/g of metals removed, respectively (Selatnia et al. 2004a, 2004b, 2004c, 2004d). This is attributed to the deprotonation of functional groups present on the cell surface, which led to negatively charged surfaces, thus improving the biosorption of metal cations. In addition, ion exchange between sodium cation and metal cations were also enhanced (Selatnia et al. 2004b).

The superiority of either heat or chemical pretreatment over the other has long been debated. In some instances, heat treatment is said to be more effective in enhancing metal removal in bacterial biosorbents followed by chemical pretreatments. In some examples, no significant increase in metal uptake was observed. Azizi et al. (2012) compared the effect of heat and chemical pretreatments on the Cd^{2+} adsorption capacity of *Oscillatoria* sp. by soaking the bacterial biomass in NaOH or HCl (chemical) or heating at 100°C in water bath and drying at 90°C. Among all treatments, heating at 100°C was the most effective (16.67 mg/g), followed by treatment with NaOH (15 mg/g), drying at 90°C (13.89 mg/g), and treatment with HCl (13.75 mg/g). Kiran et al. (2016) also conducted a similar study comparing the effects of NaOH and heat pretreatment on the cyanobacterium *Lyngbya putealis* HH-15. In their study, NaOH and heat pretreatment reduced the Cr^{6+} uptake capacity instead of enhancing the removal. As such, while chemical or heat pretreatments are generally established as effective measures to significantly enhance metal uptake, the effect is not universal to all cells and may have few exceptions.

6.3.3 Detecting Characteristics of Pretreated Cells

Pretreatments, either by heat or chemical approaches, are known to modify the surface functional groups of the bacterial cell, to expose more metal-binding sites. Heat pretreatments disrupt the cell structure causing more functional groups to be exposed, whereas chemical pretreatments alter the protonation state, causing a change in the net charges of these functional groups to attract either metal cations or anions (Selatnia et al. 2004a). These functional groups can be detected and characterized using several techniques, primarily with the use of the Fourier transform infrared spectroscopy (FTIR).

FTIR is a technique commonly used to examine the presence of functional groups on the cell surface, by studying the spectral bands and shifting of wavenumber. The significant shift in adsorption peaks of pretreated and nontreated bacterial biosorbent indicates the modification of the functional groups, whether they were significantly reduced or increased after the pretreatment (Huang et al. 2013). For example, Yap (2016) reported changes in intensity and significant shifts in adsorption peaks when cells of *Stenotrophomonas maltophilia* were pretreated with NaOH. The changes were detected at wavenumbers 1,046, 1,224, 1,390, 1,526, and 1,635 cm^{-1}, which indicate a change in carbonyl, carboxyl, and phosphate groups on the cells after pretreatment with NaOH (Figure 6.1). In some biosorbents, where NaOH treatment resulted in inferior sorption efficacy, this was attributed to the absence of amine groups, suggesting the removal of protein from the biomass surface, hence the poor metal adsorption (Kiran et al. 2016). Pretreatment with other chemical reagents would result in different effects of the functional groups, evident in the different spectra formed, for example, in the pretreatment with nitric acid. With HNO_3, slight shifts in peak wavenumber were detected compared with the nontreated biomass (Haris et al. 2018).

Pretreatment with heat damages the cell wall structure of the bacterial cells, causing cell lysis, which results in increased cell wall permeability, leading to improved metal adsorption (Huang et al. 2013). Heat pretreatment also increased the density of amino acids on the cell surface, as the intensities of the peak associated with amino and carbonyl groups on the surface of preheated bacterial cells (*E. coli*) were greater than the nontreated biomass (Ogi et al. 2016). Heat pretreatment was also capable of altering the functional groups of both live and dead cells. In a study by Li et al. (2010), significant shifts were observed at peaks associated with carbonyl, amino, carboxyl, and aromatic groups in the heat-treated cells of both live and dead cells, compared with cells that were nontreated. Therefore, it can be concluded that heat (and chemical) pretreatments can modify the functional groups present on the cell surface, making more functional groups to be available and involved in the metal adsorption process.

In addition, FTIR can also be used to study the functional groups involved in metal binding by comparing the spectra of unloaded and metal-loaded pretreated biosorbents. Yap (2016) reported an increase in peak intensity in NaOH-treated biosorbents after binding with Cu^{2+} and Pb^{2+} ions. This is due to the involvement of the respective functional groups in metal binding and the formation of new chemical bonds between the metal cations and functional groups. Functional groups of pretreated biosorbents involved in metal binding from several studies are summarized in Table 6.6. It can be concluded that the functional groups commonly involved in metal binding are hydroxyl, amine, carbonyl, phosphate, and sulfhydryl groups. Similarly, binding with arsenic ions also resulted in shifts in peaks related to carboxyl, hydroxyl, amine groups, and aromatic amino acids (Haris et al. 2018).

Aside from FTIR, scanning electron microscopy (SEM) has also been utilized to examine the interaction of metal ions with the surface of the biosorbents. The few existing reports used SEM to describe the morphology of the bacterial cells upon pretreatment. Autoclaved cells appeared longer and more slender than live cells, which are smooth and short rod in shape (Huang et al. 2013). Upon biosorption, cell surface may be covered with shiny and bulky particles of metal ions. These particles are detected as irregular globules on the cell surface (Boeris et al. 2016). Similar observations were reported by Ogi et al. (2016) in which the preheated cells appeared concave-convex shape, whereas nontreated cells were smooth. The change in the shape of cells due to heating is hypothesized to increase the surface area, thus exposing more metal-binding sites for metal uptake. Pretreatment with chemical reagents, such as HCl, gave similar conclusions. It was revealed that pretreated biosorbents improved porous structure with greater homogeneity as compared with the linear groove and projections of the nontreated bacterial biomass (Kiran et al. 2016). Haris et al. (2018) used a combination of field emission scanning electron microscope (FESEM) with energy-dispersive X-ray (EDX) analysis to visualize the surface of acid-treated bacterial biomass

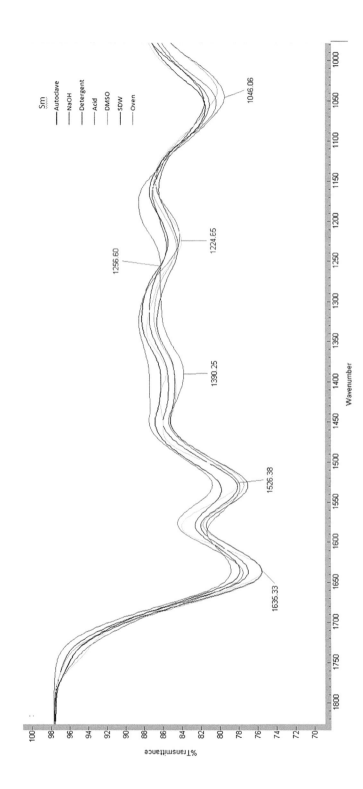

FIGURE 6.1 The FTIR spectra (range from 1,000 to 1,800 cm⁻¹) of cells of *Stenotrophomonas maltophilia* (Sm) subjected to various pretreatments. (Reprinted with permission from Yap 2016.)

TABLE 6.6

Functional Groups and Their Peaks Detected in Heat and Chemical Pretreated Biosorbents, Responsible for the Binding of Various Metal Cations

Metal Adsorbed	Hydroxyl OH (3,500–3,700)	Amine NH$_2$ (3,300–3,500)	Carbonyl C=O (1,650–1,750)	Carboxyl COOH (1,540)	Phosphate PO$_3$ (1,030–1,200)	Sulfhydryl SH (470–800)	Reference
Al^{3+}	✓				✓		Boeris et al. (2016)
Ni^{2+}, Pb^{2+}	✓	✓	✓	✓	✓	✓	Gabr et al. (2008)
Cd^{2+}	✓	✓	✓	✓	✓		Huang et al. (2013)
As^{3+}	✓	✓	✓				Haris et al. (2018)
Cu^{2+}, Pb^{2+}	✓	✓	✓	✓	✓		Yap (2016)

and to evaluate the chemical composition of the sample. FESEM results showed that the surface of the bacterial cell appeared to be more porous and deformed after the treatment with nitric acid, resulting in an increased number of available metal-binding site. The EDX analysis identified the presence of arsenic ions on the pretreated biomass upon biosorption of arsenic ions.

The other approach in characterizing pretreated biosorbents is through the use of zeta potential. Zeta potential is the electrokinetic potential measured in millivolts (mV), on the potential difference between a solid surface and its liquid medium. Zeta potential is often used to measure the net electrical charge present on the bacterial cell surface (Aryal and Liakopoulou-Kyriakides 2015). Huang et al. (2013) reported that the pretreated cells have lower zeta potential, indicating higher amounts of negative charges present on the cell surface, which essentially leads to greater Cd^{2+} adsorption. Cells, which exhibited negative zeta potential values sooner, imply greater change of overall surface charges of the cells from positive to negative, which then leads to more rapid biosorption. Ogi et al. (2016) also demonstrated similar finding, where adsorption capacity of the pretreated *E. coli* biomass toward tungsten anions (W$^-$) increased with its zeta potential. The zeta potential increased significantly at low pH region (pH 1.7–2.0) from 10 mV (60°C) to 24 mV (100°C) with the heating temperature used during pretreatment (60°C–100°C) as compared with the untreated bacterial biomass (4 mV). Hence, the improvement in adsorption of W$^-$ is attributed to the increase of positive charges on the biomass after the application of heat pretreatment. It is therefore evident that zeta potential can be used to detect and characterize heat pretreated cells.

6.4 CONCLUSIONS AND FUTURE PROSPECTS

In conclusion, heat and chemical pretreatments are able to enhance metal removal of bacterial biosorbents. Heat pretreatments work by exposing more available functional groups, whereas chemical pretreatments change the charges of the metal-binding sites causing it to be more specific toward the targeted metal cations or anions. Both of the pretreatments appeared to be equally effective if the right treatments are applied for the targeted metal ions. However, heat pretreatment is more sustainable and economical; it is also more environmentally friendly compared with the chemical pretreatments as no chemical reagents are required during the pretreatment process. This reduces cost and the likelihood of chemical residual formation. Heat pretreatment does not only increase the number of metal binding sites, but it also inactivates or kills the bacteria cells, making the bacterial biosorbents safer and more environmentally friendly as it would not cause any harm when it is applied. Future studies can focus on the technologies to incorporate these pretreated bacterial biosorbents for large-scale application.

ACKNOWLEDGMENTS

The authors wish to thank Monash University Malaysia for the facilities and support in research endeavors related to metal removal using microbial cells.

REFERENCES

Abdel-Monem, M., Al-Zubeiry, A. and Al-Gheethi, A. (2010) Biosorption of nickel by Pseudomonas cepacia 120S and Bacillus subtilis 117S, *Water Science and Technology* 61(12), 2994–3007.

Ahalya, N., Ramachandra, T. and Kanamadi, R. (2003) Biosorption of heavy metals, *Research Journal of Chemistry and Environment* 7(4), 71–79.

Al-Gheethi, A., Mohamed, R., Noman, E., Ismail, N. and Kadir, O. A. (2017) Removal of heavy metal ions from aqueous solutions using Bacillus subtilis biomass pre-treated by supercritical carbon dioxide, *CLEAN–Soil, Air, Water* 45(10), 1700356.

Alluri, H. K., Ronda, S. R., Settalluri, V. S., Bondili, J. S., Suryanarayana, V. and Venkateshwar, P. (2007) Biosorption: An eco-friendly alternative for heavy metal removal, *African Journal of Biotechnology* 6(25), 2924–2931.

Aravindhan, R., Fathima, A., Selvamurugan, M., Rao, J. R. and Balachandran, U. N. (2012) Adsorption, desorption, and kinetic study on Cr (III) removal from aqueous solution using Bacillus subtilis biomass, *Clean Technologies and Environmental Policy* 14(4), 727–735.

Aryal, M. and Liakopoulou-Kyriakides, M. (2013) Binding mechanism and biosorption characteristics of Fe (III) by Pseudomonas sp. cells, *Journal of Water Sustainability* 3(3), 117–131.

Aryal, M. and Liakopoulou-Kyriakides, M. (2014) Characterization of Mycobacterium sp. strain Spyr1 biomass and its biosorption behavior towards Cr (III) and Cr (VI) in single, binary and multi-ion aqueous systems, *Journal of Chemical Technology & Biotechnology* 89(4), 559–568.

Aryal, M. and Liakopoulou-Kyriakides, M. (2015) Bioremoval of heavy metals by bacterial biomass, *Environmental Monitoring and Assessment* 187(1), 4173.

Aryal, M., Ziagova, M. and Liakopoulou-Kyriakides, M. (2010) Study on arsenic biosorption using Fe (III)-treated biomass of Staphylococcus xylosus, *Chemical Engineering Journal* 162(1), 178–185.

Azizi, S. N., Hosseinzadeh Colagar, A. and Hafeziyan, S. M. (2012) Removal of Cd (II) from aquatic system using Oscillatoria sp. biosorbent, *The Scientific World Journal* 2012 (2012), 1–7.

Bai, J., Yang, X., Du, R., Chen, Y., Wang, S. and Qiu, R. (2014) Biosorption mechanisms involved in immobilization of soil Pb by Bacillus subtilis DBM in a multi-metal-contaminated soil, *Journal of Environmental Sciences* 26(10), 2056–2064.

Barakat, M. (2011) New trends in removing heavy metals from industrial wastewater, *Arabian Journal of Chemistry* 4(4), 361–377.

Boeris, P. S., del Rosario Agustín, M., Acevedo, D. F. and Lucchesi, G. I. (2016) Biosorption of aluminum through the use of non-viable biomass of Pseudomonas putida, *Journal of Biotechnology* 236, 57–63.

Chandra, R., Kumar, V. (2015) Biotransformation and biodegradation of organophosphates and organohalides. In: *Environmental Waste Management*, R. Chandra (ed.). CRC Press, Boca Raton, FL, pp. 475–524.

Chandra, R., Kumar, V. (2018) Phytoremediation: A green sustainable technology for industrial waste management. In: *Phytoremediation of Environmental Pollutants*, R. Chandra, N. K. Dubey, V. Kumar (eds.). CRC Press, Boca Raton, FL.

Chandra, R., Kumar, V., Singh, K. (2018) Hyperaccumulator versus nonhyperaccumulator plants for environmental waste management. In: *Phytoremediation of Environmental Pollutants*, R. Chandra, N. K. Dubey, V. Kumar (eds.). CRC Press, Boca Raton, FL, pp. 43–80.

Chandra, R., Saxena, G., Kumar, V. (2015) Phytoremediation of environmental pollutants: An eco-sustainable green technology to environmental management. In: *Advances in Biodegradation and Bioremediation of Industrial Waste*, R. Chandra (ed.). CRC Press, Boca Raton, FL, pp. 1–29.

Chen, X. C., Wang, Y. P., Lin, Q., Shi, J. Y., Wu, W. X. and Chen, Y. X. (2005) Biosorption of copper (II) and zinc (II) from aqueous solution by Pseudomonas putida CZ1, *Colloids and Surfaces B: Biointerfaces* 46(2), 101–107.

Çolak, F., Atar, N., Yazıcıoğlu, D. and Olgun, A. (2011) Biosorption of lead from aqueous solutions by Bacillus strains possessing heavy-metal resistance, *Chemical Engineering Journal* 173(2), 422–428.

Colica, G., Caparrotta, S. and De Philippis, R. (2012) Selective biosorption and recovery of Ruthenium from industrial effluents with Rhodopseudomonas palustris strains, *Applied Microbiology and Biotechnology* 95(2), 381–387.

Duruibe, J., Ogwuegbu, M. and Egwurugwu, J. (2007) Heavy metal pollution and human biotoxic effects, *International Journal of Physical Sciences* 2(5), 112–118.

Fomina, M. and Gadd, G. M. (2014) Biosorption: current perspectives on concept, definition and application, *Bioresource Technology* 160, 3–14.

Fu, F. and Wang, Q. (2011) Removal of heavy metal ions from wastewaters: A review, *Journal of Environmental Management* 92(3), 407–418.

Gabr, R., Hassan, S. and Shoreit, A. (2008) Biosorption of lead and nickel by living and non-living cells of Pseudomonas aeruginosa ASU 6a, *International Biodeterioration & Biodegradation* 62(2), 195–203.

Gadd, G. M. (1990) Heavy metal accumulation by bacteria and other microorganisms, *Experientia* 46(8), 834–840.

Gaetke, L. M. and Chow, C. K. (2003) Copper toxicity, oxidative stress, and antioxidant nutrients, *Toxicology* 189(1–2), 147–163.

Gialamouidis, D., Mitrakas, M. and Liakopoulou-Kyriakides, M. (2010) Equilibrium, thermodynamic and kinetic studies on biosorption of Mn (II) from aqueous solution by Pseudomonas sp., Staphylococcus xylosus and Blakeslea trispora cells, *Journal of Hazardous Materials* 182(1–3), 672–680.

Guo, J., Zheng, X.-d., Chen, Q.-b., Zhang, L. and Xu, X.-p. (2012) Biosorption of Cd (II) from aqueous solution by Pseudomonas plecoglossicida: Kinetics and mechanism, *Current Microbiology* 65(4), 350–355.

Haris, S. A., Altowayti, W. A. H., Ibrahim, Z. and Shahir, S. (2018) Arsenic biosorption using pretreated biomass of psychrotolerant Yersinia sp. strain SOM-12D3 isolated from Svalbard, Arctic, *Environmental Science and Pollution Research* 25(28), 27959–27970.

Hasan, H. A., Abdullah, S. R. S., Kofli, N. T. and Kamarudin, S. K. (2012) Isotherm equilibria of Mn^{2+} biosorption in drinking water treatment by locally isolated Bacillus species and sewage activated sludge, *Journal of Environmental Management* 111, 34–43.

Huang, F., Dang, Z., Guo, C.-L., Lu, G.-N., Gu, R. R., Liu, H.-J. and Zhang, H. (2013) Biosorption of Cd (II) by live and dead cells of Bacillus cereus RC-1 isolated from cadmium-contaminated soil, *Colloids and Surfaces b: Biointerfaces* 107, 11–18.

Huang, W. and Liu, Z.-m. (2013) Biosorption of Cd (II)/Pb (II) from aqueous solution by biosurfactant-producing bacteria: Isotherm kinetic characteristic and mechanism studies, *Colloids and Surfaces B: Biointerfaces* 105, 113–119.

Javanbakht, V., Alavi, S. A. and Zilouei, H. (2014) Mechanisms of heavy metal removal using microorganisms as biosorbent, *Water Science and Technology* 69(9), 1775–1787.

Joo, J.-H., Hassan, S. H. and Oh, S.-E. (2010) Comparative study of biosorption of Zn^{2+} by Pseudomonas aeruginosa and Bacillus cereus, *International Biodeterioration & Biodegradation* 64(8), 734–741.

Kapoor, A. and Viraraghavan, T. (1997) Heavy metal biosorption sites in Aspergillus niger, *Bioresource Technology*, 61(3), 221–227.

Kayalvizhi, K., Vijayaraghavan, K. and Velan, M. (2015) Biosorption of Cr (VI) using a novel microalga Rhizoclonium hookeri: Equilibrium, kinetics and thermodynamic studies, *Desalination and Water Treatment* 56(1), 194–203.

Kinoshita, H., Ohtake, F., Ariga, Y. and Kimura, K. (2016) Comparison and characterization of biosorption by Weissella viridescens MYU 205 of periodic group 12 metal ions, *Animal Science Journal* 87(2), 271–276.

Kiran, B., Rani, N. and Kaushik, A. (2016) FTIR spectroscopy and scanning electron microscopic analysis of pretreated biosorbent to observe the effect on Cr (VI) remediation, *International Journal of Phytoremediation* 18(11), 1067–1074.

Ku, Y. and Jung, I.-L. (2001) Photocatalytic reduction of Cr (VI) in aqueous solutions by UV irradiation with the presence of titanium dioxide, *Water Research* 35(1), 135–142.

Kumar, D. and Gaur, J. (2011) Metal biosorption by two cyanobacterial mats in relation to pH, biomass concentration, pretreatment and reuse, *Bioresource Technology* 102(3), 2529–2535.

Kumar V., Shahi S. K., Singh S. (2018) Bioremediation: An eco-sustainable approach for restoration of contaminated sites. In: *Microbial Bioprospecting for Sustainable Development*, J. Singh, D. Sharma, G. Kumar, N. R. Sharma (eds.). Springer, Singapore, pp. 115–136.

Li, H., Lin, Y., Guan, W., Chang, J., Xu, L., Guo, J. and Wei, G. (2010) Biosorption of Zn (II) by live and dead cells of Streptomyces ciscaucasicus strain CCNWHX 72-14, *Journal of Hazardous Materials* 179(1–3), 151–159.

Masood, F. and Malik, A. (2011) Biosorption of metal ions from aqueous solution and tannery effluent by Bacillus sp. FM1, *Journal of Environmental Science and Health, Part A* 46(14), 1667–1674.

Mejáre, M. and Bülow, L. (2001) Metal-binding proteins and peptides in bioremediation and phytoremediation of heavy metals, *Trends in Biotechnology* 19(2), 67–73.

Mohammed, A. S., Kapri, A. and Goel, R. (2011) Heavy metal pollution: Source, impact, and remedies. In: *Biomanagement of Metal-Contaminated Soils*, M. Saghir Khan, A. Zaidi, R. Goel, J. Musarrat (eds.). Springer, Dordrecht, pp. 1–28.

Morsy, F. M. (2011) Hydrogen production from acid hydrolyzed molasses by the hydrogen overproducing Escherichia coli strain HD701 and subsequent use of the waste bacterial biomass for biosorption of Cd (II) and Zn (II), *International Journal of Hydrogen Energy* 36(22), 14381–14390.

Ogi, T., Makino, T., Iskandar, F., Tanabe, E. and Okuyama, K. (2016) Heat-treated Escherichia coli as a high-capacity biosorbent for tungsten anions, *Bioresource Technology* 218, 140–145.

Oves, M., Khan, M. S. and Zaidi, A. (2013) Biosorption of heavy metals by Bacillus thuringiensis strain OSM29 originating from industrial effluent contaminated north Indian soil, *Saudi Journal of Biological Sciences* 20(2), 121–129.

Paul, M. L., Samuel, J., Chandrasekaran, N. and Mukherjee, A. (2012) Comparative kinetics, equilibrium, thermodynamic and mechanistic studies on biosorption of hexavalent chromium by live and heat killed biomass of Acinetobacter junii VITSUKMW2, an indigenous chromite mine isolate, *Chemical Engineering Journal* 187, 104–113.

Sadrzadeh, M., Mohammadi, T., Ivakpour, J. and Kasiri, N. (2009) Neural network modeling of Pb^{2+} removal from wastewater using electrodialysis, *Chemical Engineering and Processing: Process Intensification* 48(8), 1371–1381.

Selatnia, A., Bakhti, M., Madani, A., Kertous, L. and Mansouri, Y. (2004a) Biosorption of Cd^{2+} from aqueous solution by a NaOH-treated bacterial dead Streptomyces rimosus biomass, *Hydrometallurgy* 75(1–4), 11–24.

Selatnia, A., Boukazoula, A., Kechid, N., Bakhti, M. and Chergui, A. (2004b) Biosorption of Fe^{3+} from aqueous solution by a bacterial dead Streptomyces rimosus biomass, *Process Biochemistry* 39(11), 1643–1651.

Selatnia, A., Boukazoula, A., Kechid, N., Bakhti, M., Chergui, A. and Kerchich, Y. (2004c) Biosorption of lead (II) from aqueous solution by a bacterial dead Streptomyces rimosus biomass, *Biochemical Engineering Journal* 19(2), 127–135.

Selatnia, A., Madani, A., Bakhti, M., Kertous, L., Mansouri, Y. and Yous, R. (2004d) Biosorption of Ni^{2+} from aqueous solution by a NaOH-treated bacterial dead Streptomyces rimosus biomass, *Minerals Engineering* 17(7–8), 903–911.

Schleifer, K. H. and Kandler, O. (1972) Peptidoglycan types of bacterial cell walls and their taxonomic implications, *Bacteriological Reviews*, 36(4), 407.

Shahalam, A. M., Al-Harthy, A. and Al-Zawhry, A. (2002) Feed water pretreatment in RO systems: Unit processes in the Middle East, *Desalination* 150(3), 235–245.

Veglio, F. and Beolchini, F. (1997) Removal of metals by biosorption: A review, *Hydrometallurgy* 44(3), 301–316.

Veneu, D. M., Torem, M. L. and Pino, G. A. (2013) Fundamental aspects of copper and zinc removal from aqueous solutions using a Streptomyces lunalinharesii strain, *Minerals Engineering* 48, 44–50.

Vijayaraghavan, K. and Yun, Y.-S. (2008) Bacterial biosorbents and biosorption, *Biotechnology Advances* 26(3), 266–291.

Vimalnath, S. and Subramanian, S. (2018) Studies on the biosorption of Pb (II) ions from aqueous solution using extracellular polymeric substances (EPS) of Pseudomonas aeruginosa, *Colloids and Surfaces B: Biointerfaces* 172, 60–67.

Wang, J. and Chen, C. (2009) Biosorbents for heavy metals removal and their future, *Biotechnology Advances* 27(2), 195–226.

Wei, W., Wang, Q., Li, A., Yang, J., Ma, F., Pi, S. and Wu, D. (2016) Biosorption of Pb (II) from aqueous solution by extracellular polymeric substances extracted from Klebsiella sp. J1: Adsorption behavior and mechanism assessment, *Scientific Reports* 6, 31575.

Wierzba, S. and Latała, A. (2010) Biosorption lead (II) and nikel (II) from an aqueous solution by bacterial biomass, *Polish Journal of Chemical Technology* 12(3), 72–78.

Yap, L.-S. (2016) Effect of Pretreatments on Microbial Biosorbents for Metal Removal. Bachelor of Science (Honours), Monash University Malaysia.

Yap, L.-S., Lee, W.-L. and Ting, A. S. Y. (2017) Endophytes from Malaysian medicinal plants as sources for discovering anticancer agents. In: *Medicinal Plants and Fungi: Recent Advances in Research and Development*, D. C. Agrawal, H.-S. Tsay, L.-F. Shyur, Y.-C. Wu, S.-Y. Wang (eds.). Springer, Singapore, pp. 313–335.

Ye, J., Yin, H., Xie, D., Peng, H., Huang, J. and Liang, W. (2013) Copper biosorption and ions release by Stenotrophomonas maltophilia in the presence of benzo [a] pyrene, *Chemical Engineering Journal* 219, 1–9.

7 Field-Scale Remediation of Crude Oil–Contaminated Desert Soil Using Various Treatment Technologies
A Large Remediation Project Case Study

Subhasis Das, Veeranna A. Channashettar,
Nanthakumar Kuppanan, and Banwari Lal
The Energy and Resources Institute

CONTENTS

7.1   INTRODUCTION

The oily sludge wastes generated from any petroleum industry during crude oil exploration, transportation from oil production, storage, and refinery process are a worldwide problem of concern. The ever-increasing demand for petroleum products generates huge wastes, which are accumulated in the natural environment, and this problem is considered as a major threat to the environment. The crude oil contains higher concentrations of petroleum hydrocarbons (PHCs), which considered themselves hazardous to the environment and human health, and, therefore, requires remediation. Petroleum is a natural reserve found in accumulation in porous reservoir rock underground during geographical time scale (Olajire, 2014). Petroleum reserves are available in liquid form, which is called crude oil. Apart from liquid form, other formations are soil (such as coal, bitumen, and tar sands) and gaseous (such as natural gas). Crude oils are predominantly made up of higher carbon atom number (C_{5+}), and in room temperature and pressure, they form a liquid (Schlumberger, 2010). Oily waste pollution is generated through gas flaring, aboveground pipeline leakage, oily waste dumping at different dumping site, and due to oil spills. Gas flare from any oil industry located near the village or urban area causes substantial ecological damages to the land, water, and vegetation by producing soot. The soot is deposited on the building roofs and surface of the leaf in the nearby area. After raining, the soot will be washed off in a liquid form of black-ink color, which contains harmful chemicals drained to the soil affecting the fertility of the soil. On the other hand, gas pipelines have also caused damage to the agricultural soils where the pipeline goes. Both the effects are hazardous to the environment, and effects range from soil infertility to health problems, which may cause fatal to a human.

7.2   COMPOSITION OF PETROLEUM HYDROCARBON

Crude oil is a complex mixture of an indefinite number of individual chemical compounds, and the properties of these compounds differ depending on many factors, such as source, geological history, age, migration, and alteration of crude oil. The PHCs in crude oil contain a complex mixture of four fractions: saturates, aromatics, resins (N, S, O), and asphaltene (Balba et al., 1998). The saturate fractions are straight chain alkanes (all normal alkanes), branched alkanes (isoalkanes), and cycloalkanes (naphthenes). The aromatic fractions contain volatile monoaromatic hydrocarbon (benzene, toluene, xylene etc.), polyaromatic hydrocarbon (PAH) (three-ring compounds—anthracene, phenanthrene; four-ring compounds—tetracene, chrysene, triphenylene; five-ring compounds—pentacene, benzopyrene, corannulene, benzopyrene; six-ring compound—coronene; and seven-ring compounds—ovalene and benzofluorene), naphthenoaromatics, and aromatic sulfur compounds (Figure 7.1). The resin and asphaltene fractions consist of polar molecules containing nitrogen, sulfur, and oxygen. Resins are amorphous in nature and dissolved in oil, whereas asphaltenes are of a big colloidal shape and dispersed in oil. It is notable that PAH fractions associated with oil contamination, as well as known carcinogens, are suspected. The most toxic PAH compound is benzopyrene.

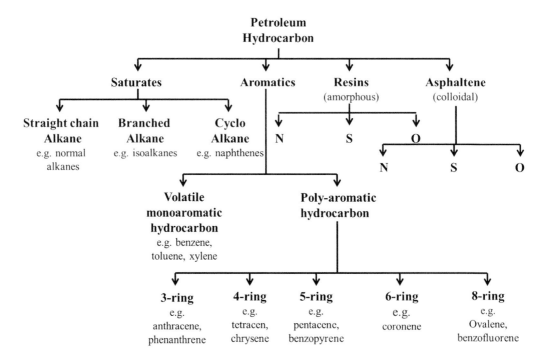

FIGURE 7.1   Composition of petroleum hydrocarbon with examples.

7.3 OILY SLUDGE CONTAMINATION SCENARIO IN KUWAIT

A huge amount of oil-contaminated soil remains unremediated in the Kuwait desert since the past few decades (Figure 7.2). The contaminated soil has the potentiality to cause pollution of underground water and to affect the health of people in the neighborhood. Kuwait had 114 km^2 of its desert severely damaged by 798 detonated oil wells by Iraqi troops (Al-Gharabally and Al-Barood, 2015). Crude oil gushed from the damaged oil well creates oil lake in the low lying desert area that contaminated over 40 km^2 land. These oil lakes are mostly dry oil materials but in some features still have semiliquid, oil/sludgy material referred to as wet oil lakes. The contaminated soil piles were further generated during a rescue operation to stop the spreading of oil. Apart from that, few more features are also contaminated during the operation of oil exploration. These features are sludge pits, where spilled oil/oil from pipeline leakages is recovered and stored for the last more than 50 years, and effluent pits, where produced water is being stored, which is one of the by-products generated along with oil and gas. The clean soil beneath the sludge pit and effluent pit is further contaminated through leaching of oil and produced water due to longer storage time. These polluted features are scattered in the desert area near to crude oil production facilities. Consequently, contaminated soil is estimated to be around 26 million m^3, posing a threat to the precious desert ecosystem, including soil, groundwater, vegetation, and wildlife (Al-Gharabally et al., 2016).

7.4 REMEDIATION STRATEGIES TO CLEAN UP CONTAMINATED SOIL

The effective remediation technology of oily sludge has become a worldwide challenge due to its hazardous nature and a sharp increase in quantity in production across the globe. During the past few decades, a variety of oily sludge treatment methods have been developed by researchers worldwide (Panno et al., 2005; Zhang et al., 2012; Okeke and Obi, 2013; Benyahia and Embaby, 2016). Few of them are commercialized and used for the treatment of large quantity of soil contaminated by crude oil. The major remediation technology is incineration, thermal desorption, chemical treatment, and bioremediation. These technologies could reduce or eliminate the hazardous constituents from oil and produce healthy soil. However, due to the recalcitrant nature of oily sludge, few technologies can reach a compromised balance between satisfying strict environmental regulations and reducing treatment costs (Hu et al. 2013). Al-Futaisi et al. (2007) proposed a three-tire oily sludge waste management strategy that includes employing technologies to reduce the quantity of oily sludge production during operation in a petroleum industry, recovery of oil from existing oily sludge, and disposing of the residues after oil recovery or oily sludge itself. The proposed first tier would help in source manipulation to prevent the oily waste generation, whereas the next two tires are more focused on effective treatment methods. Generally, a contaminated site remediation scheme is triple-stage process where the first step would be site reconnaissance and risk assessment, second is remedial options appraisal, and third is remediation and aftercare maintenance

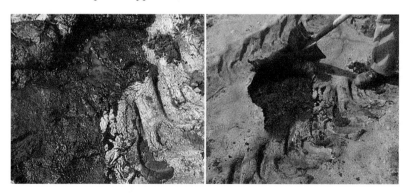

FIGURE 7.2 Crude oil contamination at Kuwait desert soil.

(Kupusamy et al. 2017). Step I aims to investigate if the land has been contaminated with oily sludge or the concentration of oily waste exceeds from primary ecotoxicity and alternate ecotoxicity remediation standard that set out as per available local or national guidelines. Following this, the site is declared to be polluted and requires the implementation of the soil management program. The management program first addresses source control to reduce PHC generation. After understanding the extent of contamination, remediation objectives could be developed in a different way ranging from developing a guideline to site-specific risk assessment. In step II, identify and evaluate possible remedial option, and develop remediation strategies. Before implementation of any technology at the field scale, the effectiveness of the selected remedial options must be checked for feasibility and treatability studies in the laboratory and/or onsite field condition. The results obtained from feasibility and treatability studies could be used for designing remedial action for any contaminated site. In the third phase, the implementation of suitable remediation strategies is carried out where the remediation technology is put into action and long-term monitoring and aftercare maintenance are implemented. Finally, ecological and toxicity assessment is once again executed to confirm if risk-based remediation of contaminated land is achieved or otherwise by the adopted remediation strategy. If risk persists, then the site has to be subjected to further remediation (Kupusamy et al. 2017; Duan et al., 2013).

7.4.1 POTENTIAL PATHWAY AND MOLECULAR MECHANISM OF CRUDE OIL–CONTAMINATED SOIL

The biodegradation process involves a series of steps: digestion, assimilation, and hydrocarbon metabolism using different enzymes, and this is selectively metabolized by bacteria, fungi, protozoa, and other organisms alone or in a consortium. The most efficient and rapid degradation of oily sludge or any other organic pollutants occurred aerobically by a group of bacteria, such as phototrophic and chemotrophic bacteria. Most of the microbial enzymes for biodegradation are encoded on plasmids, and these enzymes are contributing in various pathways for biodegradation (Varjani, 2017). Various plasmids involved in PHC degradation are Q15, OCT, TOL, NAH7, pND140, and pND160 with genes, including alkA, alkM, alkB, theA, LadA, assA1, assA2, and nahA-M (Salleh et al., 2003; Abbasian et al., 2015; Wilkes et al., 2016) The initial intracellular attack of hydrocarbon content is achieved by an oxidative process by attachment of microbial cells to the substrates and/or production of biosurfactant, emulsifier, biopolymers, solvents, gases, and acids (Banat, 1995; Pornsunthorntawee et al., 2008; Saeki et al., 2009; Varjani and Upasani, 2016). Microorganisms either catabolize PHCs available in crude oil to obtain carbon sources and energy or assimilate them into cell biomass for growth and proliferation (Leahy and Colwell, 1990). Figure 7.3 represents the schematic pathway for PHC utilization by hydrocarbon-degrading bacteria through the phototrophic anoxygenic, chemotrophic aerobic, and chemotrophic anaerobic pathway. In aerobic route of degradation, the process is carried out within the cytoplasmic membrane, and the oxidation product is transported inside the cell where potential degrading bacteria oxidize the carbon molecules through oxygenase/dehydrogenase enzyme complex and produce oxygen molecules that result in the formation of alcohol, which is more reactive. Subsequently, alcohol group is transformed to carboxylic acid through oxidation of aldehyde, which is similar to fatty acid and decomposed into an acetyl Co-A by β-oxidation and by which fatty acid is shortened to two carbon atom compounds. Other oxidation pathways are terminal oxidation and subterminal oxidation. In both the terminal oxidation pathway, oxidation of both ends of the alkane molecule takes place through w-hydroxylation where terminal methyl group of fatty acid is converted into dicarboxylic acid by β-oxidation.

Anaerobic degradation of PHC is less explored as compared with aerobic degradation (Meckenstock et al., 2016). Hence, limited information is available on genes and enzymes involved in anaerobic degradation pathway. In anaerobic route, a variety of electron acceptors such as nitrate, ferrous iron, manganese, or sulfate ions are used by the microorganisms. Wilkes et al. (2016)

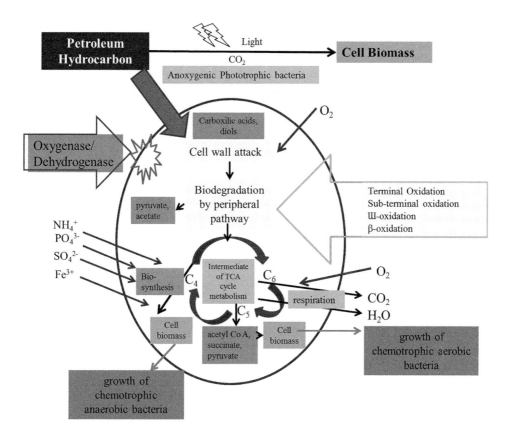

FIGURE 7.3 Schematic overview on molecular mechanism and potential pathway for petroleum hydrocarbons degradation by microorganisms.

proposed two biochemical mechanisms: addition of fumarate and carboxylation for initial activation of an alkane. Under anaerobic metabolism, aromatic fractions of PHC are first oxidized into phenol or organic acids and then transformed to long-chain volatile fatty acid, which is finally metabolized to CH_4 and CO_2 (Wilkes et al., 2016).

7.4.2 REMEDIATION OF CRUDE OIL–CONTAMINATED DESERT SOIL IN KUWAIT

7.4.2.1 Trial Experiment under Pilot-Scale Project

The journey started in 2004 with a small trial experiment conducted by TERI at Kuwait in a desert soil previously contaminated with crude oil. Extensive research had been carried out for the discovery of microorganisms (Lal and Khanna, 1996a, 1996b; Mishra et al., 2001; Bhattacharya et al., 2003; Sarma et al., 2004; Sarma et al., 2010) from the soil, which can potentially degrade all four fractions of crude oil. TERI has conducted larger field-scale bioremediation in different places where the soils were contaminated by crude oil (Mandal et al., 2012a, 2012b; 2012c). Subsequently, four bacterial strains have screened from Kuwait desert soil, and each bacterium could degrade each fraction of crude oil. These bacteria can withstand wide temperature ranges between 10°C and 50°C. Initially, 50 microbes were isolated from Kuwait desert soil contaminated by crude oil, and only 4 are highly potential for degradation of crude oil contamination. These four bacterial strains were further manipulated in laboratory condition to increase the potential for degradation of different fractions of crude oil. Subsequently, a cocktail of four bacteria is designed, and dry powdered formulation is prepared by adding inert carrier material. This cocktail formulation was scaled up further for bulk production. This product is named as "KT-Oilzapper." These KT-Oilzapper microbes are nonpathogenic in nature and not harmful to any living creatures. These cocktails were broadcasted into an experimental block (60 m × 60 m) containing approximately 1,800 m^3 of desert soil contaminated with crude oil under a demonstration project in Kuwait in the year of 2004. This bioremediation trial was continuing in this experimental block for 6 months. During 6 months, after-care maintenance had been carried out with continuous tilling and addition of water to maintain the air and moisture inside the soil for the survival of the of KT-Oilzapper microorganisms. Pilot-scale study was conducted into three blocks (blocks A, B, and C), and samples were collected from each block before application of KT-Oilzapper, which represents that baseline sample and subsequently other interim samples were collected after routine intervals in every month till the total PHC level reduced below the higher borderline of primary ecotoxicity remediation standard set by local environmental authority. Ten samples were collected from each block, and a composite sample is prepared after uniform mixing of ten samples. After collection, composite samples were tested for total PHC, water holding capacity, bulk density, organic carbon, and available phosphorus content in contaminated soil and macro- and micronutrients in the soil. After 6 months, the total PHC concentration reduced more than 90% and subsequently reached primary ecotoxicity remediation standard equivalent to 5,580 mg/kg. The concentrations of four fractions are significantly reduced in 6 months.

7.4.2.2 Large-Scale Bioremediation Project for Treatment of Crude Oil–Contaminated Soil

Based on the success achieved in the trial experiment, TERI was awarded a contract for a large-scale bioremediation project where the total volume that need to be treated was more than 2,00,000 m^3 of crude oil–contaminated soil. Mostly, the contaminated soil came from sludge pit, effluent pits, and contaminated soil piles from oil production. The baseline concentration of total PHC of the crude oil–contaminated soil ranges between 3% and 5% (30,000 and 50,000 mg/kg). TERI treated these entire contaminated soils through thermal desorption and bioremediation in 48 months. Out of 2,17,000 m^3 of crude oil–contaminated soil, approximately 1,60,000 m^3 of contaminated soil was treated by landfarming using KT-Oilzapper microbes. The end products of degradation are CO_2 and H_2O, which are not harmful to the environment. On the other hand, approximately 57,000 m^3 of contaminated soil was treated by thermal desorption, which is not at all environmentally friendly approach for the treatment of contaminated soil. Basically, thermal desorption is a widely accepted technology that provides a permanent solution at an economically competitive cost within a very short period; however, this technology is an energy-intensive process and damaging the soil properties in some extent. Moreover, thermal-based technology increases the carbon footprint in the atmosphere drastically, which causes global warming.

7.4.2.2.1 Physical Remediation of Total PHC-Contaminated Soil through Thermal Desorption

Thermal desorption is a thermally induced physical separation technique. Thermal desorption process is consisting of a two-step process, wherein the first step heat is applied to a contaminated material, such as soil, sediment, and sludge of filter cake to vaporize the contaminant into a gas stream (Gan et al., 2009; Troxler et al., 1993). The second step is treating the exhaust gas stream to prevent emissions of the volatilized contaminants to the atmosphere. In addition to volatilizing organic contaminants contained in the waste feed, moisture is volatilized and leaves with the off-gases. As a result of the thermal desorption, system also functions as a dryer. Nowadays, a variety of thermal desorption systems are used in numerous government and private remediation project across the globe. Thermal desorption units are available in two modes: continuous-feed and batch-feed mode. Continuous-feed systems are used globally for ex situ remediation of contamination (Uzgiris et al., 1995; Lighty et al. 1988). In this process, contaminated materials excavate from the contaminated site followed by some degree of civil work, including material handling, and then feed to the treatment unit. Continuous-feed thermal desorption unit is of two types: direct fired (DTU) and indirect fired (ITDU). The direct fired thermal desorption system is equipped with principal process elements, a rotary dryer, a fabric filter baghouse, and an afterburner in a sequence, whereas indirect fired thermal desorption unit comes with multiple configurations: One such system uses double shell rotary dryer with several burners mounted in the annular space between two shells. Another configuration is based on thermal screw conveyer. In direct fired thermal desorption process, no end products are generated, whereas in an indirect fired thermal process, liquid wastes are generated, which is an emulsion of oil and water and needs to be treated before discharge to any storage pit. Direct fired thermal desorption plant can only treat contaminated soil ranges between 2% and 3% (20,000–30,000 mg/kg), and feed soil needs to be uniformly mixed through all bucket for getting the best result. However, indirect fired thermal desorption unit can treat the contamination up to 7% (70,000 mg/kg).

7.4.2.2.2 Bioremediation of Total PHC-Contaminated Soil through Land Farming

A large volume of crude oil–contaminated soil received from different features, including sludge pit and effluent pits from oil production area in desert ecosystem, was characterized, excavated, transported into a bioremediation pad, measured the volume, and subsequently treated through land farming using KT-Oilzapper technology. KT-Oilzapper was grown at large scale in a 13 kl fermentor in India and transported to Kuwait at the site in a refrigerated container maintaining 2°C continuously. The duration between the production of KT-Oilzapper in India and application at the field in Kuwait was maintained 15–20 days. The desert ecosystem of Kuwait harshly damaged by Iraqi troops and, still after 28 years, unexploded ordnance are being found in several places embedded in the ground, and therefore, it is mandatory that all the features are to be cleared before starting any bioremediation work. The first step carried out for bioremediation was site characterization. This process helped preliminarily for calculating the contamination level and in the estimation of contaminated soil available at the site. Before starting site characterization, the site was protected by the installation of temporary fencing and temporary office facilities. Thereafter, radiological survey work, explosive ordnance disposal, and topographical survey work completed. Site characterization method helped in the estimation of the area and depth of contamination lying and the volume of free-phase oil on the top layer and a sludge layer beneath the free phase oil and bottom contaminated soil. Based on site characterization data and laboratory data set collected from site characterization sample, contaminated soils were excavated up to clean underneath layer and stored in a stockpile adjacent to the excavation area. The contaminated soil in stockpile was uniformly mixed, and the volume was measured before transported into a bioremediation pad. Once the contaminated soil has been removed until the gatch layer or clean soil found, then the base of the bioremediation area was prepared. The base layer comprised in certain locations with a gatch layer and other locations with a layer of compacted clean soil. After compaction of the clean base, a 10-cm-thick low contaminated soil layer was placed above the clean compacted gatch as a buffer layer before putting contaminated soil into the pad. The contaminated soil stockpiled adjacent to the pad was placed as a small heap within the bioremediation area over the underlying hard gatch layer. After placing the heap of contaminated soils in bioremediation pad area, a perimeter berm of approximately 40 cm height around the bioremediation plot was constructed keeping 2 m from the contaminated soil heaps. After placing the contaminated soil in the bioremediation area, zero time soil sampling was done before broadcasting KT-Oilzapper in the contaminated soil. After zero-day sampling, KT-Oilzapper and nutrient mixture recipe were applied to the contaminated soil and started bioremediation in that area, and then started aftercare maintenance of the sites to provide air and moisture. Tilling activity was conducted on a weekly basis to maintain air in the contaminated soil for the fast growing of the KT-Oilzapper bacteria. Water also sprinkled whenever required to maintain the moisture for growth of the KT-Oilzapper bacteria. Interim soil sampling event has done every month to monitor the reduction of the total PHC and fractionated. After 6 months, PHC concentration was reduced to below 0.5% or 5,000 mg/kg and achieved

(a)   (b)

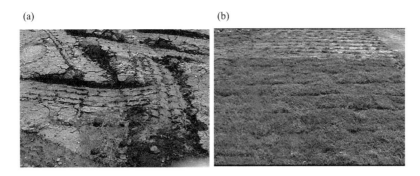

FIGURE 7.4   (a) Crude oil–contaminated site before bioremediation and (b) after bioremediation.

primary ecotoxicity remediation standard. In total, 71 batches were conducted in 3 years, and the bioremediation of 1,60,000 m³ of contaminated desert soil at a different location in Kuwait was completed. A total of 2,17,000 m³ of contaminated soils were treated through thermal desorption and bioremediation in less than 4 years (Figure 7.4). After bioremediation, treated soil was filled back into the excavated area. After successfully completed, sites were restored by mixing of customized biomanure in treated soil filled back at excavated area and amendment with some native plant varieties, including Compositae and Leguminosae family.

7.4.2.2.3   Challenges for the Large-Scale Bioremediation Project

Although land farming is considered as simplest bioremediation technology, major constraints associated with this technology at a larger scale are a requirement of large operating space, reduction of the microbial population due to heterogeneity in the environmental conditions, and the involvement of additional cost due to the huge excavation of contaminated soil in any major bioremediation project. Another real challenge in larger-scale field bioremediation is nonuniform concentration and distribution of the oily sludge into the contaminated soil, which could be possible to curb by huge excavation process that requires moderate to extensive engineering that implies more workforce and capital to succeed.

7.5   CONCLUSION

Crude oil contamination is a growing environmental problem worldwide, and field-based remediation approach is nowadays extensively used globally to tackle oil spills contamination at a larger scale. After the gulf war in 1990, crude oil–contaminated dry and wet oil lakes are still lying in the large desert area of Kuwait. Because of the Iraq invasion, Kuwait sustained significant and widespread environmental damages, including loss of habitat and disturbance to ecological equilibria. During the evacuation, the Iraqi troops set fire to oil wells in the desert oil field where a significant amount of toxic metals and carcinogenic constituents constantly released into the atmosphere for several months. In recognizing the need for large-scale environmental remediation and restoration in Kuwait, the United Nation Compensation Commission (UNCC) awarded approximately USD 3 billion in compensation to the State of Kuwait. Kuwait was awarded mainly for remediation of damages to groundwater resources, marine and coastal oil damage remediation, terrestrial resources restoration, oil lake remediation, marine shoreline reserves, and ordnance disposal site remediation. Under this compensation, 26 million m³ of surface soil and sludge need to be treated. Therefore, bioremediation technology would be expected to rise further in the next few decades for the treatment of huge turmoil in Kuwait. Only a few companies are available to offer soil remediation through bioremediation technology approaches, and this technology creates a huge market for remediation companies for the next few decades.

REFERENCES

Abbasian F, Lockington R, Mallavarapu M, Naidu R, 2015. A comprehensive review of aliphatic hydrocarbon biodegradation by bacteria. *Appl. Biochem. Biotechnol.* 1–30 DOI:10.1007/s12010-015-1603-5.

Al-Futaisi A, Jamrah A, Yaghi B, Taha R, 2007. Assessment of alternative management techniques of tank bottom petroleum sludge in Oman. *J. Hazard. Mater.* 141: 557–564.

Al-Gharabally D, Al-Barood A, 2015. Kuwait environmental remediation programme (KERP): Remediation demonstration strategy. SSPub.

Al-Gharabally D, Al-Barood A, Sudhakaran B, 2016. Kuwait environmental remediation programme (KERP): Limited site soil characterization in south east Kuwait. *IJRDO-J. Social Sci. Humanit. Res.* 1(10).

Balba MT, Al-Awadhi N, Al-Daher R, 1998. Bioremediation of oil contaminated soil: Microbiological methods for feasibility assessment and field evaluation. *J. Microbiol. Meth.* 32: 155–164.

Banat IM, 1995. Biosurfactants production and possible uses in microbial enhanced oil recovery and oil pollution remediation: A review. *Bioresour. Technol.* 51: 1–12.

Benyahia F, Embaby AS, 2016. Bioremediation of crude oil contaminated desert soil: Effect of biostimulation, bioaugmentation and bioavailability in biopile treatment systems. *Int. J. Environ. Res. Public Health.* 13: 219. DOI:10.3390/ijerph13020219.

Bhattacharya D, Sarma PM, Krishnan S, Mishra S, Lal B, 2003. Evaluation of genetic diversity among Pseudomonas citronellolis strains isolated from oily sludge-contaminated sites. *Appl. Environ. Microbiol.* 69(3): 1435–1441.

Duan L, Naidu R, Thavamani P, Meaklim J, Megharaj M, 2013. Managing long-term polycyclic aromatic hydrocarbon contaminated soils: A risk based approach. *Environ. Sci. Pollut. Res. Int.* 22(12): 8927–8941.

Gan S, Lau EV, Ng HK, 2009. Remediation of soils contaminated with polycyclic aromatic hydrocarbons (PAHs). *J. Hazard. Mater.* 172(2–3): 532–549. DOI:10.1016/j.jhazmat.2009.07.118.

Hu G, Li J, Zeng G. 2013. Recent development in the treatment of oily sludge from petroleum industry: A review. *J. Hazard. Mater.* 261: 470–490.

Kupusamy S, Thavamani P, Venkateswarlu K, Lee YB, Naidu R, Megharaj M, 2017. Remediation approaches for polyclic aromatic hydrocarbons (PAHs) contaminated soil: Technological constrains emerging trends and future directions. *Chemosphere* 168: 944–968.

Lal B, Khanna S, 1996a. Degradation of crude oil by Acinetobactercal coaceticus and Alcaligenes odorans. *J. Appl. Bacteriol.* 81: 355–362.

Lal B, Khanna S, 1996b. Mineralization of [14]C Octacosane by Acinetobactercal coaceticus S30. *Can J Microbiol* 42: 1225–1231.

Leahy JH, Colwell R, 1990. Microbial degradation of hydrocarbons in the environment. *Microbiol. Rev.* 54(3): 305–315.

Lighty JS, Pershing DW, Cundy VA, Linz DG, 1988. Characterization of thermal desorption phenomena for the clean-up of contaminated soil. *Nuclear Chemical Waste Manage.* 8(3): 225–237. DOI:10.1016/0191-815X(88)90030-7.

Mandal A, Sarma PM, Jeyaseelan CP, Channashettar VA, Singh B, Lal B, Datta J, 2012a. Large scale bioremediation of petroleum hydrocarbon contaminated waste at indian oil refineries: Casestudies. *Int. J. Life Sci. Pharma Res.* 2(4): L114–L128.

Mandal A, Sarma PM, Singh B, Jeyaseelan CP, Channashettar VA, Lal B, Datta J, 2012b. Bioremediation: An environment friendly sustainable biotechnological solution for remediation of petroleum hydrocarbon contaminated waste. *IJREISS* 2(8): 1–18.

Mandal A, Sarma PM, Singh B, Jeyaseelan CP, Channashettar VA, Lal B, Datta J, 2012c. Bioremediation: An environment friendly sustainable biotechnological solution for remediation of petroleum hydrocarbon contaminated waste. *ARPN J. Sci. Technol.* 2: 1–12.

Meckenstock RU, Boll M, Mouttaki H, Koelschbach JS, Tarouco PC, Weyrauch P, Dong X, Himmelberg AM, 2016. Anaerobic degradation of benzene and polycyclic aromatic hydrocarbons. *J. Mol. Microbiol. Biotechnol.* 26: 92–118.

Mishra S, Jyoti J, Kuhad RC, Lal B, 2001. Evaluation of inoculum addition to stimulate in situ bioremediation of oily-sludge-contaminated soil. *Appl. Environ. Microbiol.* 67(4): 1675–1681.

Okeke PN, Obi C, 2013. Treatment of oil drill cuttings using thermal desorption technique. *ARPN J. Syst. Software.* 3(7): 153–158.

Olajire AA, 2014. The petroleum industry and environmental challenges. *J. Pet. Environ. Biotechnol.* 5: 4. DOI:10.4172/2157-7463.1000186.

Panno MTD, Morelli IS, Engelen B, Berthe-Corti L, 2005. Effect of petrochemical sludge concentrations on microbial communities during soil bioremediation. *FEMS Microbiol. Ecol.* 53: 305–316.

Pornsunthorntawee O, Wongpanit P, Chavadej S, Abe M, Rujiravanit R, 2008. Structural and physicochemical characterization of crude biosurfactant produced by Pseudomonas aeruginosa SP4 isolated from petroleum contaminated soil. *Bioresour. Technol.* 99: 1589–1595.

Saeki H, Sasaki M, Komatsu K, Miura A, Matsuda H, 2009. Oil spill remediation by using the remediation agent JE1058BS that contains a biosurfactant produced by Gordonia sp. strain JE-1058. *Bioresour. Technol.* 100: 572–577.

Salleh AB, Ghazali FM, Rahman RNZA, Basri M, 2003. Bioremediation of petroleum hydrocarbon pollution. *Indian J. Biotechnol.* 2: 411–425.

Sarma PM, Bhattacharya D, Krishnan S, Lal B, 2004. Degradation of polycyclic aromatic hydrocarbons by a newlydiscovered enteric bacterium, *Leclerciaadecarboxylata. Appl. Environ. Microbiol.* 70(5): 3163–3166.

Sarma PM, Duraja P, Deshpande S, Lal B, 2010. Degradation of pyrene by an enteric bacterium, Leclerciaadecarboxylata PS4040. *Biodegradation* 21: 59–69. DOI:10.1007/s10532-009-9281-z.

Schlumberger, 2010. Schlumberger Oilfield Glossary.

Troxler WL, Cudahy JJ, Zink RP, Yezzi Jr JJ, Rosenthal SI, 1993. Treatment of nonhazardous petroleum-contaminated soils by thermal desorption technologies. *Air Waste* 43(11): 1512–1525.

Uzgiris EE, Edelstein WA, Philipp HR, Timothy Iben IE, 1995. Complex thermal desorption of PCBs from soil. *Chemosphere* 30(2): 377–387. DOI:10.1016/0045-6535(94)00404-I.

Varjani SJ, 2017. Microbial degradation of petroleum hydrocarbons. *Bioresour. Technol.* 223: 277–286.

Varjani SJ, Upasani VN, 2016. Carbon spectrum utilization by an indigenous strain of Pseudomonas aeruginosa NCIM 5514: Production, characterization and surface-active properties of biosurfactant. *Bioresour. Technol.* 221: 510–516.

Wilkes H, Buckel W, Golding BT, Rabus R., 2016. Metabolism of hydrocarbons in n-Alkane utilizing anaerobic bacteria. *J. Mol. Microbiol. Biotechnol.* 26: 138–151.

Zhang J, Li J, Thring RW, Hu S, Song X, 2012. Oil recovery from refinery oily sludge via ultrasound and freeze/thaw. *J. Hazard. Mater.* 203–204: 195–203.

8 Microbial Processes for Treatment of e-Waste Printed Circuit Boards and Their Mechanisms for Metal(s) Solubilization

Shailesh R. Dave
Loyola Centre of Research and Development

Asha B. Sodha and Devayani R. Tipre
Gujarat University

CONTENTS

8.1   INTRODUCTION

People want more and more comfort and sophistication, which leads to rapid urbanization and industrialization; these have resulted in increased demand for electronic and electrical gadgets. With fast technological advancement and development, due to the demand of people, industries are now manufacturing novel, superior, and smart electronic and electrical equipment (EEE) at an alarming rate. Moreover, the world population is also increasing so they too need technologically advanced EEE products; thus, a large number of latest products are produced, purchased, used, and discarded, resulting in the generation of huge amount of waste electronic and electrical equipment (WEEE) (Tansel 2017). Discarded, end-of-life, broken as well as obsolete electrical and electronic devices are referred as electronic waste (e-Waste) or WEEE; it is one of the fastest accumulating wastes globally (Akinseye 2013). According to Basel Action Network, a wide range of electronic appliances such as refrigerators, airconditioners, cellphones, stereo systems, computers, audio equipment, printers, and items are used in these devices and consumable for the same if discarded or they have reached their end of the life cycle are considered as e-Waste (Gaidajis et al. 2010). According to the guidelines provided by the Government of India, "E-Waste comprises of waste generated from used electronic devices and household appliances which are not fit for their original intended use and are destined for recovery, recycling or disposal. Such wastes encompass a wide range of electrical-electronic devices such as computers, hand-held cellular phones, personal stereos, including large household appliances including refrigerators, air-conditioners etc." (CPCB 2008). The electronic waste (e-Waste) normally consists of small and large equipment such as LED lamps, cell phones, smart mobile phones, superior televisions, refrigerator, printers, driers, temperature exchangers as well as advanced computing devices (Bhattacharya and Khare 2016). Due to the fast advancement in technology and tremendous market growth, most of these electronic and electric materials have a very short life span, which resulted in the fastest accumulation in e-Waste (Sodha et al. 2017). Broadly, the end-of-life EEE without the intent of reuse is considered as e-Waste. As per the information available in literature in 2014, about 41.8 million tons of e-Waste was generated globally, and as per the estimation, e-Waste is growing at the rate of 3%–5% annually, which leads to about 49.8 million tons of e-Waste generation in 2018 (Cui and Anderson 2016). E-Waste constitutes nearly 8% of municipal waste (Wei and Liu 2012). These end-of-life electronic products from the consumers and defective products at the manufacturing end itself are nonscientifically disposed as land fill or incineration or recycling ends up in e-Waste in the environment. The path of e-Waste to the environment is shown in Figure 8.1.

The large amounts of complex and diverse e-Waste generated have an adverse impact on the environment as well as the human being if they are not properly managed. Unfortunately, there is no

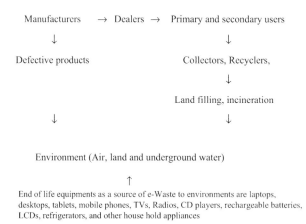

FIGURE 8.1   Sources/path of e-Waste to environment.

proper system to collect, store, transport, treat, and dispose of e-Waste. Thus, it has become a major issue of concern for the industries, government and nongovernment organization, and even the public to protect the environment from the hazardous effect of improperly managed e-Waste (Kaya 2016).

Most of us know that e-Wastes have been accumulated since many years, but their devastating environmental effects are realized recently, which has created keen awareness in the scientific community as well as in common people throughout the world, and it has forced concerned people for their proper treatment. E-Waste has been accumulated for many years; it contains large number of toxic and hazardous pollutants such as mixture of Ag, Al, As, Au, Be, Bi, Cd, Cr, Cu, Fe, Hg, Ni, Pt, Sb, Si, Zn, and rare earth metals along with flame retardants, halogens, combustible plastics, chlorofluorocarbons, polycyclic aromatic hydrocarbons, polybrominated diphenyl ethers, and dioxin-like components (Pradhan and Kumar2014; Dave et al. 2018).

PCBs are the core component of any electronic devices; they are used for the smooth, fast, and convenient functioning of small to large electronic devices. They are composed of nearly 40% metals, 30% plastics, and 30% ceramic; metal content varies from e-Waste to e-Waste depending on the brand and also on manufacturing time (Shah et al. 2014). As PCBs contain several metals as well as various organic pollutants, it is not advisable to dump them as landfills or incinerate them. However, incineration and landfilling are the common methods adapted for e-Waste management, which leads to the release of toxic gases in the atmosphere and highly harmful metals in soil and ground waters. These pollutants are accumulated and transported in plant systems, which ultimately reach to animals and humans (Awasthi et al. 2016). As per the view of the Global E-waste Monitor (Baldé et al. 2017), e-Waste shows the presence of about 60 elements of the periodic table, and many of these elements can be recovered. Due to health and environmental awareness and the presence of a large content of metals, e-Waste is considered as an alternative resource of metals or an urban mine. Thus, e-Waste management has become a vital and significant field of research throughout the world. This chapter mainly focuses on treatment of PCBs with special reference to the extraction of metals from waste PCBs using microbial technology.

8.2   WASTE ELECTRONIC AND ELECTRICAL EQUIPMENT

According to the United Nations Environment Programme (UNEP), across the globe, 20–50 million metric tons of electronic waste is generated per year with the United States being the largest producer (Pradhan and Kumar 2014). E-Waste production is increasing at the rate of 3%–5% annually. Developed countries are exporting nearly 50%–80% of their e-Waste to Asian and African countries (Dave et al. 2016a). Asian countries such as India and China are the major countries that get the e-Waste from all over the world for recycling purpose, as the developing countries have no strict legislation rules for e-Waste disposal (Dave et al. 2016a). In Asia, a total of 18.2 Mt e-Waste was generated in 2016. China generates the highest e-Waste of 7.2 Mt in Asia and in the world (Baldé et al. 2017). India is ranking 5th in terms of e-Waste production, and the production is growing by 15% every year and will reach 1.04 million tons by 2019. The adverse impact of pollutants present in e-Waste on various human organs and systems as well as on the environment is clearly stated by Kumar et al. (2013).

Metals are widely used in the manufacturing of electronic appliances due to their high chemical stability and conducting properties. According to the European Union (EU 2000), any appliance using an electric power supply that has reached its end of life would come under e-Waste. All the electronic appliances and most of the electric equipment have printed circuit boards (PCBs) as a core part, which controls the functioning of the equipment. A huge share of the precious metals is found to be present on PCBs (Sodha et al. 2017; Kiddee and Naidu 2013). On average, one metric ton of PCBs contains between 80 and 1,500 g of gold and 160 and 210 kg of copper. These amounts are 40–800 times more gold and 30–40 times more copper as compared with that are present in its ores (Isildar et al. 2018). PCBs also contain several another base, precious, and rare earth metals. Thus, PCBs are referred to as a valuable secondary source of raw materials for these metals (Sodha et al. 2017; Isildar et al. 2018). PCBs represent 3 wt% of total e-Waste. Due to such a diversified and rich source of metals, PCBs are considered as urban mines to recover metals and illustrate as an attractive fraction for the metal extraction, recycling, and recovery.

PCBs are the flat surface, on which electric components such as conductor chips and capacitors are mounted. They act as a supporting material and connect the chips using a conductive path and signals through copper sheets present on the surface of PCBs, which are mainly laminated by epoxy (Dave et al. 2016a; Hadi et al. 2015). There are different FR (flame retardant) standards used in the manufacturing of different types of PCBs. Another name for PCBs has printed wiring boards (PWBs). PCBs have many diversified physical appearances in terms of shape, size, epoxy coating, soldering, aluminum foil cover, etc. (Figure 8.2). PCBs have a wide variation in metal content and type of metals present (Dave et al. 2016a; Xue et al. 2015); some of these variations are depicted in Table 8.1. However, the content of copper present is the highest. The heterogeneous mixture of

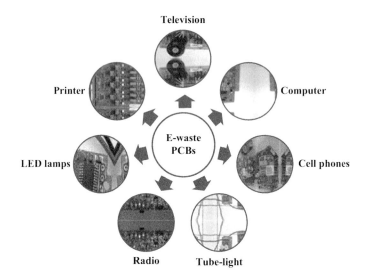

FIGURE 8.2   PCBs of various waste electrical and electron equipment.

TABLE 8.1
Metal Contents of Various Types of e-Waste PCB (Dave et al. 2018, 2016a)

Metals	Solubilized Metals in Various e-Waste PCB (mg/g)				
	Mobilephone	Computer	Television	LX	Tubelight
Cu	360.00	300.00	118.25	64.25	167.75
Zn	7.96	37.00	19.27	1.23	22.70
Ni	8.55	3.84	13.00	0.62	1.48
Al	6.66	45.93	56.27	14.28	53.22
Pb	12.07	136.5	154.80	133.70	80.50
Fe	10.50	60.12	64.00	9.46	69.75
As	4.34	7.82	5.33	5.52	1.65
Cr	0.59	1.61	1.10	0.93	1.21
Au	0.10	0.14	ND	ND	ND
Ag	0.28	0.23	0.50	0.22	0.22
Pd	0.64	0.27	0.37	0.27	0.68
Cd, Co, Se, K,	BDL	BDL	BDL	BDL	BDL

ND = not detected; BDL = below detectable limit.

toxic and heavy metals makes them a challenging task from a recovery point of view. So, in spite of high metals content and the toxic materials present in them, available environmentally friendly, economically viable separation techniques are limited for PCBs.

8.3 METHODS OF E-WASTE PROCESSING FOR METAL RECOVERY

Management of e-Waste is a critical problem; the simplest, oldest, and most common method used for disposal of e-Waste is land filling. Here, nearly 70% of e-Waste is disposed of in waste-land, which results in leaching of metals and other harmful compounds in surrounding land and underground water; such leaching causes damage to the soil and water as well as evaporation of some materials also pollutes the air (Awasthi et al. 2016; Dave et al. 2016a). This way of management is dishonored due to environmental concern. Another traditional and commonly used strategy for e-Waste management is incineration. In this method, e-Waste is allowed for combustion at an extremely high temperature to remove nonmetallic parts. The process results in the generation of poisonous gases from polychlorinated dibenzo-p dioxins/polychlorinated dibenzofurans, poly-brominated dibenzo-p dioxins, and dibenzofurans, which pollute the atmosphere. Both of these processes are responsible for the deterioration of the environment and harmful to living beings. At the same time, we lose many base and precious metals in both methods of management. So currently, some other treatment methods are practiced for e-Waste treatment. The pyrometallurgical process is a conventional, major, and quite an old approach for the recovery of nonferrous base and precious metals (Isildar et al. 2018). However, this process requires 800°C–1,350°C temperature; it is cost-intensive and also needs highly qualified persons to operate the process. Recently, the process is modified, and vacuum pyrolysis is employed, but it has the problem of using high temperature. The second widely used process is hydrometallurgy; it results in high metal recovery as compared with pyrometallurgy. In contrast to pyrometallurgy, hydrometallurgy is cheaper, generates less pollution, and can be operated even for a small scale (Cui and Anderson 2016). Apart from these two processes, multiple crushing, magnetic separation, and corona electrostatic separation are also employed for material recovery from e-Waste (Xue et al. 2015). But none of these processes has solved the problem of e-Waste processing in terms of recovery of metals, sustainability of environment, and achievability. In the past couple of decades, attention is diverted to a biological process for metal recovery from the e-Waste; the process is referred to as biohydrometallurgy or bioleaching.

8.4 METALS BIOEXTRACTION PROCESS

Bioleaching of metals is defined as mobilization of metals from their immobilized (insoluble) form from solid materials by the action of microorganisms or the metabolites produced by them in solution form (soluble). For more than 50 years, it is studied for extractions of copper from lean grade ores or tailing, and it is now applied for extraction of gold, cobalt, molybdenum, etc. It is also used for metal solubilization from concentrates (Tipre and Dave 2004; Dave and Tipre 2011). The process is mainly used for the treatment of sulfide minerals. Several dumps and reactor bioleaching processes are operating at production level for copper as well as gold recovery from the minerals (Tipre and Dave 2004). In the past 10–15 years, bioleaching is investigated for the treatment of e-Waste for metal recovery. E-Wastes are principally different from sulfide minerals. In e-Waste, metals are present in their zero valent form, whereas in the case of sulfide mineral, they are present in the form of metal sulfides along with pyrite or sulfur. Metal sulfides, pyrite, and/or sulfur serve as an energy source for iron- and sulfur-oxidizing (chemolithotrophic) organisms. During one-step process, added microorganisms (inoculum) in the leaching system utilize these as substrates and constantly generate ferric iron and/or proton as lixiviants. These lixiviants act upon metal sulfides, and metal gets solubilized. In case of e-Waste, no such substrates or energy sources are available, so it is necessary to add ferrous iron, pyrite, or sulfur in the leaching system; from the added substrate/s organism generates required lixiviants, which act upon metals and solubilized them in aqueous solution. Metal bioextraction from e-Waste is performed by one-step bioleaching, two-step bioleaching, and spent medium bioleaching process (Dave et al. 2018).

8.4.1 One-Step Bioextraction Process

In case of one-step bioleaching process, both actively growing inoculum and e-Waste are added in the suitable leaching medium; the added organisms oxidized the ferrous iron or sulfur and continuously generated ferric iron or proton, which gradually solubilizes metals. The presence of toxic organic compounds and soluble metals present or generated during the process exerts inhibitory effects on organisms (Dave et al. 2016a; Pradhan and Kumar 2012). This results in a decrease in microbial growth and activity or even inhibition of the process. Such inhibitory effects restrict the one-step bioleaching process to operate at low pulp density. As can be seen from the available literature, most of the processes are carried out in the range of 1%–5% (w/v) e-Waste or at the most at 10% (w/v) e-Waste (Liang et al. 2013). Addition of low pulp density results in a low concentration of metal extraction in the medium, and it requires a larger reactor size for the process, that is, low vessel size utilization. Moreover, in one-step process, it is difficult to satisfy the required suitable conditions simultaneously both for the growth of the organisms and for the metal extraction; this results in low metal extraction and slow rate also.

8.4.2 Two-Step Bioextraction Process

In the case of two-step bioleaching process, desired organisms are grown in their respective medium under their optimum growth conditions as the first step of the process. When the desired growth of the organism and lixiviant concentration (ferric iron, proton, cyanide) achieved, e-Waste is added as the second step of the process. In this process, organisms are grown separately in the absence of e-Waste, so the toxicity problem of e-Waste is eliminated, and moreover, desired optimum conditions for the growth can be provided to generate concentrated lixiviants at a faster rate. In the second step, optimum condition of leaching is provided, so it results in better metal extraction as compared with the one-step process. Two-step bioleaching process is found to be more attractive, as it results in faster and higher metal extraction (Sand et al. 2001). Further more, it is also possible to run the process at high pulp density in contrast to one-step bioleaching process. During the second step, generated ferric iron/organic acids/hydrocyanide (HCN) is consumed in metal extractions. In case of ferric iron due to metal extraction, it is converted to ferrous iron. The produced ferrous iron gets re-biooxidized, and ferric iron is produced, which once again serves as a lixiviant for solubilization of metals. In the case of two-step process, a large quantity of ferric iron/proton/cyanides is available when e-Waste is added, which immediately reacts with e-Waste, and metals get solubilized at the possible fastest rate.

8.4.3 Spent Medium Process

In the case of the spent medium process, as described in a two-step process, organisms are allowed to grow under optimum conditions to get maximum growth and lixiviants. After the growth phase, organisms are removed, and supernatant, which is devoid of microbes referred to as spent medium, is separated followed by addition of e-Waste in the collected spent medium. Produced ferric iron/organic acid/HCN solubilizes the metals. In the case of gold extraction, spent medium process is found to be more useful, as it results in higher gold extractions as compared with the two-step process. In this process, as organisms are removed from the spent medium, more oxygen is available for gold leaching as compared with the two-step process. In case of HCN production, the organisms produce HCN in late logarithmic or early stationary phase of their growth, and if the organisms are not removed, they metabolized available HCN into β-cyanoalanine, so the amount of available HCN decreases, which results in low gold extraction in the two-step process as compared with this process. Such negative effect may not be occurring in case of ferric iron or proton production system; thus, when iron or sulfur oxidizers are used, two-step process gives better metal extraction as compared with spent medium process. On the other hand, use of spent medium method may be preferred for gold extraction if the cyanogenic bacterial process is used (Dave et al. 2018, 2016a).

8.5 FACTORS INFLUENCING E-WASTE BIOTREATMENTS

The bioleaching process is mainly governed by the production of ferric iron/protons/organic acids and HCN by the organisms. Productions of all these lixiviants are dependent on the growth of the organisms, so factors affecting the growth are playing a significant role in metal extractions. The major factors that affect the growth and thus metal ion extraction are pH, temperature, dissolved oxygen, pulp density, the concentration of substrate, type of the culture, medium, particle size of the e-Waste, and addition of surfactants and complexing agents (Shah et al. 2015). Organisms used in bioleaching processes are aerobic organisms; thus, aeration or agitation plays a significant role in metal extractions. The cyanogenic mediated process is operating at alkaline pH, whereas in case of fungi, mild acidic pH is preferred, and in case of iron and sulfur oxidizers, pH below 2.0 results in more metal extraction as compared with pH above 2.0. Application of high pulp density is desired for economic extractions of metals, but it exerts inhibitory effects on the organisms in one-step process, which restrict the process above certain pulp density. Most of the reported e-Waste bioleaching processes have used less than 10% w/v pulp density. Shah et al. (2015) and Ilyas et al. (2014) have studied the process at 15% w/v and higher pulp density. As compared with the one-step process, two-step bioleaching or spent medium bioleaching process can be operated at higher pulp density. The positive influence of the addition of a surfactant and complexing agent for metal extraction from e-Waste is highlighted by Lan et al. (2009). Shah et al. (2015) have also reported the effect of various chelating agents and pointed out the benefit of ethylenediamine disuccinic acid (EDDS) as compared with use of ethylene diamine tetraacetic acid (EDTA). The added chelating agent forms complexation of ferricchelatemetal that helps in keeping a more ferric iron in solution, thus resulting in better extractions (Vlivainio 2010). In most of the bioleaching processes, metal extraction is reported to be inversely proportional to the particle size of the e-Waste used, but in some experiments with our experience, the results were different, and in case of certain e-Waste, critical selection of particle size is essential. Irrespective of the e-Waste or the type of process used, the metal extraction is directly proportional to a number of lixiviants produced. Thus, it is preferred to keep these lixiviants at the desired optimum concentration to achieve the best possible metals extractions (Shah et al. 2015; Pant et al. 2012).

In a one-step process, if the total pulp density is not added as one lot, but it is added in fractions as two/three/four lots (fed-batch process), which shows better metal extractions as compared with addition in one lot (Shah et al. 2015). Normally, use of the small size of e-Waste is preferred for high metal extraction, but it requires grinding, which increases process cost, generates air pollution, makes solid separation difficult after leaching process, and generates more sludge. In several e-Wastes, when PCBs are used without grinding with pretreatment, it gives more metal extractions and reduces many of the indicated problems of small size application (unpublished data). When large PCBs are used without grinding, after the leaching process, the PCBs glass epoxy plates can be used for various applications. Use of membrane-based techniques in bioleaching process for e-Waste may reduce the problems arising due to use of smaller particles of e-Waste. The authors have studied that cheaper membranes prepared from clothes for entrapping ground waste during bioleaching showed better metal extraction and very easy separation of leached e-Waste solids from the solution (Dave et al. 2016b).

8.6 ROLE OF MICROORGANISMS IN METAL EXTRACTION

Microorganisms are playing a significant role in the extraction of metals from tailings, ores, concentrates, and e-Waste, and most of these processes are working at ambient temperature and pressure with mild conditions; thus, they are easier to be regulated and maintained (Erüst et al. 2013). Colmer and Hinkle isolated acidophilic autotrophic iron- and sulfur-oxidizing bacteria from acid mine drainage in 1947 and provided the first scientific evidence on the role of microorganisms in metal solubilization (Dave and Tipre 2011). In the early 1950s, Colmer was the first to extract copper from mine dump at Kennecott Copper Corporation through the action of microorganisms (Zimmerley et al. 1958). Copper, gold, and cobalt are extracted from their respective minerals at their industrial scale applying microbial technologies. Apart from copper and gold leaching, now encouraging results are also available for extraction of Co, Ga, Mo, Ni, Pb, Zn, and platinum group of metals by biomining (Lee and Pandey 2012). Use of molecular biology techniques has expanded our knowledge on detection and identification of organisms as well as in understanding the role of metal-microbial interaction in metal extraction and recovery. Because of advancement in molecular biology techniques, today, evidence is available that microorganisms belonging to all the three domains of life that are archaea, bacteria, and eukaryote have potential role and application in biomining (Ehrlich 2001; Vera et al. 2013). The types of organisms playing significant roles in metal extraction are iron and sulfur oxidizers, cyanogenic organisms, and organic acid producers.

8.6.1 Iron and Sulfur Oxidizer

The most widely studied iron and sulfur oxidizers for sulfide minerals' leaching are *Acidithiobacillus ferrooxidans*, *Acidithiobacillus thiooxidans*, and *Leptospirillum ferrooxidans*. However, microorganisms responsible for ferrous iron, sulfur, or reduced sulfur compound oxidations are found in four phyla, namely, Proteobacteria, Nitrospirae, Firmicutes, and Actinobacteria. Phylum Proteobacteria include *Acidithiobacillus*, *Acidiphilum*, *Acidiferrobacter*, and *Ferrovum*. *Leptospirillum* belongs to Nitrospirae. Species of *Alicyclobacillus* and *Sulfobacillus* are included in phylum Firmicutes; on the other hand, *Ferrimicrobium*, *Acidimicrobium*, and *Ferrithrix* are Actinobacteria. This group of bioleaching organisms is also found in the domain Archaea, which includes *Sulfolobus*, *Acidianus*, *Metallosphaera*, and *Sulfurisphaera* as a member of Crenarchaeota. Two species *Ferroplasma acidiphilum* and *Ferroplasma acidarmanus* belong to Euryarchaeota (Dave et al. 2018; Dave and Tipre 2011; Vera et al. 2013). With response to the optimum temperature of growth, these organisms are referred to as mesophiles, moderate thermophiles, and thermophiles. There is no precise boundary of temperature for the growth; for example, *Acidimicrobium ferrooxidans* and *Acidithiobacillus caldus* grow between temperature 25°C and 55°C, which is the temperature range of neither mesophiles nor moderate thermophiles. When substrates for growth are considered, they are grouped as iron oxidizers, sulfur oxidizers, and the third group, which oxidizes both iron and sulfur. Among these organisms, *A. ferrooxidans* is highly diverse and lives aerobically by utilizing both iron and reduced inorganic sulfur compounds. It also functions in the absence of air and shows anaerobic growth by the oxidation of hydrogen and sulfur coupled with the reduction of ferric iron (Johnson 2012). When the ferric/ferrous ratio is high in the medium *L. ferrooxidans* and *Leptospirillum ferriphilum* over grow *A. ferrooxidans* and play a dominant role in ferric iron generation leads to generate high ferric iron containing lixiviant, these organisms play a predominant role in the extraction of metals from waste PCBs.

8.6.2 HCN Producers

Since 1898, the cyanidation process has been known for the extraction of gold and silver from gold-containing minerals (Smith and Mudder 1991). Several bacteria and fungi are reported for HCN production; among these few organisms capable of HCN production are reported for gold extractions from e-Waste. Bacteria such as *Chromobacteriun violaceum*, *Pseudomonas fluorescens*, *Pseudomonas plecoglossicida*, and *Pseudomonas aeruginosa* are reported for gold solubilization due to HCN production. Some members from fungi such as *Marasmius oreades* and species of *Clitocyde* and *Polysphrous* are reported for HCN production, so they could be used for gold and silver extraction from e-Waste (Dave et al. 2018, 2016a).

8.6.3 Organic Acid Producers

Fungi such as *Aspergillus niger* and *Penicillium simplicissimum* are reported to grow at low pH and produce organic acids, such as citric acid, oxalic acid, malic acid, and gluconic acid. These acids serve as lixiviant and help in solubilization of various base metals from e-Waste. Considering the ability of such fungi, which modify metal speciation and increase the mobility of the metals, they could offer a potential alternative method of metal extraction from e-Waste (Dave et al. 2018, 2016a).

8.7 MECHANISMS OF METAL EXTRACTIONS FROM E-WASTE

8.7.1 Ferric and Proton Generation

The bioleaching of metal sulfides by *A. ferrooxidans* was initially described by two mechanisms: direct and indirect mechanism. However, it is accepted that direct mechanism does not exist and sole mechanisms of metal extractions are referred as an indirect mechanism, now which is more clearly explained by the largely acceptable terminology, contact, and noncontact mechanisms (Rohwerder and Sand 2007). For the contact mechanisms, the production of extracellular polymeric substances (EPSs) is very essential for the contact and filling up the space between the cell wall and the surface of the substrate. In e-Waste, specifically, PCB metals are not present in the form of metal sulfides, but they are present at zero valent metal, as for example, Cu^0, Zn^0, Ni^0, etc. Iron-oxidizing organisms oxidize ferrous sulfate to ferric sulfate (Eq. 8.1). The biologically generated ferric iron reacts with Cu^0, Zn^0, and Ni^0 present on e-Waste and converts them to Cu^{2+}, Zn^{2+}, and Ni^{2+} as shown in Eqs. (8.2–8.4). This step results in the reduction of ferric iron to ferrous iron again, and the processed copper, zinc, and nickel are converted to their respective sulfates. As sulfates of these metals are water soluble, they get solubilized. The delta energy showed in various equations indicates the thermodynamic feasibility of the reactions (Eqs. 8.2–8.4).

$$4Fe^{2+} + O_2 \xrightarrow{\text{Fe-oxidizer}} 4Fe^{3+} + H_2O \tag{8.1}$$

$$2Fe^{3+} + Cu^0 \rightarrow 2Fe^{2+} + Cu^{2+} \qquad \Delta G = -82.9\,\text{kJ/mol} \tag{8.2}$$

$$2Fe^{3+} + Zn^0 \rightarrow 2Fe^{2+} + Zn^{2+} \qquad \Delta G = -295.4\,\text{kJ/mol} \tag{8.3}$$

$$2Fe^{3+} + Ni^0 \rightarrow 2Fe^{2+} + Ni^{2+} \qquad \Delta G = -196.6\,\text{kJ/mol} \tag{8.4}$$

On the other hand, when sulfur or reduced sulfur compounds oxidizing organisms are present, they stimulate the reaction (Eq. 8.5) and produce protons. In the presence of protons, copper, zinc, and nickel get solubilized. Solubilization of copper from PCBs in the absence of iron takes place through oxidation of elemental sulfur, which provides required protons for copper solubilization. However, in this case, the presence of molecular oxygen is required (Eq. 8.6). The most interesting and useful part is that all these reactions are taking place at ambient temperature and pressure under mild acidic environment, which leads to eco-friendly environment-sustainable microbial technology for metal extraction from waste PCBs (Dave et al. 2018).

$$S^0 + 1.5O_2 + H_2O \xrightarrow{\text{microbes}} SO_4^{2-} + 2H^+ \tag{8.5}$$

$$2Cu^0 + 4H^+ + O_2 \rightarrow 2Cu^{2+} + 2H_2O \tag{8.6}$$

8.7.2 HCN Generation

Cyanide forms a water-soluble complex with gold, and it is responsible for gold extraction. The use of biogenic cyanide provides an alternative, eco-friendly gold extraction process, which can be referred alkaline bioleaching or heterotrophic bioleaching process. HCN is generated by diverse heterotrophic organisms. The enzyme involved for gold recovery is called as HCN synthase. Normally, HCN is produced between the end of an exponential phase and early stationary phase during the growth of *C. violaceum* and at the beginning of the stationary phase by *Pseudomonas*. In the case of *Chromobacterium*, the produced HCN is detoxified during the late stationary phase by transforming it to β-cyanoalanine (Dave et al. 2018, 2016a; Knowles and Bunch 1989; Brandl et al. 2008). The dissolution of gold can be expressed as following anodic (Eq. 8.7) and cathodic (Eq. 8.8) reactions. The overall reaction is known as Elsner's equation (Eq. 8.9).

$$4Au + 8CN^- \rightarrow 4Au(CN)^{2-} + 4e^- \tag{8.7}$$

$$O_2 + 2H_2O + 4e^- \rightarrow 4OH^- \tag{8.8}$$

$$4Au + 8CN^- + O_2 + 2H_2O \rightarrow 4Au(CN)^{2-} + 4OH \tag{8.9}$$

In case of PCBs, before giving the cyanogenic bacterial treatment, it is essential to treat PCBs with iron and sulfur bacteria for removal of copper and other metals; otherwise, these metals interfere in the process of gold extractions by consuming the HCN. The presence of oxygen plays an important role in the dissolution of gold, which is normally consumed rapidly during the growth phase of organisms, resulting in a high decrease in solubilization of gold.

8.7.3 Organic acid Production

Several fungi are found to metabolize carbohydrate present in the medium, and as a metabolic product, they generate various organic acids found in Krebs cycle (Eq. 8.10);these produced acids can attack mineral surface by acidolysis and solubilized the metals as shown in Eq. 8.11 (Saidan et al. 2012) (43). As compared with iron and sulfur bacteria, metal leaching is possible by these fungi at less acidic pH, which further minimizes environmental risk (Dave et al. 2018). Addition of some of these organic acids during metal mobilization by iron-oxidizing bacteria favors better metal solubilization and keeps ferric iron for more time in solution form.

$$\text{Sugar} \xrightarrow{\text{Aspergillus niger}} \text{orgainc acids} \tag{8.10}$$

$$Cu^0 + \text{citric acid} \rightarrow \text{copper citrate} + H^+ \tag{8.11}$$

8.8 BIORECOVERY OF METALS FROM LEACHATE

Bioleaching of e-Waste generates metal-loaded aqueous solutions. These metals are positively charged, and bacterial cell wall is negatively charged; thus, almost all bacterial biomasses are used as biosorbent for the sorption of metals. Bacterial, yeast, fungal, and plant biomass alive or dead are used for metal recovery from solution. Several mechanisms and many types of biomasses are involved in metal recovery from metal-loaded solution. It is a very wide topic, and it cannot be covered in this chapter. This topic needs a separate chapter to give proper justification.

8.9   FUTURE ASPECTS

Biomining is a well-known process for extraction of metals from ore and concentrates. It is a safe, green environmentally friendly technology. However, use of this process is not well documented in e-Waste bioleaching field as compared with other disciplines, mainly because the handling of a live system is difficult as compared with chemical and mechanical treatments. The prevailing conventional e-Waste treatment methods adopted are harmful to the environment and found to be toxic to human as well as surrounding other living creatures. Biohydrotechnology seems to be safe and acceptable technique as compared with other methods of e-Waste treatments available; nevertheless, the gaps in the study still prevail and need to be connected. The major problems are that this method can tackle a small amount of e-Waste, the organisms used in the process fail to remain viable for a longer time due to the toxicity of e-Waste, and the amount of lixiviant generated is not sufficient as well as takes a longer time for the production, so there is an urgent need to focus on the development of more efficient strain, search of new strain, process optimization, scale-up of the current process for e-Waste management, and treatment. Thus, there is a need for the development of hybrid technology, which requires more attention to various interdisciplinary subjects, such as microbiology, metallurgy, engineering, and biotechnology.

8.10   CONCLUSIONS

Generation and accumulation of e-Waste is a global serious problem, and there is an urgent need to tackle diversified toxic materials present in e-Waste for their efficient remediation from the sites where they are disposed or treat them to minimize their effect on the environment. Microbiological hybrid processes have a bright future for the development of an environmentally safe, sustainable, and cost-effective process for e-Waste treatment. For the efficient metal extraction, the prime requirement is the removal of organic pollutants and coating of color epoxy layers present on metals on PCBs by pretreatment to make metals easily available for the microbial attack, so they can be extracted at a much faster rate and in more amounts. The efficiency of biological treatments would be enhanced by the combination of different methods and the use of nanoparticles or with some nonpolluting biodegraded agents. Public awareness for the 3R strategy of reducing, reusing, and recycling of e-Waste should be focused more enthusiastically, and there is also a need to bridge the knowledge gap through further efforts.

ACKNOWLEDGMENTS

The authors acknowledge the financial support from the UGC New Delhi, India, for Emeritus Professor Fellowship to S.R. Dave. We are also thankful to the Department of Science and Technology, New Delhi, to Inspire fellowship to A.B. Sodha. All the authors acknowledge the Department of Biotechnology, New Delhi, for the project grant (102/IFD/SAN/3407/2016–2017).

REFERENCES

Akinseye VO, Electronic waste components in developing countries: harmless substances or potential carcinogen. *Annu Rev Res Biol* 3: 131–147 (2013).

Awasthi A, Zeng X, Li J, Environmental pollution of electronic waste recycling in India: A critical review. *Environ Poll.* 211: 259–270 (2016).

Baldé CP, Forti V, Gray V, Kuehr R, Stegmann P, The Global E-waste Monitor - 2017, United Nations University (UNU), International Telecommunication Union (ITU) and International Solid Waste Association (ISWA), Bonn/Geneva/Vienna (2017).

Bhattacharya A, Khare SK, Sustainable options for mitigation of major toxicants originating from electronic waste. *Curr Sci* 111: 1946–1953 (2016).

Brandl H, Lehmann S, Faramarzi MA, D. Martinelli, Biomobilization of silver, gold, and platinum from solid waste materials by HCN-forming microorganisms. *Hydrometallurgy* 94: 14–17 (2008).

Cui H, Anderson CG, Literature review of hydrometallurgical recycling of printed circuit boards (PCBs). *J Adv Chem Eng* 6:142 (2016). doi:10.4172/2090-4568.1000142.

Dave SR, Shah MB, Tipre DR, E-waste: metal pollution Threat or Metal Resource? *J Adv Res Biotechnol* 1: 1–14 (2016a).

Dave SR, Sodha AB, Tipre DR, Microbial technology for metal recovery from e-waste printed circuit boards. *J Bacteriol Mycol Open Access* 6: 241–247 (2018).

Dave SR, Tipre DR, Chapter 6: Bioleaching of metals from sulphidic minerals. In: *Environmental Security, Human and Animal Health*, Garg S.R. (ed.). IBDC Publishers, Lucknow, pp. 71–94 (2011).

Dave SR, Tipre DR, Shah MB, Bioleaching process for metal extraction from E-waste at high pulp density using membrane technology. Application filed no. 201621015364 A; Controller General of Patents, Designs and Trade Marks, India. The Patent Office Journal 25/2016: 27445 (2016b).

Ehrlich HL, Past, present and future of biohydrometallurgy. *Hydrometallurgy* 59: 127–134 (2001).

Erüst C, Akcil A, Gahan CS, Tuncuk A, Deveci H, Biohydrometallurgy of secondary metal resources: a potential alternative approach for metal recovery. *J Chem Tech Biotech* 88: 2115–2132 (2013).

EU, European Union Waste Electronic and Electrical Waste (WEEE) Directive, Brussels (2000).

Gaidajis G, Angelakoglou K, Aktsoglou, D, E-Waste: Environmental problems and current management. *J Eng Sci Technol* 3: 193–199 (2010).

Hadi P, Xu M, Lin CSK, Hui- C-W, McKaya G, Waste printed circuit board recycling techniques and product utilization. *J Hazard Mater* 283: 234–243 (2015).

Ilyas S, Lee J, Kim B, Bioremoval of heavy metals from recycling industry electronic waste by a consortium of moderate thermophiles: process development and optimization. *J Clean Prod* 70: 194–202 (2014).

Isildar A, Rene ER, van Hullebusch ED, Lens PN, Electronic waste as a secondary source of critical metals: management and recovery technologies. *Resour Conserv Recycl* 135: 296–312 (2018).

Johnson DB, Geomicrobiology of extremely acidic subsurface environments. *FEMS Microbiol Ecol* 81: 2–12 (2012).

Kaya M, Recovery of metals and non-metals from electronic waste by physical and chemical recycling process. *Waste Manage* 57: 64–90 (2016).

Kiddee P, Naidu R, Electronic waste management approaches: an overview. *Waste Manage*. 33: 1237–1250 (2013).

Knowles CJ, Bunch AW, 1986. Microbial cyanide metabolism. *Adv Microbial Physio* 27: 73–111 (1989).

Kumar S, Singh R, Singh D, Prasad R, Yadav T, Electronics-waste management. *Int J Environ Eng Manag* 4: 389–396 (2013).

Lan Z, Hu Y, Qin W, Effect of surfactant OPD on the bioleaching of marmatite. *Miner Eng* 22: 10–23 (2009).

Lee JC, Pandey BD, Bio-processing of solid wastes and secondary resources for metal extraction – a review. *Waste Manage* 32: 3–18 (2012).

Liang G, Ting J, Liu W, Zhou Q, Optimizing mixed culture of two acidophiles to improve copper recovery from printed circuit boards (PCBs). *J Hazard Mater* 250–251: 238–245 (2013).

Ministry of Environment and Forests, Central Pollution Control Board Delhi, India (2008) Guidelines for environmentally sound management of e-waste.

Pant D, Joshi D, Upreti MK, Kotnala RK, Chemical and biological extraction of metals present in e-waste: a hybrid technology. *Waste Manage* 32: 575–583 (2012).

Pradhan J, Kumar S, Metals bioleaching from electronic waste by *Chromobacterium violaceum* and *Pseudomonas* sp. *Waste Manage Res* 30: 1151–1159 (2012).

Pradhan JK, Kumar S, Informal e-waste recycling: environmental risk assessment of heavy metal contamination in Mandoli industrial area, Delhi, India, *Environ Sci Pollut Res* (2014). DOI:10.1007/s11356-014-2713-2.

Rohwerder T, Sand W, Oxidation of inorganic sulphur compounds in acidophilic prokaryotes. *Eng Life Sci* 7: 301–309 (2007).

Saidan M, Brown B, Valix M, Leaching of electronic waste using biometabolised acids. *Chinese J Chem Eng* 20: 530–534 (2012).

Sand W, Gehre T, Jozsa PG, Schippers A, Biochemistry of bacterial leaching-direct vs indirect bioleaching. *Hydrometallurgy* 59: 159–175 (2001).

Shah M, Tipre D, Dave S, Chemical and biological processes for multi-metal extraction from waste printed circuit boards of computers and mobile phones. *Waste Manage Res* 32: 1134–1141 (2014). DOI: 10.1177/0734242X14550021.

Shah M, Tipre D, Purohit M, Dave S, Development of two-step process for enhanced biorecovery of Cu-Zn-Ni from computer printed circuit boards. *J Biosci Bioeng* 120: 167–173 (2015).

Smith A, Mudder T, Chemistry and treatment of cyanidation wastes. Mining Journal Books Ltd. (UK), 345 (1991).

Sodha AB, Qureshi SA, Khatri BR, Tipre DR, Dave SR, Enhancement in iron oxidation and multi metal extraction from waste television printed circuit boards by iron oxidizing *Leptospirillum ferriphilum* isolated from coal sample. *Waste Biomass Valor* (2017). doi:10.1007/s12649-017-0082-z.

Tansel B, From electronic consumer products to e-waste: Global outlook, waste, quantities, recycling challenges. *Environ Inter* 98: 35–45 (2017).

Tipre DR, Dave SR, Bioleaching process for Cu-Pb-Zn bulk concentrate at high pulp density. *Hydrometallurgy* 75: 37–43 (2004).

Vera M, Schippers A, Sand W, Progress in bioleaching: Fundamentals and mechanisms of bacterial metal sulfide oxidation - Part A. *Appl Microbiol Biotechnol* 97: 7529–7541 (2013).

Vlivainio K, Effects of iron (III) chelates on the solubility of heavy metals in calcareous soils. *Environ Pollut* 158: 3194–3200 (2010).

Wei L, Liu Y, Present status of e-waste disposal and recycling in China. *Procedia Environ Sci* 16: 506–514 (2012).

Xue M, Kendall A, Xu Z, Schoenung JM, Waste management of printed wiring boards: a life cycle assessment of the metals recycling chain from liberation through refining. *Environ Sci Technol* 49: 940–947 (2015).

Zimmerley SR, Wilson DG, Prater JD, Cyclic leaching process employing iron oxidizing bacteria. *US Patent* 2: 829–964 (1958).

9 Application of Thermostable Enzymes for Retaining Sustainable Environment

Improvement of the Enzymatic Activities of Two Thermophilic Archaeal Enzymes without Decreasing Their Stability

Yutaka Kawarabayasi
National Institute of Advanced Industrial
Science and Technology (AIST)

CONTENTS

9.1   INTRODUCTION

At present for the production of useful chemical compounds, they are mainly synthesized through a chemical process utilizing high-pressure and toxic heavy metal ions, which cause consumption of a large amount of energy and pollution of the environment. Therefore, this production process is not convenient for retaining environment to the next generation. Until now, the development of industry and economics are mainly due to this production system. However, now we, especially natural scientists, must consider our next generation to remain a sustainable society. To this purpose, chemical production system must be changed to the more sustainability-friendly system.

Meanwhile, most organisms are keeping their life using the most economic and environmentally friendly system. Within our life system, a huge number of enzymes are working for retaining our liability. As features of natural enzymes, it is well known that enzyme can only catalyze a specific substrate, process reaction under a mild condition, produce only specific product even though the product has chirally different forms such as D-amino acid and L-amino acid, and process reaction without using the toxic metal ions. From these reasons, enzymes are thought to be one of the most eco-friendly catalysts of natural chemicals. Among these natural enzymes, thermostable enzymes are expected to be utilized for production of useful compounds at the industrial and massive production level because high stability of the thermostable enzymes is convenient for its industrial usage. For thermostable enzymes, two different origin enzymes are present, one is a bacterial enzyme and the other is an archaeal enzyme.

Archaea was established as the third domain of microorganisms separated from bacteria by sequence feature of the 16S rRNA sequences around 40 years ago [1]. Archaea are known to possess unusual properties such as monolayer structure of cell membrane, eukaryote-like transcription regulation, and bacteria-like metabolic pathway. Many Archaeal species indicate thermophilic feature because they are isolated from geo- or hydrothermal environments. These thermophilic Archaea possess thermostable or thermophilic proteins and enzymes, which are thought to be suitable for the application. Previous experimental analyses of these archaeal thermophilic enzymes indicate that they are stable under the condition with an organic solution and metal ions as well as high temperature. However, the Archaeal enzymes exhibit less activity than that of the mesophilic enzymes. Therefore, improvement of thermophilic enzymes from Archaea is required for their usage in the application. For this purpose, it has been attempted to increase enzymatic activities of the Archaeal thermophilic enzymes, and successful result is obtained.

In this chapter, improvement of two archaeal thermophilic enzymatic activities, for their utilization in the application, is described. Success examples of different types of introduction of substitutional and truncating mutations, and also their success cases are summarized.

9.2 THE BIFUNCTIONAL ST0452 PROTEIN IDENTIFIED FROM *Sulfolobus tokodaii*

The ST0452 gene was detected in the genomic data of an acidothermophilic archaeon *Sulfolobus tokodaii* strain 7 [2,3], which was isolated from hot springs in Japan and grows optimally at around 80°C and pH 4, as the longest gene among the homologous genes predicted to encode sugar-1-phosphate nucleotidylyltransferase (Sug-1-P NTase) [4]. When heterologously expressed in *Escherichia coli*, the ST0452 protein exhibited glucose-1-phosphate thymidylyltransferase (Glc-1-P TTase) activity from 65°C to 100°C and absolute stability at high temperature, with a half-life of 180 min at 80°C and 60 min at 95°C [4]. Analysis of the effect of metal ions on ST0452 Glc-1-P TTase activity showed cations to be functionally essential. Higher enzymatic activity was detected in the presence of 2 mM Co^{2+} or Mn^{2+}, which were, respectively, 2.43 and 1.64 times higher than that detected in the presence of 2 mM Mg^{2+} [4]. Analysis of substrate specificity also indicated that the ST0452 protein possessed 4.76 times higher *N*-acetylglucosamine-1-phosphate uridyltransferase (GlcNAc-1-P UTase) activity than its Glc-1-P TTase activity [4].

Owing to the presence of 24 repeats of a signature motif sequence at the C-terminus of the ST0452 protein, amino-sugar-1-phosphate acetyltransferase (amSug-1-P AcTase) activity was analyzed. Experimental analysis indicated that the ST0452 protein possessed glucosamine-1-phosphate acetyltransferase (GlcN-1-P AcTase) activity [5]. Analyses of the effect of metal ions on the ST0452 acetyltransferase activity established that metal ions were not essential for its GlcN-1-P AcTase activity. Furthermore, analysis of the substrate specificity on the ST0452 amSug-1-P AcTase activity showed that the ST0452 protein could not catalyze glucosamine-6-phosphate (GlcN-6-P) as a substrate, predicting that in *S. tokodaii* only contains the bacterial-type UDP-GlcNAc biosynthetic pathway [5]. Substrate specificity analysis also indicated that, in addition to GlcN-1-P AcTase activity, the ST0452 protein possesses the unique and novel galactosamine-1-phosphate acetyltransferase (GalN-1-P AcTase) activity [5]. In addition to the novel acetyltransferase activity, the ST0452 protein also exhibits *N*-acetylgalactosamine-1-phosphate uridyltransferase (GalNAc-1-P

UTase) activity [5]. From these two observations, a novel UDP-GalNAc biosynthetic pathway was predicted to be present in this thermophilic archaeon [5]. As indicated, the ST0452 protein possesses useful Sug-1-P NTase and unique amSug-1-P AcTase activities. Attempts have, therefore, been made to increase these activities.

9.3 ENHANCEMENT OF THE GlcNAc-1-P UTase ACTIVITY ENCODED IN THE N-TERMINUS OF THE ST0452 PROTEIN

The thermostability of the ST0452 protein makes this protein suitable for industrial applications; however, increasing its GlcNAc-1-P UTase activity would improve its suitability for such applications. To improve ST0452 GlcNAc-1-P UTase activity without diminishing its thermostability, amino acid residues at its reaction center were altered by targeted substitution. This could be achieved since the reaction center is enclosed within a pocket-shaped location, for the mutation of which should not change the overall protein structure. Targeted substitution was therefore used to introduce Ala into nine amino acid positions: 9th Gly, 13th Arg, 23rd Lys, 80th Thr, 97th Tyr, 99th Asp, 146th Glu, 147th Lys, and 208th Asp (shown by asterisks in Figure 9.1). Nine mutant ST0452 proteins were designated as D9A, R13A, K23A, T80A, Y97A, D99A, E146A, K147A, and D208A, respectively, and expressed in *E. coli*. Mutant proteins remaining in a soluble fraction after 30 min treatment at 80°C (data not shown) indicated that the targeted substitutions introduced at the residues in the ST0452 Sug-1-P NTase reaction center did not reduce their thermostability [6]. Therefore, the GlcNAc-1-P UTase activity of all mutant ST0452 proteins was subsequently measured and compared with that of the wild-type ST0452 protein. Two mutants, K23A and D99A, completely lost GlcNAc-1-P UTase activity by the introduction of substitution to Ala, indicating that the 23rd Lys and 99th Asp play essential roles in the ST0452 GlcNAc-1-P UTase activity. Conversely, as shown in Figure 9.2, four of the nine mutant ST0452 proteins (G9A, T80A, Y97A, and K147A) exhibited higher GlcNAc-1-P UTase activity than that of the wild-type ST0452 protein under optimized reaction condition, 100 µM UTP and 10 mM GlcNAc-1-P [6].

The ST0452 GlcNAc-1-P UTase activity was successfully increased by single amino acid substitution within the protein's reaction center; however, the highest activity of the mutant protein was still lower than the activity of a similar *E. coli* enzyme, GlmU [6]. To overcome this limitation, combined two single mutations were introduced into the single ST0452 protein's reaction center [7]. Six double-mutant ST0452 proteins were constructed, expressed in *E. coli*, and were found to remain as soluble proteins after 20 min treatment at 80°C (data not shown). Although thermostability of these double-mutant ST0452 proteins was not affected, their GlcNAc-1-P UTase activities

```
              *   *                             *              * *
ST0452     9 GSGERLEPITHTRPK 23    73 QKDDIKGT 80    94 LIIYGDL 100
EcGlmU    14 GKGTRMY--S-DLPK 25    76 QAE-QLGT 82    99 LMLYGDV 106


               * *                            *
ST0452   144 IIE-KPEIPPSN 154      202 EGY-WMDIG 210
EcGlmU   152 IVEHKDATDEQR 163      222 EVEG-VNNR 229


                #  #                          #        #   #
ST0452   307 PHLSYVG 313          324 GAGTLIA(NLRFDEKEVKVN 342
EcGlmU   362 GHLTYL  367          379 GAGTIT(CNYDGANKF-K   394
```

FIGURE 9.1 Sequence alignment of the highly conserved domains between the ST0452 protein and the *Escherichia coli* GlcNAc-1-P UTase. EcGlmU indicates the GlcNAc-1-P UTase from *E. coli* (NC_000913). The amino acid residues chosen for construction of the substitution mutant proteins within the N- and C-terminus are indicated by asterisks (*) and sharps (#), respectively. The numerals indicate the coordinates of the two ends of each domain relative to the N-terminus of each protein.

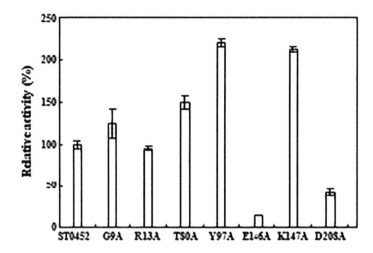

FIGURE 9.2   GlcNAc-1-P UTase activity of mutant ST0452 proteins under optimized reaction conditions. GlcNAc-1-P UTase activity of each mutant protein was measured in a standard reaction solution containing 100 μM UTP plus 10 mM GlcNAc-1-P. Relative activity is expressed as a percentage of the activity detected for the wild-type ST0452 protein under the same reaction condition. Assays were carried out in triplicate.

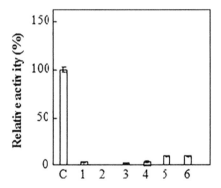

FIGURE 9.3   GlcNAc-1-P UTase activity of wild-type and double-mutant ST0452 proteins. Enzymatic activity was measured for 2 min at 80°C, with relative activity expressed as a percentage of the wild-type ST0452 activity. Lane C: the wild-type ST0452 protein; Lane 1: G9A/T80A; Lane 2: G9A/Y97A; Lane 3: G9A/K147A; Lane 4: T80A/Y97A; Lane 5: T80A/K147A; Lane 6: Y97A/K147A. Assays were carried out in triplicate.

were drastically reduced to 15% or less of the wild-type ST0452 GlcNAc-1-P UTase activity (Figure 9.3) [7]. These results indicated that introduction of two single mutants was not appropriate for the GlcNAc-1-P UTase activity of the ST0452 protein. Improving the ST0452 GlcNAc-1-P UTase activity thus required an alternative approach.

To further increase ST0452 GlcNAc-1-P UTase activity, site saturation mutagenesis of the 97th Tyr residues was undertaken because replacement with Ala at this residue showed the highest specific activity of all nine single mutants constructed (Figure 9.2) [7]. All mutant ST0452 proteins with a saturation substitution at the 97th position Tyr were successfully expressed in *E. coli* and were remained within the soluble fraction after 20 min treatment at 80°C (data not shown), revealing that these mutations did not make a serious effect on these protein's overall structures. Of 19 mutant proteins constructed, 14 (from N to G) exhibited higher GlcNAc-1-P UTase activity than that of the wild-type ST0452 protein (Figure 9.4). Of these 14 mutant proteins, six (N, H, Q, S, L, and M; shown as hatched bars in Figure 9.4) also exhibited increased Glc-1-P UTase activity. The highest

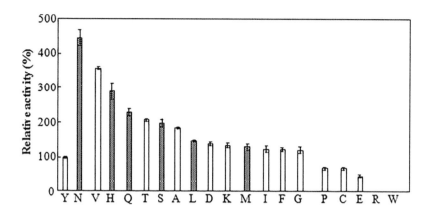

FIGURE 9.4 GlcNAc-1-P UTase activity of wild-type and mutant ST0452 proteins. Relative GlcNAc-1-P UTase activity is expressed as a percentage of the wild-type ST0452 activity. Lane Y: the wild-type ST0452 protein; Lane N: Y97N; Lane V: Y97V; Lane H: Y97H; Lane Q: Y97Q; lane T: Y97T; Lane S: Y97S; Lane A; Y97A; Lane L: Y97L; Lane D: Y97D; Lane K: Y97K; Lane M: Y97M; Lane I: Y97I; Lane F: Y97F; Lane G: Y97G; Lane P: Y97P; Lane C: Y97C; Lane E: Y97E; Lane R: Y97R; Lane W: Y97W. Mutant proteins with increased Glc-1-P UTase activity are highlighted by hatched bars. Assays were carried out in triplicate.

activity, 4.4 times higher than that of the wild-type ST0452 protein, was observed for the Y97N mutant protein. The Y97V mutant protein showed the second highest GlcNAc-1-P activity, but increased Glc-1-P UTase activity was not observed for this mutant protein. Kinetic analysis of these two mutant proteins indicated that increased GlcNAc-1-P UTase activity was mainly due to an increase in their *kcat* values (data not shown) [7]. From these results, it is proposed that thermophilic enzymes from archaea are not fully optimized and can thus be improved by introducing amino acid substitution mutations within the enzyme's reaction center.

9.4 ENHANCEMENT OF GlcN-1-P AcTase ACTIVITY LOCATED IN THE C-TERMINUS OF THE ST0452 PROTEIN

Similar to GlcNAc-1-P UTase activity located in the N-terminal region of the ST0452 protein, increasing of the acetyltransferase activity encoded in the C-terminus of the ST0452 protein is thought to be suitable for downstream application in producing modified sugar molecules. It was therefore attempted to increase amSug-1-P AcTase activity located in the C-terminal domain of the ST0452 protein without reducing its thermostability. Comparison of the amino acid sequences of the ST0452 C-terminal domain with the C-terminal domain of *E. coli* GlmU protein, the 3D structure of which is already determined, predicted amino acid residues with important roles for the ST0452 amSug-1-P AcTase activity (shown by sharps in Figure 9.1). Five mutant ST0452 proteins, containing Ala substitutions for the residues predicted to be important for amSug-1-P AcTase activity, were constructed (H308A, Y311A, N331A, K337A, and K340A) [8]. Three of the five mutant ST0452 proteins (Y311A, K337A, and K340A) exhibited increased GlcN-1-P AcTase activity (Figure 9.5), ranging from 1.18 to 1.47 times higher than that of the wild-type ST0452 protein. These observations indicated that substitution of two residues (H308A and N331A) slightly inhibits the GlcN-1-P AcTase activity of the ST0452 protein. However, a higher increase of the ST0452 amSug-1-P AcTase activity is required for its intended application.

To determine the region essential for catalyzing ST0452 GlcN-1-P AcTase activity, a series of mutant ST0452 proteins with C-terminal truncation were constructed [8]. Mutant ST0452 proteins with 21-, 31-, 41-, 71-, and 121-residue-long truncations were expressed as an insoluble form in *E. coli* and thus could not be used for further activity analysis. Mutant ST0452 proteins with 51- and 171-residue-long truncations were initially expressed as a soluble form in *E. coli* but changed to an insoluble form after only 5 min treatment at 70°C, indicating that mutant protein's thermostability

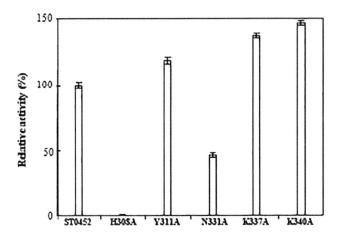

FIGURE 9.5 GlcN-1-P AcTase activity of wild-type and mutant ST0452 proteins. Relative GlcN-1-P AcTase activity is expressed as a percentage of the wild-type ST0452 activity. Assays were carried out in triplicate.

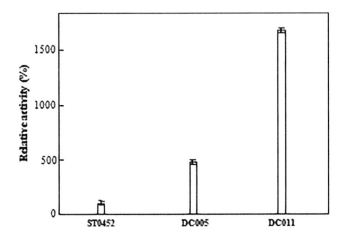

FIGURE 9.6 GlcN-1-P AcTase activity of wild-type and mutant ST0452 proteins containing C-terminal truncation. Relative GlcN-1-P AcTase activity is expressed as a percentage of the wild-type ST0452 activity. Assays were carried out in triplicate.

was reduced by C-terminal truncation. Only two truncated mutant ST0452 proteins, with deletions of 5 or 11 residues from the C-terminus, could, therefore, be used for activity analysis. Surprisingly, the specific activities of these two truncated proteins exhibited 4.8 and 16.9 times higher GlcN-1-P AcTase activities than that of the wild-type ST0452 protein (Figure 9.6) [8]. Further kinetic analysis of these mutant proteins indicated that their increase of the GlcN-1-P AcTase activity was mainly due to increase of *kcat* values [8]. This observation suggests that the C-terminal 11-residue-long region of the ST0452 protein plays an important role in regulation of its GlcN-1-P AcTase activity.

9.5 THE BIFUNCTIONAL PH0925 PROTEIN IDENTIFIED FROM *Pyrococcus horikoshii* OT3

In most organisms, compatible solutes, low-molecular-weight organic metabolites, are accumulated for resistance and adjustment to changing of the external osmotic pressure or increasing of the temperature in the surrounding environment. From genomic data, only species within

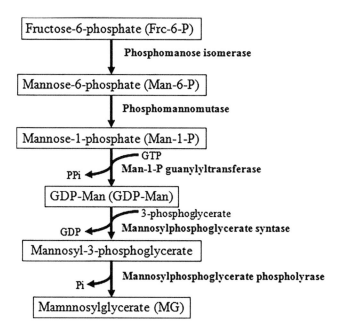

FIGURE 9.7   Putative *Pyrococcus horikoshii* mannosylglycerate biosynthetic pathway. The enzymatic name catalyzing each reaction is shown in bold characters.

the genera *Pyrococcus* and *Thermococcus* possess a four-gene cluster for the biosynthesis of mannosylglycerate, found as the main compatible solute in these thermophilic archaeal species, from fructose-6-phosphate (Frc-6-P) (Figure 9.7). In *Pyrococcus horikoshii* OT3 [9,10], mannosylglycerate was found as the main compatible solute in response to increasing external osmotic pressure [11]. Within the four-gene cluster, the PH0925 gene was predicted to encode a bifunctional protein with phosphomannose isomerase (PMI) and mannose-1-phosphate guanylyltransferase (Man-1-P GTase) activities by similarity. These two activities correspond to the first and third reaction of the mannosylglycerate biosynthetic pathway from Frc-6-P in this archaeon. The PH0925 protein exhibits multiple Sug-1-P NTase activities, which utilizes all four NTPs and deoxy-GTP when Man-1-P is used as the substrate, but catalyzes only GTP and deoxy-GTP when Glc-1-P or glucosamine-1-phosphate (GlcN-1-P) is used as the substrate at an optimum reaction temperature of 95°C [12]. These thermostable enzymatic activities were expected to be useful for the production of nucleotide sugar molecules, substrate molecules constructing artificial polysaccharides. It was therefore attempted to increase the Sug-1-P NTase activity of the PH0925 protein without decreasing its thermostability.

9.6   ENHANCEMENT OF Man-1-P GTase ACTIVITY LOCATED IN THE N-TERMINUS OF THE PH0925 PROTEIN

For the analysis of PH0925 Sug-1-P NTase activity, expression vectors encoding mutant PH0925 proteins with N- or C-terminal truncations were constructed [12]. Mutant PH0925 proteins with N-terminal truncation were expressed as an insoluble form in *E. coli*, revealing that the N-terminal region of the PH0925 protein is essential for construction of the correct structure of the protein. Seven mutant PH0925 proteins with truncations from C-terminus (DN450, DN350, DN345, DN328, DN321, DN310, and DN271 with 14, 114, 119, 136, 143, 154, and 193 residues deleted from the C-terminus, respectively) were successfully expressed as soluble protein in *E. coli*. All C-terminal truncated mutant PH0925 proteins did not exhibit any PMI activity, indicating that its PMI activity

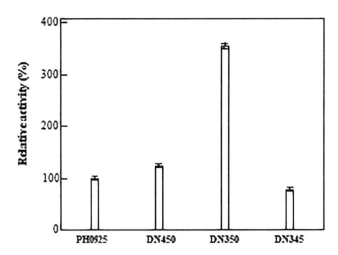

FIGURE 9.8 Man-1-P GTase activity of wild-type and mutant PH0925 proteins with C-terminal truncations. Relative Man-1-P GTase activity is expressed as a percentage of the wild-type PH0925 activity. Assays were carried out in triplicate.

is encoded at the C-terminal domain of the PH0925 protein. Furthermore, this indicates that the most C-terminal 14 residues are essential for the PMI activity of the PH0925 protein.

For these mutant PH0925 proteins with the C-terminal truncations, Man-1-P GTase activity was measured, since this was the highest activity displayed in the wild-type PH0925 protein. Four truncation mutants (DN328, DN321, DN310, and DN271) exhibited less than 2% of the activity detected in the wild-type PH0925 protein, whereas the remaining three truncation mutant PH0925 proteins (DN450, DN350, and DN345) exhibited Man-1-P GTase activity more consistent with that of the wild-type PH0925 protein. Surprisingly, DN450 and DN350 respectively exhibited 1.25 and 3.5 times higher Man-1-P GTase activities compared with that of the wild-type PH0925 protein (Figure 9.8) [12]. These observations suggest that the 114-residue-long C-terminal regions strongly suppress the Sug-1-P NTase activity encoded in the N-terminal region of the PH0925 protein. In addition, it was suggested that the five residues in positions 345–350 played an essential role in PH0925 Man-1-P GTase activity. It was thus established that the mutant PH0925 protein D350 was useful for the production of GDP-Man, an important nucleotide sugar molecules.

9.7 CONCLUSION

Enzymes or proteins isolated from thermophilic archaea show absolute thermostability, remaining in a soluble form even after 20 min treatment at 80°C. This feature makes these proteins suitable for application in the efficient production of the targeted chemical compounds. Improving the thermostability of enzymes isolated from mesophilic microorganisms with useful activities, on the other hand, has proved to be difficult. Therefore, increasing the enzymatic activity of the thermostable enzymes will expand the application areas of these natural enzymes, and it is useful for the establishment of sustainable society. The results summarized in this chapter would provide important information regarding artificial improvement of the natural thermostable enzymes from archaea and materials or systems to maintain the present environment for a long time.

ACKNOWLEDGMENT

This work was partially supported by the Institute for Fermentation, Osaka (IFO).

REFERENCES

1. Woese, C.R., Mangrum, L.J. & Fox, G.E. Archaebacteria. *J. Mol. Evol.* 1978, 11, 245–252.
2. Suzuki, T., Iwasaki, T., Uzawa, T., Hara, K., Nemoto, N., Kon, T., Ueki, T., Yamagishi, A. & Oshima, T. *Sulfolobus tokodaii* sp. nov. (f. *Sulfolobus* sp. strain 7), a new member of the genus *Sulfolobus* isolated from Beppu Hot Springs, Japan. *Extremophiles* 2002, 6, 39–44.
3. Kawarabayasi, Y., Hino, Y., Horikawa, H., Jin-no, K., Takahashi, M., Sekine, M., Baba, S., Ankai, A., Kosugi, H., Hosoyama, A., Fukui, S., Nagai, Y., Nishijima, K., Otsuka, R., Nakazawa, H., Takamiya, M., Kato, Y., Yoshizawa, T., Tanaka, T., Kudoh, Y., Yamazaki, J., Kushida, N., Oguchi, A., Aoki, K., Masuda, S., Yanagii, M., Nishimura, M., Yamagishi, A., Oshima, T. & Kikuchi, H. Complete genome sequence of an aerobic thermoacidophilic crenarchaeon, *Sulfolobus tokodaii* strain 7. *DNA Res.* 2001, 8, 123–140.
4. Zhang, Z., Tsujimura, M., Akutsu, J., Sasaki, M., Tajima, H. & Kawarabayasi, Y. Identification of an extremely thermostable enzyme with dual sugar-1-phosphate nucleotidylyltransferase activities from an acidothermophilic archaeon, *Sulfolobus tokodaii* strain 7. *J. Biol. Chem.* 2005, 280, 9698–9705.
5. Zhang, Z., Akutsu, J. & Kawarabayasi, Y. Identification of novel acetyltransferase activity on the thermostable protein ST0452 from *Sulfolobus tokodaii* strain 7. *J. Bacteriol.* 2010, 192, 3287–3293.
6. Zhang, Z., Akutsu, J., Tsujimura, M. & Kawarabayasi, Y. Increasing in archaeal GlcNAc-1-P uridyltransferase activity by targeted mutagenesis while retaining its extreme thermostability. *J. Biochem.* 2007, 141, 553–562.
7. Honda, Y., Zang, Q., Shimizu, Y., Dadashipour, M., Zhang, Z. & Kawarabayasi, Y. Increasing the thermostable sugar-1-phosphate nucleotidylyltransferase activities of archaeal ST0452 protein through site-saturation mutagenesis of the 97th amino acid position. *Appl. Environ. Microbiol.* 2017, 83, e02291-16.
8. Zhang, Z., Shimizu, Y. & Kawarabayasi, Y. Characterization of the amino acid residues mediating the unique amino-sugar-1-phosphate acetyltransferase activity of the archaeal ST0452 protein. *Extremophiles* 2015, 19, 417–427.
9. González, J.M., Masuchi, Y., Robb, F.T., Ammerman, J.W., Maeder, D.L., Yanagibayashi, M., Tamaoka, J. & Kato, C. *Pyrococcus horikoshii* sp. nov., a hyperthermophilic archaeon isolated from a hydrothermal vent at the Okinawa Trough. *Extremophiles* 1998, 2, 123–130.
10. Kawarabayasi, Y., Sawada, M., Horikawa, H., Haikawa, Y., Hino, Y., Yamamoto, S., Sekine, M., Baba, S., Kosugi, H., Hosoyama, A., Nagai, Y., Sakai, M., Ogura, K., Otsuka, R., Nakazawa, H., Takamiya, M., Ohfuku, Y., Funahashi, T., Tanaka, T., Kudoh, Y., Yamazaki, J., Kushida, N., Oguchi, A., Aoki, K. & Kikuchi, H. Complete sequence and gene organization of the genome of a hyper-thermophilic archaebacterium, *Pyrococcus horikoshii* OT3. *DNA Res.* 1998, 5, 55–76.
11. Empadinhas, N., Marugg, J., Borges, N., Santos, H. & da Costa, M.S. Pathway for the synthesis of mannosylglycerate in the hyperthermophilic archaeon *Pyrococcus horikoshii*. Biochemical and genetic characterization of key enzymes. *J. Biol. Chem.* 2001, 276, 43580–43588.
12. Akutsu, J., Zhang, Z., Morita, R. & Kawarabayasi, Y. Identification and characterization of a thermostable bifunctional enzyme with phosphomannose isomerase and sugar-1-phosphate nucleotidylyltransferase activities from a hyperthermophilic archaeon, *Pyrococcus horikoshii* OT3. *Extremophiles* 2015, 19, 1077–1085.

10 Metagenomics
A Genomic Tool for Monitoring Microbial Communities during Bioremediation of Environmental Pollutants

Gaurav Saxena, Narendra Kumar,
Nandkishor More, and Ram Naresh Bharagava
Babasaheb Bhimrao Ambedkar University (A Central University)

CONTENTS

10.1 INTRODUCTION

Rapid industrialization and urbanization around the world have led to the recognition and understanding of the relationship between environmental contamination and public health. Industries are the key players in the national economy of many developing countries, unfortunately; however, these are also the major polluters in the environment. Among the different sources of environmental pollution, industrial wastewater discharged from different industries is considered as the major source of environmental pollution (soil and water). Industries use a variety of chemicals for the processing of raw materials to obtain a good quality of products within a short period of time and in an economic way. To obtain the good quality of products within a short period of time, industries generally use cheap and poorly or nonbiodegradable chemicals, and their toxicity is usually ignored. However, there are many reports available in the public domain, which confirm the presence of a variety of highly toxic chemicals in industrial wastewaters.

Industrial wastewaters contain a variety of organic and inorganic pollutants, which causes serious environmental pollution and health hazards in living beings (Goutam et al. 2018; Bharagava et al. 2017a,b,c; Saxena and Bharagava 2017; Saxena et al. 2016; Maszenan et al. 2011; Megharaj et al. 2011). The wastewater generated from pollution causing industries is characterized by a high chemical oxygen demand (COD), biological oxygen demand (BOD), total dissolved solids (TDS), total suspended solids (TSS), and a variety of recalcitrant organic and inorganic pollutants (Goutam et al. 2019; Bharagava et al. 2019; Gautam et al. 2017; Saxena and Bharagava 2015; Saxena et al. 2015). Organic pollutants include phenols, chlorinated phenols, endocrine-disrupting chemicals, azodyes, polyaromatic hydrocarbons, polychlorinated biphenyls, and pesticides, whereas inorganic pollutants include a variety of toxic heavy metals such as cadmium (Cd), chromium (Cr), arsenic (As), lead (Pb), and mercury (Hg). The high concentration and poor biodegradability of recalcitrant organic pollutants and nonbiodegradable nature of inorganic metal pollutants in industrial wastewaters pose a major challenge for environmental safety and human health protection, and thus, it is required to adequately treat the industrial wastewaters before its final disposal in the environment.

Recently, bioremediation is emerging as an eco-friendly waste management method that utilizes inherent potential of microbes such as algae, fungi, bacteria, and plants, to remove the organic and inorganic pollutants from industrial wastes whereas biodegradation is the degradation of highly toxic complex organic pollutants into less toxic forms (Saxena and Bharagava 2016, 2019; Bharagava and Saxena 2019a,b; Saxena et al. 2019a,d,e,f; Chandra et al. 2015; Chandra and Kumar 2015). It involves the detoxification and mineralization, where the waste is converted into inorganic compounds, such as carbon dioxide, water, and methane (Reshma et al. 2011). But if the pollutants are persistent in nature, their biodegradation often proceeds through multiple steps utilizing different enzymes or microbial populations. The process of bioremediation depends on the metabolic potential of microbes to degrade/detoxify or transform the pollutants, which is also dependent on both the pollutants accessibility and bioavailability (Antizar-Ladislao 2010; Kumar et al. 2018). Bioremediation can be applied as in situ (waste to be treated at the site) or ex situ (waste to be treated elsewhere) remediation technology. However, the in situ bioremediation is usually preferred over ex situ bioremediation due to the high cost associated with excavation of contaminated sites to be treated. Furthermore, in situ bioremediation involves following three strategies (Maszenan et al. 2011): attenuation (natural process of degradation and can be monitored by decrease in pollutants concentration with increasing time), biostimulation (intentional stimulation of pollutants degradation by addition of water, nutrient, and electron donors or acceptors), and bioaugmentation (addition of laboratory-grown microbes with potential for degradation).

Microbes play an important role in the degradation and detoxification of organic and inorganic pollutants from industrial wastewaters (Mulla et al. 2019; Saxena et al. 2019b,c; Saxena and Bharagava 2017; Maszenan et al. 2011; Megharaj et al. 2011; Chandra and Kumar 2017a; Kumar and Chandra 2018). However, sometimes, the bioremediation of pollutants takes place at a very slow rate that might be due to the inability of the native microbial community to degrade/detoxify the pollutants of high concern. To overcome this, exogenous or nonnative microbes or their enzyme can be added to augment the system to enhance the degradability. But this approach can be the most invasive, as a nonnative organism added to an ecosystem may not be able to survive in the contaminated system and persist long after the contaminant has been removed altering the ecosystem (Techtmann and Hazen 2016; Chandra and Kumar 2017b). Hence, it is important to understand the microbial communities involved in bioremediation of environmental pollutants because microbes are the drivers of bioremediation and shifts in their composition and activity may impact the fate of pollutants in the environment (Techtmann and Hazen 2016). Metagenomics approach is applied to better study the microbial community composition and activity during bioremediation in a contaminated environment. Thus, the objective of this chapter is to provide the basic knowledge of metagenomic approaches and their applications to better understand the microbial community structure and functions during bioremediation of environmental pollutants in a contaminated matrix.

10.2 CONVENTIONAL GENOMICS APPROACHES

Molecular techniques are primarily used for the monitoring of microbial communities and to determine the efficacy of bioremediation processes. A number of culturable and nonculturable molecular techniques are currently applied to determine the microbial community composition and structure at contaminated sites. The culture-dependent molecular techniques include various molecular methods, such as amplified ribosomal DNA restriction analysis (ARDRA), randomly amplified polymorphic DNA analysis (RAPD), amplified fragment length polymorphisms (AFLP), length heterogeneity PCR (LH-PCR), single-strand conformation polymorphism (SSCP), automated ribosomal intergenic spacer analysis (ARISA), denaturing gradient gel electrophoresis (DGGE)/temperature gradient gel electrophoresis (TGGE), and terminal-restriction fragment length polymorphism (T-RFLP) (Desai et al. 2011). A short description of conventional molecular techniques with their merits and demerits is presented in Table 10.1. However, a brief description of some of the conventional molecular techniques is provided below:

DGGE and TGGE are based on the principle of amplifying rRNA or functional gene PCR products obtained from community DNA using primers containing a 50 bp GC clamp and their separation on polyacrylamide gels having chemical or temperature-based denaturing gradients (Nocker et al. 2007). In many studies, the use of DGGE has been reported in monitoring of microbial communities, and their functional genes at the sites contaminated with various environmental pollutants (Wakase et al. 2007), such as a shift in microbial community structure, were detected in multimetal-contaminated soils (Li et al. 2006), and abundances of dsrB (dissimilatory sulfite reductase b-subunit) genes were also assessed at an in situ metal precipitation site using DGGE fingerprinting technique (Geets et al. 2006). However, DGGE or TGGE has some limitations, as these are labor-intensive and often less reproducible in terms of band pattern and intensity detection obtained after electrophoretic separation (Desai et al. 2011). Furthermore, other associated limitations are as follows (Rastogi and Sani 2011): (i) limited sequence information, (ii) different DNA fragments may have similar melting points, (iii) number of different DNA fragments, which can be separated by polyacrylamide gel electrophoresis (PAGE), and (iv) sequence heterogeneity among multiple rRNA operons of one bacterium, leading to multiple bands in DGGE, which might overestimate the diversity. DGGE analysis has been also used to screen the unique clones in clone libraries based on distinct patterns and determine the number of operational taxonomic units (OTUs) (Rastogi and Sani 2011).

RAPD and DNA amplification fingerprinting (DAF) techniques are the important molecular techniques and utilize PCR amplification with a short primer (usually ten nucleotides), which anneals randomly at multiple sites on the genomic DNA under low annealing temperature, typically 35°C (Rastogi and Sani 2011). These methods generate PCR amplicons of various lengths in a single reaction that are separated on an agarose or polyacrylamide gel depending on the genetic complexity of the microbial communities. Due to the high speed and ease of use, RAPD/DAF has been used extensively in fingerprinting of overall microbial community structure and closely related bacterial species and strains (Rastogi and Sani 2011). RAPD and DAF are highly sensitive to experimental conditions (e.g., annealing temperature, $MgCl_2$ concentration) and quality and quantity of template DNA and primers. Hence, the primers and reaction conditions need to be evaluated to compare the relatedness between microbial communities and obtain the most discriminating patterns between species or strains. A RAPD profiling study was used with 14 random primers to assess the changes in microbial diversity in soil samples that were treated with pesticides (triazolone) and chemical fertilizers (ammonium bicarbonate) (Yang et al. 2000). RAPD fragment richness data demonstrated that pesticide-treated soil maintained an almost identical level of diversity at the DNA level as the control soil (i.e., without contamination). In contrast, chemical fertilizer caused a decrease in DNA diversity as compared with the control soil.

ARDRA is based on DNA sequence variations present in the PCR-amplified 16S rRNA genes. The PCR product amplified from environmental DNA is generally digested with tetracutter restriction endonucleases (e.g., Alu I and Hae III), and restricted fragments are resolved on agarose or polyacrylamide gels. ARDRA provides little or no information about the type of microorganisms present in the sample, but this method is still useful for rapid monitoring of microbial communities over time or to compare microbial diversity in response to changing environmental conditions. ARDRA is also used for the identification of unique clones and to estimate

TABLE 10.1

Description of Conventional Molecular Techniques Used for Microbial Characterization

Molecular Technique	Merits	Demerits
Ribotyping	Highly reproducible; classify isolates from multiple sources	Complex, expensive; labor-intensive; geographically specific; database required; variation in methodology
Amplified ribosomal DNA restriction analysis	Culture-independent technique and suitable for analysis of a variety of microbes	Not quantitative and require DNA extraction and PCR biases
Ribosomal RNA intergenic spacer analysis	Culture-independent technique, suitable for analysis of a variety of microbes, and give remarkable heterogeneity in length and sequence among bacteria	Not quantitative and require DNA extraction and PCR biases
Pulse-field gel electrophoresis	Extremely reproducible and highly sensitive to point genetic difference	Long assay time, too sensitive for broadly discriminate source, limited simultaneous processing, and required database
Denaturing gradient gel electrophoresis	Culture-independent technique, suitable for analysis of a variety of microbes, and use rRNA gene sequence heterogeneity	DNA extraction and PCR biases
Terminal restriction fragment length polymorphism analysis	Fast, semiquantitative, culture-independent technique, and suitable for analysis of a variety of microbes	DNA extraction and PCR biases
Fluorescent in situ hybridization	Quantitative and directly visualize the microbial cells including nonculturables	Inactive cells may not be detected
Quantitative PCR	Culture-independent technique and suitable for analysis of a variety of microbes	Expensive equipment; technically demanding
Repetitive DNA sequences	Simple and rapid	Reproduce a concern; cell culture required; large database required; variability increases as database increases
Length heterogeneity PCR	Culture-independent technique	Expensive equipment; technically demanding
Multiplex PCR	Fast and simultaneous detection of several target microorganisms	Combination of primer pairs must function in a single PCR reaction
Nucleic acid microarrays	High-throughput design with wider applications	Low sensitivity and processing complexities for environmental samples
Host-specific 16S rDNA	Does not require culturing or a database; an indicator of recent pollution	Only tested on human and cattle markers; limited simultaneous processing; expensive equipment; technically demanding; little information about the survival of *Bacteroides* spp. in the environment
On-chip technology	Combination of PCR with nucleic acid hybridization on a single chip and less interference between parallel reactions	Integration and packaging

Source: Adopted from Saxena et al. (2015).

the OTUs in environmental clone libraries based on the restriction profiles of clones (Rastogi and Sani 2011). However, the major limitation of ARDRA is that the restriction profiles generated from complex microbial communities are sometimes too difficult to resolve by agarose/PAGE (Rastogi and Sani 2011).

T-RFLP is almost similar to ARDRA, but it uses one 5¢ fluorescently labeled primer during the PCR reaction. The resulting PCR products are digested with restriction enzyme(s), and terminal restriction fragments (T-RFs) are separated on an automated DNA sequencer (Thies 2007). Only the terminally fluorescent labeled restriction fragments are detected and thus simplify the banding pattern and allow the analysis of complex microbial communities. Community diversity is also estimated by analyzing the size, numbers, and peak heights of resulting T-RFs. Each T-RF is assumed to represent a single OTU or ribotype. With recent developments in bioinformatics, several Web-based T-RFLP analysis programs have been developed, which enable researchers to rapidly assign putative identities based on a database of fragments produced by known 16S rDNA sequences. Similar to ARDRA, a T-RFLP pattern is characteristic of the restriction enzyme(s) used, and more than two enzymes should typically be applied. One pitfall of T-RFLP method is that it underestimates community diversity because only a limited number of bands per gel (generally <100) can be resolved, and different bacterial species can share the same T-RF length (OTU overlap or OTU homoplasy) (Rastogi and Sani 2011). Nonetheless, the method provides a robust index of community diversity, and T-RFLP results are generally very well correlated with the results from clone libraries (Fierer and Jackson 2006). Fierer and Jackson (2006) applied the T-RFLP technique to understand the biogeographical patterns in soil bacterial communities and to investigate the biotic and abiotic factors that shape the composition and diversity of bacterial communities. They collected 98 soil samples from across North and South America representing a wide range of temperature, pH, and other geographical conditions. Their results demonstrated that bacterial diversity was higher in neutral soils as compared with acidic soils and was unrelated to factors, such as site temperature, latitude, and other variables that typically act as good predictors of animal and plant diversity.

LH-PCR analysis is also similar to T-RFLP method, but it further detects the amplicon length variations, which are raised after restriction digestion, whereas in LH-PCR, different microorganisms are discriminated based on natural length polymorphisms that occur due to mutation within genes (Mills et al. 2007). Amplicon LH-PCR interrogates the hypervariable regions present in 16S rRNA genes and produces a characteristic profile. LH-PCR utilizes a fluorescent dye–labeled forward primer, and a fluorescent internal size standard is run with each sample to measure the amplicon lengths in base pairs. The intensity (height) or area under the peak in the electropherogram is proportional to the relative abundance of that particular amplicon. The advantage of using LH-PCR over the T-RFLP is that the former does not require any restriction digestion and PCR products can be directly analyzed by a fluorescent detector; however, the limitations of LH-PCR technique include inability to resolve complex amplicon peaks and underestimation of diversity, as phylogenetically distinct taxons may produce same-length amplicons (Mills et al. 2007). LH-PCR was used in combination with fatty acid methyl ester (FAME) analysis to investigate the microbial communities in soil samples that differed in terms of type and/or crop management practices (Ritchie et al. 2000). LH-PCR results strongly correlated with FAME analysis and were highly reproducible and successfully discriminated different soil samples. The most abundant bacterial community members, based on cloned LH-PCR products, were members of the β-Proteobacteria, Cytophaga-Flexibacter-Bacteroides, and the high G+C content gram-positive bacterial group (Rastogi and Sani 2011).

Ribosomal intergenic spacer analysis (RISA) involves PCR amplification of a portion of the intergenic spacer region (ISR) present between the small (16S) and large (23S) ribosomal subunits (Rastogi and Sani 2011). The ISR contains significant heterogeneity in both length and nucleotide sequence. By using primers annealing to conserved regions in the 16S and 23S rRNA genes, RISA profiles can be generated from most of the dominant bacteria existing in an environmental sample. RISA provides a community-specific profile, with each band corresponding to at least one organism in the original community (Rastogi and Sani 2011). The automated version of RISA is known as ARISA and involves the use of a fluorescence-labeled forward primer, and ISR fragments are detected automatically by a laser detector. ARISA allows simultaneous analysis of many samples; however, the technique has been shown to overestimate microbial richness and diversity (Rastogi and Sani 2011). Ranjard et al. (2001) evaluated ARISA to characterize the bacterial communities from four types of soil differing in geographic origins, vegetation cover, and physicochemical properties. ARISA profiles generated from these soils were distinct and contained several diagnostic peaks with respect to size and intensity. Their results demonstrated that ARISA is a very effective and sensitive method for detecting differences between complex bacterial communities at various spatial scales (between- and within-site variability).

DNA microarrays have been used primarily to provide a high-throughput and comprehensive view of microbial communities in environmental samples. The PCR products amplified from total environmental DNA are directly hybridized to known molecular probes, which are attached to the microarrays (Gentry et al. 2006). After the fluorescently labeled PCR amplicons are hybridized to the probes, positive signals are scored by the use of confocal laser scanning microscopy. The microarray technology allows samples to be rapidly evaluated with replication, which is a significant advantage in microbial community analyses. In general, the hybridization signal intensity on microarrays is directly proportional to the abundance of the target organism. Cross-hybridization is a major limitation of microarray technology, particularly when dealing with environmental samples (Rastogi and Sani 2011). In addition, the microarray is not useful in identifying and detecting novel prokaryotic taxa. The ecological importance of a genus could be completely ignored if the genus does not have a corresponding probe on the microarray. DNA microarrays used in microbial ecology could be classified into two major categories depending on the probes: (i) 16S rRNA gene microarrays (PhyloChip) and (ii) functional gene arrays (FGAs).

Quantitative PCR (Q-PCR), or real-time PCR, has been used in microbial investigations to measure the abundance and expression of taxonomic and functional gene markers (Bustin et al. 2005; Smith and Osborn 2009). Unlike traditional PCR, which relies on endpoint detection of amplified genes, Q-PCR uses either intercalating fluorescent dyes such as SYBR Green or fluorescent probes (TaqMan) to measure the accumulation of amplicons in real time during each cycle of the PCR. The software records the increase in amplicon concentration during the early exponential phase of amplification, which enables the quantification of genes (or transcripts) when they are proportional to the starting template concentration. When Q-PCR is coupled with a preceding reverse transcription (RT) reaction, it can be used to quantify gene expression (RT-Q-PCR). Q-PCR is highly sensitive to starting template concentration and measures template abundance in a large dynamic range of around six orders of magnitude. Several sets of 16S and 5.8S rRNA gene primers have been designed for rapid Q-PCR-based quantification of soil bacterial and fungal microbial communities (Fierer et al. 2005). Q-PCR has also been successfully used in environmental samples for quantitative detection of important physiological groups of bacteria such as ammonia oxidizers, methane oxidizers, and sulfate reducers by targeting amoA, pmoA, and dsrA genes, respectively (Foti et al. 2007). Kolb et al. (2003) estimated the abundance of total methanotrophic population and specific groups of methanotrophs in a flooded rice field soil by Q-PCR assay of the pmoA genes. The total population of methanotrophs was found to be 5×10^6 pmoA molecules g^{-1}, and *Methylosinus* (2.7×10^6 pmoA molecules g^{-1}) and *Methylobacter/Methylosarcina* groups (2.0×10^6 pmoA molecules g^{-1}) were the dominant methanotrophs. The *Methylocapsa* group was below the detection limit of Q-PCR (1.9×10^4 pmoA molecules g^{-1}).

Fluorescence in situ hybridization (FISH) enables in situ phylogenetic identification and enumeration of individual microbial cells by whole cell hybridization with oligonucleotide probes (Rastogi and Sani 2011). A large number of molecular probes targeting 16S rRNA genes have been reported at various taxonomic levels (Rastogi and Sani 2011). The FISH probes are generally 18–30 nucleotides long and contain a fluorescent dye at the 5¢ end that allows detection of probe bound to cellular rRNA by epifluorescence microscopy. In addition, the intensity of fluorescent signals is correlated to cellular rRNA contents and growth rates, which provide insight into the metabolic state of the cells. FISH can be combined with flow cytometry for a high-resolution automated analysis of mixed microbial populations. The FISH method was used to follow the dynamics of bacterial populations in agricultural soils treated with s-triazine herbicides (Caracciolo et al., 2010). A variety of molecular probes were used to target specific phylogenetic groups of bacteria such as α, β, λ, and δ subdivisions of Proteobacteria and Planctomycetes. Results demonstrated that α-Proteobacteria populations diminished sharply after 14 days of incubation in treated soil compared with control soil with no s-triazine treatment. In contrast, β-Proteobacteria populations remained higher than that of the control soils throughout the incubation period (70 days). Other bacterial groups, e.g., α-Proteobacteria and Planctomycetes, were not significantly affected by the presence of the herbicide. Low signal intensity, background fluorescence, and target inaccessibility are commonly encountered problems in FISH analysis. In the past few years, extensive improvements have been made to solve some of these problems, which include the use of brighter fluorochromes, chloramphenicol treatment to increase the rRNA content of active bacterial cells, hybridization with probes carrying multiple fluorochromes, and signal amplification with reporter enzymes (Rogers et al. 2007). In a modified FISH method known as catalyzed reporter deposition (CARD) FISH, the hybridization signal is enhanced through the use of tyramide-labeled fluorochromes (Pernthaler et al. 2002). This allows the accumulation of several fluorescent probes at the target site, which ultimately increases the signal intensity and sensitivity. Li et al. (2008) developed an advanced imaging technique by combining FISH to secondary-ion mass spectrometry (SIMS). In principle, the technique uses 16S rRNA probes for in situ hybridization; however, the probes are labeled with a stable isotope or element (e.g., fluorine or bromine atoms) rarely present in biomass. Once the probe is hybridized, the microbial identities of stable isotope-labeled cells are simultaneously determined in situ by NanoSIMS imaging. With next-generation SIMS instruments, the spatial resolution of ~50 nm (NanoSIMS) was achieved, which allowed quantifying the isotopic composition at the single-cell level.

The above-discussed molecular methods are culture dependent, as these need microbial culture for characterization and identification, and thus are not efficient to discover the uncultivable microbial diversity (Desai et al., 2011). To overcome the limitations of conventional molecular techniques, the use of metagenomic approaches is increasing, which advances our knowledge regarding the full-scale characterization of microbial community composition, structure, and activity during bioremediation at a contaminated site.

10.3 METAGENOMICS: AN OVERVIEW

Metagenomics (also known as ecological genomics, community genomics, or environmental genomics) is a discipline that uses genomic methods to analyze natural ecological communities, namely, the collective genomes in an environmental community (Uhlik et al., 2013; Handelsman et al., 1998, Riesenfeld et al., 2004) (Figure 10.1). The major goal of metagenomics is to explicate the genomes of uncultured microbes, thereby permitting investigation of the broad diversity of taxonomically and phylogenetically relevant genes, individual catabolic genes, and whole operons (Schmeisser et al., 2007). Metagenomics itself was initially recognized for its potential to aid in the discovery of novel biomolecules for biotechnological applications (Riesenfeld et al., 2004). Although the basic concept of metagenomics was first introduced at the end of the past century (Handelsman et al., 1998), early forms of metagenomics had begun to emerge previously, with one example being the phylogenetic analysis of environmental microbial communities (Pace et al., 1985).

The approach introduced by Handelsman et al. (1998) involves extraction of the metagenome (genomic DNA from all organisms inhabiting the environment), its fragmentation, cloning, transformation, and subsequent screening of the constructed metagenomic library. The primary aim is to screen environmental communities for a specific biological activity and identify genes or gene clusters associated with it, also referred to as function-based screening (Yun and Ryu, 2005).

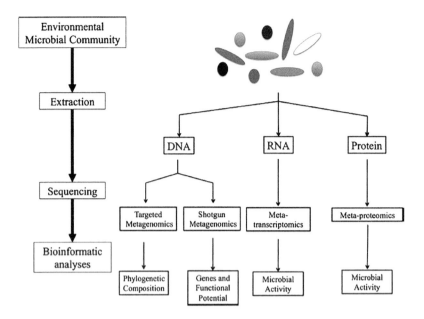

FIGURE 10.1 Metagenomic approaches to understand the microbial community structure, composition, and activity in environment. (Adapted from Techtmann and Hazen 2016.)

The advent of high-throughput next-generation DNA sequencing (e.g., 454 pyrosequencing, Illumina (Solexa) sequencing, SOLiD sequencing) gave rise to another approach to metagenomics, sequence-based screening. This method was first demonstrated by environmental genome shotgun sequencing of the Sargasso Sea (Venter et al., 2004), showing the potential of revealing the vast phylogenetic and metabolic diversity of microbial communities (Yun and Ryu, 2005).

Ma et al. (2015) employed MiSeq Illumina high-throughput sequencing to identify the microbial community composition and structure of coal mine wastewater treatment plants (WWTPs) in China. The most abundant phylum was *Proteobacteria* ranging from 63.64% to 96.10%, followed by *Bacteroidetes* (7.26%), *Firmicutes* (5.12%), *Nitrospira* (2.02%), *Acidobacteria* (1.31%), *Actinobacteria* (1.30%), and *Planctomycetes* (0.95%). At the genus level, *Thiobacillus* and *Comamonas* were the two primary genera in all sludges; other major genera included *Azoarcus, Thauera, Pseudomonas, Ohtaekwangia, Nitrosomonas*, and *Nitrospira*. Most of these core genera were closely related to aromatic hydrocarbon degradation and denitrification processes (Figure 10.2).

In addition to the sequence-based screening of environmental metagenomic libraries, direct pyrosequencing of environmental communities is possible, bypassing cloning completely (Edwards et al., 2006). Hu et al. (2012) employed 454-pyrosequencing technology to investigate the microbial communities in 12 municipal WWPTs with different treatment processes. In total, 202,968 effective sequences of the 16S rRNA gene were generated from 16 samples that widely represented the diversity of the microbial communities. While Proteobacteria was found to be the dominant phylum in some samples, in other samples, it was Bacteroidetes (Figure 10.3a,b).

In addition to high-throughput sequencing, metagenomic analyses have recently been performed with the use of high-throughput microarrays. These have been used to analyze microbial communities and monitor environmental biogeochemical processes. GeoChip microarrays (He et al., 2007, 2010) currently contain 83,992 50-metric sequences covering approximately 152,414 genes encoding for enzymes responsible for biogeochemical (C, N, P, S) cycling, metabolic processes, heavy metal resistance, antibiotic resistance, degradation of pollutants, and gyrB genes (Hazen et al., 2010, Lu et al., 2012). Marker gyrB encoding for the gyrase β-subunit is used instead of the more common 16S rRNA genes, as probes for 16S rRNA usually do not provide resolution below genus level. gyrB

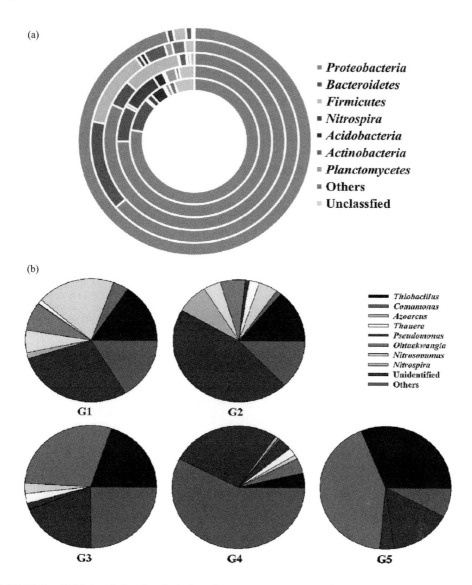

FIGURE 10.2   (a) Major phyla of each sludge of wastewater treatment plants (sequence percentage >1% in at least one sludge; inner to outer: G1–G5). (b) Major genera of each sludge of wastewater treatment plants (sequence percentage >1% in at least two sludges). (Adapted from Ma et al. 2015.)

can be used to differentiate even closely related species (He et al., 2010). GeoChip microarrays can be therefore used to study structure, dynamics, and potential metabolic activity of microbial communities and their variations depending on different stimuli (Uhlik et al., 2013). Another type of microarray valuable to microbial ecology and contaminant biodegradation is the PhyloChip, which is used for high-throughput phylogenetic analyses of microbial communities (Brodie et al., 2006), and has been used for a variety of applications, including assessing microbial community responses to petroleum contamination (Hazen et al., 2010; DeAngelis et al., 2011).

   With the first applications of metagenomic techniques, it became apparent that they enable the discovery of genomic and metabolic diversity that had not been previously imagined (Schloss and Handelsman, 2005). As research progressed, however, the main drawbacks of metagenomics were realized: the inability to link specific functions to individual populations and to achieve

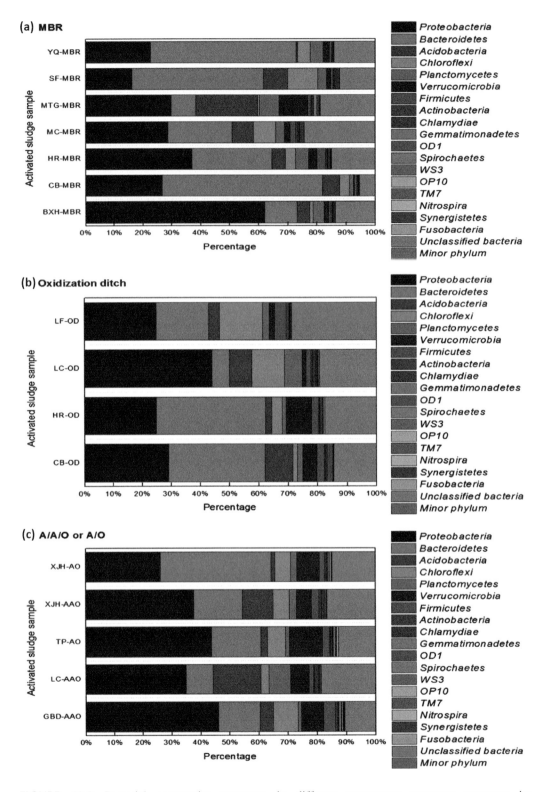

FIGURE 10.3 Bacterial community structures in different wastewater treatment processes by 454-pyrosequencing analysis. (Adapted from Hu et al. 2012.)

full-sequence coverage in more complex communities (Vieites et al., 2009). By combining metagenomic techniques with stable isotope probing (SIP), these limitations can be significantly reduced (Uhlik et al., 2013).

10.4   SCHEME OF METAGENOMICS

Metagenomic approaches often take two forms: targeted metagenomics or shotgun metagenomics. In targeted metagenomics—or microbiomics—the diversity of a single gene is probed to identify the full complement of sequences of a particular gene in an environment, and it is most often employed to investigate both the phylogenetic diversity and relative abundance of a particular gene in a sample (Techtmann and Hazen 2016). This approach is regularly used to investigate the diversity of small subunit rRNA sequences (16S/18S rRNA) in a sample. Microbial ecologists routinely use small subunit rRNA sequencing to understand the taxonomic diversity of an environment. It can also be applied as a tool to investigate the impact of environmental contaminants in altering the microbial community structure. To perform targeted metagenomics, environmental DNA is extracted, and the gene of interest is PCR-amplified using primers designed to amplify the greatest diversity of sequences for that gene of interest. These amplified genes are then sequenced using next-generation sequencing. Next-generation sequencing results in thousands of small subunit rRNA read per sample and can probe hundreds of samples simultaneously. Targeted metagenomics captures the diversity of single gene of interest but is limited by the universality of the PCR primers chosen for the analysis (Shakya et al., 2013; Parada et al., 2015; Klindworth et al., 2013; Prosser, 2012). Furthermore, various bioinformatics analyses have the potential to skew the overall diversity estimates (Rinke et al., 2013). The strength of targeted metagenomics is that it provides a fairly comprehensive catalog of the microbial taxa present in a set of samples and allows for in-depth comparison of shifts in microbial diversity before and after a perturbation.

In shotgun metagenomics, the total genomic complement of an environmental community is probed through genomic sequencing. In this approach, environmental DNA is extracted and then fragmented to prepare sequencing libraries. These libraries are then sequenced to determine the total genomic content of that sample. Shotgun metagenomics is a powerful technique where the functional potential of a microbial community can be identified. Shotgun metagenomics is often most limited by the depth of sequencing. Gaining a complete inventory of the genes in an environmental sample often requires extremely deep sequencing. Good coverage of the entire genomic content of every organism in the community is required for a comprehensive analysis of the functional potential of a community. Oftenly, shotgun metagenomics heavily samples the dominant microbes in a community and only sparsely covers the genomic content of the low abundance members of that community. Furthermore, analysis of metagenomic sequencing data can be very complex, as it involves accurately annotating diverse gene sequences, many of which have no homologs in the current sequence databases (Delmont et al., 2012). The goal of many studies is to link a functional gene with a taxonomic classification using a phylogenetic anchor. This can often be difficult with metagenomic sequencing unless sufficient sequencing depth is achieved and the reads can be accurately assembled into sufficiently long contigs. Many computational approaches have sought to assemble metagenomic sequences into complete genomes to gain a more complete understanding of the functional potential of particular species within a community. Several recent reviews have sought to summarize the key steps in metagenomics and the many potential pitfalls in these techniques (Faust et al., 2015; Mason et al., 2014; Moran et al., 2013).

In addition to sequencing-based approaches, several microarray-based techniques have been developed (Hazen et al., 2013). PhyloChip and GeoChip are the two most commonly used microarray technologies. PhyloChip is a 16S rRNA-based microarray able to probe the diversity of 10,993 subfamilies in 147 phyla (Hazen et al., 2010). GeoChip is a functional gene microarray able to probe the diversity of 152,414 genes from 410 gene categories (Zhou et al., 2013). Microarray techniques are not dependent on the depth of sequencing to provide comprehensive insights into the microbial

TABLE 10.2
Description of Metagenomic Sequencing Methods

Sequencing Method	Comments
BAC-based sequencing	Old method of genomic era where bacterial artificial chromosomes (BACs) were used (scrupulously used between the year 1995–2002)
Whole-genome sequencing also called as shotgun sequencing	This technology first time boosted the speed of sequencing genome and was being meticulously used from 1999 to 2006 to sequence the whole genome through the various project, e.g., the Human Genome Project
Roche/454/Pyro/FLX Sequencing	Very rapid parallel sequencing method, being used for metagenome since 2005 onward
Illumina/Solexa Genome Analyzer	Very rapid parallel sequencing method, being used for metagenome since 2006 onward, utilizes a sequencing-by-parallel-synthesis approach
Applied Biosystems SOLiD™ Sequencer	Uses an emulsion PCR approach with small magnetic beads to amplify the fragments for sequencing
Chromatin immunoprecipitation sequencing	Sequences and maps "reactome" than "genome"
Helicos Heliscope™ and Pacific Biosciences SMRT	Recently announced massive parallel sequencing and analysis platform

Source: Adopted from Patake and Patake (2011); Mardis (2008); Warren et al. (2007).

community (Hazen et al., 2013). They also have the advantage of providing rigorous annotation for the various taxa/genes present on the chip alleviating the limitation of the need for good homologs in the database to achieve accurate classification. Microarray-based approaches are, however, limited in that only the genes on the chip can be detected, thus limiting the potential for discovery of new genes or pathways in a sample. Microarray-based approaches are often a helpful complement to sequencing-based approaches as an additional line of evidence. The brief description of metagenomic approaches can be found in Table 10.2.

10.5 APPLICATIONS OF METAGENOMICS IN CHARACTERIZATION OF MICROBIAL COMMUNITIES DURING BIOREMEDIATION OF ENVIRONMENTAL POLLUTANTS

Metagenomic approaches play an important role in understanding the microbial community composition and dynamics in an ecological matrix. Metagenomics also advances our understanding of microbial degradation and detoxification of organic and inorganic pollutants at the contaminated site. Metagenomic approaches may be useful to find out the potential microbial degrader for the bioremediation of a specific pollutant or catabolic gene responsible for the degradation and detoxification of the specific pollutant (Table 10.3). Metagenomics is also used for comparing the microbial functional diversity at different contaminated sites impacted by a specific pollutant. Furthermore, advances in sequencing technologies such as next-generation sequencing (NGS) also allow us to investigate deeper and deeper layers of the microbial communities and are vital in presenting an unbiased view of phylogenetic composition and functional diversity of environmental microbial communities (Zwolinski, 2007).

The availability of whole-genome sequences from several environmental microorganisms pertinent to bioremediation has been useful to determine the gene pool of enzymes involved in degradation of anthropogenic pollutants (Galvao et al., 2005). Metagenomics surmounts the major limitations of cultivation-dependent studies, as it involves extraction of nucleic acids directly from environmental samples, which theoretically embodies the entire set of microbial

TABLE 10.3

Discovery of Pollutant-Degrading Microbes and Catabolic Enzymes Using Sequence-Based Metagenomic Approaches

Ecosystem	Pollutant	Category	Function	Method	Reference
Petroleum-contaminated water aquifers	Petroleum hydrocarbon	Mixture of hydrocarbons	Profiling of microbial diversity and adaptation to degradation of hydrocarbons	Illumina MiSeq platform	Kachienga et al. (2018)
Soil	Polychlorinated biphenyls	POP, adverse health effects	Metagenomic analysis of a biphenyl-degrading soil bacterial consortium revealing the metabolic roles of specific populations	Whole-genome shotgun Illumina sequencing	Garrido-Sanz et al. (2018)
Activated sludge	Sulfadiazine	Sulfa drug (antibiotic: bacteriostatic)	Revealing the role of sulfadiazine degrader, *Arthrobacter* sp. D2, and another *Pimelobacter bacterium* in degradation of sulfadiazine	Illumina HiSeq-2500 platform	Deng et al. (2018)
Soil	Aromatic compounds and pesticides	Insecticides, detergents, dyes, and many miscellaneous chemicals	Discovery of multifunctional genes and enzymes involved in xenobiotics degradation	Metatranscriptomics	Singh et al. (2018)
Wastewater	The precursor to PCBs; biphenyl; phenol; o-toluidine	Toxic effect on neurosystem; dye precursor	Discovery of oxygenases involved in the metabolism of the aromatic compound degradation	Functional annotation and comparative metagenomics	Jadeja et al. (2014)
Methane-rich arctic ocean sediments	Organic matter	Organic matter	Discovery of genes for organic matter degradation (mineralization)	Tag pyrosequencing and qPCR	Kirchman et al. (2014)
Soil	Polycyclic aromatic hydrocarbon	Used to make dyes, plastics, and pesticides	Revealing novel polycyclic aromatic hydrocarbon dioxygenases involved in degradation	454 pyrosequencing (Roche Biosciences), as well as Illumina technology	Chemerys et al. (2014)

(Continued)

TABLE 10.3 (Continued)

Discovery of Pollutant-Degrading Microbes and Catabolic Enzymes Using Sequence-Based Metagenomic Approaches

Ecosystem	Pollutant	Category	Function	Method	Reference
Oil sands tailings	Petroleum hydrocarbon	Mixture of hydrocarbons	Revealing alkane-degrading sulfate-reducing bacterium *Desulfatibacillum alkenivorans* AK-01 and detection of putative hydrocarbon succinate synthase genes (e.g., assA, bssA, and nmsA) implicated in anaerobic hydrocarbon degradation	454 pyrosequencing, Illumina paired-end sequencing, and 16S rRNA amplicon pyrotag sequencing	Tan et al. (2013)
Surface	Microcystin	Cyanobacterial harmful blooms (CyanoHABs) that produce microcystins	Metagenomic identification of bacterioplankton taxa and pathways involved in microcystin degradation	Pyrosequencing	Mou et al. (2013)
Ground water	Naphthalene/PAH	Mothball, fumigant	Discovery of 1,2-dihydroxynaphthalene-degrading enzymes (biodegradation)	PCR amplification and functional annotation	Wang et al. (2012)
Soil	3-Chlorobenzoate	Pesticide precursor	Discovery of dioxygenases involved in pesticide degradation	PCR and metagenome walking	Morimoto and Fujii (2009)
Soil	Polychlorinated biphenyl	Dielectric, coolant fluid; POP	Discovery of biphenyl dioxygenase involved in degradation of polychlorinated biphenyl	Pyrosequencing/functional annotation	Iwai et al. (2009)

community genomes present in a given ecosystem (Desai et al., 2011). Over the past few years, metagenomics-based methods have been useful to determine novel gene families and/or microbes involved in bioremediation of xenobiotics. Recently, DNA microarrays have been applied to microbial ecological research (Bae and Park, 2006); these microbial ecological microarrays have been useful in monitoring microbial communities and the efficacy of several bioremediation processes. In general, application of these "omics" techniques in bioremediation research has significantly aided in characterizing or monitoring pollutant-biodegrading microbial populations and identification of novel biodegradation pathways.

Many studies have reported the construction and screening of metagenomic libraries to identify genes involved in bioremediation. Martin et al. (2006) constructed metagenomic libraries to decipher ecological and metabolic functions of microbial communities involved in enhanced biological phosphate removal (EPBR) systems by a dominant polyphosphate-accumulating microorganism (POA) known as Candidatus *Accumulibacter phosphates*. Transcriptomic or metatranscriptomic tools are used to gain functional insights into the activities of environmental microbial communities by studying their mRNA transcriptional profiles (McGrath et al., 2008). Jennings et al. (2009) performed a transcriptomic analysis on a cis-dichloroethene (cDCE)-assimilating *Polaromonas* sp. JS666 strain in order to identify the genes upregulated by cDCE using DNA microarrays. Proteomics-based investigations have been useful in determining changes in the composition and abundance of proteins, as well as in the identification of key proteins involved in the physiological response of micro-organisms when exposed to anthropogenic pollutants. Kim et al. (2006) analyzed aromatic hydrocarbon catabolic pathways in *Pseudomonas putida* KT 2440 using a combined proteomic approach based on 2DE/MS and cleavable isotope-coded affinity tag analysis. Beyond genomics, transcriptomics, and proteomics, the research forefronts are now expanding toward the global analysis of the entire repertoire of cellular metabolites within a microbial cell; this newly introduced approach is known as "metabolomics." Several studies have recently applied microbial metabolome analysis to study biodegradation of anthropogenic pollutants. Recently, Keum et al. (2009) described studies on comparative metabolome analysis of *Sinorhizobium* sp. C4 during the degradation of phenanthrene. Recently, Tang et al. (2009) performed a fluxomics analysis on *Shewanella* sp. known to have cometabolic pathways for bioremediation of toxic metals, radionuclides, and halogenated organic compounds.

10.6   COMPUTATIONAL TOOLS IN METAGENOMIC STUDY

Several in silico software, pipelines, web resources, and algorithms are used to interpret or correlate molecular and -omics data (Desai et al., 2011). Some software, which are used in the metagenomic study, are summarized in Table 10.4. However, bioinformatic resources exclusively committed to bioremediation are still scanty. The University of Minnesota Biocatalysis/Biodegradation Database (UMBBD) has been the pioneer and most prominent web resource enlisting 187 pathways, 1287 reactions, 1195 compounds, 833 enzymes, 491 microorganism entries, and 259 biotransformation rules encompassing microbial bioremediation. UMBBD is freely available online at (http://umbbd. msi.umn.edu/) (Ellis et al., 2006). MetaRouter is yet another system, which is employed for maintaining heterogeneous information related to bioremediation and biodegradation in a framework that allows updating query modification (Desai et al., 2011). The core of the system is a relational database where the information on chemical compounds, reactions, enzymes, and organisms is stored in an integrated framework. It permits to find pathways between one compound and another (or between two sets of compounds). The representation of possible pathways can be restricted according to a number of criteria such as the length of the pathway, the required enzymes present in a given organism(s), and the intermediate compounds having a range of values for a given property (e.g., highly soluble). The system can be accessed and administrated through a web interface (Pazos et al., 2005). The full-featured MetaRouter analysis is freely available on the web (http://www.pdg. cnb.uam.es/biodeg_net/MetaRouter/).

TABLE 10.4

Computational and Bioinformatics Software Used for Metagenomic Analysis

Name of Software	Web Server	Application	Reference
UniFrac	http://bmf.colorado.edu/unifrac/	An online tool for comparing microbial diversity	Lozupone et al. (2006)
JANE	http://jane.bioapps.biozentrum. uniwuerzburg.de	Mapping of ESTs and variable length prokaryotic genome sequence reads on related template genomes	Liang et al. (2009)
WebCARMA	http://webcarma.cebitec. unibilefeld.de	A web application for the functional and taxonomical classification of unassembled metagenomic reads	Gerlach et al. (2009)
Prodigal	http://compbio.ornl.gov/prodial/	Prokaryotic gene recognition and translation initiation site identification	Hyatt et al. (2010)
MEGAN	http://www.ab.informatik. unituebingen.de/software/megan	Illumina sequencing metagenome reads analysis tool	Mitra et al. (2010)
MG-RAST	http://metagenomics.anl.gov/	Metagenome annotation server	Aziz (2010)
CAMERA	http://camera.calit2.net	Metagenomic database server, which contains sequences from environmental samples collected during the global ocean sampling	Maumus et al. (2009)
FUNGIpath	http://www.fungipath.upsud.fr	Database and tool serve for fungal orthology and metagenomics	Grossetete et al. (2010)
envDB	http://metagenomics.uv.es/ envDB/	Database and tool server for environmental distribution of prokaryotic taxa	*Tamames et al. (2010)*

Source: Adopted from Patake and Patake (2011).

10.7   CONCLUSION

Microbes play an important role in the degradation and detoxification of a variety of organic and inorganic pollutants and also help in the biogeochemical cycling of minerals in the ecosystem, and thus, the knowledge of microbes in contaminated ecological matrix is required to better understand the mechanism of bioremediation of a specific pollutant and find out the key enzyme of catabolic gene involved. To understand this, metagenomics is increasingly used to unfold the microbial community structure, composition, and functions in a contaminated environment. Metagenomics is quicker, accurate, and highly efficient genomic approaches that overcome the limitations of the conventional molecular techniques. With the rapid decrease in the sequencing prices, the NGS technologies are also becoming popular among the molecular biologists, as these techniques are easy to handle and require the use of various computational and bioinformatic tools that accurately interpret data and provide the same in a very informative way. Furthermore, the use of metagenomics in the monitoring of environmental microorganisms will open up a new vista for various applications at commercial enterprises.

ACKNOWLEDGMENTS

The authors are highly thankful to the University Grant Commission (UGC) and Department of Science and Technology (DST), Government of India (GOI), New Delhi, India, for providing financial support to our research works.

REFERENCES

Antizar-Ladislao, B. (2010) Bioremediation: working with bacteria. *Elements* 6:389–394.

Aziz, R.K. (2010) Subsystems-based servers for rapid annotation of genomes and metagenomes. *BMC Bioinformatics* 11(4):2–3.

Bae, J.-W., Park, Y.-H. (2006) Homogeneous versus heterogeneous probes for microbial ecological microarrays. *Trends Biotechnol.* 24:318–323.

Bharagava, R.N., Chowdhary, P., Saxena, G. (2017a) Bioremediation: an ecosustainable green technology: its applications and limitations. In: Bharagava, R.N. (ed.) *Environmental Pollutants and their Bioremediation Approaches.* 1st ed. CRC Press, Taylor & Francis Group, Boca Raton, FL, pp. 1–22. DOI:10.1201/9781315173351-2.

Bharagava, R.N., Saxena, G. (2019a) *Bioremediation of Industrial Waste for Environmental Safety: Biological Agents and Methods for Industrial Waste Management.* 1st ed. Springer Nature, Singapore. DOI:10.1007/978-981-13-3426-9.

Bharagava, R.N., Saxena, G. (2019b) Progresses in Bioremediation Technologies for Industrial Waste Treatment and Management: Challenges and Future Prospects. In: Bharagava, R.N., Saxena, G. (eds.) *Bioremediation of Industrial Waste for Environmental Safety: Biological Agents and Methods for Industrial Waste Management.* 1st ed. Springer Nature, Singapore. DOI: 10.1007/978-981-13-3426-9_21.

Bharagava, R.N., Saxena, G., Chowdhary, P. (2017b) Constructed wetlands: an emerging phytotechnology for degradation and detoxification of industrial wastewaters. In: Bharagava, R.N. (ed.) *Environmental Pollutants and their Bioremediation Approaches.* 1st ed. CRC Press, Taylor & Francis Group, Boca Raton, FL, pp. 397–426. DOI:10.1201/9781315173351-15.

Bharagava, R.N., Saxena, G., Mulla, S.I. (2019) Introduction to industrial wastes containing organic and inorganic pollutants and bioremediation approaches for environmental management. In: Saxena, G., Bharagava, R.N. (eds.) *Bioremediation of Industrial Waste for Environmental Safety: Volume I: Industrial Waste and its Management.* Springer Nature, Singapore. DOI:10.1007/978-981-13-1891-7_1.

Bharagava, R.N., Saxena, G., Mulla, S.I., Patel, D.K. (2017c) Characterization and identification of recalcitrant organic pollutants (ROPs) in tannery wastewater and its phytotoxicity evaluation for environmental safety. *Arch. Environ. Contam. Toxicol.* DOI:10.1007/s00244-017-0490-x.

Brodie, E.L., Desantis, T.Z., Joyner, D.C., Baek, S.M., Larsen, J.T., Andersen, G.L., Hazen, T.C., Richardson, P.M., Herman, D.J., Tokunaga, T.K., Wan, J.M., Firestone, M.K. (2006) Application of a high-density Oligonucleotide microarray approach to study bacterial population dynamics during uranium reduction and reoxidation. *Appl. Environ. Microbiol.* 72:6288–6298.

Bustin, S.A., Benes, V., Nolan, T., Pfaffl, M.W. (2005) Quantitative real-time RT-PCR – a perspective. *J. Mol. Endocrinol.* 34:597–601.

Caracciolo, A.B., Bottoni, P., Grenni, P. (2010) Fluorescence in situ hybridization in soil and water ecosystems: a useful method for studying the effect of xenobiotics on bacterial community structure. *Toxicol. Environ. Chem.* 92:567–579.

Chandra, R., Kumar, V. (2015) Biotransformation and biodegradation of organophosphates and organohalides. In: Chandra, R. (ed.) *Environmental Waste Management.* CRC Press, Boca Raton, FL, pp. 475–524.

Chandra, R., Kumar, V. (2017a) Detection of androgenic-mutagenic compounds and potential autochthonous bacterial communities during in situ bioremediation of post methanated distillery sludge. *Frontiers Microbiol.* 8:887.

Chandra, R., Kumar, V. (2017b) Detection of Bacillus and Stenotrophomonas species growing in an organic acid and endocrine-disrupting chemicals rich environment of distillery spent wash and its phytotoxicity. *Environ. Monitoring Assess.* 189:26.

Chandra, R., Saxena, G., Kumar, V. (2015) Phytoremediation of environmental pollutants: an eco-sustainable green technology to environmental management. In: Chandra, R. (ed.) *Advances in Biodegradation and Bioremediation of Industrial Waste.* 1st ed. CRC Press, Taylor & Francis Group, Boca Raton, FL, pp. 1–30. DOI:10.1201/b18218-2.

Chemerys, A., Pelletier, E., Cruaud, C. et al. (2014) Characterization of novel polycyclic aromatic hydrocarbon dioxygenases from the bacterial metagenomic DNA of a contaminated soil. *Appl. Environ. Microbiol.* 80:6591–6600.

DeAngelis, K.M., Wu, C.H., Beller, H.R., Brodie, E.L., Chakraborty, R., DeSantis, T.Z., Fortney, J.L., Hazen, T.C., Osman, S.R., Singer, M.E., Tom, L.M., Andersen, G.L. (2011) PCR amplification-independent methods for detection of microbial communities by the high-density microarray PhyloChip, *Appl. Environ. Microbiol.* 77:6313–6322.

Delmont, T.O., Simonet, P., Vogel, T.M. (2012) Describing microbial communities and performing global comparisons in the 'omic' era. *ISME J.* 6:1625–1628.

Deng, Y., Wang, Y., Mao, Y., Zhang, T. (2018) Partnership of Arthrobacter and Pimelobacter in aerobic degradation of sulfadiazine revealed by metagenomics analysis and isolation. *Environ. Sci. Technol.* 52:2963–2972.

Desai, C., Pathak, H., Madamwar, D. (2011) Advances in molecular and "-omics" technologies to gauge microbial communities and bioremediation at xenobiotic/anthropogen contaminated sites. *Bioresour. Technol.* 101:1558–1569.

Edwards, R.A., Rodriguez-Brito, B., Wegley, L., Haynes, M., Breitbart, M., Peterson, D.M., Saar, M.O., Alexander, S., Alexander, E.C., Rohwer, F. (2006) Using pyrosequencing to shed light on deep mine microbial ecology. *BMC Genomics* 7:57.

Ellis, L.B.M., Roe, D., Wackett, L.P. (2006) The University of Minneso biocatalysis/biodegradation database: the first decade. *Nucleic Acids Res.* 34:D517–D521.

Faust, K., Lahti, L., Gonze, D., de Vos, W.M., Raes, J. (2015) Metagenomics meets time series analysis: unraveling microbial community dynamics. *Curr. Opin. Microbiol.* 25:56–66.

Fierer, N., Jackson, J.A., Vilgalys, R., Jackson, R.B (2005) Assessment of soil microbial community structure by use of taxon-specific quantitative PCR assays. *Appl. Environ. Microbiol.* 71:4117–4120.

Fierer, N., Jackson, R.B. (2006) The diversity and biogeography of soil bacterial communities. *Proc. Natl. Acad. Sci. USA* 103:626–631.

Foti, M., Sorokin, D.Y., Lomans, B., Mussman, M., Zacharova, E.E., Pimenov, N.V., Kuenen, J.G., Muyzer, G. (2007) Diversity, activity, and abundance of sulfate-reducing bacteria in saline and hypersaline soda lakes. *Appl. Environ. Microbiol.* 73:2093–3000.

Galvao, T.C., Mohn, W.W., Lorenzo, V. (2005) Exploring the microbial biodegradation and biotransformation gene pool. *Trends Biotechnol.* 23:497–506.

Garrido-Sanz, D., Manzano, J., Martín, M., Redondo-Nieto, M., Rivilla, R. (2018) Metagenomic analysis of a biphenyl-degrading soil bacterial consortium reveals the metabolic roles of specific populations. *Front. Microbiol.* 9:232.

Gautam, S., Kaithwas, G., Bharagava, R.N., Saxena, G. (2017) Pollutants in tannery wastewater, pharmacological effects and bioremediation approaches for human health protection and environmental safety. In: Bharagava, R.N. (ed.) *Environmental Pollutants and their Bioremediation Approaches.* 1st ed. CRC Press, Taylor & Francis Group, Boca Raton, FL, pp. 369–396. DOI:10.1201/9781315173351-14.

Geets, J., Borremans, B., Diels, L., Springael, D., Vangronsveld, J., vanderLelie, D., Vanbroekhoven, K. (2006) DsrB gene-based DGGE for community and diversity surveys of sulfate-reducing bacteria. *J. Microbiol. Methods* 66:194–205.

Gentry, T.J., Wickham, G.S., Schadt, C.W., He, Z., Zhou, J. (2006) Microarray applications in microbial ecology research. *Microb. Ecol.* 52:159–175.

Gerlach, W., Unemann, S., Tille, F., Goesmann, A., Stoye, J. (2009) WebCARMA: a web application for the functional and taxonomic classification of unassembled metagenomic reads. *BMC Bioinformatics* 10:430–439.

Goutam, S.P., Saxena, G., Roy, D., Yadav, A.K., Bharagava, R.N. (2019) Green synthesis of nanoparticles and their applications in water and wastewater treatment. In: Saxena, G., Bharagava, R.N. (eds.) *Bioremediation of Industrial Waste for Environmental Safety: Volume I: Industrial Waste and Its Management.* Springer Nature, Singapore. DOI:10.1007/978-981-13-1891-7_17.

Goutam, S.P., Saxena, G., Singh, V., Yadav, A.K., Bharagava, R.N. (2018) Green synthesis of TiO_2 nanoparticles using leaf extract of Jatropha curcas L. for photocatalytic degradation of tannery wastewater. *Chem. Eng. J.* 336:386–396. DOI:10.1016/j.cej.2017.12.029.

Grossetete, S. et al. (2010) Fungi path: a tool to assess fungal metabolic pathways predicted by orthology. *BMC Genomics* 11:81–95.

Handelsman, J., Rondon, M.R., Brady, S.F., Clardy, J., Goodman, R.M. (1998) Molecular biological access to the chemistry of unknown soil microbes: a new frontier for natural products. *Chem. Biol.* 5:R245–R249.

Hazen, T.C., Dubinsky, E.A., DeSantis, T.Z., Andersen, G.L., Piceno, Y.M., Singh, N., Jansson, J.K., Probst, A., Borglin, S.E., Fortney, J.L., Stringfellow, W.T., Bill, M., Conrad, M.E., Tom, L.M., Chavarria, K.L., Alusi, T.R., Lamendella, R., Joyner, D.C., Spier, C., Baelum, J., Auer, M., Zemla, M.L., Chakraborty, R., Sonnenthal, E.L., D'haeseleer, P., Holman, H.Y.N., Osman, S., Lu, Z.M., Van Nostrand, J.D., Deng, Y., Zhou, J.Z., Mason, O.U. (2010) Deep-sea oil plume enriches indigenous oil-degrading bacteria. *Science* 330:204–208.

Hazen, T.C., Rocha, A.M., Techtmann, S.M. (2013) Advances in monitoring environmental microbes. *Curr. Opin. Biotechnol.* 24:526–533.

He, Z., Deng, Y., Van Nostrand, J.D., Tu, Q., Xu, M., Hemme, C.L., Li, X., Wu, L., Gentry, T.J., Yin, Y., Liebich, J., Hazen, T.C., Zhou, J. (2010) GeoChip 3.0 as a high-throughput tool for analyzing microbial community composition, structure and functional activity. *ISME J.* 4:1167–1179.

He, Z., Gentry, T.J., Schadt, C.W., Wu, L., Liebich, J., Chong, S.C., Huang, Z., Wu, W., Gu, B., Jardine, P., Criddle, C., Zhou, J. (2007) GeoChip: a comprehensive microarray for investigating biogeochemical, ecological and environmental processes. *ISME J.* 1:67–77.

Hu, M., Wang, X., Wen, X., Xia, Y. (2012) Microbial community structures in different wastewater treatment plants as revealed by 454-pyrosequencing analysis. *Bioresource Technol.* 117:72–79.

Hyatt, D. et al. (2010) Prodigal: prokaryotic gene recognition and translation initiation site identification. *BMC Bioinformatics* 11:119–129.

Imelfort, M., Parks, D., Woodcroft, B.J., Dennis, P., Hugenholtz, P., Tyson, G.W. (2014) GroopM: an automated tool for the recovery of population genomes from related metagenomes. *Peerj.* DOI:10.7717/peerj.603.

Iwai, S., Chai, B., Sul, W.J., Cole, J.R., Hashsham, S.A., Tiedje, J.M. (2009) Gene-targeted metagenomics reveals extensive diversity of aromatic dioxygenase genes in the environment. *ISME J.* 4(2):279–285.

Jadeja, N.B., More, R.P., Purohit, H.J., Kapley, A. (2014) Metagenomic analysis of oxygenases from activated sludge. *Bioresour. Technol.* 165:250–256.

Jennings, L.K., Chartrand, M.M., Lacrampe-Couloume, G., Lollar, B.S., Spain, J.C., Gossett, J.M. (2009) Proteomic and transcriptomic analyses reveal genes upregulated by cis-dichloroethene in polaromonas JS666. *Appl. Environ. Microbiol.* 11:3733–3744.

Kachienga, L., Jitendra, K., Momba, M. (2018) Metagenomic profiling for assessing microbial diversity and microbial adaptation to degradation of hydrocarbons in two South African petroleum-contaminated water aquifers. *Scientific Rep.* 8:7564.

Keum, Y.S., Seo, J.S., Qing, X.L., Kim, J.H. (2009) Comparative metabolomic analysis of Sinorhizobium sp. C4 during the degradation of phenanthrene. *Appl. Microbiol. Biotechnol.* 80:863–872.

Kim, Y.H., Cho, K., Yun, S.H., Kim, J.Y., Kwon, K.H., Yoo, J.S., Kim, S.I (2006) Analysis of aromatic catabolic pathways in Pseudomonas putida KT 2440 using a combined proteomic approach: 2DE/MS and cleavable isotope-coded affinity tag analysis. *Proteomics* 6:1301–1318.

Kirchman, D.L., Hanson, T.E., Cottrell, M.T., Hamdan, L.J. (2014) Metagenomic analysis of organic matter degradation in methane-rich Arctic Ocean sediments. *Limnol. Oceanogr.* 59(2):548–559.

Klindworth, A., Pruesse, E., Schweer, T., Peplies, J., Quast, C., Horn, M., Glockner, F.O. (2013) Evaluation of general 16S ribosomal RNA gene PCR primers for classical and next-generation sequencing-based diversity studies. *Nucleic Acids Res.* DOI:10.1093/nar/gks808

Kolb, S., Knief, C., Stubner, S., Conrad, R (2003) Quantitative detection of methanotrophs in soil by novel pmoA-targeted real-time PCR assays. *Appl. Environ. Microbiol.* 69:2423–2429.

Kumar, V., Chandra, R. (2018) Bacteria-assisted phytoremediation of industrial waste pollutants and ecorestoration. In: Chandra, R., Dubey, N.K., Kumar, V. (eds.) *Phytoremediation of Environmental Pollutants.* CRC Press, Boca Raton, FL, pp. 159–200.

Kumar, V., Shahi, S.K., Singh, S. (2018) Bioremediation: an eco-sustainable approach for restoration of contaminated sites. In: Singh, J., Sharma, D., Kumar, G., Sharma, N.R. (eds.) *Microbial Bioprospecting for Sustainable Development.* Springer, pp. 115–136.

Li, T., Wu, T.D., Mazéas, L., Toffin, L., Guerquin-Kern, J.L., Leblon, G., Bouchez, T (2008) Simultaneous analysis of microbial identity and function using NanoSIMS. *Environ. Microbiol.* 10:580–588.

Li, Z., Xu, J., Tang, C., Wu, J., Muhammad, A., Wang, H (2006) Application of 16S rDNA-PCR amplification and DGGE fingerprinting for detection of shift in microbial community diversity in Cu-, Zn-, and Cd-contaminated paddy soils. *Chemosphere* 62:1374–1380.

Liang, C., Schmid, A., Lopez-Sanchez, M.J., Moya, A., Gross, R., Bernhardt, J., Dandekar, T.J. (2009) Efficient mapping of prokaryotic ESTs and variable length sequence reads on related template genomes. *BMC Bioinformatics* 10:391–413.

Lozupone, C., Hamady, M., Knight, R. (2006) UniFrac—an online tool for comparing microbial community diversity in a phylogenetic context. *BMC Bioinformatics* 7:371–384.

Lu, Z.M., Deng, Y., Van Nostrand, J.D., He, Z.L., Voordeckers, J., Zhou, A.F., Lee, Y.J., Mason, O.U., Dubinsky, E.A., Chavarria, K.L., Tom, L.M., Fortney, J.L., Lamendella, R., Jansson, J.K., D'Haeseleer, P., Hazen, T.C., Zhou, J.Z. (2012) Microbial gene functions enriched in the Deepwater Horizon deep-sea oil plume. *ISME J.* 6:451–460.

Ma, Q., Qu, Y.-Y., Zhang, X.-Y. (2015) Identification of the microbial community composition and structure of coal-mine wastewater treatment plants. *Microbiol. Res.* 175:1–5.

Mardis, E.R. (2008) Next-generation DNA sequencing methods. *Annu. Rev. Genomics Hum. Genet.* 9:387–402.

Martin, H.G., Ivanova, N., Kunin, V., Warnecke, F., Barry, K.W., McHardy, A.C., Yeates, C., He, S., Salamov, A.A., Szeto, E., Dalin, E., Putnam, N.H., Shapiro, H.J., Pangilinan, J.L., Rigoutsos, I., Kyrpides, N.C., Blackall, L.L., McMahon, K.D., Hugenholtz, P. (2006) Metagenomic analysis of two enhanced biological phosphorus removal (EBPR) sludge communities. *Nat. Biotechnol.* 24:1263–1269.

Mason, O.U., Scott, N.M., Gonzalez, A., Robbins-Pianka, A., Baelum, J., Kimbrel, J., Bouskill, N.J., Prestat, E., Borglin, S., Joyner, D.C., Fortney, J.L., Jurelevicius, D., Stringfellow, W.T., Alvarez-Cohen, L., Hazen, T.C., Knight, R., Gilbert, J.A,, Jansson J.K. (2014) Metagenomics reveals sediment microbial community response to deepwater horizon oil spill. *ISME J.* 8:1464–1475.

Maszenan, A.M., Liu, Y., Ng, W.J (2011) Bioremediation of wastewaters with recalcitrant organic compounds and metals by aerobic granules. *Biotechnol. Adv.* 29:111–123.

Maumus, F., Allen, A.E., Mhiri, C., Hu, H., Jabbari, K., Vardi, A., Grandbastien, M., Bowler, C. (2009) Potential impact of stress activated retrotransposons on genome evolution in a marine diatom. *BMC Genomics* 10:624–62.

McGrath, K.C., Thomas-Hall, S.R., Cheng, C.T., Leo, L., Alexa, A., Schmidt, S., Schenk, P.M. (2008) Isolation and analysis of mRNA from environmental microbial communities. *J. Microbiol. Methods* 75:172–176.

Megharaj, M., Ramakrishnan, B., Venkateswarlu, K., Sethunathan, N., Naidu, R. (2011) Bioremediation approaches for organic pollutants: a critical perspective. *Environ. Int.* 37:1362–1375.

Mills, D.K., Entry, J.A., Gillevet, P.M. (2007) Assessing microbial community diversity using amplicon length heterogeneity polymerase chain reaction. *Soil Sci. Soc. Am. J.* 71:572–578.

Mitra, S., Schubach, M., Huson, D.H. (2010) Short clones or long clones? A simulation study on the use of paired reads in metagenomics. *BMC Bioinformatics* 11(1):512–522.

Moran, M.A., Satinsky, B., Gifford, S.M., Luo, H.W., Rivers, A., Chan, L.K., Meng, J., Durham, B.P., Shen, C., Varaljay, V.A., Smith, C.B., Yager, P.L., Hopkinson, B.M. (2013) Sizing up metatranscriptomics. *ISME J.* 7:237–243.

Morimoto, S., Fujii, T., 2009. A new approach to retrieve full lengths of functional genes from soil by PCR-DGGE and metagenome walking. *Appl. Microbiol. Biotechnol.* 83(2):389–396.

Mou, X., Lu, X., Jacob, J., Sun, S., Heath, R. (2013) Metagenomic identification of bacterioplankton taxa and pathways involved in microcystin degradation in Lake Erie. *PLoS ONE* 8(4):e61890.

Mulla, S.I., Ameen, F., Talwar, M.P., Eqani, S.A.M.A.S., Bharagava, R.N., Saxena, G., Tallur, P.N., Ninnekar, H.Z. (2019) Organophosphate pesticides: impact on environment, toxicity, and their degradation. In: Saxena, G., Bharagava, R.N. (eds.) *Bioremediation of Industrial Waste for Environmental Safety: Volume I: Industrial Waste and Its Management.* Springer Nature, Singapore. DOI:10.1007/978-981-13-1891-7_14.

Nocker, A., Burr, M., Camper, A (2007) Genotypic microbial community profiling: a critical technical review. *Microb. Ecol.* 54:276–289.

Pace, N.R., Stahl, D.A., Lane, D.J., Olsen, G.J. (1985) Analyzing natural microbial populations by rRNA sequences. *ASM News* 51:4–12.

Parada, A.E., Needham, D.M., Fuhrman, J.A. (2015) Every base matters: assessing small subunit rRNA primers for marine microbiomes with mock communities, time series and global field samples. *Environ. Microbiol.* 18:1403–1404.

Patake, R.S., Patake, G.R. (2011) A Mini review on metagenomics and its implications in ecological and environmental biotechnology. *Univ. J. Environ. Res. Technol.* 1:1–6.

Pazos, F., Guijas, D., Valencia, A., De Lorenzo, V. (2005) MetaRouter: bioinformatics for bioremediation. *Nucleic Acids Res.* 33:D588–D592.

Pernthaler, A., Pernthaler, J., Amann, R. (2002) Fluorescence in situ hybridization and catalyzed reporter deposition for the identification of marine bacteria. *Appl. Environ. Microbiol.* 68:3094–3101.

Prosser, J.I. (2012) Ecosystem processes and interactions in a morass of diversity. *FEMS Microbiol. Ecol.* 81:507–519.

Ranjard, L., Poly, F., Lata, J.C., Mougel, C., Thioulouse, J., Nazaret, S. (2001) Characterization of bacterial and fungal soil communities by automated ribosomal intergenic spacer analysis fingerprints: biological and methodological variability. *Appl. Environ. Microbiol.* 67:4479–4487.

Rastogi, G., Sani, R.K. (2011) Molecular techniques to assess microbial community structure, function, and dynamics in the environment. In: Ahmad, I. et al. (eds.) *Microbes and Microbial Technology: Agricultural and Environmental Applications.* DOI:10.1007/978-1-4419-7931-5_2.

Reshma, S.V., Spandana, S., Sowmya, M. (2011) *Bioremediation Technologies.* World Congress of Biotechnology, India.

Riesenfeld, C.S., Schloss, P.D., Handelsman, J. (2004) Metagenomics: genomic analysis of microbial communities. *Annu. Rev. Genet.* 38:525–552.

Rinke, C., Schwientek, P., Sczyrba, A., Ivanova, N.N., Anderson, I.J., Cheng, J.F., Darling, A., Malfatti, S., Swan, B.K., Gies, E.A., Dodsworth, J.A., Hedlund, B.P., Tsiamis, G., Sievert, S.M., Liu, W.T., Eisen, J.A., Hallam, S.J., Kyrpides, N.C., Stepanauskas, R., Rubin, E.M., Hugenholtz, P., Woyke, T. (2013) Insights into the phylogeny and coding potential of microbial dark matter. *Nature* 499:431–437.

Ritchie, N.J., Schutter, M.E., Dick, R.P., Myrold, D.D (2000) Use of length heterogeneity PCR and fatty acid methyl ester profiles to characterize microbial communities in soil. *Appl. Environ. Microbiol.* 66:1668–1675.

Rogers, S.W., Mooreman, T.B., Onge, S.K. (2007) Fluorescent in situ hybridization and microautoradiography applied to ecophysiology in soil. *Soil Sci. Soc. Am. J.* 71:620–631.

Saxena, G., Bharagava, R.N. (2015) Persistent organic pollutants and bacterial communities present during the treatment of tannery wastewater. In: Chandra, R. (ed.) *Environmental Waste Management.* 1st ed. CRC Press, Taylor & Francis Group, Boca Raton, FL, pp. 217–247.

Saxena, G., Bharagava, R.N. (2016) Ram Chandra: advances in biodegradation and bioremediation of industrial waste. *Clean Tech. Environ. Policy* 18:979–980.

Saxena, G., Bharagava, R.N. (2017) Organic and inorganic pollutants in industrial wastes, their ecotoxicological effects, health hazards and bioremediation approaches. In: Bharagava, R.N. (ed.) *Environmental Pollutants and their Bioremediation Approaches.* 1st ed. CRC Press, Taylor & Francis Group, Boca Raton, FL, pp. 23–56.

Saxena, G., Bharagava, R.N. (2019) *Bioremediation of Industrial Waste for Environmental Safety: Industrial Waste and Its Management.* 1st ed. Springer Nature, Singapore. DOI:10.1007/978-981-13-1891-7.

Saxena, G., Bharagava, R.N., Kaithwas, G., Raj, A. (2015) Microbial indicators, pathogens and methods for their monitoring in water environment. *J. Water Health* 13:319–339. DOI:10.2166/wh.2014.275.

Saxena, G., Chandra, R., Bharagava, R.N. (2016) Environmental pollution, toxicity profile and treatment approaches for tannery wastewater and its chemical pollutants. *Rev. Environ. Contam. Toxicol.* 240:31–69. DOI:10.1007/398_2015_5009.

Saxena, G., Purchase, D., Mulla, S.I., Saratale, G.D., Bharagava, R.N. (2019a) Phytoremediation of heavy metal-contaminated sites: eco-environmental concerns, field studies, sustainability issues and future prospects. *Rev. Env. Contam. Toxicol.* DOI:10.1007/398_2019_24.

Saxena, G., Kishor, R., Bharagava, R.N. (2019b) Application of microbial enzymes in degradation and detoxification of organic and inorganic pollutants. In: Saxena, G., Bharagava, R.N. (eds.) *Bioremediation of Industrial Waste for Environmental Safety: Volume I: Industrial Waste and Its Management.* Springer Nature, Singapore. DOI:10.1007/978-981-13-1891-7_3.

Saxena, G., Purchase, D., Bharagava, R.N. (2019c) Environmental hazards and toxicity profile of organic and inorganic pollutants of tannery wastewater and bioremediation approaches. In: Saxena, G., Bharagava, R.N. (eds.) *Bioremediation of Industrial Waste for Environmental Safety: Volume I: Industrial Waste and Its Management.* Springer Nature, Singapore. DOI:10.1007/978-981-13-1891-7_18.

Saxena, G., Kishor, R., Purchase, D., Bharagava, R.N. (2019d) Chandra R, Dubey NK, Kumar V: Phytoremediation of environmental pollutants. *Environmental Earth Sciences* 78:418.

Saxena, G., Kishor, R., Saratale, G.D., Bharagava, R.N. (2019e) Genetically Modified Organisms (GMOs) and Their Potential in Environmental Management: Constraints, Prospects, and Challenges. In: Bharagava, R.N., Saxena, G. (eds.) *Bioremediation of Industrial Waste for Environmental Safety: Industrial Waste and Its Management.* Springer Nature, Singapore. DOI: 10.1007/978-981-13-3426-9_1.

Saxena, G., Goutam, S.P., Mishra, A., Mulla, S.I. Bharagava, R.N. (2019f) Emerging and Ecofriendly Technologies for the Removal of Organic and Inorganic Pollutants from Industrial Wastewaters. In: Bharagava, R.N., Saxena, G. (eds.) *Bioremediation of Industrial Waste for Environmental Safety: Biological Agents and Methods for Industrial Waste Management.* Springer Nature, Singapore. DOI: 10.1007/978-981-13-3426-9_5.

Schloss, P.D., Handelsman, J. (2005) Metagenomics for studying unculturable microorganisms: cutting the Gordian knot. *Genome Biol.* 6:229.

Schmeisser, C., Steele, H., Streit, W.R. (2007) Metagenomics, biotechnology with non-culturable microbes. *Appl. Microbiol. Biotechnol.* 75:955–962.

Shakya, M., Quince, C., Campbell, J.H., Yang, Z.M.K., Schadt, C.W., Podar, M. (2013) Comparative metagenomic and rRNA microbial diversity characterization using archaeal and bacterial synthetic communities. *Environ. Microbiol.* 15:1882–1899.

Sharon, I., Banfield, J.F. (2013) Genomes from metagenomics. *Science* 342:1057–1058.

Singh, D.P., Prabha, R., Gupta, V.K., Verma, M.K. (2018) Metatranscriptome analysis deciphers multifunctional genes and enzymes linked with the degradation of aromatic compounds and pesticides in the wheat rhizosphere. *Front. Microbiol.* 9:1331.

Smith, C.J., Osborn, A.M. (2009) Advantages and limitations of quantitative PCR (Q-PCR)-based approaches in microbial ecology. *FEMS Microbiol. Ecol.* 67:6–20.

Tamames et al. (2010) Environmental distribution of Prokaryotic Taxa. *BMC Microbiol.* 10:85–98.

Tan, B., Dong, X., Sensen, C.W., Foght, J. (2013) Metagenomic analysis of an anaerobic alkane-degrading microbial culture: potential hydrocarbon-activating pathways and inferred roles of community members. *Genome* 56:599–611.

Tang, Y.J., Martin, H.G., Dehal, P.S., Deutschbauer, A., Llora, X., Meadows, A., Arkin, A., Keasling, J.D (2009) Metabolic flux analysis of Shewanella spp. reveals evolutionary robustness in central carbon metabolism. *Biotechnol. Bioeng.* 102:1161–1169.

Techtmann, S.M., Hazen, T.C. (2016) Metagenomic applications in environmental monitoring and bioremediation. *J. Ind. Microbiol. Biotechnol.* DOI:10.1007/s10295-016-1809-8.

Thies, J.E. (2007) Soil microbial community analysis using terminal restriction fragment length polymorphisms. *Soil Sci. Soc. Am. J.* 71:579–591.

Uhlik, O., Leewis, M-C., Strejcek, M., Musilova, L., Mackova, M., Leigh, M.B., Macek, T. (2013) Stable isotope probing in the metagenomics era: a bridge towards improved bioremediation. *Biotechnol. Adv.* 31(2):154–165.

Venter, J.C., Remington, K., Heidelberg, J.F., Halpern, A.L., Rusch, D., Eisen, J.A., Wu, D., Paulsen, I., Nelson, K.E., Nelson, W., Fouts, D.E., Levy, S., Knap, A.H., Lomas, M.W., Nealson, K., White, O., Peterson, J., Hoffman, J., Parsons, R., Baden-Tillson, H., Pfannkoch, C., Rogers, Y.H., Smith, H.O. (2004) Environmental genome shotgun sequencing of the Sargasso Sea. *Science* 304:66–74.

Vieites, J.M., Guazzaroni, M.E., Beloqui, A., Golyshin, P.N., Ferrer, M. (2009) Metagenomics approaches in systems microbiology. *FEMS Microbiol. Rev.* 33:236–255.

Wakase, S., Sasaki, H., Itoh, K., Otawa, K., Kitazume, O., Nonaka, J., Satoh, M., Sasaki, T., Nakai, Y. (2007) Investigation of the microbial community in a microbiological additive used in a manure composting process. *Biores. Technol.* 99:2687–2693.

Wang, Y., Chen, Y., Zhou, Q., Huang, S., Ning, K., Xu, J., et al. (2012) A culture-independent approach to unravel uncultured bacteria and functional genes in a complex microbial community. *PLoS One* 7(10):e47530.

Warren, R.L., Sutton, G.G., Jones, S.J.M. and Holt, R.A. (2007) Assembling millions of short DNA sequences using SSAKE. *Bioinformatics* 23(4):500–501.

Yang, Y., Yao, J., Hu, S., Qi, Y. (2000) Effects of agricultural chemicals on DNA sequence diversity of soil microbial community: a study with RAPD marker. *Microb. Ecol.* 39:72–79.

Yun, J., Ryu, S. (2005) Screening for novel enzymes from metagenome and SIGEX, as a way to improve it. *Microb. Cell Fact.* 4:8.

Zhou, A.F., He, Z.L., Qin, Y.J., Lu, Z.M., Deng, Y., Tu, Q.C., Hemme, C.L., Van Nostrand, J.D., Wu, L.Y., Hazen, T.C., Arkin, A.P., Zhou, J.Z. (2013) StressChip as a high-throughput tool for assessing microbial community responses to environmental stresses. *Environ. Sci. Technol.* 47:9841–9849.

Zwolinski, M.D. (2007) DNA sequencing: strategies for soil microbiology. *Soil Sci. Soc. Am. J.* 71:592–600.

11 Microbial Capacities for Utilization of Nitroaromatics

Asifa Qureshi, Hitesh Tikariha, and Hemant J. Purohit
CSIR-National Environmental Engineering Research Institute

CONTENTS

11.1 INTRODUCTION

Nitroaromatic compounds (NACs), namely, nitrobenzene, p-nitrophenol (PNP), nitrotoluenes (TNT, DNT, NT), are critical environmental pollutants reported because of their toxicity to many living organisms (Kumari et al. 2017, Saha et al. 2014). The nitro substituents on NACs possess the electron-withdrawing character, which causes resistance toward biodegradation. Therefore, the oxidative attack by bacterial oxygenases becomes difficult. Still, several potent bacteria have adapted to utilize NACs using oxygenases. Increase in the number of nitro groups and electron-withdrawing substituents on aromatic ring causes an increase in recalcitrant character forcing the nitroaromatics to be utilized by partial reduction mechanism.

Microorganisms use oxidative and reductive degrading pathways to convert NACs entirely to CO_2 and H_2O or partially to an organic compound. It is the genetic machinery present in the microbes that guide the conversion of simple products from NACs. While aerobic bacteria use aerobic and partial reductive catabolic pathways systems, anaerobic bacteria were capable of using only the reductive mechanism to utilize NACs.

NACs and their substituted derivatives have been reported to be efficiently degraded by microorganisms, but still, pollutants persist in the environment because microbial strains have been reported to be useful as bioremediation agents at controlled laboratory conditions (Qureshi et al. 2009, 2012, Ghosh et al. 2017, Singh et al. 2015). Researchers have reported that bacteria could grow only under the influence of pH, temperature, oxygen, moisture, the appropriate level of nutrients, the bioavailability of contaminants, and the presence of other toxic compounds (Yadav et al. 2015, Saha et al. 2017).

Thus, exploiting the maximum potential of bacteria to degrade pollutants in the contaminated environment remains a challenge today, which could be accomplished in the future by understanding the microbial capacities of utilization based on "-omics" analysis.

11.2 WHY THE STUDY ON NITROAROMATIC DEGRADATIVE PATHWAY IMPORTANT?

A recent literature survey has reported that nitro-/amino-/chloroaromatic containing industrial wastewaters should be uncontaminated before being discharged into the environment (Chatterjee et al. 2017, Purohit et al. 2016) to reduce pollution levels in the environment, so knowledge of biodegradation capacities of bacteria at genetic levels and their regulatory pathways becomes imperative. Reports on NAC biodegradation indicate that such microbes could serve as a biocatalyst for degradation (Begum and Arundhati 2016, Thangaraj et al. 2008).

11.3 CHALLENGES IN NITROAROMATIC BIODEGRADATION

Environmental sustainability is one of the significant challenges and drivers toward the development of the world economy. Globally, one of the goals followed for sustainable urbanization and industrialization process includes responsible production and consumption, but discharge and disposal of hazardous wastes containing organic and inorganic pollutants remains a big issue.

With the listing of nitroaromatic compounds in the U.S. Environmental Protection Agency's hazardous category, there is a need to protect the environment from such hazardous chemicals. The production of nitroaromatics derived from anthropogenic sources such as pharmaceutical, defense, and agricultural sectors could not be restricted. Instead, the ways to counteract their effect need to be proposed and practiced throughout the world. One of the best strategies involved in the biodegradation of the NACs involves the use of microbes. It is envisaged from the researcher's inputs toward biodegradation of some nitroaromatics, that is, by proper use of microbes and their field applications (bioremediation), mitigation of nitroaromatics could find promiscuity. For filling this gap, it becomes imperative to understand various biodegradative mechanisms, capacities, and genetics of bacteria responsible for biodegradation of a wide array of nitroaromatics, insights of which have been described in this chapter.

Currently, genomics, proteomics, metabolomics, and their combinations are reported as emerging powerful tools for understanding microbial capacities and evaluating biodegradation of aromatics and substituted aromatic compounds (Tikariha et al. 2016, Kapley and Purohit 2009, Qureshi et al. 2009).

11.4 MOLECULAR MECHANISMS OF BIODEGRADATIVE PATHWAYS OF AROMATIC AND NITROAROMATICS

11.4.1 Aromatic Biodegradative Pathways

A common biodegradative pathway of aromatics and substituted aromatic compounds such as NACs is illustrated in Figure 11.1. In general, the aerobic biodegradation proceeds via two phases: first, ring cleavage of an aromatic compound by a specific ring modification reactions occurs, leading to the generation of a dihydroxylated benzene ring or substituted dihydroxylated benzene ring. Fission of this aromatic ring in the second phase of degradation with subsequent reactions leads to the

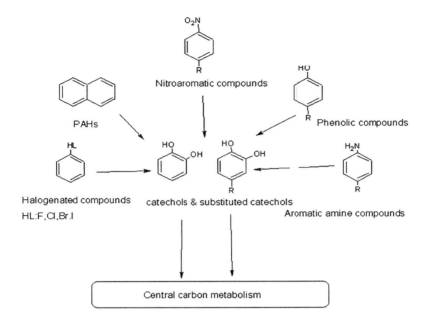

FIGURE 11.1 Aromatic compound biotransformations/biodegradation by bacteria. Substituted derivatives, viz., nitroaromatics, phenolics, polyaromatic hydrocarbons, aromatic amines, and halogenated compounds degrade via catechols and substituted catechols as intermediates, which enter into central carbon metabolism.

central carbon intermediates. There are various dioxygenases responsible for catalyzing the step of ring fission. Depending on the type of cleavage, it can be known as either ortho- or beta-ketoadipate pathway (from the fact that beta-ketoadipate is a crucial intermediate of ortho-cleavage) when the cleavage of the ring occurs between the hydroxyl groups (intradiol cleavage) or when adjacent to one of the hydroxyls gets cleaved, it is known as the meta pathway (estradiol cleavage).

Major pathways reported for catabolism of aromatic compounds in bacteria have revealed that initial conversion steps were carried out by different enzymes, but the compounds get transformed into a limited number of derived metabolites, such as protocatechuates and catechols (Figure 11.1). These intermediate metabolites get channeled into central metabolic routes via different enzymes in the ring cleavage pathway. This generalized scheme of catabolic pathways for aromatic compounds suggested that microorganisms have extended their substrate range and capacities by developing "peripheral" enzymes, which were able to transform initial substrates into one of the central intermediates such as catechol, resorcinol, and benzoquinones.

11.4.2 NACs Biodegradative Pathways

Several researchers have reported degradative pathways of nitroaromatic pollutants and their intermediates (Tiwari et al. 2017, Ghosh et al. 2017, Qureshi and Purohit, 2002, Min et al. 2017). Studies have shown the bacterial degradation of toxic compound, viz., 4-chloro-3-nitrophenol (4C3NP) and formation of 4-chlororesorcinol intermediate in the degradation pathway of 4C3NP, and also 2-chloro-4-aminophenol degradation. A 4C3NP-mineralizing bacterium, *Pseudomonas* sp., has been isolated from wastewaters collected from a chemically contaminated area.

11.4.2.1 Nitrobenzene as Model NAC

Nitrobenzene as one of the representative NACs has been reported here for understanding the biodegradative pathways being adopted by different bacteria such as *Pseudomonas pseudoalcaligene, Commamonas* sp., *Pseudomonas mendocina,* and *Pseudomonas putida* (Figure 11.2).

(a)
(d)
(b)
(c)
2-amino muconic semialdehyde
2-amino phenol
Hydroxylam inobenzene
Nitrosobenzene
Nitrobenzene
3-nitrocatechol
Ring cleavage and formation of muconic semialdehyde derivative
Catechol
4-nitrocatechol
2-hydroxymuconic semialdehyde

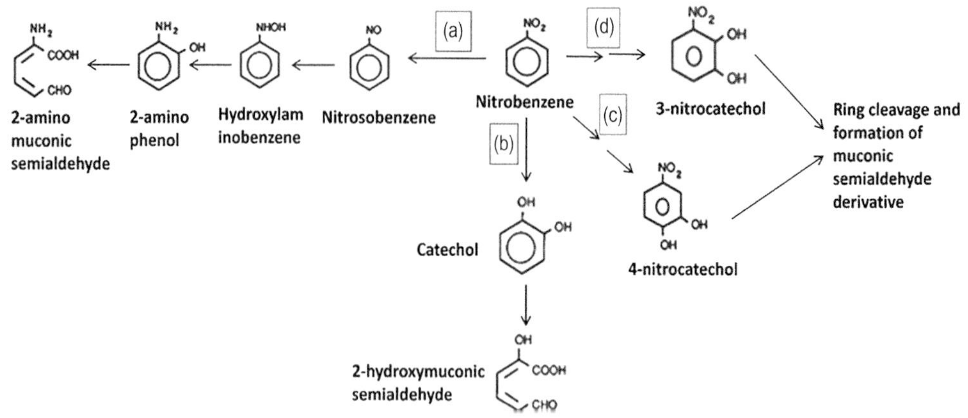

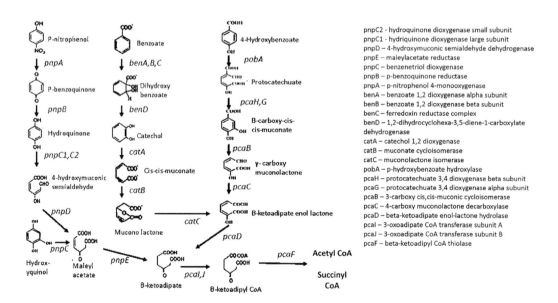

FIGURE 11.2 Degradation scheme adopted by different bacteria. Pathway A is followed by *Pseudomonas pseudoalcaligene*, B by *Comamonas* ssp., C by *Pseudomonas mendocina*, and D by *Pseudomonas putida*. All pathways lead to formation of catechol derivative, which then leads to ring cleavage and then proceeds to form maleylacetate and kedoadipate and then moves on to tricarboxylic acid cycle.

11.4.2.2 Nitrophenol as Model NAC

Other pollutants such as 4-nitrobenzoate, 2-nitrobenzoate, 2-nitrotoluene, nitrobenzene, *p*-nitroaniline, and PNP were reported to be used as growth substrates by several bacteria (Yanzhen et al. 2016, Tikariha et al. 2016, Qureshi et al. 2007). para-Nitrophenol as a representative model compound for nitroaromatic pollutants has been shown in Figure 11.3 for understanding biodegradative pathways, in bacteria such as *Pseudomonas* species.

P-nitrophenol
pnpA
P-benzoquinone
pnpB
Hydroquinone
pnpC1,C2
4-hydroxymuconic semialdehyde
pnpD
pnpC
Hydroxyquinol
Maleyl acetate
pnpE
Benzoate
benA,B,C
Dihydroxy benzoate
benD
Catechol
catA
Cis-cis-muconate
catB
Mucono lactone
catC
4-Hydroxybenzoate
pobA
Protocatechuate
pcaH,G
B-carboxy-cis-cis-muconate
pcaB
γ-carboxy muconolactone
pcaC
B-ketoadipate enol lactone
pcaD
B-ketoadipate
pcaI,J
B-ketoadipyl CoA
pcaF
Acetyl CoA
Succinyl CoA
pnpC2 - hydroquinone dioxygenase small subunit
pnpC1 - hydriquinone dioxygenase large subunit
pnpD – 4-hydroxymuconic semialdehyde dehydrogenase
pnpE – maleylacetate reductase
pnpC – benzenetriol dioxygenase
pnpB – p-benzoquinone reductase
pnpA – p-nitrophenol 4-monooxygenase
benA – benzoate 1,2 dioxygenase alpha subunit
benB – benzoate 1,2 dioxygenase beta subunit
benC – ferredoxin reductase complex
benD – 1,2-dihydrocyclohexa-3,5-diene-1-carboxylate dehydrogenase
catA – catechol 1,2 dioxygenase
catB – muconate cycloisomerase
catC – muconolactone isomerase
pobA – p-hydroxybenzoate hydroxylase
pcaH – protocatechuate 3,4 dioxygenase beta subunit
pcaG – protocatechuate 3,4 dioxygenase alpha subunit
pcaB – 3-carboxy cis,cis-muconic cycloisomerase
pcaC – 4-carboxy muconolactone decarboxylase
pcaD – beta-ketoadipate enol-lactone hydrolase
pcaI – 3-oxoadipate CoA transferase subunit A
pcaJ – 3-oxoadipate CoA transferase subunit B
pcaF – beta-ketoadipyl CoA thiolase

FIGURE 11.3 Degradation pathways adopted by bacteria (*Pseudomonas species*) for p-nitrophenol (PNP) utilization indicating their capacity at molecular level.

TABLE 11.1

Primary source of Nitroaromatic Compounds in the Environment

S.No.	Nitroaromatic Compounds	Sources
1	p-Nitrophenol	Intermediate in the synthesis of insecticides such as parathion and methyl parathion and also in drug synthesis
2	2-Nitrophenol	As an intermediate in the synthesis of pigments, rubber chemicals, and dyestuffs
3	m-Nitrophenol	Widely used as an intermediate in the preparation of dyes, pigments, lumber preservatives, photographic chemicals, and pesticides
4	Nitrotoluenes	Used in manufacturing of dyestuff and also as an intermediate for explosive, plastics, and pharmaceuticals
5	Other nitrophenol derivatives	Pesticides such as carbofuran, fluorodifen, nitrogen, and bifenox
6	Dinitrophenols	Pesticides such as ovicides, insecticides, herbicides, and fungicides
7	Substituted nitrobenzenes and nitropyridines	Synthesis of indoles from which drugs and agrochemicals are manufactured
8	Nitrobenzene	For the production of aniline, which on downstream used in drug, pesticides, and explosives manufacturing
9	Nitrobenzene, 3-nitrotoluene, 1- and 2-nitronaphthalene, 3-nitrobiphenyl	Generated in the environment from hydrocarbons released during combustion and fossil fuel burning process

11.4.2.3 Nitrotoluene as NAC

NACs trinitrotoluene (TNT) and dinitrotoluene (DNT) are known to be as recalcitrant and toxic. Their complete degradation has been mainly reported mainly under anoxic conditions following partial and complete reductive pathways (Liu et al. 2017, Thijs et al. 2014) (Table 11.1).

11.5 MICROBES ASSOCIATED WITH BIODEGRADATION OF NITROAROMATICS IN THE ENVIRONMENT AND THEIR ADAPTIVE FEATURES

Many species of bacteria have proven their ability as potent candidate for NAC degraders, which mainly include *Pseudomonas, Burkholderia, Rhodococcus, Roseivirga,* and *Shewanella oneidensis* (Tikariha et al. 2016, Sengupta et al. 2015, Yanzhen et al. 2016, Xu et al. 2016, Selvaratnam et al. 2016, Liu et al. 2017, Min et al. 2016, Ghosh et al. 2017) (Table 11.1, Figure 11.4). Most of these bacteria belong to phyla Proteobacteria and class Gammaproteobacteria, viz., *Pseudomonas, Burkholderia, Bradyrhizobium,* Cupriviadus, *Shewanella, Raoultella.* Only a few exceptions such as *Roseivirga* belonging to Actinobacteria and Rhodococcus belonging to phyla Bacteriodetes exist. Thus, nitroaromatic-utilizing genera fall in those groups of bacteria, which are observed mostly in soil and often associated with the degradation of a varied range of aromatic compounds.

11.5.1 NAC-UTILIZING MICROBES

NAC-degrading bacterial genomes harbor specific genes, such as oxygenases, for opening the aromatic ring, which then goes down to the central aromatic pathway (Figure 11.2). Moreover, the genetic mapping of the bacteria mentioned above has not revealed any highly specialized feature that could be solely present in nitroaromatic-utilizing bacteria. However, like other aromatic-degrading bacteria, they mostly have a broad set of genes for the utilization of other

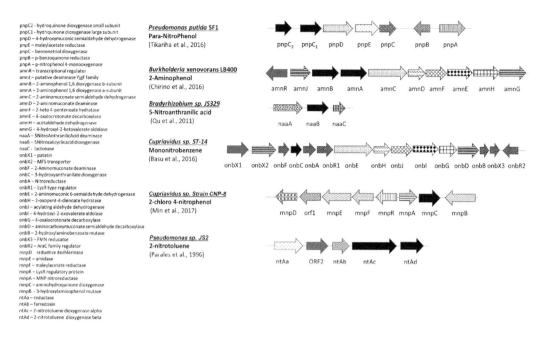

pnpC2 - hydroquinone dioxygenase small subunit
pnpC1 - hydroquinone dioxygenase large subunit
pnpD - 4-hydroxymuconic semialdehyde dehydrogenase
pnpE - maleylacetate reductase
pnpC - benzenetriol dioxygenase
pnpB - p-benzoquinone reductase
pnpA - p-nitrophenol 4-monooxygenase
amnR - transcriptional regulator
amnJ - putative deaminase YjgF family
amnB - 2-aminophenol 1,6 dioxygenase b-subunit
amnA - 2-aminophenol 1,6 dioxygenase a-subunit
amnC - 2-aminomuconate semialdehyde dehydrogenase
amnD - 2-aminomuconate deaminase
amnF - 2-keto 4-pentenoate hydratase
amnE - 4-oxalocrotonate decarboxylase
amnH - acetaldehyde dehydrogenase
amnG - 4-hydroxyl-2-ketovalerate aldolase
naaA - 5NitroAnthranilicAcid deaminase
naaB - 5Nitrosalicylicacid dioxygenase
naaC - lactonase
onbX1 - patatin
onbX2 - MFS transporter
onbF - 2-Aminomuconate deaminase
onbC - 3-hydroxyanthranilate dioxygenase
onbA - Nitroreductase
onbR1 - LysR type regulator
onbE - 2-aminomuconic 6-semialdehyde dehydrogenase
onbH - 2-oxopent-4-dienoate hydratase
onbJ - acylating aldehyde dehydrogenase
onbI - 4-hydroxy-2-oxovalerate aldolase
onbG - 4-oxalocrotonate decarboxylase
onbD - aminocarboxymuconate semialdehyde decarboxylase
onbB - 2-hydroxylaminobenzoate mutase
onbX3 - FMN reducatse
onbR2 - AraC family regulator
mnpD - reductive dechlorinase
mnpE - amidase
mnpF - maleylacetate reductase
mnpR - LysR regulatory protein
mnpA - MNP nitroreductase
mnpC - aminohydroquinone dioxygenase
mnpB - 3-hydroxylaminophenol mutase
ntAa - reductase
ntAb - ferredoxin
ntAc - 2-nitrotoluene dioxygenase alpha
ntAd - 2-nitrotoluene dioxygenase beta

FIGURE 11.4 Gene cluster for various nitroaromatic compounds degradation present in various bacteria. The arrow represents genes and shows their direction of transcription. The size is just representative; it does not correlate exactly with actual size of genes. The solid black arrow represents dioxygenases.

aromatic compounds (Tikariha et al. 2016). Also, a closer look at the various categories of genes present in aromatic compounds–degrading bacteria has shown lots of heavy metal stress–resistant genes, antibiotic resistance, and genes for oxidative stress. Bacterial ability to survive at higher concentration of NACs seems to be unlikely, as these compounds act as uncouplers of oxidative phosphorylation and have a toxic effect on the bacterial cells (Yanzhen et al. 2016). Moreover, the concentration of nitroaromatics detected in water bodies or soil is very low (less than 1 ppm in most cases) and can be easily degraded by bacteria reported till date, which have shown the ability to utilize even up to 50 ppm.

Bacteria from contaminated sites have evolved to utilize many of these nitroaromatics as a sole source of carbon for their growth. In-depth analyses of the degrading strains have led to the identification and characterization of the enzymes in degradative pathways and genes encoding these enzymes (Table 11.2), which primarily governs the utilization capacities in bacteria (Tikariha et al. 2016, Ju and Parales 2010) by expressing multiple degradative pathways.

11.5.2 Gene Clusters in Nitroaromatic-Degrading Bacteria

High-throughput DNA sequencing, gene expression data as well as functional (proteomics) data have emerged and have provided a coherent approach to explore the capabilities of microorganisms (Figure 11.4). Gene technology, combined with a knowledge of degradative pathways, and microbial physiology have improved our understanding of catabolic activities for pollutants such as pesticides, insecticides, chloro-/nitro-/amino-substituted aromatics (Qureshi et al. 2001, Chirino et al. 2013, Ghosh et al. 2017).

11.5.3 Regulation of NACs Degradation in Bacteria

Nitroaromatic catabolic operons are controlled by LysR type of transcriptional regulators (LTTRs), which are the largest family of prokaryotic regulatory proteins identified so far (Tikariha et al. 2016).

TABLE 11.2

Nitroaromatic Compound Utilizing Capacity in Bacteria Illustrated by Whole-Genome Data Submitted by Researchers at Genbank NCBI

S. No.	Bacteria	Genome Accession Number	Nitroaromatic Compound Utilized by Bacteria	Operons Present	References
1	*Pseudomonas putida* SF1	NZ_ LDPF00000000.1	p-Nitrophenol	pnp gene cluster	Tikariha et al. 2016
2	*Pseudomonas fragi* P121	NZ_CP013861.1	Aniline, nitrobenzene	lysR gene regulator	Yanzhen et al. 2016
3	*Rhodococcus* sp. WB1	NZ_CP015529.1	Toluene and nitrotoluene degradation	Tod gene cluster	Xu et al. 2016
4	*Raoultella ornithinolytica* TNT	NZ_ JHQH00000000.1	Trinitrotoluene (TNT)	Nitroreductase A, B, and NEM reductase genes present	Thijs et al. 2014
5	*Roseivirga* sp. *strain* D-25	NZ_JSVA00000000.1	2,4,6-Trinitrotoluene	lysR gene regulator	Selvaratnam et al. 2016
6	*Shewanella oneidensis MR1*	NC_004347.2	2, 6-Dinitrotoluene	cymA and nfnB for anaerobic reduction	Liu et al. 2017
7	*Paraburkholderia xenovorans* LB400	NC_007951.1	2-Aminophenol	amn gene cluster	Chirino et al. 2013
8	*Rhodococcus imtechensis RKJ300*	NZ_AJJH00000000.1	2-Chloro-4-nitrophenol (2C4NP) and PNP	mnp gene cluster	Min et al. 2016, 2017

Other regulators for catabolic operons comprise IclR type for ortho-cleavage pathways and AraC/ XylS type of regulator for meta-cleavage pathways. GntR-type regulators control the catabolic operons by repression of genes.

Bacteria sense environmental signals by two-component signal transduction system. Some two-component signal regulatory systems are known to control the expression of catabolic pathways and control the global cellular process.

LysR is a regulatory protein, which is involved in the expression of control of pathways for degradation of NACs such as 2-nitrotoluene (in *Acidovorax* species), where 2,4- and 2,6-dinitrotoluene act as inducers, and 4-nitrophenol, where it exists in tetrameric forms (Figure 11.5).

XylR/NtrC-type transcriptional regulators were also found in the regulation of catabolic pathways of nitroaromatics. These regulators activate RNA polymerase (RNAP) holding the alternative sigma factor rho54. Rho54 RNAP holoenzyme forms a stable complex with −12 and −24 promoters but is unable to start transcription without further activators such as NACs. However, studies show that 3-nitrotoluene can also bind XylR without mediating activation. It thus demonstrates that diversity in control/regulatory mechanisms operate during the utilization of compounds as carbon and energy source and contribute for enhancing utilization capacities of microbes.

Aromatic compounds degradation pathway is also often associated with the diverse array of regulatory genes such as LTTRs. For example, LTTRs regulate a single target operon such as catR that controls catBCA expression for catechol metabolism in *P. putida*.

11.5.4 Genomes of NACs-Degrading Bacteria

Whole-genome sequences (WGSs) of many of the nitroaromatic-degrading bacteria since the past decade showcase their genetic capacities toward utilization of NACs (Table 11.2).

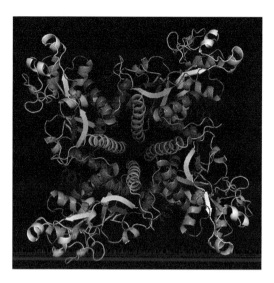

FIGURE 11.5   Quaternary structure of homo-tetrameric p-nitrophenol regulator protein.

11.5.5   Genome Plasticity: Nitroaromatic-Degrading Bacterial Genome Diversification

Bacteria in their genome have inherent capacity to evolve through mutations, rearrangements, or horizontal gene transfer (HGT) (Juhas et al. 2009). The genome has a set of genes for carrying out essential metabolic functions known as core genes and another set of genes called as accessory genes acquired by HGT that are beneficial under specific environmental conditions. The transfer of genes by HGT plays a vital part in diversification and adaptation of microorganisms in varied stress condition reflecting a more vivid scenario of genome plasticity.

Moreover, genomic islands can also be responsible for the evolution of a broad spectrum of bacteria due to their role in the dissemination of different genes such as antibiotic resistance and virulence genes, which might lead to the generation of new catabolic genes. This new gene can then intermingle with the existing machinery, or it proceeds to the formation of whole new metabolic pathways.

11.6   EVOLUTION OF NEW BIODEGRADATIVE PATHWAYS AND NETWORKING OF GENES/OPERONS

Ecosystem biology has come up to conceptualize the networking of the evolving microbiota in a niche. It postulates that different microbes in a community together form a community pathway to utilize a complex metabolite. In such scenario, whole cascade of enzymes for biodegradation of nitroaromatic cannot be found in a single microbe and thus require the development of consortia to mitigate the challenges. The implication of "-omics"-based approach toward the evaluation of microbial capacities could become an evolving tool for future domains of research and would serve as a scope toward mitigation of pollutants.

11.7   CONCLUSION

Nitroaromatics are major pollutants released in the environment and pose toxic effects to ecosystems. In the present report, microbial utilization capacities for nitroaromatics under aerobic condition have been discussed based on information encoded on their genomes. Bacterial strains such as *Pseudomonas, Burkholderia, Rhodococcus, Roseivirga,* and *Shewanella oneidensis* have capacities to degrade various NACs based on sharing of degradative pathway enzymes and genes. The requirement of triggering/inducing factor for expressing the degradative genes remains critical element to exploit the utilization capacities of microbes in case of NACs maximally.

ACKNOWLEDGEMENTS

All the authors are thankful to the Director of CSIR-NEERI for inspiration and providing infrastructural facilities (NEERI/KRC/2019/JAN/EGBD/3). Mr. Hitesh Tikariha acknowledges the Senior Research Fellowship (SRF) from the University Grants Commission (UGC) of India for carrying out his work. Grants from project DST/IS-STAC/CO2-SR-107/11(G) are also acknowledged.

REFERENCES

Begum, S.S. and Arundhati, A., 2016. A study of bioremediation of methyl parathion in vitro using potential Pseudomonas sp. isolated from agricultural soil, Visakhapatnam, India. *International Journal of Current Microbiology and Applied Sciences, 5*(2), pp. 464–474.

Chatterjee, S., Deb, U., Datta, S., Walther, C. and Gupta, D.K., 2017. Common explosives (TNT, RDX, HMX) and their fate in the environment: Emphasizing bioremediation. *Chemosphere, 184*, pp. 438–451.

Chirino, B.S., Agullo, E., Gonzaĺez, L., Seeger M., 2013. Genomic and functional analyses of the 2-aminophenol catabolic pathway and partial conversion of its substrate into picolinic acid in Burkholderiaxenovorans LB400. *PLoS ONE, 8*(10), p. e75746. DOI:10.1371/journal.pone.0075746.

Ghosh, S., Qureshi, A. and Purohit, H.J., 2017. Enhanced expression of catechol 1,2 dioxygenase gene in biofilm forming Pseudomonas mendocina EGD-AQ5 under increasing benzoate stress. *International Biodeterioration and Biodegradation, 118*, pp. 57–65.

Ju, K.-S. and Parales, R.E., 2010. Nitroaromatic compounds, from synthesis to biodegradation. *Microbiology and Molecular Biology, 74*(2), pp. 250–272.

Juhas, M., vander Meer, J.R., Gallard, M., Harding, R.M., Hood, D.W. and Crook, D.W., 2009. Genome islands: tools of bacterial horizontal gene transfer and evolution *FEMS Microbiol Reviews, 33*(2), pp. 376–393.

Kapley, A. and Purohit, H. J., 2009. Diagnosis of treatment efficiency in industrial wastewater treatment plants: a case study at a refinery ETP. *Environmental Science & Technology, 43*(10), pp. 3789–3795.

Kivisaar, M., 2009. Degradation of nitroaromatic compounds: a model to study evolution of metabolic pathways. *Molecular Microbiology, 74*(4), pp. 777–781.

Kumari, A., Singh, D., Ramaswamy, S. and Ramanathan, G., 2017. Structural and functional studies of ferredoxin and oxygenase components of 3-nitrotoluene dioxygenase from Diaphorobater sp. strain DS2. *PLoS ONE, 12*(4), p. e0176398.

Liu, D.F., Min, D., Cheng, L., Zhang, F., Li, D.B., Xiao, X., Sheng, G.P. and Yu, H.Q., 2017. Anaerobic reduction of 2, 6-dinitrotoluene by Shewanellaoneidensis MR-1: Roles of Mtr respiratory pathway and NfnB. *Biotechnology and Bioengineering, 114*(4), pp. 761–768.

Min, J., Chen, W., Wang, J. and Hu, X., 2017. Genetic and biochemical characterization of 2-Chloro-5-nitrophenol degradation in a newly isolated bacterium, Cupriavidus sp. strain CNP-8. *Frontiers in Microbiology, 8*, p. 1778.

Min, J., Zhang, J.J. and Zhou, N.Y., 2016. A two-component para-nitrophenol monooxygenase initiates a novel 2-chloro-4-nitrophenol catabolism pathway in RKJ300. *Applied and Environmental Microbiology, 82*(2), pp. 714–723.

Purohit, H.J., Kapley, A., Khardenavis, A., Qureshi, A. and Dafale, N.A., 2016. Chapter three-insights in waste management bioprocesses using genomic tools. *Advances in Applied Microbiology, 97*, pp. 121–170.

Qureshi, A., Kapley, A., Purohit, H.J., 2012. Degradation of 2,4,6-Trinitrophenol (TNP) by Arthrobacter sp. HPC1223 isolated from effluent treatment plant. *Indian Journal of Microbiology, 52*(4), pp. 642–647. DOI:10.1007/s12088-012-0288-5.

Qureshi, A., Mohan, M., Kanade, G.S., Kapley, A. and Purohit, H.J., 2009. In situ bioremediation of organochlorine-pesticide-contaminated microcosm soil and evaluation by gene probe. *Pest Management Science, 65*(7), pp. 798–804.

Qureshi, A., Prabu, S.K. and Purohit, H.J., 2001. Isolation and Characterisation of *Pseudomonas* strain for degradation of 4-nitrophenol. *Microbes and Environment, 16*(1), pp. 49–52.

Qureshi, A. and Purohit, H.J., 2002. Isolation of bacterial consortia for degradation of p-nitrophenol from agricultural soil. *Annals in Applied Biology, 140*, pp. 159–162.

Qureshi, A., Verma, V., Kapley, A. and Purohit, H.J., 2007. Degradation of 4-nitroaniline by Stenotrophomonas strain HPC 135' *International Biodeterioration and Biodegradation, 60*, pp. 215–218.

Saha, S., Badhe, N., Pal, S., Biswas, R. and Nandy, T., 2017. Carbon and nutrient-limiting conditions stimulate biodegradation of low concentration of phenol. *Biochemical Engineering Journal, 126*, pp. 40–49.

Saha, S.P., Banik, S.P., Majumder, A., Noor, A., Biswas, K., Hasan, N., Banerjee, N., Saha, P., Das, R., Halder, S. and Parveen, S., 2014. Bioremediation of methyl parathion by bacterial strains isolated from fresh vegetables. *Journal of Environment and Sociobiology, 11*(1), pp. 43–56.

Selvaratnam, C., Thevarajoo, S., Ee, R., Chan, K.G., Bennett, J.P., Goh, K.M. and Chong, C.S., 2016. Genome sequence of Roseivirga sp. strain D-25 and its potential applications from the genomic aspect. *Marine Genomics, 28*, pp. 29–31.

Sengupta, K., Maiti, T.K. and Saha, P., 2015. Degradation of 4-nitrophenol in presence of heavy metals by a halotolerant Bacillus sp. strain BUPNP2, having plant growth promoting traits. *Symbiosis, 65*(3), pp. 157–163.

Singh, D., Mishra, K. and Ramanthan, G., 2015. Bioremediation of nitroaromatic compounds. In *Wastewater Treatment Engineering*. IntechOpen publisher. DOI:10.5772/61253.

Thangaraj, K., Kapley, A. and Purohit, H.J., 2008. Characterization of diverse Acinetobacter isolates for utilization of multiple aromatic compounds. *Bioresource Technology, 99*(7), pp. 2488–2494.

Thijs, S., Van Hamme, J., Gkorezis, P., Rineau, F., Weyens, N. and Vangronsveld, J., 2014. Draft genome sequence of Raoultellaornithinolytica TNT, a trinitrotoluene-denitrating and plant growth-promoting strain isolated from explosive-contaminated soil. *Genome Announcements, 2*(3), pp. e00491–14.

Tikariha, H., Pal, R.R., Qureshi, A., Kapley, A. and Purohit, H.J., 2016. In silico analysis for prediction of degradative capacity of Pseudomonas putida SF1. *Gene, 591*(2), pp. 382–392.

Tiwari, J., Naoghare, P., Sivanesan, S. and Bafana, A., 2017. Biodegradation and detoxification of chloronitroaromatic pollutant by Cupriavidus. *Bioresource Technology, 223*, pp. 184–191.

Xu, Y., Yu, M. and Shen, A., 2016. Complete genome sequence of the polychlorinated biphenyl degrader Rhodococcus sp. WB1. *Genome Announcements, 4*(5), p. e00996-16.

Yadav, T.C., Pal, R.R., Shastri, S., Jadeja, N.B. and Kapley, A., 2015. Comparative metagenomics demonstrating different degradative capacity of activated biomass treating hydrocarbon contaminated wastewater. *Bioresource Technology, 188*, pp. 24–32.

Yanzhen, M., Yang, L., Xiangting, X. and Wei, H., 2016. Complete genome sequence of a bacterium Pseudomonas fragi P121, a strain with degradation of toxic compounds. *Journal of Biotechnology, 224*, pp. 68–69.

12 Methane Monooxygenases
Their Regulations and Applications in Biofuel Production

Dipayan Samanta
South Dakota School of Mines and Technology

Tanvi Govil and David R. Salem
South Dakota School of Mines and Technology
Composite and Nanocomposite Advanced Manufacturing
Center – Biomaterials (CNAM-Bio Center)

Lee R. Krumholz
University of Oklahoma

Robin Gerlach
Montana State University

Venkata Gadhamshetty
South Dakota School of Mines and Technology

Rajesh K. Sani
South Dakota School of Mines and Technology and
Composite and Nanocomposite Advanced Manufacturing
Center–Biomaterials (CNAM-Bio Center)

CONTENTS

12.1 INTRODUCTION

Environmental deterioration is one of the most severe issues worldwide (Mosier et al., 1988). The National Oceanic and Atmospheric Administration (NOAA) documented that the global level of methane had exceeded 1,860.7 ppb in 2018 (https://www.esrl.noaa.gov/gmd/aggi/aggi.html). The Intergovernmental Panel on Climate Change (IPCC) has projected that the global temperature may rise by 7.2°F over the next ten decades (https://www.climatecentral.org/news/ipcc-predictions-then-versus-now-15340). This rise in methane level leads to global warming due to the strong heat-absorbing properties of this gas, posing a menace to the environment (Wuebbles and Hayhoe, 2002). Landfills waste, coal mines, and agricultural and human waste are major anthropogenic sources of methane emission in the atmosphere (Karakurt et al., 2012). Furthermore, the use of fossils fuels such as coal, oil, and natural gas also significantly contribute toward increasing the levels of methane in the environment (Perera, 2017). Additionally, industrialization and urbanization make a minor contribution toward the increase in the level of methane in the atmosphere (Sapart et al., 2012). Methane is also a precursor for liquid fuels, which can be a conceptual framework toward the circumvention of vented methane (Fox, 1993). However, the oxidation of methane is quite challenging in the field of chemical catalytic conversion due to its high C–H bond energy (104 kcal/mol, Sen, 1998), thus yielding only partial oxidation (Iglesia et al., 2001). Microbially mediated oxidation of methane has been a focus of research in the past decade. There is a continual search toward exploring novel microorganisms that can convert methane into value-added products. Among the microbes reported, the methylotrophic and methanotrophic genera make use of a fascinating enzymatic methane conversion machinery and serve as the major sink for methane (Dalton, 1980).

Methylotrophs are a diverse group of microorganisms that can utilize C1 compounds (e.g., methane, methanol) as the carbon and energy source for their growth and survival; these organisms also often degrade a variety of contaminants (e.g., trichloroethylene [TCE], aromatic halides, and aliphatic halides) (Bowman et al., 1991; Semrau, 2011). Methanotrophs, a unique and ubiquitous group of Gram-negative bacteria, are the subset of methylotrophs that utilize methane as a sole source of carbon for their energy fulfillment (Hanson and Hanson, 1996). Methanotrophs belonging to the class I and class X category are γ-Proteobacteria, whereas the class II methanotrophs are α-Proteobacteria (Dispirito et al., 1991). There are distinguishable features between these methanotrophs in terms of cell morphology and membrane arrangement. Moreover, because of their physiologically unique and adaptable nature, they can be found in a broader range of environmental conditions, including a wide range of pH, temperature, oxygen concentration, salinity, and heavy metal concentrations (Hanson and Hanson, 1996; Pandey et al., 2014). Furthermore, the structural and functional feature of these bacteria allows them to adapt to a range of environmental and sometimes extreme conditions (Kwon et al., 2008). Their adaptability in these extreme conditions makes them promising candidates for environmental bioremediation. Methanotrophs have been shown to degrade and cooxidize a variety of organic pollutants and heavy metals, respectively, due to the presence of a broad-spectrum oxidative enzyme, methane monooxygenase (MMO) (Green and Dalton, 1989). However, the mechanisms of degradation and conversion of methane and other aliphatic or aromatic halides by methanotrophs are not well understood. Therefore, this chapter is concerned with (i) computational modeling of MMO subunits, for which structures are not available, and molecular docking simulation between the subunits and TCE; (ii) mutations in the particulate methane monooxygenase (pMMO) active site and its regulation of activity; and (iii) applications of MMOs in biofuels and value-added products.

12.2 METHANE MONOOXYGENASE

Methanotrophs employ a distinct broad-spectrum methane-oxidizing enzyme, MMO, which mediates the facile conversion of methane into methanol under ambient conditions (Que and Tolman, 2008). MMO has been shown to catalytically oxidize a broader range of substrates, including aromatic compounds, e.g., halogenated benzenes, toluene, and styrene as well as aliphatic hydrocarbons with up to eight carbons. Unlike other microbes, they are documented to degrade halogenated compounds, chlorinated hydrocarbons, and aromatic halides (Bouwer and Zehnder, 1993). MMOs exist in two forms: membrane-associated (particulate; pMMO) and cytoplasmic (soluble; sMMO) as shown in Figure 12.1. Both pMMO and sMMO are involved in the oxidation of methane at metal centers having a complex, multisubunit scaffold, but with distinguishable structure, active sites, and reaction mechanisms (Lawton and Rosenzweig, 1982; Sazinsky and Lippard, 2015). However, the relatively high bond dissociation energy of C–H (104 kcal/mol) in methane hinders its selective oxidation and is thus challenging (Sen, 1998). This motivated substantial interest in understanding how MMOs carry out this difficult reaction, such as how methane is stereospecific to the active sites and how it bonds with the metal center.

Figure 12.2 shows the overall biochemical pathway describing the oxidation of methane to methanol than to formaldehyde.

The oxidation of methane to methanol by MMO is the first step in the methane oxidation pathway. The first product (methanol) is further oxidized to formaldehyde by methanol dehydrogenase. The formaldehyde is oxidized to formic acid by formate dehydrogenases. The assimilation of formaldehyde takes place via the serine pathway or the ribulose monophosphate

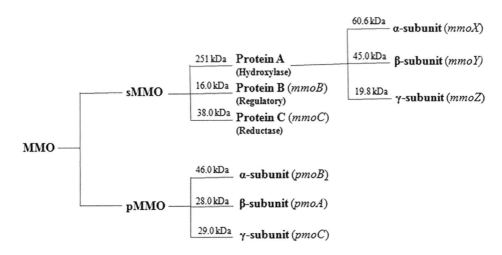

FIGURE 12.1 Subunits of MMOs (e.g., for *Methylococcus capsulatus* BATH).

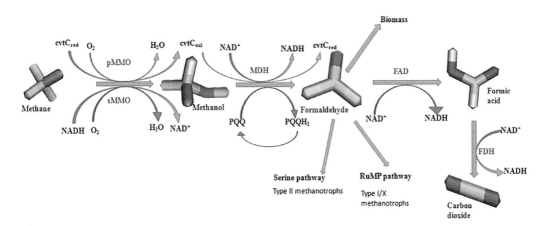

FIGURE 12.2 Biochemistry of methane oxidation.

(RUMP) pathway, where it gets converted into biomass (assimilatory pathway) or carbon dioxide (dissimilatory pathway). The pathway is dependent on the type of Proteobacteria, whether it is γ-Proteobacteria (type I/X; RUMP pathway) or α-Proteobacteria (type II; serine pathway) (Zehnder and Brock, 1979).

12.2.1 Structural Analyses of MMOs

12.2.1.1 Soluble Methane Monooxygenase

sMMO consists of three protein components: a hydroxylase (MMOH), a reductase (MMOR), and a regulatory protein (MMOB) (Green and Dalton, 1985). MMOH is a 251-kDa dimer of three subunits in $\alpha_2\beta_2\gamma_2$ configuration with α (60.6 kDa), β (45.0 kDa), and γ (19.8 kDa). Each α subunit contains a carboxylate group and a hydroxo-bridged dinuclear iron center, where the activation of oxygen and hydroxylation of methane takes place (Lipscomb, 1994). MMOR is a 38-kDa iron-sulfur flavoprotein that helps in the shuttling of electrons from NADH through its FAD and Fe-S cofactors to the hydroxylase active site. MMOB is a 16-kDa protein containing no prosthetic groups and modulates MMO reactivity by forming specific complexes with hydroxylase that affect the structure and reactivity of the diiron site indirectly. The weight percent of this protein's components varies from one methanotroph species to another (Whittington and Lippard, 2014; Walters et al., 1999).

The oxidation of methane and reduction of atmospheric oxygen at the carboxylate-bridged diiron active site of the soluble MMO is the first obvious step in methane-to-carbon dioxide or biomass conversion (Rosenzweig et al., 1995; Kopp and Lippard, 2002). The addition of MMOB to the other sMMO components affects both the rate and regioselectivity of hydroxylase-catalyzed reactions. Furthermore, it also interferes with the redox potentials of the MMOH diiron center (Sullivan et al., 1998). Literature has documented the genes encoding the sMMO proteins. Reviewing the identified and sequenced data from various methanotrophic genera, such as *Methylococcus capsulatus* (BATH), *Methylosinus trichosporium* OB3b, *Methylocystis* sp. strain M, *Methylocystis* sp. strain W114, and *Methylomonas* sp. strains KSPIII and KSWIII, has shown that sMMO genes are clustered on a 5.5-kb operon, comprising *mmoX*, *mmoY*, *mmoB*, *mmoZ*, *orfY*, and *mmoC*, which code, respectively, for MMOH (α), MMOH (β), MMOB, MMOH (γ), an unidentified functional protein (*OrfY*, 12 kDa), and MMOR (Murrell et al., 2000).

12.2.1.2 Particulate Methane Monooxygenase

The expression of pMMO in almost all methanotrophs except the *Methylocella* and *Methyloferula* genera has already been documented in the literature (Dedysh et al., 2000; Vorobev et al., 2010). Understanding of the structural integrity, the metal nucleated active site(s), and the mechanism of the formation of oxidized metal complexes in pMMO is still hypothetical and controversial (Culpepper and Rosenzweig, 2012). Detailed characterization of pMMO has been hampered by its instability and by difficulties in purification (Lieberman et al., 2003). Despite noteworthy progress in pMMO experimental data reported in the literature, its mechanism of formation of metal complexes in the active site has not yet been established. This is mainly due to lack of knowledge of the bonding nature of metal complexes present in the active site of pMMO.

pMMO consists of three subunits, namely, α (46 kDa), β (28 kDa), and γ (29 kDa). The crystal structure of pMMO of *M. trichosporium* OB3b (PDB 3CHX) shows a total of 15 chains. Each subunit contains three chains, namely, A, E, I for α, B, F, J for β, and C, G, K for γ. The remaining six chains consist of unknown metals in their metallic site. Though their location is shown in the crystal structure of 3CHX, they are not located within any of the annotated subunits of pMMO. Therefore, their function and properties are not clear and cannot be explained from the function and properties of pMMO subunits. The metallic site inside each chain of the subunits of pMMO differs in the number of copper atoms (mononucleated, dinucleated, or trinucleated copper). There are no detected zinc sites in *M. trichosporium* OB3b, but experiments showed the presence of these in *M. trichosporium* BATH (Lontoh, 2000).

The crystal structure of pMMO in *M. trichosporium* OB3b did not show any trinucleated copper clusters, which became a thought-provoking observation regarding the metal content in pMMO. Moreover, the conserved domain analysis of pMMO subunits has shown that the three genes *pmoA*, *pmoB*, and *pmoC* encode for the β, α, and γ subunits, respectively. The genes encoding the three-subunit enzyme, pMMO, are clustered on the chromosome in the order *pmoCAB*. There are two copies of these genes present in the genome of *M. capsulatus* BATH, *M. trichosporium* OB3b, and *Methylocystis* sp. strain M. The existence of a third copy of *pmoC* is also discussed in the literature. The *pmoCAB* gene copies are functionally essential for maximal pMMO activity (Gilbert et al., 2000). The genetic level study of pMMO is limited to a very few methanotrophic genera, which restricts proper understanding of the mechanism (Figure 12.3).

The amino acid FASTA sequence of the *pmoB* gene (414-amino-acid residue) of *M. capsulatus* BATH was retrieved from UniProt(ID: G1UBD1) to predict the transmembrane helixes and signal peptides. The website TMHMM was used to determine the number and position of transmembrane helices. There are three transmembrane helixes present within the structure at position 12–34, 189–206, and 235–254. The signal peptides for G1UBD1 were evaluated using SignalP-5.0. The probability scale of SignalP-5.0 shows that there is a likelihood of 0.8968, 0.0789, and 0.0152 for this protein to be a signal peptide, TAT signal peptide, and lipoprotein signal peptide, respectively.

12.2.2 MODE OF ACTION OF MMOS TOWARD POLLUTANTS

Under low copper-biomass ratio conditions, the sMMO is expressed in methanotrophs that have the genes for it (Murrell et al., 2000). The list of various methanotrophs, which express sMMO, pMMO, or both, is shown in Table 12.1.

sMMO has attracted considerable attention due to greater stability and easier purification relative to pMMO. sMMO has also been shown to possess broad substrate specificity, including aliphatic

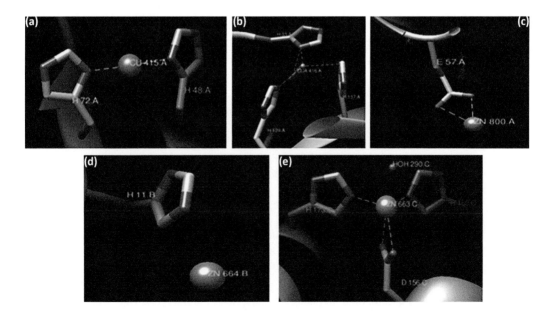

FIGURE 12.3 Arrangement of atoms in pMMO in *Methylococcus capsulatus* BATH. (a–c) Chain A with different metallic centers: (a) mononucleated copper center bonded with two amino acids, (b) mononucleated copper center with three amino acids, and (c) zinc center with two bonds with an amino acid; (d) free zinc center in chain B; (e) Zn center in chain C with four bonds with three amino acids.

TABLE 12.1
List of Methanotrophic Genera Expressing pMMO, sMMO, or Both pMMO and sMMO

Genus	pMMO	sMMO	References
α-Proteobacteria			
Methylocapsa	+	−	Dedysh et al. 2002
Methylocella	−	+	Dedysh et al. 2004
Methylocystics	+	Varies	Im et al. 2011
Methyloferula	−	+	Dedysh and Dunfield 2015
Methylosinus	+	+	Bosma and Janssen 1998
Verrucomicrobia			
Methyloacidiphilum	+	−	Hou et al. 2008
β-Proteobacteria			
Candidatus Accumulibacter	+	−	
Hydrogenophaga	+	−	
γ-Proteobacteria			
Methylobacter	+	−	Smith et al. 1997
Methylocaldum	+	−	Islam et al. 2016
Methylococcus	+	+	Colby et al. 1970
Methylohalobius	+	−	Wang et al. 2014
Methylomicrobium	+	−	Gilman et al. 2017
Methylomonas	+	Varies	Koh et al. 1993
Methylosarcina	+	−	Oswald et al. 2017
Methylosoma	+	−	Semrau et al. 2014
Methanosphaera	+	−	Bang et al. 2014
Methanothermus	+	−	Hirayama et al. 2011
Methylovulum	+	+	Iguchi et al. 2010

TABLE 12.2
Degradation of Waste Products and the Components of Degradation by MMO

Compound	Acidic Products	Volatile Products	References
Trichloroethylene	Glyoxylate	Choral	Little et al. 1988
	Dichloroacetate	CO	
	Formate	-	
Vinylidene chloride	Glycolate	Dichloroacetaldehyde	Janssen et al. 1988
Trifluoroethylene	Glyoxylate	-	Sirajuddin and Rosenzweig 2015
	Difluoroacetate	-	
Chlorotrifluoroethylene	Oxalate	-	Jiang et al. 2010
Tribromoethylene	Formate	Bromal	Smith and Dalton 2004
Trichloroethylene epoxide	Glyoxylate	Chloral	Little et al. 1988
	Formate	CO	

and aromatic hydrocarbons and their halogenated derivatives. On the other hand, pMMO has a narrow spectrum of carbon substrate specificity (alkanes and alkenes). sMMO is capable of oxidizing chlorinated, fluorinated, and brominated alkenes (Burrows et al., 1984) and has unique oxidizing properties (Table 12.2). The rate of oxidation of chloroalkanes by MMO is 7,000-fold higher than that of methane oxidation (Fox et al., 1990).

The rate and range of the degradation of the pollutants are highly dependent on the form of MMO expressed. Studies have shown that at relatively high initial concentrations of chlorinated ethenes, the faster degradation by sMMO-expressing methanotrophs led to the rapid formation of toxic products, resulting in reduced methanotrophic growth. This factor is contrasting when compared with pMMO-expressing methanotrophs.

Semrau (2011) reported a delta model, which is based on Michaelis-Menten kinetics and can predict whether we can use sMMO- or pMMO-expressing methanotrophs under different concentrations of pollutant. This helps us in the selection of methanotrophic genera for the degradation of pollutants with particular concentration level. At high delta value (>0.4), we can use pMMO-expressing methanotrophs, whereas at low delta value (≤0.4), we can use sMMO-expressing methanotrophs. The algorithm of this model determines this based on the difference between the rate of growth substrate turnover and the sum of the rate(s) of competing pollutant degradation. Therefore, the delta model depends on the concentration of the growth substrate and competing for pollutant. The ratio varies from 1 to less than zero. The values of V_G and V_{P_i} also depend on the concentration of growth substrate and competing pollutants. When the value is less than zero, methanotrophs grow with limited associated pollutant degradation. Moreover, the value decreases significantly with an increase in pollutant concentration in the sMMO-expressing methanotroph when compared with the other one (Semrau, 2011).

$$\left(\Delta = \frac{V_G - \sum_{i=1}^{n} V_{P_i}}{V_g} = \frac{\dfrac{V_m^G \times S^G}{K_s^G + S^G} - \sum_{i=1}^{n} \dfrac{V_m^{P_i} X P_i}{K_s^{P_i} + P_i}}{\dfrac{V_m^s \times S^G}{K_s^G + S^G}} \right)$$

V_G is the growth substrate turnover;
V_{P_i} is the rate of competing for pollutant concentration;
K_s is the rate constant of the reaction between the substrate and the enzyme;
S^G is the entropy change during substrate conversion;
V_m is the maximum velocity of the reaction; and
$K_s^{P_i}$ is the rate constant of the reaction between the competing pollutant and the enzyme.

12.2.3 SURFACE HYDROPHOBICITY ANALYSIS OF MMOs

Hydrophobicity has been one of the imperative physiochemical parameters to describe the surface attributes of any membrane-bound protein. The hydrophobicity is a measure of the hydrophobic or hydrophilic notch of any residue side chain. Moreover, this parameter is also valuable to predict the ability of a ligand to be transportable through the tunnel of the membrane-bound protein. To predict the bottlenecks of the tunnel in several MMOs, hydropathy calculations were performed and analyzed using the Kyte-Doolittle algorithm through MOLE 2.0 (Kyte and Doolittle, 1982; Sehnal et al., 2013). We evaluated the change in bottleneck-free radius and actual radius with the tunnel length to ensure the TCE compatibility. The mutability parameter measured using the MOLE2.0 provides us with an idea about the high or low propensity of a surface residue to be mutated (in vitro) without affecting the physiochemical properties of the transmembrane, determined by high or low probability values. The physicochemical surface property indices for *M. capsulatus* BATH were calculated using boundary residues, interior residues, and all residues. Each parameter of all residues is evaluated as a sum of the respective parameter of boundary and interior residues. Hydropathy of pMMO subunits in *M. capsulatus* BATH is shown in Figure 12.4, whereas the hydropathy parameters for α-, β-, and γ-subunits of pMMO are shown in Table 12.3.

With an adjustment in bottleneck tunnel radius to 1.25 Å, we estimated the number of tunnels influenced by the active site neighbor amino acid residues (H33, H137, and H139). The number

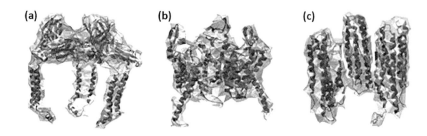

FIGURE 12.4 Hydropathy of pMMO subunits in *Methylococcus capsulatus* BATH: (a) *pmoB*, (b) *pmoA*, and (c) *pmoC*.

TABLE 12.3
Surface Hydrophobicity Analysis of pMMO from *Methylococcus capsulatus* BATH

Types of Residues	Subunits	Number of Residues	Charge (C)	Tunnel Parameter					
				Hydropathy (Scale of MOLE 2.0)	Hydrophobicity (Scale of MOLE 2.0)	Polarity (Debye)	Mutability (%)	Depth (Å)	Volume (Å³)
All residues	α (pmmoB)	66	2	0.01	−0.29	6.57	77	27	2010
	β (pmmoA)	33	1	0.13	−0.19	4.37	73	17	771
	γ (pmmoC)	21	−3	0.46	0.02	5.54	72	15	558

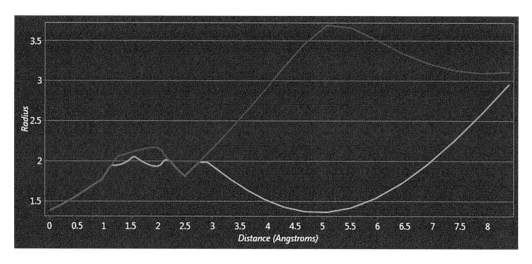

FIGURE 12.5 Analysis of bottleneck of *pmoB* gene in *Methylococcus capsulatus* BATH. Green, free radius; Orange, actual radius.

of tunnels estimated is one with a length of 8.39 Å. The bottleneck was found at 4.5 Å with a free radius of less than 1.5 Å, as shown in Figure 12.5. Figure 12.5 shows the change in the actual radius (orange) and free radius (green) with the tunnel length. The free radius describes the elastic range of the tunnel to which it can deform when a substrate is passing through it. This shows that the substrate TCE can travel through the transmembrane tunnel and cavity.

12.3 MODELING OF UNKNOWN STRUCTURES AND MOLECULAR DOCKING

The FASTA sequences of the subunits of sMMO (*mmoX, mmoY, mmoZ, mmoB,* and *mmoC*) and pMMO (*pmoA, pmoB,* and *pmoC*) were retrieved from the UNIPROT with IDs: P22869, P18798, P11987, P18787, P22868, G1UBD1, Q607G3, and Q60C16, respectively. NCBI Blast analyses were performed to find the best homology with good query coverage. sMMO (PDB #1YEW) and pMMO (PDB #3RGB) showed 100% homology with the targeted sequences and were used as templates to model each individual subunit. The modeling of the structures was performed using the homology modeling technique in an offline software modeler 9.13. Prediction of the unknown structures involved the energy minimization steps and was optimized using CHARM22_PROT and CHARM22_CHAR force-field algorithm in VEGAZZ 2.0.8 and Nanoscale Molecular Dynamics (NAMD). The motive of this modeling was to generate individual subunit structures and further compare the binding energies between whole wild-type (WT) pMMO and individual subunits to determine the site of catalysis (Figure 12.6).

Docking investigations were performed with pMMO and the substrate (TCE), using Chimera 1.9 with AutoDock Vina 4.0 algorithm (Trott and Olson, 2010). The pMMO structure of *M. capsulatus* BATH (PDB #3RGB) showed nine chains (A, E, I for α-subunit, B, F, J for β-subunit, and C, G, K for γ-subunit). The chain A of both WT and mutant-type (MT) *pmoB* gene was processed with the addition of polar hydrogens and Kollmann's partial atomic charge (gasteiger charge) (Sadowski and Gasteiger, 1993) to minimize the energy of the structure. The protein structures with minimized energy were then saved in PDBQ file format. The binding energies of the WT and MT proteins with TCE were observed, such that the entire ligand-binding site or the dinucleated copper site was covered within the grid box. The dimensions of the grid box were set as $15 \times 10 \times 15$ Å

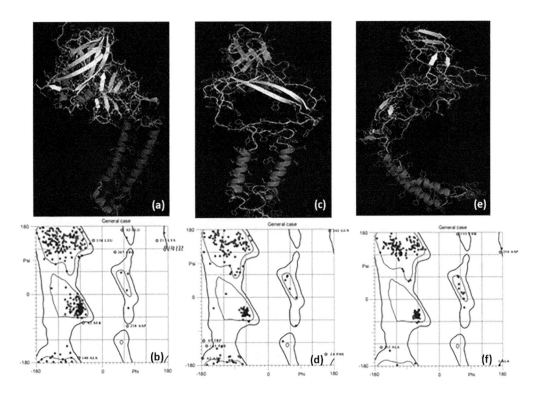

FIGURE 12.6 Modeled structures and validation of pMMO subunits in *Methylococcus capsulatus* BATH: (a and b) *pmoB*; (c and d) *pmoA*; (e and f) *pmoC*.

TABLE 12.4

Binding Affinities between the Subunits of sMMO and pMMO of *Methylococcus capsulatus* BATH and the Respective Substrate, Root Mean Square Deviation (RMSD), and Ramachandran Statistics of the Modeled Subunits of pMMO

Genes	Change in Binding Energies (kcal/mol)	Difference in Change in Binding Energies (kcal/mol)	RMSD (with the Template Used for Modeling, Å)	Ramachandran Statistics	
				Residues in Favorable Region (%)	Residues in Outlier Region (%)
pmoA	−5.2	+4.3%	5.074	84.1	7.1
pmoB	−6.2	+2.1%	6.045	88.4	3.5
pmoC	−5.1	+3.1%	2.980	84.9	6.9
pMMO (whole structure)	−5.9	+3.8%	-	-	-

depending on all the distance of the ligand from the protein along with all three directions. The docking study revealed that the active site may be present in the *pmoB* gene on the basis of binding affinity; e.g., pMMO and pmoB had similar binding energies (−5.9 and −6.2 kcal/mol, respectively). The binding energies of *pmoA* and *pmoC* are −5.2 and −5.1 kcal/mol, respectively, and are significantly different from the WT pMMO (Table 12.4).

12.4 MUTATIONS IN MMO AND ITS REGULATION OF ACTIVITY

The 2.8 Å crystal structure of *M. capsulatus* BATH pMMO was retrieved from the protein databank with PDB ID #3RGB (Berman et al., 2002). The protein structure was analyzed in an offline tool Chimera 1.9 (Pettersen et al., 2004), and the amino acid residues were visualized in PyMol 2.2.3 (DeLano, 2002). In Chimera 1.9, the chain A of the *pmoB* was selected, and the dinucleated copper site was observed, and the neighboring amino acid residues were noted. Swiss PDB viewer (SPDBV 4.10) (Guex and Peitsch, 1997) was used to mutate the selected neighbor amino acid residues and predict the changes in the number of hydrogen bonds that accounts for protein stability (Hubbard, 2001). The PyMOL was used to determine the minimum strain energy of the possible rotamers of each amino acid substitution. To mutate the conserved predicted amino acids, the neighbor amino acids were chosen with 4 Å and focused. The hydrogen bonds were displayed, and the distances were calculated. The mutated options permitted us to select the intended WT residue within the structure and choose the mutant residue rotamer. The strain energy of the possible number of rotamers was noted using the mutagenesis option available in PyMOL.

The dicopper metal site in chain A is found to be surrounded with H33, H137, and H139 residues. There are bonded contacts of this dinucleated copper with the above three neighbor amino acid residues within 4 Å radius of the metallic center. Furthermore, these residues have nonbonded contacts and salt bridges with other chains present in the structure of pMMO of *M. capsulatus* BATH (PDB #3RGB), but hydrogen bonding is absent between these residues of chain A and other chains. A mono zinc site is present within chain A, where there is a contact between the Zn atom and the H11 and G284 residues. The bonded contact is in the form of intramolecular hydrogen bonding with H11 and G284 distance, 3.19 and 2.90 Å respectively. Therefore, mutations at amino acid residues H33, H137, and H139 were performed using Swiss-PDB viewer (SPDV version r4.10). The results had been analyzed using online mutation servers MutPred (http://mutpred.mutdb.org/) and PolyPhen 2.0 (http://genetics.bwh.harvard.edu/pph2/) to confirm the variation in transmembrane properties, which are shown in Table 12.5.

TABLE 12.5

Mutational Effects on the Physiological Performance of the *pmoB* Gene

Mutations	Physiochemical Properties
H33S, H33G	Gain of Helix
	Gain of B-factor
	Gain of Methylation at K36
	Gain of Ubiquitylation at K36
	Gain of pyrrolidone carboxylic acid at Q38
H137W, H137T, H137S, H137K, H137G, H137Q	Loss of allosteric site at W136
	Gain of strand
	Gain of catalytic site at H139
H139 mutations	Negative effects on physiological parameters

We considered several parameters that make the substitution suitable without significantly changing the physiochemical properties of the transmembrane particulate enzyme. Moreover, the strain energy was also considered since it signifies the steric hindrance caused by the overlapping of Van der Waal surfaces. The less the strain energy of the rotamers of the substituted amino acid residue, the more the stability of the structure and the more the binding energy (Loftsson and Brewster, 1996) The Van der Waal electron clouds are shown for WT and each of the mutants in Figure 12.7. The mutational study compared the substitution of different amino acids and determined the most probable substitution at the metallic site of the enzyme. The statistics of various substitutions that include root mean square deviation (RMSD), binding affinity, and strain energy are shown in Table 12.6. While scanning for hydrogen bonds (H-bonds) within 4 Å in WT *pmoB*

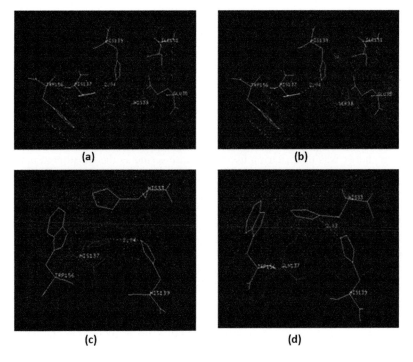

FIGURE 12.7 Mutational study in *Methylococcus capsulatus* BATH: (a) H33 wildtype; (b) H33S mutant; (c) H137 wild type; and (d) H137Q mutant.

TABLE 12.6
Screening of Mutants of *pmoB* on Basis of Binding Affinities

Mutations	Change in Binding Energy (kcal/mol)	Difference in Change in Binding Energy (kcal/mol)
H33S	−7.1	+10.1%
H33G	−5.2	+7.8%
H137W	−6.2	+9.1%
H137T	−7.5	+5.8%
H137K	−6.1	+7.6%
H137G	−7.1	+6.7%
H137Q	−6.9	+5.7%
H137S	−6.2	+4.1%

gene chain A, at H33, there is a presence of 2.94 Å H-bond between the H137 and H139 residues. The H-bond remained unhampered when H33 was substituted with SER (H33S). While at position H137, there is a presence of 2.94 Å H-bond with H139, which became a 2.83 Å H-bond when substituted with GLN (H137Q). This mutation leads to either formation of new hydrogen bonds or shortening the distance of hydrogen bonds, which ultimately resulted in greater stability of the protein secondary backbone, whereas when H33 is mutated with H33G, there was a gain of the helix, B-factor, methylation at K36, ubiquitylation at K36, and pyrrolidone carboxylic acid at Q38. The gain of helix and introduction of covalent forces within the secondary backbone allows residues to be in the favorable region (pro-proline) in the Ramachandran plot, whereas loss of helix results in the formation of rigid bonds and torsion angles, which leads to poor geometry of the secondary structure. B-factor measures the agitation of atoms in polypeptide backbones and side chains due to kinetic energy and thermal motion. Therefore, gain in B-factor denotes greater fluctuation of atoms within the polypeptide chain, which leads to the existence of a greater number of rotamers (good for protein health). Methylation at K36 is a useful parameter since methylation in any amino group will lead to the loss of hydrogen and covalent bond existing between the residues where methylation occurs. Ubiquitylation at K36 is also an important parameter in determining whether the mutation is causing the breakdown of amino acids or not. Pyrrolidone carboxylic acid at Q38 is known as a measurement parameter for membrane barrier and is useful only for the measurement of any transmembrane protein. Gain in pyrrolidone carboxylic acid at Q38 leads to the generation of multipass channels through the transmembrane protein, which allows greater diffusion of any substrates.

The binding energy was the criterion used to choose serine as the substituted residue. The substitution at the H137 site leads to loss of allosteric site at W136 and the gain of the strand and catalytic site at H139, which made H137W, H13T, H137S, H137K, H137G, and H137Q substitutions promising toward increasing the stability of the protein. But later when evaluated with H-bonds, only H137W and H137Q increased the stability. Other substitution led to the loss of H-bonds, which in turn hampers the stability of the protein. In a further study, we docked the mutated protein with the substrate to investigate the most probable enzyme-substrate complex with more values for the change in Gibb's energy parameter. This screening result can help us to proceed with less number of mutations (hence can save time) in further experimental studies in wet lab with an objective to increase the uptake rate of methane by tailoring pMMO genes.

The suitable mutants discussed above with minimum strain analysis and structure validation also coincide with the results of binding energies. H33S is the best-suited mutant since the mutant enzyme-substrate complex will hold strongly. The kinetics of this will unravel the enzyme complex lock and key pattern in the near future. The leucine gate is nearly unaffected due to mutations, and there is no significant negative change in the bonding energies between the WT and MT.

Though later, docking investigations showed that TCE was deeply seated in the active site when mutation had been performed. But the structure of the enzyme-substrate complex is still not revealed. The H137 mutations, screened among the scanned amino acid residues, are not promising with respect to physiochemical transmembrane properties (loss in the strand) in increasing the uptake rate of methane in methanotrophs.

12.5 APPLICATIONS OF MMOS IN PRODUCTION OF BIOFUEL AND VALUE-ADDED PRODUCTS

The recent focus on the food versus fuel debate has largely caused a decline in interest in converting edible oils to biofuels, e.g., biodiesel. Alternatively, an increase in interest in nonedible lipid sources from microorganisms, where the primary class of lipids is triglycerides, has been revealed. Methanotrophs with pMMO are among a very small subset of microorganisms that have a metabolic potential for the accumulation of lipids suitable for liquid fuels (Haynes and Gonzalez, 2014; Fei et al., 2014). The fatty acids in lipids accumulated by methanotrophs are saturated or monounsaturated with different positions of the double bond and the C14–C18 chain lengths. The intact phospholipid profiles (IPPs) of methanotrophs have revealed phosphatidylglycerol (PG), phosphatidylethanolamine (PE) as well as its derivatives phosphatidylmethylethanolamine (PME) and phosphatidyldimethylethanolamine (PDME), as the major classes of phospholipids with predominantly C16:1 and C18:1 fatty acid (Fang et al., 2010). These fatty acids can be an ideal source for diesel production via a hydrotreating process with respect to the projected fuel properties (Fei et al., 2014). The Advanced Research Projects Agency-Energy (ARPA-E) issued a $30 million Funding Opportunity Announcement for Reducing Emissions using Methanotrophic Organisms for the Transportation Energy (REMOTE) in early 2013 (DOE, 2013) to fund research on the development of bioconversion technologies to convert methane into liquid fuels. The goal of the REMOTE project is to identify transformational concepts for cost-effective, one-step conversion of methane-to-liquid transportation fuels with low capital expenditure and flexible deployment to access remote, flared, or pipeline gases (Fei et al., 2014). Thus, methanotrophs producing MMOs have potential as "cell factories" to produce biofuels from methane.

Reports on industrial conversion with these microorganisms are still limited and subject to several limitations, including (i) inhibition on cell growth by H_2S (or other compounds) when CH_4 in biogas is used; (ii) high costs of growth medium required for the conversions; and (iii) limited gas (CH_4)-liquid mass transfer. To date, only one methanotrophic bacterium, *Methylocystis parvus* OBBP, expressing MMO, has been reported to be able to grow using biogas (Criddle et al., 2010). Exploiting this bacterium's monooxygenase machinery could serve to sustainably process methane to biofuel.

The use of methanotrophic bacteria as biocatalysts for synthetic chemistry and bioremediation has long been investigated due to the unique catalytic properties of the pMMO and sMMO systems (Guo et al., 2017). There is an increasing interest in the development of biological routes for the synthesis of biofuels and value-added products such as methanol, biodiesel, single-cell protein (SCP), biopolymers, pigments, organic acids, ectoine, farnesene, vitamin B12, and exopolysaccharides from methane (Strong et al., 2016).

   i. Single-cell protein:
       The method for the production of SCPs from methane was developed as early as in the 1970s. In 2010, significant progress was made in support of industrial SCP production from the methane fraction of natural gas using methanotrophic bacterial consortia, to be used as an amino acid–balanced feed for fish and other animals (Hamer, 2010). Presently, SCP derived from a mixed culture dominated by *M. capsulatus* BATH is legally approved by the European Union for use in protein nutritional feeds for salmon and livestock

(e.g., pigs, poultry, and cattle) (Strong et al., 2016). This methanotrophic conversion process has the advantage in that it controls greenhouse gas emissions from perturbed natural and engineered aquatic ecosystems, besides being critical to the overall economic viability of the process.

ii. Isoprenoid compounds and carotenoid pigments:

Methanotrophs are known to accumulate both isoprenoid compounds and carotenoid pigments of various carbon lengths. Many patents have been filed in the United States regarding the production of isoprene (2-methyl-1,3-butadiene) from methane gas or methanol (Cheng et al., 2013; SongKwang et al., 2016). In 2003, under a patent filed by Dupont, genes were isolated from *Methylomonas* sp. strain 16a, encoding the isoprenoid biosynthetic pathway. A methylotroph, *Methylomicrobium alcaliphilum*, was transformed with a gene encoding an isoprene synthase from *Ipomoea batatas*, to accumulate isoprene from methane as the sole source of carbon (SongKwang et al., 2016). Similarly, the methanotroph, *Methylomonas* sp., has been genetically engineered to knock out the native carotenoid pathway of the organism, leading to the production of pink-pigmented C_{30} diapocarotenoids, thereby increasing the available carbon flux directed toward C_{40} carotenoids of interest (Donaldson et al., 2013). DuPont reported the engineering of *Methylomonas* sp.16a to produce the carotenoid astaxanthin from methane (Guo et al., 2017). Both these pigment types, isoprene and carotenoids, are of commercial interest, whereas isoprene is used in the manufacture of polyisoprene, and various copolymers with isobutylene, butadiene, styrene, or other monomers are most notably used commercially in synthetic rubber for tires. Carotenoids usage as antioxidants is valued in decreasing the risk of disease, particularly certain cancers and eye disease (Johnson, 2002). Thus, genetically engineered methanotrophs can overproduce naturally occurring metabolites or nonnative compounds, including such molecules as carotenoids and isoprene discussed above as well as 1,4 butanediol and farnesene (Strong et al., 2016).

iii. Osmoprotectants:

Moderately halotolerant methanotrophs and methylotrophs have been shown to accumulate osmoprotectants. 5-Oxoproline (pyrrolidone carboxylic acid), ectoine (1,4,5,6-tetrahydro-2-methyl-4-pyrimidine carboxylate), and its derivative hydroxyectoine stabilizes enzymes, nucleic acids, and DNA-protein complexes and protects cells from the destructive effects of various physicochemical factors (e.g., temperature, dehydration, UV radiation, and chemotherapeutic agents) (Khmelenina et al., 2015). *M. alcaliphilum* 20Z and other strains of the species accumulate ectoine up to 20% of cell weight and quickly and stably grow on methane or high methanol concentrations under a wide range of salinity (0%–10% NaCl) and pH (pH 7–10) (Cantera et al., 2017; Khmelenina et al., 1999). As ectoine is a valuable product that retails at approximately $1000/kg and can be used as a moisturizer in cosmetics, the potential to generate these osmoprotectants from methylotrophs is promising.

iv. Lactic acid:

Lactic acid, which can be produced chemically or biologically, is a widely used chemical compound in the cosmetics industry, food industry, and pharmaceutical and chemical industries. Recently, increased attention has been directed to the use of lactic acid to produce polylactic acid (PLA), which is a raw material used in the manufacturing of bioplastics that offers a more sustainable alternative to petrochemical resources. A report by Hernard et al. (2016) suggests that methanotrophs (e.g., *Methylomicrobium buryatense*) possessing MMO can carry out biocatalysis of methane to lactate (Henard et al., 2016). A US patent filed in 2014 by Calysta Inc. also suggests that lactic acid may be overproduced (Saville et al., 2014). Here, two different methanotrophs (*M. capsulatus* BATH and a *Methylomicrobium* sp.) were engineered to overexpress lactate dehydrogenase, which can convert pyruvate into lactate—a chemical precursor for polymer synthesis. On a commercial scale, Calysta Inc. has now partnered with Natureworks with the intention of commercializing polylactic acid from the methane-derived lactate monomer (Haynes and Gonzalez, 2014).

v. Methanobactin:

Methanobactins are high-affinity copper-binding peptides synthesized by methanotrophs. They are used as the pMMO that requires copper for catalytic activity (Bowman, 2006). Because of their copper chelating functionality, methanobactins protect methanotrophs against copper toxicity and can extract copper from insoluble minerals (Balasubramanian and Rosenzweig, 2006). Also, with its ability to bind other metals such as gold, iron, nickel, zinc, cobalt, cadmium, mercury, and uranium, methanobactins could be used in environmental remediation (solubilization or immobilization of many metals in situ) and metal recovery from mine leachates or even therapeutics (Strong et al., 2016).

vi. Carbohydrates:

In addition to the synthesis of SCP, carotenoids, and isoprenes, methanotrophic cells can further build the oxidation products of methane (i.e., methanol and formaldehyde) into complex molecules such as carbohydrates and lipids. For example, under high methane flux rates accompanied by nitrogen or oxygen deprivation, methanotrophs are known to produce exopolysaccharides (Chiemchaisri et al., 2001; Wilshusen et al., 2004). Huq et al. (1978) characterized exopolysaccharides produced from a methylotrophic enrichment culture on methane as a sole carbon source and found glucose, galactose, mannose, fucose, and rhamnose, were the significant components (Huq et al., 1978).

vii. Biopolymers:

In addition to the production of exopolysaccharides from methylotrophs, as described above, it has been shown that methanotrophic genera such as *Methylocella, Methylocapsa*, and *Methylocystis* with the serine C1 metabolism accumulate large amounts of the biopolymer and can be promising for the production of biodegradable plastics such as polyhydroxyalkanoate (PHA) from methane. Approximately 70% PHA accumulation (by weight) in a type II methanotroph *M. parvus* OBBP has been demonstrated (Asenjo and Suk, 1986). Myung et al. (2017) reported the first methanotrophic synthesis of PHAs that contain repeating units beyond 3HB and 3HV, including poly(3-hydroxybutyrate-*co*-4-hydroxybutyrate) (P(3HB-*co*-4HB)), poly(3-hydroxybutyrate-*co*-5-hydroxyvalerate-*co*-3-hydroxyvalerate) (P(3HB-*co*-5HV-*co*-3HV)), and poly(3-hydroxybutyrate-*co*-6-hydroxyhexanoate-*co*-4-hydroxybutyrate) (P(3HB-*co*-6HHx-*co*-4HB)) by a pure culture of *M. parvus* OBBP when the primary substrate was CH_4 and the corresponding ω-hydroxyalkanoate monomers were added as cosubstrates (Myung et al., 2017). Commercially, the use of methane to generate PHA is being investigated by a number of US, Russian, and Indian companies because of the potential to lower production costs for PHB compared with other carbon feedstocks (Khosravi-Darani, 2013). NewLight Technologies Inc. has developed a commercial process to produce PHB-based products using a proprietary methanotroph (http://newlight.com). In summary, methanotrophic conversion of CH_4 to fuels and value-added products offers a means to reduce GHG emissions instead of squandering this high-volume, high-energy gas. However, to date, advances in methane biocatalysis have been constrained by the low productivity and limited genetic tractability of natural methane-consuming microbes. The intensive research committed to these unique methanotrophic bacteria reflects the great potential envisaged for biologically generating valuable commercial products from them.

12.6   SUMMARY AND CONCLUSIONS

The structural integrity of pMMO was discussed in detail, taking into consideration the operon architecture. Modeling and simulation were demonstrated, providing an understanding of the intriguingly conserved amino acid residue–bound copper sites in pMMO genes. The in silico mutation in *pmoB*e elucidated the candidate amino acid residues responsible for methane oxidation. To show the possible presence of exogenous ligands and to reveal which metal center activates the oxygen molecule, further experimental studies can be performed. The crystal structure of pMMO in *M. trichosporium* OB3b did not show any trinucleated copper clusters. Therefore, using approaches described in this study, investigations of pMMOs structures from different organisms could reveal the unique set of amino acid ligands (metallic centers) for the controversial trinuclear copper cluster. This may aid the quest for understanding the mechanism of methane hydroxylation by copper and potentially allows to enhance the production of biofuels and value-added products using methanotrophs.

ACKNOWLEDGMENTS

This work was supported by the National Science Foundation in the form of BuG ReMeDEE initiative (Award #1736255) and the Department of Chemical and Biological Engineering at the South Dakota School of Mines and Technology.

REFERENCES

Asenjo, J.A., & Suk, J.S. "Microbial conversion of methane into poly-β-hydroxybutyrate (PHB): growth and intracellular product accumulation in a type II methanotroph." *Journal of Fermentation Technology*, 64 no 4 (1986), 271–278.

Balasubramanian, R., & Rosenzweig, A.C. "Copper methanobactin: a molecule whose time has come." *Current Opinion in Chemical Biology*, 12 no 2 (2006), 245–249.

Bang, C., Weidenbach, K., Gutsmann, T., Heine, H., & Schmitz, R.A. "The intestinal archaea Methanosphaera stadtmanae and Methanobrevibacter smithii activate human dendritic cells." *PLoS One*, 9 no 6 (2014), e99411.

Berman, H.M., Westbrook, J., Feng, Z., Gilliland, G., Bhat, T.N., Weissig, H., Shindyalov, I.N., & Bourne, P.E. "The Protein Databank." *Nucleic Acids Research*, 28 (2002), 235–242. doi:10.1093/nar/28.1.235.

Bosma, T., & Janssen, D.B. "Conversion of chlorinated propanes by *Methylosinus trichosporium* OB3b expressing soluble methane monooxygenase." *Applied Microbiology and Biotechnology*, 50 no 1 (1998), 105–112.

Bowman, J.P., Sly, L.I., & Hayward, A.C. "Contribution of genome characteristics to assessment of taxonomy of obligate methanotrophs." *International Journal of Systematic and Evolutionary Microbiology*, 41 no 2 (1991), 301–305.

Bowman, J. "The methanotrophs—the families methylococcaceae and methylocystaceae." In: M. Dworkin, S. Falkow, E. Rosenberg, K.-H. Schleifer, E. Stackebrandt (Eds.) *The Prokaryotes: Volume 5: Proteobacteria: Alpha and Beta Subclasses* (pp. 266–289). Springer, New York, 2006.

Bouwer, E.J., & Zehnder, A.J. "Bioremediation of organic compounds—putting microbial metabolism to work." *Trends in biotechnology*, 11 no 8 (1993), 360–367.

Burrows, K.J., Cornish, A., Scott, D., & Higgins, I.J. "Substrate specificities of the soluble and particulate methane mono-oxygenases of Methylosinustrichosporium OB3b." *Microbiology*, 130 no 12 (1984), 3327–3333.

Cantera, S., Lebrero, R., Rodríguez, E., García-Encina, P.A., & Muñoz, R. "Continuous abatement of methane coupled with ectoine production by Methylomicrobium alcaliphilum 20Z in stirred tank reactors: a step further towards greenhouse gas biorefineries." *Journal of Cleaner Production*, 152 (2017), 134–141.

Cheng, Q., Koffas, M., Norton, K.C., Odom, J.M., Picataggio, S.K., Rouviere, P.E., Schenzle, A., & Jean-Francois, T. "Genes involved in isoprenoid compound production." US6660507B2, Du Pont (2013).

Chiemchaisri, W., Wu, J.S., V& isvanathan, C. "Methanotrophic production of extracellular polysaccharide in landfill cover soils." *Water Science and Technology*, 43 no 6 (2001), 151–158.

Colby, J., Stirling, D.I., & Dalton, H.O.W.A.R.D. "The soluble methane mono-oxygenase of *Methylococcus capsulatus* (Bath). Its ability to oxygenate n-alkanes, n-alkenes, ethers, and alicyclic, aromatic and heterocyclic compounds." *Biochemical Journal*, 165 no 2 (1970), 395–402.

Criddle, C.S., Hart J.R., Wu W.M., Sundstrom E.R., Morse M.C., & Billington S.L. "Production of PHA using biogas as feedstock and power Source." (Ed.) U.S.P. Application, Vol. Pub. No.: US20130071890 A1 (2010).

Culpepper, M.A., & Rosenzweig, A.C. "Architecture and active site of particulate methane monooxygenase." *Critical Reviews in Biochemistry and Molecular Biology*, 47 no 6 (2012), 483–492.

Dalton, H. "Oxidation of hydrocarbons by methane monooxygenases from a variety of microbes." *Advances in Applied Microbiology*, 26 (1980), 71–87.

Dedysh, S.N., Liesack, W., Khmelenina, V.N., Suzina, N.E., Trotsenko, Y.A., Semrau, J.D., & Tiedje, J.M. "Methylocellapalustris gen. nov., sp. nov., a new methane-oxidizing acidophilic bacterium from peat bogs, representing a novel subtype of serine-pathway methanotrophs." *International Journal of Systematic and Evolutionary Microbiology*, 50 no 3 (2000), 955–969.

Dedysh, S.N., Khmelenina, V.N., Suzina, N.E., Trotsenko, Y.A., Semrau, J.D., Liesack, W., & Tiedje, J.M. "Methylocapsa acidiphila gen. nov., sp. nov., a novel methane-oxidizing and dinitrogen-fixing aci-dophilic bacterium from Sphagnum bog." *International Journal of Systematic and Evolutionary Microbiology*, 52 no 1 (2002), 251–261.

Dedysh, S.N., Berestovskaya, Y.Y., Vasylieva, L.V., Belova, S.E., Khmelenina, V.N., Suzina, N.E., & Zavarzin, G.A. "Methylocella tundrae sp. nov., a novel methanotrophic bacterium from acidic tundra peatlands." *International Journal of Systematic and Evolutionary Microbiology*, 54 no 1 (2004), 151–156.

Dedysh, S.N., & Dunfield, P.F. "Methyloferula." In: Bergey's Manual of Systematics of Archaea and Bacteria, (2015), 1–5.

DeLano, W.L. Pymol: an open-source molecular graphics tool. *CCP4 Newsletter On Protein Crystallography*, 40 (2002), 82–92.

DiSpirito, A.A., Gulledge, J., Shiemke, A.K., Murrell, J.C., Lidstrom, M.E., & Krema, C.L. (1991). "Trichloroethylene oxidation by the membrane-associated methane monooxygenase in type I, type II and type X methanotrophs." *Biodegradation*, 2 no 3 (1991), 151–164.

DOE. "Reducing emissions using methanotrophic organisms for transportation energy (REMOTE)." DE-FOA-0000881 (2013).

Donaldson, G.K., Hollands, K., & Picataggio, S.K. "Biocatalyst for conversion of methane and methanol to isoprene." Vol. 20150225743, DU PONT (2013).

Fang, J., Barcelona, M.J., & Semrau, J.D. "Characterization of methanotrophic bacteria on the basis of intact phospholipid profiles." *FEMS Microbiology Letters*, 189 no 1 (2010), 67–72.

Fei, Q., Guarnieri, M.T., Tao, L., Laurens, L.M.L., Dowe, N., & Pienkos, P.T. "Bioconversion of natural gas to liquid fuel: opportunities and challenges." *Biotechnology Advances*, 32 no 3 (2014), 596–614.

Fox III, J.M. "The different catalytic routes for methane valorization: an assessment of processes for liquid fuels." *Catalysis Reviews—Science and Engineering*, 35 no 2 (1993), 169–212.

Fox, B.G., Borneman, J.G., Wackett, L.P., & Lipscomb, J.D. "Haloalkene oxidation by the soluble methane monooxygenase from Methylosinus trichosporium OB3b: mechanistic and environmental implica-tions." *Biochemistry*, 29 no 27 (1990), 6419–6427.

Gilbert, B., McDonald, I.R., Finch, R., Stafford, G.P., Nielsen, A.K., & Murrell, J.C. "Molecular analysis of the pmo (particulate methane monooxygenase) operons from two type II methanotrophs." *Applied and Environmental Microbiology*, 66 no 3 (2000), 966–975.

Gilman, A., Fu, Y., Hendershott, M., Chu, F., Puri, A.W., Smith, A.L., & Lidstrom, M.E. "Oxygen-limited metabolism in the methanotroph Methylomicrobium buryatense 5GB1C." *PeerJ*, 5 (2017), e3945.

Green, J., & Dalton, H. "Protein B of soluble methane monooxygenase from Methylococcus capsulatus (Bath). A novel regulatory protein of enzyme activity." *Journal of Biological Chemistry*, 260 no 29 (1985), 15795–15801.

Green, J., & Dalton, H. "Substrate specificity of soluble methane monooxygenase. Mechanistic implications." *Journal of Biological Chemistry*, 264 no 30 (1989), 17698–17703.

Guex, N. and Peitsch, M.C. "SWISS-MODEL and the Swiss-Pdb viewer: an environment for comparative protein modeling." *Electrophoresis*, 18 (1997), 2714–2723.

Guo, W., Li, D., He, R., Wu, M., Chen, W., Gao, F., Zhang, Z., Yao, Y., Yu, L., & Chen, S. "Synthesizing value-added products from methane by a new Methylomonas." *Journal of Applied Microbiology*, 123 no 5 (2017), 1214–1227.

Hamer, G. "Methanotrophy: from the environment to industry and back." *Chemical Engineering Journal*, 160 no 2 (2010), 391–397.

Hanson, R.S., & Hanson, T.E. "Methanotrophic bacteria." *Microbiological Reviews*, 60 no 2 (1996), 439–471.

Haynes, C.A., & Gonzalez, R. "Rethinking biological activation of methane and conversion to liquid fuels." *Nature Chemical Biology*, 10 no 5 (2014), 331–339.

Henard, C.A., Smith, H., Dowe, N., Kalyuzhnaya, M.G., Pienkos, P.T., & Guarnieri, M.T. "Bioconversion of methane to lactate by an obligate methanotrophic bacterium." *Scientific Reports*, 6 (2016), 21585.

Hirayama, H., Suzuki, Y., Abe, M., Miyazaki, M., Makita, H., Inagaki, F., & Takai, K. "Methylothermus subterraneus sp. nov., a moderately thermophilic methanotroph isolated from a terrestrial subsurface hot aquifer." *International Journal of Systematic and Evolutionary Microbiology*, 61 no 11 (2011), 2646–2653.

Hou, S., Makarova, K.S., Saw, J.H., Senin, P., Ly, B.V., Zhou, Z., & Wolf, Y.I. "Complete genome sequence of the extremely acidophilic methanotroph isolate V4, Methylacidiphilum infernorum, a representative of the bacterial phylum Verrucomicrobia." *Biology Direct*, 3 no 1 (2008), 26.

Hubbard, R.E. "Hydrogen bonds in proteins: role and strength" (2001). https://doi.org/10.1002/9780470015902. a0003011.pub2.

Huq, M., Ralph, B., Rickard, P. "The extracellular polysaccharide of a methylotrophic culture." *Australian Journal of Biological Sciences*, 31 no 3 (1978), 311–316.

Iglesia, E., Spivey, J.J., Fleisch, T.H., Schmidt, L.D., Bell, A.T., Zhang, Y., ... Ma, D. (2001). *Natural Gas Conversion VI*. Elsevier, Amsterdam.

Iguchi, H., Yurimoto, H., & Sakai, Y. "Soluble and particulate methane monooxygenase gene clusters of the type I methanotroph Methylovulum miyakonense HT12." *FEMS Microbiology Letters*, 312 no 1 (2010), 71–76.

Im, J., Lee, S.W., Yoon, S., DiSpirito, A.A., & Semrau, J.D. "Characterization of a novel facultative Methylocystis species capable of growth on methane, acetate and ethanol." *Environmental Microbiology Reports*, 3 no 2 (2011), 174–181.

Islam, T., Torsvik, V., Larsen, Ø., Bodrossy, L., Øvreås, L., & Birkeland, N.K. "Acid-tolerant moderately thermophilic methanotrophs of the class Gammaproteobacteria isolated from tropical topsoil with methane seeps." *Frontiers in Microbiology*, 7 (2016), 851.

Janssen, D.B., Grobben, G., Hoekstra, R., Oldenhuis, R., & Witholt, B. "Degradation of trans-1, 2-dichloroethene by mixed and pure cultures of methanotrophic bacteria." *Applied Microbiology and Biotechnology*, 29 no 4 (1988), 392–399.

Jiang, H., Chen, Y., Jiang, P., Zhang, C., Smith, T.J., Murrell, J.C., & Xing, X.H. "Methanotrophs: multifunctional bacteria with promising applications in environmental bioengineering." *Biochemical Engineering Journal*, 49 no 3 (2010), 277–288.

Johnson, E.J. The role of carotenoids in human health. *Nutrition in Clinical Care*, 5 no 2 (2002), 56–65.

Karakurt, I., Aydin, G., & Aydiner, K. "Sources and mitigation of methane emissions by sectors: a critical review." *Renewable Energy*, 39 no 1 (2012), 40–48.

Khmelenina, V.N., Kalyuzhnaya, M.G., Sakharovsky, V.G., Suzina, N.E., Trotsenko, Y.A., & Gottschalk, G. "Osmoadaptation in halophilic and alkaliphilic methanotrophs." *Archives of Microbiology*, 172 no 5 (1999), 321–329.

Khmelenina, V.N., Rozova, O.N., But, S.Y., Mustakhimov, I.I., Reshetnikov, A.S., Beschastnyi, A.P., & Trotsenko, Y.A. "Biosynthesis of secondary metabolites in methanotrophs: biochemical and genetic aspects (Review)." *Applied Biochemistry and Microbiology*, 51 no 2 (2015), 150–158.

Koh, S.C., Bowman, J.P., & Sayler, G.S. "Soluble methane monooxygenase production and trichloroethylene degradation by a type I methanotroph, Methylomonas methanica 68-1." *Applied and Environmental Microbiology*, 59 no 4 (1993), 960–967.

Kopp, D.A., & Lippard, S.J. "Soluble methane monooxygenase: activation of dioxygen and methane." *Current Opinion in Chemical Biology*, 6 no 5 (2002), 568–576.

Kwon, M., Ho, A., & Yoon, S. "Novel approaches and reasons to isolate methanotrophic bacteria with biotechnological potentials: recent achievements and perspectives." *Applied Microbiology and Biotechnology* (2008), 1–8.

Kyte, J., & Doolittle, R.F. "A simple method for displaying the hydropathic character of a protein." *Journal of Molecular Biology*, 157 no 1 (1982), 105–132.

Lawton, T.J., & Rosenzweig, A.C. "Methane-oxidizing enzymes: an upstream problem in biological gas-to-liquids conversion." *Journal of the American Chemical Society*, 138 no 30 (1982), 9327–9340.

Lontoh, S.T. "Substrate oxidation by methanotrophs expressing particulate methane monooxygenase (pMMO): a study of whole-cell oxidation of trichloroethylene and its potential use for environmental remediation" (Doctoral dissertation) (2000).

Loftsson, T., & Brewster, M.E. "Pharmaceutical applications of cyclodextrins. 1. Drug solubilization and stabilization." *Journal of Pharmaceutical Sciences*, 85 no 10 (1996), 1017–1025.

Lieberman, R.L., Shrestha, D.B., Doan, P.E., Hoffman, B.M., Stemmler, T.L., & Rosenzweig, A.C. "Purified particulate methane monooxygenase from Methylococcus capsulatus (Bath) is a dimer with both mononuclear copper and a copper-containing cluster." *Proceedings of the National Academy of Sciences*, 100 no 7 (2003), 3820–3825.

Lipscomb, J.D. "Biochemistry of the soluble methane monooxygenase." *Annual Reviews in Microbiology*, 48 no 1 (1994), 371–399.

Little, C.D., Palumbo, A.V., Herbes, S.E., Lidstrom, M.E., Tyndall, R.L., & Gilmer, P.J. "Trichloroethylene biodegradation by a methane-oxidizing bacterium." *Applied and Environmental Microbiology*, 54 no 4 (1988), 951–956.

Mosier, A.R., Duxbury, J.M., Freney, J.R., Heinemeyer, O., Minami, K., & Johnson, D.E. "Mitigating agricultural emissions of methane." *Climatic Change*, 40 no 1 (1988), 39–80.

Murrell, J.C., McDonald, I.R., & Gilbert, B. "Regulation of expression of methane monooxygenases by copper ions." *Trends in Microbiology*, 8 no 5 (2000), 221–225.

Myung, J., Flanagan, J.C.A., Waymouth, R.M., & Criddle, C.S. "Expanding the range of polyhydroxyalkanoates synthesized by methanotrophic bacteria through the utilization of omega-hydroxyalkanoate co-substrates." *AMB Express*, 7 no 1 (2017), 118.

Oswald, K., Graf, J.S., Littmann, S., Tienken, D., Brand, A., Wehrli, B., & Schubert, C.J. "Crenothrix are major methane consumers in stratified lakes." *The ISME Journal*, 11 no 9 (2017), 2124.

Pandey, V.C., Singh, J.S., Singh, D.P., & Singh, R.P. "Methanotrophs: promising bacteria for environmental remediation." *International Journal of Environmental Science and Technology*, 11 no 1 (2014), 241–250.

Perera, F. "Pollution from fossil-fuel combustion is the leading environmental threat to global pediatric health and equity: solutions exist." *International Journal of Environmental Research and Public Health*, 15 no 1 (2017), 16.

Pettersen, E.F., Goddard, T.D., Huang, C.C., Couch, G.S., Greenblatt, D.M., Meng, E.C., & Ferrin, T.E. "UCSF Chimera—a visualization system for exploratory research and analysis." *Journal of Computational Chemistry*, 25 no 13 (2004), 1605–1612.

Que Jr, L., & Tolman, W.B. (2008). "Biologically inspired oxidation catalysis." *Nature*, 455 no 7211 (2008), 333.

Rosenzweig, A.C., Nordlund, P., Takahara, P.M., Frederick, C.A., & Lippard, S.J. "Geometry of the soluble methane monooxygenase catalytic diiron center in two oxidation states." *Chemistry & Biology*, 2 no 6 (1995), 409–418.

Sadowski, J., & Gasteiger, J. "From atoms and bonds to three-dimensional atomic coordinates: automatic model builders." *Chemical Reviews*, 93 no 7 (1993), 2567–2581.

Sapart, C.J., Monteil, G., Prokopiou, M., Van de Wal, R.S.W., Kaplan, J.O., Sperlich, P., ... Blunier, T. "Natural and anthropogenic variations in methane sources during the past two millennia." *Nature*, 490 no 7418 (2012), 85.

Saville, R.M., Lee, S., Regitsky, D.D., Resnick, S.M., & Silverman, J. "Compositions and methods for biological production of lactate from c1 compounds using lactate dehydrogenase transformants." Patent, CALYSTA Inc (2014).

Sazinsky, M.H., & Lippard, S.J. "Methane monooxygenase: functionalizing methane at iron and copper." In: *Sustaining Life on Planet Earth: Metalloenzymes Mastering Dioxygen and Other Chewy Gases* (pp. 205–256). Springer, Cham, 2015.

Sehnal, D., Vařeková, R.S., Berka, K., Pravda, L., Navrátilová, V., Baňáš, P., ... Koča, J. "MOLE 2.0: advanced approach for analysis of biomacromolecular channels." *Journal of Cheminformatics*, 5 no 1 (2013), 39.

Semrau, J. "Bioremediation via methanotrophy: overview of recent findings and suggestions for future research." *Frontiers in Microbiology*, 2 (2011), 209.

Semrau, J.D., DiSpirito, A.A., & Yoon, S. "Methanotrophs and copper." *FEMS Microbiology Reviews*, 34 no 4 (2014), 496–531.

Sen, A. "Catalytic functionalization of carbon–hydrogen and carbon–carbon bonds in protic media." *Accounts of Chemical Research*, 31 no 9 (1998), 550–557.

Sirajuddin, S., & Rosenzweig, A.C. "Enzymatic oxidation of methane." *Biochemistry*, 54 no 14 (2015), 2283–2294.

Smith, K.S., Costello, A.M., & Lidstrom, M.E. "Methane and trichloroethylene oxidation by an estuarine methanotroph, Methylobacter sp. strain BB5.1." *Applied and Environmental Microbiology*, 63 no 11 (1997), 4617–4620.

Smith, T.J., & Dalton, H. "Biocatalysis by methane monooxygenase and its implications for the petroleum industry." *Petroleum Biotechnology, Developments and Perspectives*, 151, (2004), 177–192.

Song, J.Y., Kuk Cho, K., Sung Lee, K., Hwa La, Y., & Kalyuzhnaya, M. "Method for producing isoprene using recombinant halophilic methanotroph." Vol. US20170211100A1, SK Innovation Co Ltd San Diego State University Research Foundation (2016).

Strong, P. J., Xie, S., & Clarke, W. P. "Methane as a resource: can the methanotrophs add value?." *Environmental science & technology*, 49 no 7 (2015), 4001–4018.

Sullivan, J.P., Dickinson, D., & Chase, H.A. "Methanotrophs, Methylosinus trichosporium OB3b, sMMO, and their application to bioremediation." *Critical Reviews in Microbiology*, 24 no 4 (1998), 335–373.

Trott, O., & Olson, A.J. "AutoDock Vina: improving the speed and accuracy of docking with a new scoring function, efficient optimization and multithreading." *Journal of Computational Chemistry*, 31 (2010), 455–461.

Vorobev, A.V., Baani, M., Doronina, N.V., Brady, A.L., Liesack, W., Dunfield, P.F., & Dedysh, S.N. "Methyloferula stellata gen. nov., sp. nov., an acidophilic, obligately methanotrophic bacterium that possesses only a soluble methane monooxygenase." *International Journal of Systematic and Evolutionary Microbiology*, 61 no 10 (2010), 2456–2463.

Walters, K.J., Gassner, G.T., Lippard, S.J., & Wagner, G. "Structure of the soluble methane monooxygenase regulatory protein B." *Proceedings of the National Academy of Sciences*, 96 no 14 (1999), 7877–7882.

Wang, P.L., Chiu, Y.P., Cheng, T.W., Chang, Y.H., Tu, W.X., & Lin, L.H. "Spatial variations of community structures and methane cycling across a transect of Lei-Gong-Hou mud volcanoes in eastern Taiwan." *Frontiers in Microbiology*, 5 (2014), 121.

Whittington, D.A., & Lippard, S.J. "Crystal structures of the soluble methane monooxygenase hydroxylase from Methylococcus capsulatus (Bath) demonstrating geometrical variability at the dinuclear iron active site." *Journal of the American Chemical Society*, 123 no 5 (2014), 827–838.

Wilshusen, J.H., Hettiaratchi, J.P., De Visscher, A., & Saint-Fort, R. "Methane oxidation and formation of EPS in compost: effect of oxygen concentration." *Environmental Pollution* 129 no 2 (2004), 305–314.

Wuebbles, D.J., & Hayhoe, K. "Atmospheric methane and global change." *Earth-Science Reviews*, 57 no 3–4 (2002), 177–210.

Zehnder, A.J., & Brock, T.D. "Methane formation and methane oxidation by methanogenic bacteria." *Journal of Bacteriology*, 137 no 1 (1979), 420–432.

13 Plant Growth–Promoting Rhizobacteria (PGPR) and Bioremediation of Industrial Waste

Sangeeta Yadav, Kshitij Singh, and Ram Chandra
Babasaheb Bhimrao Ambedkar University (A Central University)

CONTENTS

13.1 INTRODUCTION

Plant-associated bacteria can be classified into beneficial, deleterious, and neutral groups on the basis of their effects on plant growth. Beneficial free-living soil bacteria are usually referred to as plant growth–promoting rhizobacteria (PGPR). PGPR are a group of bacteria that can be found in the rhizosphere. The term "plant growth–promoting bacteria" refers to bacteria that colonize the roots of plants (rhizosphere) that enhance plant growth. Rhizosphere is the soil environment where the plant root is available and is a zone of maximum microbial activity, resulting in a confined nutrient pool in which essential macro-and micronutrients are extracted. The microbial population present in the rhizosphere is relatively different from that of its surroundings due to the presence of root exudates that function as a source of nutrients for microbial growth. PGPR include bacteria that inhabit the rhizosphere, improve plant health, and may also enhance plant growth. The term PGPR was coined by Kloepper and coworkers in 1980, although PGPR was first mentioned in 1978 by the same author in the Proceedings of the Fourth International Congress of Bacterial Plant Pathogens, conducted in France. This zone is rich in nutrients when compared with the bulk soil due to the accumulation of a variety of plant exudates, such as carbohydrates and amino acids, providing a rich source of energy and nutrients for bacteria. Generally, 10–100 times higher bacterial count is present than the bulk soil, which shows high diversity present in rhizosphere. It is well established that only 1%–2% of bacteria promote plant growth in the rhizosphere. The microbial-colonizing rhizosphere includes bacteria, fungi, acticomycetes, protozoa, and algae. However, bacteria are the most abundant microbial present in the rhizosphere. The genera such as *Azospirillum*, *Azotobacter*, *Acetobacter diazotrophicus*, and *Azoarcus* include bacterial species that have the ability to fix nitrogen biologically. Besides biological nitrogen fixation (BNF), phosphate solubilization is also an important phenomenon in the rhizosphere that enhances the nutrient availability to the host plant. Direct and indirect mechanisms are followed by PGPR for plant growth. The direct mechanism includes secretion of phytohormones, i.e., auxin, and decrease of plant ethylene levels and facilitates the uptake of nutrients from the environment through nitrogen fixation. Direct plant growth promotion by PGPR involves either providing plants with microbe-oriented compounds or helping in the absorption of several nutrients from the environment that are essential for plant growth, whereas indirect mechanism of PGPR prevents the deleterious effects of one or more phytopathogenic organisms. This includes production of antagonistic substances for inhibition of pathogenic microbes, which is a type of biological control. Among these mechanisms, PGPR adopted one or more for plant growth. PGPRs control the detrimental effects of pathogenic agents on plants to reduce the impact of diseases by producing growth inhibitors, i.e., antibiotics, bacteriocins, siderophores, induction of systemic resistance, competition for nutrients, and niches and lytic enzymes, or by increasing natural resistance of host plant. Antagonistic activity of PGPR is regulated by several mechanisms, including competition, parasitism, and siderophores or antibiotics production. The mechanism for biocontrol by PGPR is induced systemic resistance (ISR) that manipulates the physical and biochemical properties of the host plant for controlling plant diseases. The direct mechanisms are biofertilization, stimulation of root growth, rhizoremediation, and plant stress control.

PGPR and their interactions with plants are exploited commercially and hold great promise for sustainable agriculture. Applications of these associations have been investigated in maize, wheat, oat, barley, peas, canola, soy, potatoes, tomatoes, lentils, radicchio, and cucumber (Gray and Smith, 2005). PGPR are agriculturally important bacteria having specific symbiotic relationships with plants. PGPR serve as one of the active ingredients in biofertilizer formulation. Based on the interactions with plants, PGPR can be separated into symbiotic bacteria, whereby they live inside plants and exchange metabolites with them directly and free-living rhizobacteria, which live outside plant cells. Symbiotic bacteria mostly reside in the intercellular spaces of the host plant, but there are certain bacteria that are able to form mutualistic interactions with their hosts and penetrate plant cells. In addition to that, a few are capable of integrating their physiology with the plant, causing the formation of specialized structures. Rhizobia, the famous mutualistic symbiotic bacteria, could establish symbiotic associations with leguminous crop plants, fixing atmospheric nitrogen for the plant in specific root structures known as nodules. PGPR can also be termed as plant health–promoting rhizobacteria (PHPR) or nodule-promoting rhizobacteria (NPR) and are attached with the rhizosphere that is an important ecological environment of soil for plant-microbe interactions. Various species of rhizobacteria belonging to the genera *Alcaligenes, Arthrobacter, Azospirillum, Azotobacter, Bacillus, Bradyrhizobium, Burkholderia, Enterobacter, Flavobacterium, Klebsiella, Mesorhizobium, Pseudomonas, Rhodococcus, Streptomyces, Serratia*, etc. have been reported to promote plant growth and antagonize plant pathogens.

Industrial wastewater is a global issue for safe disposal. High organic and inorganic pollution loads along with potential toxic elements such as Zn^{2+}, Cu^{2+}, Pb^{2+}, Cd^{2+}, and As^{3+} are usually found in industrial wastewater (Chandra and Kumar, 2017a,b; Kumar and Chandra, 2019; Chandra et al., 2018c). This has led to a search for sustainable methods for the remediation of contaminated environments. Therefore, it is important to develop methods to remediate the heavy metal entry of toxic elements into the food chain. Various engineering methods (excavation, landfill, thermal treatment, leaching, and electroreclamation) presently being used are not fully satisfactory, as they destroy the biotic and abiotic components of the soil, and are also technically difficult and expensive to use. Phytoremediation for in situ remediation for industrial wastewater being seen as an efficient, sustainable, and cost-effective remediation technique compared with conventional physical-chemical techniques (Chandra et al., 2015; Chandra and Kumar, 2017c, 2018; Chandra et al., 2018a,b). During phytoremediation, biodegradation of pollutants is promoted by the synergy between plants and the microorganisms present in the rhizosphere, the region of soil that is directly influenced by root secretions. Successful phytoremediation is dependent on the survival and growth of plants on contaminated sites, as well as the ability of the rhizosphere to support an active soil microbial population (Chandra and Kumar, 2015; Kumar and Chandra, 2018). Phytoremediation is defined as the use of plants to destroy, sequester, and remove toxic pollutants from the environment. However, this method also has many drawbacks. Therefore, phytoremediation associated with rhizospheric microorganisms has emerged as an acceptable agronomic remediation technology. The relationships that exist between plants and microbes in the rhizosphere play a key role in enhancing the efficacy of phytoremediation through a process known as "bioassisted phytoremediation." In the soil, microorganisms present in and around the roots are called PGPR; they use many types of mechanisms to promote plant growth and minimize stress. PGPR are helpful for plant growth enhancement and bioremediation of contaminated soil through sequestering or degrading heavy metals and other toxicants. Bioremediation is, therefore, an option that offers the possibility to destroy or render harmless, various contaminants using natural biological activity. PGPR assist phytoremediation directly or indirectly through several mechanisms, such as increased nutrient uptake, suppressing pathogens by producing antibiotics and siderophores or bacterial and fungal antagonistic substances (hydrogen cyanide (HCN)), phytohormone production, and nitrogen. This chapter focuses on the role of PGPR in remediation of industrial waste water. Many plants are shown to accumulate metals and biotransformation of recalcitrant compounds. During this process, how plant-bacteria partnerships can be applied for mitigating environmental pollution is highlighted in this chapter. This chapter mainly focuses on antagonistic features of PGPR and their beneficial effects on the agricultural system. Role of PGPR in agriculture is well documented, but the involvement in the bioremediation of heavy metals and industrial waste is so less. Some PGPR can also help plants withstand abiotic stresses including contamination by heavy metals or other pollutants; certain are even able to increase the capacity of plants to sequester heavy metals (Tak et al., 2013). Therefore, utilizing PGPR is a new and promising approach for improving the success of remediation or phytoremediation of contaminated soils.

13.2 RHIZOSPHERE MICROBIAL ECOLOGY

Microbial ecology of the rhizosphere refers to the study of the interactions of microorganisms with each other and the environment surrounding the plant root. Lorenz Hiltner first coined the term "rhizosphere" to describe the plant-root interface. Hiltner described the rhizosphere as the area around a plant root that is inhabited by a unique population of microorganisms influenced by the root exudates released from plant roots. The rhizosphere region is a highly favorable habitat for the proliferation, activity, and metabolism of numerous microorganisms. The rhizosphere is the physical location in soil where plants and microorganisms interact. The ecology of microorganisms in the rhizosphere is also the study of structure and function. An understanding of the basic principles of rhizosphere microbial ecology, including the function and diversity of the microorganism that reside there, is necessary before soil microbial technologies can be applied to the rhizosphere.

The close proximity of plant roots to the soil provides abundant simple sugars and amino acids that sustain large populations of microorganisms. The diffusion of gases through roots creates habitats rich in oxygen, which further supports zones of high microbial activity within the soil matrix. The interactions of plants, soil, and microorganisms within the rhizosphere heavily influence the biogeochemical cycling of carbon, nitrogen, and phosphorus through ecosystems. Microorganisms of rhizospheres are fungi, bacteria, nematodes, and archaea (Figure 13.1). Fungi are eukaryotic microorganisms that inhabit the rhizosphere in diverse forms. Arbuscular mycorrhizal fungi (AMFs) and ectomycorrhizal fungi (EcMs) are both symbiotic fungi that form mutualistic relationships with plants. AMFs enter plant roots and extend their hyphae into the cell membranes. Highly branched structures termed arbuscules or balloon-like vesicles are formed within the roots and aid in nutrient exchange. In contrast, EcMs form a thick sheath around roots, with a network of hyphae (termed the

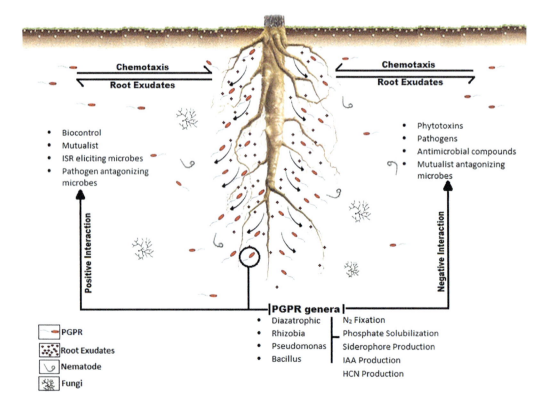

FIGURE 13.1 Ecology and biodiversity of PGPR living in the rhizosphere.

Hartig net) penetrating between epidermal and cortical cells. Whereas AMFs are ubiquitous across most plant families, the EcMs are associated primarily with trees and shrubs.

Other nonsymbiotic fungi that inhabit the rhizosphere can serve a commensalism or antagonistic relationship. Saprotrophic fungi found in the rhizosphere can compete with mycorrhizal fungi for access to plant carbon, without providing direct benefits to the plant. Saprotrophs play a critical role in the decomposition of cellulose and lignin into reduced forms, providing other microorganisms and plants with the necessary substrates to carry out metabolic functions. Antagonistic fungi show pathogenic traits that inhibit root growth and seed germination; sometimes, they kill the adult plant. Fungal pathogens play an important role in maintenance of plant populations by influencing plant competition and reproductive fitness. Bacteria are the most abundant microorganisms in the rhizosphere, occupying 1 g of soil with up to half a billion individual cells. Bacterial diversity in the rhizosphere is also high, with at least 2,000–5,000 bacterial species inhabiting a single gram. Bacteria perform a wide range of functions in the rhizosphere, including mediation of biogeochemical cycles, acquisition of nutrients, and protection of the host plant from antagonistic microbial attacks, maintenance of plant populations, and production of secondary metabolites. Chemical transformations of molecules to bioavailable forms, such as conversion of organic nitrogen to ammonium and ammonium to nitrate, are carried out by saprotrophic and ammonia-oxidizing bacteria, respectively. Nitrogen-fixing bacteria are able to fix atmospheric N_2 into ammonia, which can be taken up directly by plants and microorganisms. Nitrogen fixers are found as symbiotic inhabitants on plant roots and as free-living bacteria within the rhizosphere.

Rhizospheric bacteria also help maintain plant populations by serving as antagonistic or growth-promoting organisms. The antagonistic bacteria keep plant populations in check by suppressing plant growth, seedling elongation, and seed germination. PGPR are common in the rhizosphere, providing plants with metabolites, nutrients, and antibiotics that protect the plants from pathogenic attack and enhance plant growth.

13.2.1 Basic of PGPR

They are an important group of microorganisms used in biofertilizer. Biofertilization accounts for about 65% of the nitrogen supply to crops worldwide. PGPR have different relationships with different species of host plants. The two major classes of relationships are rhizospheric and endophytic. Rhizospheric relationships consist of the PGPR that colonize the surface of the root, or superficial intercellular spaces of the host plant, often forming root nodules. The dominant species found in the rhizosphere is a microbe from the genus *Azospirillum*. Endophytic relationships involve the PGPR residing and growing within the host plant in the apoplastic space. PGPR have the following inherent specialties:

1. They must be proficient to colonize the root surface.
2. They must survive, multiply, and compete with other microbiota.
3. They must promote plant growth.

PGPR have the diverse functional role in maintaining the soil fertility as well as assisting phytoremediation to plants; they fix atmospheric nitrogen; produce phytohormones and siderophores; solubilize phosphate, potassium, and zinc; alleviate the various stress by secreting ACC (1-aminocyclopropane-1 carboxylate) deaminase enzyme; and control disease by suppressing or killing the phytopathogens. Several PGPR and their role are shown in Table 13.1 and Figure 13.2. Kloepper and Schroth (1978) said that soil containing competitive microflora, which exerts a beneficial effect on plant growth, is termed as PGPR. In accordance with Vessey (2003), soil bacterial species burgeoning in plant rhizosphere, which grow in, on, or around plant tissues, stimulate plant growth by a plethora of mechanisms and are collectively known as PGPR. Alternatively, Somers et al. (2004) classified PGPR based on their functional activities as (i) biofertilizers (increasing the

TABLE 13.1

PGPR Parameters Responsible for Plant-Assisted Phytoremediation

PGPR	PGPR Parameters	References
Pseudomonas aeruginosa	IAA, siderophores, HCN, ammonia, exopolysaccharides, phosphate solubilization	Ahemad and Khan (2010d, 2012d, 2011a,d)
Rhizobium sp. *(pea)*	IAA, siderophores, HCN, ammonia, exopolysaccharides	Ahemad and Khan (2010c, 2011c, 2012b)
Pseudomonas putida	IAA, siderophores, HCN, ammonia, exopolysaccharides, phosphate solubilization	Ahemad and Khan (2011b, 2012a,c)
Enterobacter asburiae	IAA, siderophores, HCN, ammonia, exopolysaccharides, phosphate solubilization	Ahemad and Khan (2010a,b)
Pseudomonas sp. A3R3	IAA, siderophores	Ma et al. (2011a)
Psychrobacter sp. SRS8	Heavy metal mobilization	Ma et al. (2011b)
Azospirillum amazonense	IAA, nitrogenase activity	Rodrigues et al. (2008)
Pseudomonas sp.	ACC deaminase, IAA, siderophores	Poonguzhali et al. (2008)
Burkholderia	ACC deaminase, IAA, siderophores, heavymetal solubilization, phosphate solubilization	Jiang et al. (2008)
Pseudomonas jessenii	ACC deaminase, IAA, siderophores, heavymetal solubilization, phosphate solubilization	Rajkumar and Freitas (2008)
Pseudomonas sp.	ACC deaminase, IAA, siderophores, heavymetal solubilization, phosphate solubilization	Rajkumar and Freitas (2008)
Azotobacter sp., *Mesorhizobium* sp.	IAA, siderophores, antifungal activity, ammonia	Ahmad et al. (2008)
Bradyrhizobium sp.	IAA, siderophores, HCN, ammonia	Wani et al. (2007a)
Pseudomonas, Bacillus	Phosphate solubilization, IAA, and siderophores	Wani et al. (2007b)
Klebsiella oxytoca	IAA, phosphate solubilization, nitrogenase activity	Jha and Kumar (2007)
Pseudomonas fluorescens	Induced systemic resistance, antifungal activity	Saravanakumar et al. (2007)
Brevibacillus spp.	Zn resistance, IAA	Vivas et al. (2006)
Bacillus sp.	P-solubilization	Canbolat et al. (2006)
Pseudomonas putida	Siderophores, Pb and Cd resistance	Tripathi et al. (2005)
Variovorax paradoxus, *Rhodococcus* sp., *Flavobacterium*	IAA and siderophores	Belimov et al. (2005)
Pseudomonas fluorescens	IAA, siderophores, antifungal activity	Dey et al. (2004)
Rhizobium, Bradyrhizobium	HCN, siderophores, IAA, P-solubilization	Deshwal et al. (2003)
Mesorhizobium, Bradyrhizobium sp.	Siderophores	Khan et al. (2002)
Azotobacter chroococcum	P-solubilization	Kumar et al. (2001)
Kluyvera ascorbata	Siderophores	Burd et al. (2000)
Rhizobium cicero	Siderophopres	Berraho et al. (1997)
Rhizobium leguminosarum	Cytokinin	Noel et al. (1996)
Rhizobium, Bradyrhizobium	P-solubilization	Abd-Alla (1994)

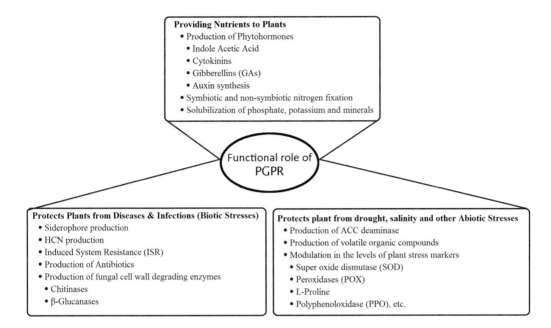

FIGURE 13.2 Schematic representation of functional role of PGPR.

availability of nutrients to plant), (ii) phytostimulators (plant growth promotion, generally through phytohormones), (iii) rhizoremediators (degrading organic pollutants), and (iv) biopesticides (controlling diseases, mainly by the production of antibiotics and antifungal metabolites). Furthermore, in most studied cases, a single PGPR will often reveal multiple modes of action, including biological control (Kloepper, 2003). Generally, PGPR can be separated into extracellular (ePGPR), existing in the rhizosphere, on the rhizoplane, or in the spaces between cells of the root cortex and intracellular (iPGPR), which exist inside root cells, generally in specialized nodular structures. Some examples of ePGPR are *Agrobacterium, Arthrobacter, Azotobacter, Azospirillum, Bacillus, Burkholderia, Caulobacter, Chromobacterium, Erwinia, Flavobacterium, Micrococcous, Pseudomonas, and Serratia*. Similarly, some examples of the iPGPR are *Allorhizobium, Azorhizobium, Bradyrhizobium, Mesorhizobium,* and Rhizobium of the family Rhizobiaceae. Most of rhizobacteria belonging to this group are gram-negative rods with lower proportion of gram-positive rods, cocci or pleomorphic. Moreover, numerous actinomycetes are also one of the major components of rhizosphere microbial communities showing plant growth beneficial traits. Among them, *Micromonospora* sp. (gram-positive, spore-forming, generally aerobic and form a branched mycelium), *Streptomyces* spp., *Streptosporangium* sp., and *Thermobifida* sp. have shown an enormous potential as biocontrol agents against different root fungal pathogens.

13.2.2 ECOLOGY AND BIODIVERSITY OF PGPR LIVING IN THE RHIZOSPHERE

The rhizosphere (soil zone influenced by roots) is a complex environment that consists of diverse bacterial populations, which have an important role in biogeochemical cycling of organic matter and mineral nutrients. It harbors a wide variety of bacteria species, and the compositions of bacterial communities differ according to root zone, plant species, plant phenological phase, stress, and disease events. The presence of rhizobacteria in the rhizosphere can have a neutral, detrimental, or beneficial effect on plant growth. The presence of neutral rhizobacteria in the rhizosphere probably has no effect on plant growth. Deleterious rhizobacteria are presumed to adversely affect plant growth and development not only through the production of metabolites such as phytotoxins or phytohormones but also through competition for nutrients or inhibition of the beneficial effects of mycorrhizae. Plant-friendly bacteria residing in rhizosphere, which exert beneficial effect on it, are called as PGPR. Currently, there are many bacterial genera that include PGPR among them, revealing a high diversity in this group. A discussion of some of the most abundant genera of PGPR follows to describe the genetic diversity and ecology of PGPR. A number of diazotrophic bacteria are present in PGPR. A number of diazotrophic PGPR participate in interactions with C3 and C4 crop plants. The mechanisms involved have a significant plant growth–promoting potential, retaining more soil organic-N and other nutrients in the plant-soil system and reducing the need for fertilizer N and P. Plant-associated nitrogen-fixing bacteria have been considered as one of the suitable alternatives for inorganic nitrogen fertilizer for promoting plant growth and yield. Though a variety of nitrogen-fixing bacteria such as *Acetobacter, Arthrobacter, Azoarcus, Azospirillum, Azotobacter, Bacillus, Beijerinckia, Derxia, Enterobacter, Herbaspirillum, Klebsiella, Pseudomonas,* and *Zoogloea* have been isolated from the rhizosphere of various plants species such as sugarcane, rice, sorghum, corn, other cereals, coffee bean, and pineapple. *Azoarcus* has recently gained attention due to its great potential in degradation of some contaminants, commonly habitating in soil. The most typical characteristic of these genera, which particularly distinguish them from other species, is their potential to grow in carboxylic acids or ethanol instead of sugars, with their optimum growth temperature ranging between 37°C and 42°C. Diazotrophic bacteria have also been growing in the endophytic compartment (inside the plant between the living cells) of some plant species and other grasses. It has been split into three different genera (*Azovibrio, Azospira,* and *Azonexus*). Beside diazotrophic bacteria, *Bacillus* sp. is also the most important genus present in PGPR. *Bacillus* sp. is the most efficient PGPR that enhances the plant growth by producing a vast variety of substances. They are having potent plant growth–promoting traits such as IAA production, phosphate solubilization, and nitrogen fixation, and biocontrol attributes such as production of HCN, siderophore, hydrolytic enzymes, and antibiotics have been isolated from different plants. Study reveals that 95% of gram-positive soil bacilli belong to the genus *Bacillus. Bacillus* sp. is able to form endospores that allow them to survive for extended periods under adverse environmental conditions. Moreover, nitrogen-fixing PGPR collectively known as rhizobia are also come under PGPR. They are capable of colonizing the rhizosphere of nonhost plants (nonlegumes), thus living within plant tissues as endophytes. Due to these properties and their ability to secrete phytohormones and siderophores and solubilize insoluble phosphates, besides eliciting plant defense reactions against phytopathogens, rhizobia have been placed along the organisms with high potential to act as PGPR. The information available on rhizobial application and the number of rhizobia stored in different culture collection centers around the world may provide an important microbiological resource to reduce the use of expensive synthetic fertilizers and pesticides in agricultural practices. Furthermore, among gram-negative soil bacteria, *Pseudomonas* is the most abundant genus in the rhizosphere, and the PGPR activity of some of these strains has been known for many years, resulting in a broad knowledge of the mechanisms involved. The ecological diversity of this genus is enormous, since individual species has been isolated from a number of plant species in different soils throughout the world. *Pseudomonas*strains show high versatility in their metabolic capacity. Antibiotics, siderophores, or HCN are among the metabolites generally released by these strains. These metabolites strongly affect the environment, both because they inhibit growth of other deleterious microorganisms and because they increase nutrient availability for the plant.

13.3   ROLE OF PGPR IN BIOREMEDIATION OF INDUSTRIAL WASTE

The contamination of soil and water with various pollutants is escalating day by day due to excessive industrialization. Technologies of remediation and bioremediation are continuously being improved using genetically modified microorganisms or those naturally occurring, to clean residues and contaminated areas from toxic organics. Research has shown that among the different microbes, use of PGPR for bioremediation activity is gaining importance due to their differential abilities to degrade and detoxify contaminants and their positive effects on plant growth promotion. To avoid industrial waste problems, bioremediation via PGPR is getting more consideration due to ecofriendly nature, less expense, and proven efficiency in comparison with physical or chemical remediation methods. Improving growth of plants and conquering the metal toxicity can be enhanced by association of PGPR. These microbes colonize the root or inhabit near root surfaces and involve in mechanisms for plant prevention from toxicity through secretion and production of several regulatory compounds such as phytohormones, siderophores, an metal-binding proteins. Although the extensive use of PGPR for the environmental remediation with plants emerged as a promising field, only very few field studies have been reported. Some organic contaminants can persist in the environment for a long time and bring great threat to human health. Such contaminants mainly include total petroleum hydrocarbons and polycyclic aromatic hydrocarbons (PAHs) coming from the exploration and consumption of fossil fuel, polychlorinated biphenyls (PCBs) widely used in the industrial process and are most degradation-resistant and other chlorinated aromatics used as PCB replacement such as polychlorinated terphenyls (PCTs), halogenated compounds such as perchloroethylene (PCE) and trichloroethylene (TCE), and pesticides such as atrazine and bentazon. Heavy metals are the primary inorganic contaminants of industrial waste, which include cadmium, chromium, copper, lead, mercury, nickel, and zinc.

For example, the soil and PGPR have been reported in bioremediation of sugarcane molasses-based anaerobically digested distillery effluent. It is dark brown due to high concentration of complex organic compounds and heavy metals, which do not alter even after long extended aeration. On land, it causes inhibition of seed germination and depletion of vegetation by reducing the soil alkalinity and manganese availability. Recently, Chandra et al. (2017) reported potential native weeds and grasses for heavy metal phytoextraction during in situ phytoremediation of distillery waste. In aquatic environment, it reduces sunlight penetration and decreases both photosynthetic activity and dissolved oxygen content damaging both aquatic fauna and flora. Disposal of distillery spent wash on land is equally harmful, causing a reduction in soil alkalinity, inhibition of seed germination, and damage to vegetation. Bioremediation of pollutants is shown in Table 13.2.

TABLE 13.2
List of PGPR during Bioremediation of Industrial Waste

PGPR	Type of Waste and Plants	Results
Rhizobium sp., *Microbacterium* sp.	Chromium, *Pisum sativum*	Improved the nitrogen (54%) concentration in *P. sativum* Decreased Chromium toxicity
Brevundimonas diminuta, Alcaligenes faecalis	Mercury, *Scripus mucronatus*	Increase phytoremediation by *Scripus mucronatus*, Decrease toxicity in soil
Bradyrhizobium japonicum CB1809	Arsenic, *Helianthus annuus*	Excess of plant biomass of *Helianthus annuus* and *Triticuma estivum*, Growth in high arsenic concentration
Bacillus megaterium	Lead, *Brassica napus*	Soil pollution decreased by *Brassica napus*
Bacillus, Staphylococcus, Aerococcus	Chromium, cadmium, copper, lead, and zinc, *Prosopis juliflora* and *Lolium mltiforum*	Improve the efficiency of phytoremediation by *Prosopis juliflora* and *Lolium mltiforum*; Tolerate high conc of and Zinc Chromium (up to 3,000 mg/l)
Bacillus sp. PSB10	Cr, *Cicer arietinum*	Prominently enhanced growth, nodulation, seed yield, and grain protein. Reduced the uptake of Cr in roots, shoots, and grains
Bacillus subtilis SJ-101	Ni, *Brassica juncea*	Acilitated Ni accrual

(Continued)

TABLE 13.2 (*Continued*)

List of PGPR during Bioremediation of Industrial Waste

PGPR	Type of Waste and Plants	Results
Bacillus weihenstephanens	Ni, Cu, Zn *Helianthus annuus*	Improved plant biomass and the accretion of Zn and Cu in the root and shoot systems, also enhanced the concentrations of soluble Ni, Zn, and Cu in soil with their metal mobilizing potential
Bradyrhizobium sp.750, *Pseudomonas* sp., *Ochrobactrum cytisi*	Cu, Cd, Pb *Lupinus luteus*	Increased plant biomass, phytostabilization
Brevibacillus sp.	Zn, *Trifolium repens*	Improved plant growth and nutrition and reduced Zn content in plant tissues
Kluyvera ascorbata SUD165	Ni, Pb Zn, *Brassica napus, Solanum lycopersicum*	No augmentation of metal uptake in comparison withnoninoculated plants. Reduction in decrease of metal stress
Mesorhizobium sp. RC3	Cr (VI), *Cicer arietinum*	Increased the nodules number, dry matter content, grain protein, and seed yield. N_2 in roots and shoots improved by 46% and 40%
Microbacterium oxydans AY509223 (RS)	Ni, *Alyssum murale*	Aided in phytoextraction of Ni
Ochrobactrum sp., *Bacillus cereus*	Cr (VI), *Vigna radiata*	In seedlings, Cr toxicity was lowered by reduction of Cr (VI) to Cr (III)
*Pseudomonas aerugino sa*strain MKRh3	Cd, *Vigna munga*	Plants elaborated lessened accretion, increased rooting and stimulated plant growth
Pseudomonas aeruginosa, Pseudomonas fluorescens, Ralstonia metallidurans	Pb, Cr, *Zea mays*	Supported plant growth, aided soil metal mobilization, and increased Pb and Cr uptake
Pseudomonas putida KNP9	Cd, Pb, *Vigna radiata*	Reduction of Cd and Pb uptake and enhanced plant growth
Pseudomonas sp.	Cr, Cd, Ni, *Glycine max, Vigna radiata, Tanacetum vulgare*	Improvement of plant growth in all the species under the applied metal stress
Psychrobacter sp. SRS8	Ni, *Helianthus annuus, Ricinus communis*	Enhanced plant growth and Ni accretion in both plant species with improved plant biomass, content of proteins and chlorophyll
Rhizobium sp. RP5	Ni, *Pisum sativum*	Improved the dry biomass, nodule numbers, seed yield, and grain protein
Sinorhizobium sp. Pb002	Pb, *Brassica juncea*	Effectiveness of Pb phytoextraction improved
Variovox paradoxus, Rhodococcus sp.	Cd, *Brassica juncea*	Elongation of root stimulated
Xanthomonas sp. RJ3, *Azomonas* sp. RJ4, *Pseudomonas* sp. RJ10, *Bacillus* sp. RJ31	Cd, *Brassica napus*	Cd accretion elevated and stimulation of growth of plant

Phragmites communis growing in temperate climatic conditions is the most commonly accepted wetland plant for the decolorization and detoxification of industrial effluents. The bacterial pretreatment of industrial effluents mediates the degradation and transformation of organic and inorganic pollutants and becomes easily bioavailable to wetland plant roots and rhizosphere microbes utilizing these biotransformed products as carbon, nitrogen, and energy source (Calheiros et al., 2009; Chandra et al., 2008). Mechanisms of organic and heavy metals removal by PGPR are shown in Figure 13.3.

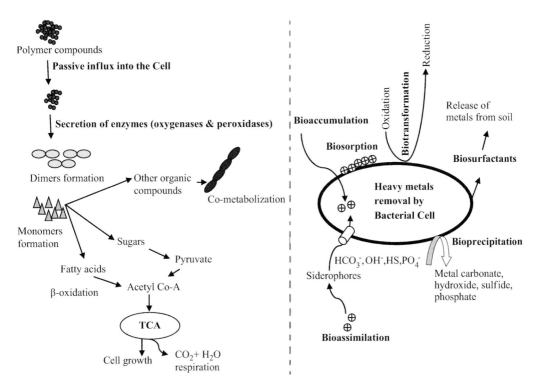

FIGURE 13.3 Mechanism of organic and heavy metals removal by PGPR.

The heterotrophs isolated from *Cyperus papyrus*r hizosphere contaminated with distillery wastewaters showed that upper zone (i.e., from 0–5 cm depth) was dominated by *Aeromonas hydrophilla, Pasteurella* sp., *Actinobacillus equuli, Aeromonas salmonicida, Kingella* sp., *Vibrio, Propionibacterium* sp., *Corynebacterium* sp., *and Staphylococcus* sp.; middle zone (i.e., 5–10 cm depth) by *Enterobacteria* sp., *Vibrio* sp., *Aeromonas salmoricida, Pasteurella* sp., *Actinobacillus* sp., and *Micrococcus* sp.; and the deeper zone by strict/obligatory anaerobes, i.e., methanogens and sulfur-reducing bacteria (SRB) (Chandra and Chaturvedi, 2002). Methanogens utilize hydrogen as substrate produced by acetate oxidizers, e.g., *Aerobacter aerogens, Alcaligens faecalis, Bacillus cereus, Bacillus megaterium, Bacillus subtilis, Micrococcus luteus, Nocardia* sp., *Streptomyces bikinensis, and Sarcina cooksoni.* Similarly, bacteria were isolated and characterized from the rhizosphere of *Phragmites australis* growing in distillery effluent-contaminated site, and out of 22 isolates from upper, middle, and lower regions, 16 were facultative anaerobes and 6 were obligate aerobes, which were mostly present in the lower region of roots (Chandra and Chaturvedi, 2002). These rhizosphere bacteria were found capable for the bioremediation of distillery wastewater-contaminated sites.

The persistent organic pollutants (POPs) and consequent environmental problems have brought the possibility of long-term environmental disasters into the public compunction. Therefore, various strategies are being developed, and further research is currently underway to develop means of sustaining the environment. Diverse catabolic pathways for degradation of complex pollutants depend on enzymatic activities of microbes for transformation and degradation of environmental pollutants into less toxic or harmless component such as CO_2 and H_2O. Transfer of electrons from electron donors to electron acceptors is a very essential step for any metabolic pathways. The electron donors act as food for microbes. Microorganisms can degrade pollutants in presence and absence of oxygen as aerobic and anaerobic degradation, respectively. Pollutants that have tendency to donate electron may be degraded in presence of oxygen by aerobic microbes, whereas those contaminants that are poor electron donors may degrade under anaerobic conditions. Many redox reactions are also involved in immobilization of trace elements found in the contaminated sites. Change in oxidation potential of chemicals such as metals will change the solubility and toxicity. Biosorption (microbial or plant cell–mediated) and biotransformation (enzymes- or metabolites-mediated) are probably the most widely explored biological metal and organic hazardous pollutants removal strategies of PGPR. In addition, PGPR are also playing a very important role in phytodegradation of organic and inorganic pollutants. Root exudates play an important role in bacterial QS and biofilm formation, since they can chemotactically attract rhizobia toward plants, adhere and colonize on legume roots as well as regulate expression of rhizobial nodulation genes [nod and rhizosphere-expressed (rhi), which promotes the growth of plants and their remediation potential]. Overall, mobilization of metals during phytodegradation takes place through the acidification, protonation, and chelation, whereas immobilization takes place through precipitation, complexation, and alkalinization. The PGPR can be used to give increased crop yields through a range of plant growth mechanisms. The eco-friendly approaches inspire a wide range of exploitation of beneficial agriculturally important bacteria and have led to improved nutrient uptake and good plant health. PGPR also play a major role in improving soil fertility, plant health, and remediation of pollutants. However, the microbial inoculants industry would also benefit greatly by developing PGPR to achieve results in novel agricultural endeavors. Agriculture is one of the most important sectors of the developing countries. The Indian economy, especially, is highly dependent on agriculture. In modern agriculture, the widespread use of chemicals during the past three decades has been a subject of public concern due to potential harmful effects on the environment as well as human and animal health. Stimulated by increasing demand and by the awareness of the environmental and human health damage by industrial waste and overuse of pesticides and fertilizers, agricultural practice is moving to a more sustainable and environmentally friendly approach worldwide.

13.3.1　PGPR-Assisted Phytoremediation of Heavy Metals

The fast industrialization and modernization all around the world leads to an unfortunate consequence: the production and release of considerable amounts of toxic wastes to the environment. Ecosystems have been contaminated with heavy metals and organic pollutants because of various human and natural activities. Because of the high cost of chemical technology, phytoremediation, which uses plants to treat metal-contaminated sites, is believed to be more environmentally friendly and economically cheap cleanup strategy than other conventional remediation strategies. However, environmental stresses inhibit the growth and development of plants, potentially reducing the efficiency of phytoremediation. The relationships that exist between plants and microbes in the rhizosphere play a key role in enhancing the efficacy of phytoremediation through a process known as "PGPR-assisted phytoremediation."

Soil microorganisms, which live in close association with plants, support the establishment and growth of plants on heavy metal–contaminated soils by producing plant growth hormones, inducing siderophores, solubilizing phosphorus, and, with a host of enzymatic activities, consequently altering the bioavailability of metals. Soil microbes are thought to exert positive effects on plant health via mutualistic relationships between them. However, microbes are sensitive to pollution and depletion of microbial populations both in terms of diversity, and biomass often occurs in such contaminated soils. Such naturally occurring rhizobacteria could assist phytoremediation both indirectly by increasing the overall fertility of the contaminated soil and enhancing plant growth through nutrient uptake and control of pathogenity and also directly catabolizing certain organics and/or intermediate partly oxidized biodegradation products. Soil microorganisms are also able to lower the level of the growth-inhibiting stress hormone, ethylene, within a plant growing in soils contaminated with heavy metals. Through these mechanisms, they may facilitate plant growth and thus increase the efficiency of phytoremediation. Apart from this, *Pseudomonas aeruginosa*, used as potent bioinoculant, express PGPR activity and biofilm and biosurfactant production and comprise key role in metal phytoextraction process. PGPR can positively influence plant growth and development in two different ways: indirectly or directly. The indirect promotion of plant growth occurs when these bacteria decrease or prevent some of the deleterious effects of a phytopathogenic organism. Bacteria can directly promote plant growth by providing the plant with a compound that is synthesized by the bacterium or by facilitating the uptake of nutrients from the environment by the plant. Plant growth–promoting bacteria may fix atmospheric nitrogen and supply it to plants; synthesize siderophores that can solubilize and sequester iron from the soil and provide it to plant cells; synthesize several different phytohormones including auxins and cytokinins, which can enhance various stages of plant growth; have mechanisms for the solubilization of minerals such as phosphorus, which then become more readily available for plant growth; and contain enzymes that can modulate plant growth and development. A particular bacterium may affect plant growth and development using any one, or more, of these mechanisms and a bacterium may utilize different mechanisms under different conditions. For example, bacterial siderophore synthesis is likely to be induced only in soils that do not contain sufficient levels of iron. Similarly, bacteria do not fix nitrogen when sufficient fixed nitrogen is available.

13.4 ROLE OF PGPR IN BIOFILM FORMATIONS AND ITS IMPORTANCE IN PLANT HEALTH

Beneficial PGPR also play a key role in agricultural and wastewater management through quorum sensing (QS) in their biofilm mode. Microbial biofilm formation is the primary step of bioremediation of any hazardous environmental pollutants. Biofilms are structured microbial communities in which the microbial cells irreversibly attach to a surface and become embedded in a matrix of extracellular polymeric substances, changes in cellular biochemistry, and social interactions that facilitate the exchange of metabolites, signal molecules, genetic material between cell and cell, sorption, and ability to immobilize pollutants. Biofilms are clusters of microbial cells that are attached to living or nonliving surface. They occur in nearly every moist environment where enough nutrient flow is available and surface attachment can be accomplished. Biofilm formation initiated when free-floating bacteria come in contact with suitable living or nonliving surface. This first step of attachment occurs when the microorganisms produce an adhesive substance known as an extracellular polymeric substance. An exopolysaccharides (EPS) is composed of sugars, proteins, and nucleic acids (extracellular DNA). Due to EPS, the microorganisms in a biofilm are stick together, which leads to create a bulbous and complex 3D structure known as biofilm. The main steps in the biofilm formations are attachment, aggregation, maturation, and dispersal. The final stage of biofilm growth cycle is seeding dispersal. In seeding dispersal process, the cells within a biofilm can leave the fold and establish themselves on a new surface either through a clump of cells that break away or individual cells that burst out from the biofilm and explore out a new biofilm. A biofilm can be formed by a single bacterial species, although it can also be formed by many species of bacteria, fungi, algae, and protozoa. Biofilm matrix is a major part of biofilm besides water and microbial cells. Biofilm matrix is a complex of secreted polymers, absorbed nutrients and metabolites, cell lysis products, and even particulate material and detritus from the surrounding environment. Approximately 97% of the biofilm matrix is either water, which is bound to the capsules of microbial cells, or solvent, the physical properties of which (such as viscosity) are driven by the solutes dissolved in it. The diffusion processes that occur within the biofilm matrix are reliant on the water-binding capacity and mobility of the biofilm. Biofilm formation is governed by a process known as QS. QS is a communication process between cells, in which bacteria secrete and sense the specific chemicals (autoinducer) and regulate gene expression in response to population density, whereas quorum quenching (QQ) blocks QS system and inhibits gene expression mediating bacterial behaviors. The QS-induced processes, such as biofilm formation, competence, sporulation, and antibiotic production, have been widely documented in plant-microbe interactions. N-acyl-L-homoserine lactones (AHLs) are an autoinducer as QS signals in many gram-negative bacteria (*Rhizobium radiobacter*, *Erwinia carotovora*,

TABLE 13.3

Summary of Known Biofilm-Forming Rhizosphere Bacteria

PGPR	Type of Waste
Acinetobacter calcoaceticus	P23 root colonization of duckweed
Azorhizobium brasilense	Root colonization of wheat
Azorhizobium caulinodans	Root colonization of rice
Bacillus amyloliquefaciens S499	Root colonization of tomato, maize, and *Arabidopsis thaliana*
Bacillus polymyxa	Root colonization of cucumber
Cyanobacteria spp.	Enhanced mixed-species biofilm formation *with Rhizobium, Azotobacter*, and *Pseudomonas* spp.
Klebsiella pneumoniae	Root colonization of wheat
Pantoea agglomerans	Root colonization of chickpea and wheat
Rhizobium leguminosarum bv. *viciae* 3841	Root colonization of various legumes. Nitrogen fixation and desiccation tolerance
Rhizobium (Sinorhizobium) sp. strain NGR234	Root colonization of various legumes and competitive colonization in the rhizosphere of cowpea. Nitrogen fixation

and *P. aeruginosa*) and PGPR (*Gluconacetobacter diazotrophicus* and *Burkholderia graminis*), which can be used to control a broad range of bacterial traits (such as symbiosis, biofilm formation, conjugation, motility, sporulation, virulence, pathogenicity, competence, and antibiotic production). Biofilm-forming rhizosphere bacteria are shown in Table 13.3.

Most of the gram-negative bacteria have LuxI/R QS systems for AHL production and expression of specific genes. Bacterial AHLs can be also recognized for modulation of plant growth homeostasis and defense response tissue specific gene expression. Specific gene expression by plants due to biofilm-forming bacteria plays a very important role in bioremediation of organic and inorganic wastes. Different types of AHLs are secreted by different species of microorganisms. In opposite, the AHL mimic compounds, i.e., furanones secreted by some bacteria and higher plants such as rice barrel clover and soybean, can disrupt the QS-regulated behaviors among bacterial population. Furanones and other AHL mimic compounds can antagonize AHL-type behaviors by binding to the AHL receptor, i.e., LuxR due to their structural similarities to bacterial AHLs, therefore affecting bacterial AHL signaling. Plants may also adopt AHL mimics to communicate with specific bacteria to protect them from pathogens. Moreover, QS systems in gram-positive bacteria differ from the AHL-mediated QS in gram-negative bacteria. In gram-positive bacteria, a peptide known as autoinducing peptide (AIP) is used as auto-inducer signal in place of AHL signal used in gram-negative bacteria. However, most of the bacteria of gram-negative and gram-positive groups have a common QS system mediated by autoinducer-2 (AI-2, i.e., furanone). This type of QS is known as hybrid QS. Comparison of free-living planktonic bacteria and their biofilm counterparts performance for remediation of pollutants or toxins showed different result, and biofilms showed more remediation. However, survival of free-living planktonic bacteria in the stress environment is less due to the low metabolic activity, decreased protection, and low bioavailability of the pollutants in the water phase causing insignificant transformation by planktonic bacteria. In contrast, attached and sessile microorganisms located in biofilm communities provide structure and protection because of their growth in a self-produced and complex polymeric matrix. In addition to genetic diversification in single species biofilms, biofilms can have different species of both aerobic and anaerobic organisms allowing them to survive in the presence of different nutrients by different metabolization process. This diversity and metabolic range makes biofilm-forming bacteria more potential in bioremediation. In the environment, indigenous biofilms forming microorganisms perform bioremediation mainly in the soils, which is a part of nutrient cycle and a part of the global self-purification system. Interestingly, QS in bacterial populations are capable of regulating a diversity of physiological processes, including bioluminescence, motility, symbiosis, plasmid transfer, antibiotic production, virulence factors, and biofilm formation. The common characteristic is that each of these processes is executed only if the bacterial cell population density is sufficiently high to ensure the success of the communication to each other. Most rhizobial species have been found to produce one or more signaling molecules associated with a QS system. Plant-associated PGPR strain *Bacillus amyloliquefaciens* SQR9 showed biofilm formation through the mechanism as shown in Figure 13.4. Bacteria in the rhizosphere sense root exudate components released by plants through methyl-accepting proteins, activate their motility-related genes (e.g., fla-cheoperon), and then swim to the root surface for attachment. At the same time, the genes involved in metabolism (e.g., fbaB, sucC) and transport (e.g., glcU) of various substrates are also induced. Activation of several NRPS/PKS genes related to antibiotic production (e.g., srf) also takes place to outcompete other microbes in the struggle for access to the root surface and to form biofilms. Thereafter, regulation of genes related to biofilm formation in cells attached to the root surface stimulates bacterial aggregation, thus allowing effective colonization and establishing a rhizospheric competition with soil pathogens. Finally, stimulation of the NRPS/PKS and plant growth–promoting (e.g., alsS, alsD) genes contributes to pathogen biocontrol and growth stimulation, respectively (Figure 13.4). Thus, root exudates can activate the rhizosphere adaptation and survival elements of SQR9, which in turn exerts beneficial biocontrol and growthpromotion effects, resulting in a mutually beneficial relationship between plant and PGPR strain. Similarly, biofilms are used in treating waste water, heavy metals, POPs, hydrocarbons, and explosives such as TNT and radioactive substances such as uranium into simple and less harmful compounds.

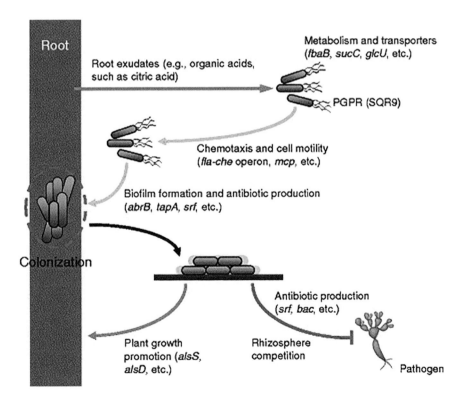

FIGURE 13.4 Biofilm formation of PGPR and rhizospheric interactions of the PGPR strain, plant, and pathogens.

Plant health can be improved through PGPR as a biofertilizer. Biofertilizers can be defined as products that contain living microorganisms; when applied to seeds, plant surfaces, and soil, they colonize in the rhizosphere or interior of the plant and promote plant growth by increasing the supply or availability of primary nutrients to the host plant. Biofertilizer is a mixture of live or latent cells encouraging nitrogen fixing, phosphate solubilizing, sulfate solubilization either by replacing soil nutrients or by making nutrients more available to plants or by increasing plant access to nutrients. PGPR are applied in the agricultural fields as replacement to conventional fertilizers. Biofertilizers are gaining impetus due to the maintenance of soil health, minimizing environmental pollution and cutting down the use of chemicals in the agriculture. However, it has been reported that PGPR have been used worldwide as biofertilizers, contributing to increased crop yields and soil fertility. Hence, with the potential contribution of the PGPR, this leads to sustained agriculture and forestry. PGPRs, particularly N_2-fixing, phosphate, and potassium solubilizers, are recommended as a sustainable solution to improve plant nutrient uptake and crop production.

13.5   MOLECULAR MECHANISM OF PGPR FOR SURVIVAL AND BIOREMEDIATION OF INDUSTRIAL WASTE

Plant growth–promoting bacteria (PGPB) also include various strains of rhizobia that form nodules on the roots of specific plants (legumes) and endophytes that can exist within the interior tissues of a plant. Today, several PGPB have been commercialized as either biocontrol agents or biofertilizers.

The PGPB can be classified into three dominant groups of microorganisms: arbuscular mycorrhizal fungi (AMF), PGPR, and nitrogen-fixing rhizobia, which are deemed to be beneficial to plant growth and nutrition. A thorough understanding of the PGPR action mechanisms is fundamental to manipulating the rhizosphere in order to maximize the processes within the system that strongly influences plant productivity. PGPR action mechanisms have been grouped traditionally into direct and indirect mechanisms. Direct mechanisms are those that occur inside the plant and directly affect the plant's metabolism, whereas indirect mechanisms are those that happen outside the plant. Direct mechanisms include those that affect the balance of plant growth regulators, either because the microorganisms themselves release growth regulators that are integrated into the plant or because the microorganisms act as a sink of plant-released hormones. The indirect mechanisms are those that improve nutrient availability to the plant; inhibition of microorganisms that have a negative effect on the plant (niche exclusion); and free nitrogen fixation in the rhizosphere, which improves nitrogen availability. Details are described below and shown in Figure 13.5. Use of PSB not only promotes the plant growth but also facilitates the bacterial-assisted phytodegradation of complex wastewater release by various industries. The knowledge of PGPR and their mechanisms and ecology in the rhizosphere will play a vital role in their use in development of sustainable technology for industrial wastewater management and in agriculture.

13.5.1   PGPR Using Direct Mechanisms

The list of direct mechanisms used by PGPR is substantial. Some have been included here, with the most relevant being discussed in detail:

- Free nitrogen fixation
- Phosphate solubilization
- Production of hormones
- Antagonistic bacteria
- Production of siderophores
- Production of volatile organic compounds

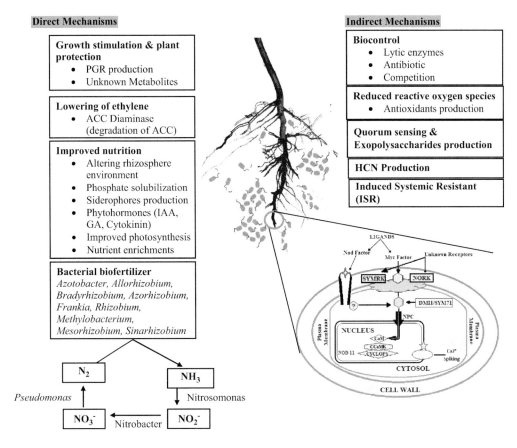

FIGURE 13.5 Direct and indirect mechanisms of PGPR.

13.5.1.1 Free Nitrogen Fixation

Atmospheric nitrogen or molecular dinitrogen (N_2) is relatively inert: it does not easily react with other chemicals to form new compounds. The fixation process frees nitrogen atoms from their triply bonded diatomic form, $N{\equiv}N$, to be used in other ways. Although 78% of the atmospheric air is N, this gaseous form is unavailable for direct assimilation by plants. Currently, a variety of industrial N fertilizers are used for enhancing agricultural productivity. BNF, discovered by Beijerinck in 1901, is carried out by a specialized group of prokaryotes. These organisms utilize the metalloenzymes called nitrogenasesto catalyze the conversion of N_2 to ammonia (NH_3). These enzymes contain iron, often with a second metal, usually molybdenum but sometimes vanadium. Microorganisms that can fix nitrogen are prokaryotes (both bacteria and archaea, distributed throughout their respective domains) called diazotrophs. Nitrogen is required for biosynthesis of the basic building blocks of plants, animals, and other life forms, e.g., nucleotides for DNA and RNA, the coenzyme nicotinamide adenine dinucleotide for its role in metabolism (transferring electrons between molecules), enzymes, proteins, chlorophyll, and hence it is a vital element for plant growth. It is also, indirectly, relevant to the manufacture of all chemical compounds that contain nitrogen, which includes explosives, most pharmaceuticals and dyes. Nitrogen fixation is carried out naturally in the soil by a wide range of nitrogen-fixing bacteria, including *Azotobacter*. Some nitrogen-fixing bacteria have symbiotic relationships with some plant groups, especially legumes.

BNF is a high-cost process in terms of energy. The free-living organisms that live in the rhizosphere do not establish a symbiotic relation with the plant. Although they do not penetrate the plants tissues, a very close relationship is established; these bacteria live sufficiently close to the root such that the atmospheric nitrogen fixed by the bacteria that is not used for their own benefit is taken up by the plant, forming an extra supply of nitrogen. This relationship is described as an unspecific and loose symbiosis. Free nitrogen-fixing bacteria belong to a wide array of taxa; among the most relevant bacterial genera are *Azotobacter, Azospirillum, Herbaspirillum, Burkholderia,* and *Bacillus.* This technique showed that the benefits of free nitrogen-fixing bacteria are due more to the production of plant growth regulators than to the nitrogen fixation.

Azotobacteraceae is the most representative of bacterial genera able to perform free nitrogen fixation. The effect of *Azotobacter* and *Azospirillum* is attributed not only to the amounts of fixed nitrogen but also to the production of plant growth regulators [indole acetic acid, gibberellic acid (GA), cytokinins, and vitamins], which result in additional positive effects to the plant. Application of inoculants in agriculture has resulted in notable increases in crop yields, especially in cereals, where *Azotobacter chroococcum* and *Azospirillum brasilense* have been very important. These two species include strains capable of releasing substances such as vitamins and plant growth regulators, which have a direct influence on plant growth. The amount of nitrogen from free fixation available to the plant is low because it is used efficiently by the bacteria. Three strategies have been proposed to address this low-yield problem: (i) glutamine synthase bacterial mutants, (ii) formation of paranodules, and (iii) facilitating the penetration of plant tissues by nitrogen-fixing bacterial endophytes that enhance colonization in a low competition niche. Nitrogenase, a major enzyme involved in the nitrogen fixation, has two components: (i) dinitrogenase reductase, the iron protein, and (ii) dinitrogenase (metal cofactor). The iron protein provides the electrons with a high reducing power to dinitrogenase, which in turn reduces N_2 to NH_3. Depending on the availability of metal cofactor, three types of nitrogen-fixing systems have been identified (i) Mo-nitrogenase, (ii) V-nitrogenase, and (iii) Fe-nitrogenase. Legume-rhizobia symbiosis is a cheaper source of N and an effective agronomic practice ensuring adequate supply of N than the application of fertilizer-N. Furthermore, ammonia is converted to nitrite and nitrate by nitrifying bacteria such as nitrosomonas and nitrobacter during nitrification, which is an important process in nitrogen cycle. Therefore, a reduced rate or inhibition of nitrification provides enough time to plant for assimilation of fixed N. Plants also produce secondary metabolites such as phenolic acids and flavonoids for inhibiting nitrification. The natural ability of plants to suppress nitrification is not currently recognized or utilized in agricultural production. However, they have no effects on other soil microbial community.

In the plant-microbial world, the association of PGPR with plants can be exploited for benefits not only for the associated organisms but also for the ecosystem. Plant health can be improved through PGPR as a biofertilizer. Biofertilizers can be defined as products that contain living microorganisms; when applied to seeds, plant surfaces, or soil, they colonize the rhizosphere or interior of the plant and promote plant growth by increasing the supply or availability of primary nutrients to the host plant. Biofertilizer is a mixture of live or latent cells encouraging nitrogen-fixing, phosphate solubilizing, or cellulolytic microorganisms used for applications to soil, seed, roots, or composting areas with the purpose of increasing the quantity of those mutualistic beneficial microorganisms and accelerating those microbial processes, which enhance the nutrient status of the plants by either replacing soil nutrients or by making nutrients more available to plants or by increasing plant access to nutrients. They are applied in the agricultural fields as replacement to conventional fertilizers. Biofertilizers are gaining impetus due to the maintenance of soil health, minimizing environmental pollution and cutting down the use of chemicals in the agriculture.

13.5.1.2 Phosphate Solubilizers

After nitrogen, phosphorus (P) is the most limiting nutrient for plant growth. Phosphorus (P) is a macronutrient that is essential for plant growth and development. It is a component of biological molecules, such as DNA, RNA, ATP, and phospholipids, and on a macrolevel, it affects root development, stalk and stem strength, crop maturity, and nitrogen fixation in legumes. It exists in both inorganic (bound, fixed, or labile) and organic (bound) forms, and the concentration depends on the parental material. Some P minerals complexed with calcium, aluminum, or iron have faster dissolution rates that are dependent on the pH of the surrounding soil and on the size of the particles. Higher soil pH (basic) causes aluminum and iron-complexed P to become more soluble, whereas lower soil pH values (acidic) promote the solubility of calcium-complexed P since the concentration of available P in soil is lower than what is found in healthy plant tissues.

The industrial solid or liquid wastes also contain phosphorus in complex form. This complex form of phosphorus is not easily taken by plants. Hence, phosphate-solubilizing bacteria (PSB), which are part of PGPR, play a significant role in releasing P from its insoluble complexes to a form that is more readily usable by plants. Soil microbes help in P release to the plants that absorb only the soluble P such asmonobasic ($H_2PO_4^-$) and dibasic ($H_2PO_4^{2-}$) forms as shown in Figure 13.6. The inorganic forms of P can be solubilized by microorganisms that secrete low-molecular-weight organic acids to dissolve phosphate-complexed minerals and chelate cations that partner with P ions (PO_4^{3-}) to release P directly into the surrounding soil. P solubilizing bacteria, which dissolves various sparingly soluble P sources such as $Ca_3(PO_4)_2$ and $Zn_3(PO_4)_2$ through lowering pH of the rhizosphere soil and making P available for plant uptake. Use of PSB, either in conjunction with or as a replacement for expensive and environmentally damaging fertilizers, would be a better alternative for sustainable agriculture. The P content in industrial waste is also present in complex form. Moreover, P content of agricultural soil is typically in the range of 0.01–3.0 mg P/L representing a small portion of plant needs. The rest must be obtained from the solid phase through intervention of biotic and abiotic processes where the phosphate-solubilizing activity of the microbes has a role to play. Among bacteria, the most efficient Phosphate-solubilizing microorganism (PSM) belong to genera *Bacillus*, *Rhizobium*, and *Pseudomonas*. Within *Rhizobia*, two species nodulating chickpea, *Mesorhizobium cicero* and *Mesorhizobium mediterraneum*, are known as good phosphate solubilizers. Rhizobia, including *Rhizobium leguminosarum*, *Rhizobium meliloti*, *M. mediterraneum*, *Bradyrhizobium* sp., and *Bradyrhizobium japonicum* are the potential *P solubilizers*. Some bacterial strains are found to possess both solubilization and mineralization capacity. PSB poses several mechanisms to solubilize phosphate as shown in Figure 13.6 (i) by releasing organic acids and affecting the mobility of phosphorus by means of ionic interactions; (ii) by means of phosphatases that help to unbind the phosphate groups from organic matter; (iii) by lowering the rhizospheric pH, PSB dissolves the soil phosphate through production of low-molecular-weight organic acids, biotic

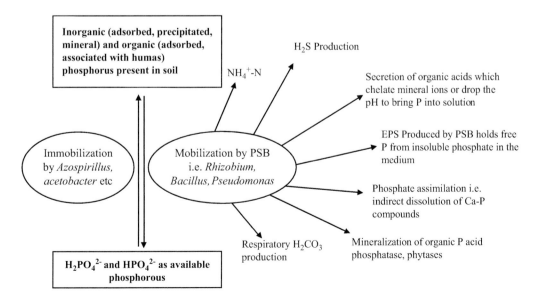

FIGURE 13.6 Mechanism of phosphate solubilization by PSB.

production of proton/bicarbonate release (anion/cation balance), and gaseous (O_2/CO_2) exchanges; (iv) enzymes of microbial origin, such as acid phosphatases, phosphohydrolases, phytase, phosphonoacetate hydrolase, D-a-glycerophosphatase, and C-P lyase; (v) EPS production; and (vi) phosphate assimilation, i.e., indirect dissolution of Ca-P compounds.

13.5.1.3 Production of Hormones

The success of remediation through plant is dependent on the potential of the plants to yield high biomass and withstand under stress condition. Plant hormones are chemical messengers that influence the plant's ability to react to its environment. These are naturally organic compounds that are effective at very low concentration and are mostly synthesized in certain parts of the plant and transported to another location. Plant hormones, also referred to as phytohormones, influence physiological processes at low concentrations. Auxin is a class of plant hormones important in the promotion of lateral root formation. Increased lateral root formation leads to an enhanced ability to take up nutrients and pollutants by the plant. Indole-3-acetic acid (IAA) is the most common, naturally occurring, plant hormone of the auxin class. It is the best known of the auxins. IAA is the foremost phytohormone that accelerates plant growth and development by improving root/shoot growth and seedling vigor. IAA is involved in phototropism and geotropism, cell division, vascular bundle formation, vascular tissue differentiation, apical dominance, root initiation (lateral and adventitious), stem and root elongation, and an essential hormone for nodule formation.

Other classes of plant hormones, which can enhance plant growth and removal of pollutants along with growth, include abscisic acid (ABA), ethylene, cytokinins, and gibberellins. The concentration of plant-synthesized auxin determine sits effect on the stimulation or inhibition of plant growth. Plant's auxin concentrations may be either suboptimal or optimal so that the addition of bacterial auxin that can be taken up by the plant may change the hormone level in the plant to either optimal or supraoptimal. Thus, bacterial IAA produced by a PGPR may either stimulate root development in cases where the plant's concentration is suboptimal or inhibit root development in cases where the auxin level is already optimal. Most auxin/IAA is synthesized from the amino acid tryptophan present in plant root exudates at varying low concentrations based on the plant's genotype. IAA appears to be synthesized by at least three different biosynthetic pathways with each pathway being named for a key intermediate within the pathway. These pathways include the indole pyruvic acid (IPyA) pathway, the indole acetamide (IAM) pathway, the indole acetaldoxime (IAOx)/indoleacetonitrile (IAN) pathway, the indole acetaldehyde (IAH) pathway, and the tryptamine pathway. It should be noted that various PGPR can have one, two, or even three functional IAA biosynthesis pathways, suggesting that the synthesis of IAA is clearly very important in the life and functioning of the bacterium.

However, relatively few studies determine the role of exogenous ABA in plant-microbe interactions and whether bacterial ABA influences ABA status of plants under salt stress. However, PGPR modulate ABA biosynthesis and ABA-mediated signaling pathways that may contribute to the enhanced growth of salt-stressed plants. The transport of abscisic acid can occur in both xylem and phloem tissues and can also be translocated through parenchyma cells. The movement of abscisic acid in plants does not exhibit polarity like auxins. Abscisic acid was reported to stimulate the stomatal closure, inhibit shoot growth while not affecting or even promoting root growth, induce seeds to store proteins and in dormancy, induce gene transcription for proteinase inhibitors, and thereby provide pathogen defense and counteract with gibberellins.

Furthermore, ethylene has also been recognized as a stress hormone. Ethylene in low levels has been observed to promote growth, but at moderate to high levels, it may inhibit root elongation. In plants, 1-aminocyclopropane-1-carboxylate (ACC) and 5'-deoxy-5'methylthioadenosine (MTA) are converted to ACC by ACC synthase. A number of PGPB have been found to contain the enzyme ACC deaminase, which cleaves and sequesters the plant ethylene precursor ACC and thus lowers the level of ethylene in a developing or stressed plant. ACC deaminase is a member of a large group of enzyme that utilizes vitamin B6 and is considered to be under tryptophan synthase family. This enzyme can diminish or prevent some of the harmful effects of the high ethylene levels. Representation of the stimulation of plant growth by a PGPR-containing ACC deaminase is shown in Figure 13.7. The ACC deaminase acts on ACC, an immediate ethylene precursor in higher plants, degrading this chemical to α-ketobutyrate and ammonium. Under such condition, in order to maintain the equilibrium between the rhizosphere and root interior ACC levels, the plants release more ACC through exudation, and thus, results decrease in the production of stress ethylene. Rhizosphere bacteria with ACC deaminase activity belonging to the genera, *Achromobacter, Azospirillum, Bacillus, Enterobacter, Pseudomonas*, and *Rhizobium* have been isolated from different soils. The presence of PGPB thereby moderates concentration of ACC so that it does not reach a level where it begins to impair root growth. Ethylene is essential for the growth and development of plants, but it has different effects on plant growth depending on its concentration in root tissues. At high concentrations, it can be harmful, as it induces defoliation and cellular processes that lead to inhibition of stem and root growth as well as premature senescence, all of which lead to reduced crop performance. Under stress conditions such as those generated by various industries, salinity, drought, water logging, heavy metals, pathogenicity, organic and inorganic load, and the endogenous level of ethylene is significantly increased, which negatively affects the overall plant growth.

In addition to improving plant's nutrient uptake and growth, the plant-associated microbes alleviate heavy metal toxicity or any types of stress by reducing stress ethylene production. By the inoculation of rhizobia-producing ACC deaminase, the plant ethylene levels are lowered and result in longer roots providing relief from stresses, such as heavy metals, pathogens, drought, radiation, and salinity. IAA-producing bacteria are reported to produce high levels of ACC and known to inhibit ethylene levels.

Furthermore, cytokinins enhance cell division, root development, and root hair formation and are also involved in the processes such as photosynthesis or chloroplast differentiation. They are also known to induce opening of stomata, suppress auxin-induced apical dominance, and inhibit senescence of plant organs, especially in leaves. Cytokinins are widely distributed in algae, bacteria, and higher plants. PGPR such as *Azospirillum* and *Pseudomonas* sp. produce cytokinins and gibberellins. They are produced in the root tips and transported through the xylemto the shoot by translocation. Cytokinin content and plant growth have been increased with the inoculation

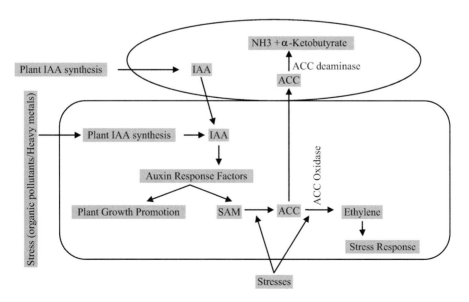

FIGURE 13.7 Schematic representation of the stimulation of plant growth by a PGPR-containing ACC deaminase.

of lettuce with *B. subtilis*. Rhizobium strains are also reported as the potent producers of cytokinins. Cytokinin production has been reported in various PGPR, such as *Arthrobacter giacomelloi*, *A. brasilense*, *B. japonicum*, *Bacillus licheniformis*, *P. fluorescens*, and *Paenibacillus polymyxa*. Plant responses to exogenous applications of cytokinin result in either one of the following effects: (i) enhanced cell division; (ii) enhanced root development; (iii) enhanced root hair formation; (iv) inhibition of root elongation; and (v) shoot initiation and certain other physiological responses. However, gibberellins are also synthesized by higher plants, fungi, and bacteria. They are involved in several plant developmental processes, including cell division and elongation, seed germination, stem elongation, flowering, fruit setting, and delay of senescence in many organs of a range of plant species. They can also regulate root hair abundance and hence promote the root growth. Gibberellins include a large group of tetracyclic diterpenoid carboxylic acids having either C20 or C19carbonskeletons. PGPR production of GAs has been observed in the following genera: *Achromobacter xylosoxidans*, *Gluconobacter diazotrophicus*, *Acinetobacter calcoaceticus*, *Rhizobia*, *Azotobacter* spp., *Bacillus* spp., *Herbaspirillum seropedicae*, and *Azospirillum* spp. The biochemistry of gibberellins in bacteria is similar to that of plants with some small differences. The absence of GAs is readily observed as a reduction of the lateral root number and length. As is the case for cytokinins, most of the currently attributed functional role of bacterially produced gibberellins in plant growth promotion is due to plant's response to the exogenic addition of purified gibberellins to growing plants.

13.5.1.4 Antagonistic Behavior of PGPR

Antagonistic potential of PGPR can be accomplish as biopesticides on commercial scale for sustainable agriculture system. Moreover, industrial waste is the major source of pathogenic bacteria. Hence, antagonistic behavior of PGPR can be a good mechanism for remediation of industrial waste. But limited researches have been done in this direction. One of the major disadvantages of chemical pesticides is that it is not easily broken down into simple and safer constituents and remained intact over a long time in the environment, causing soil pollution. Synthetic pesticides are nontargeted in nature as they affect the broad spectrum of microbe including plant beneficial microbe. PGPR as biopesticide is an appealing alternative to chemical pesticide. Rhizobacteria can inhibit the growth of several phytopathogens in different ways, competing for space and nutrients, producing bacteriocins, lytic enzymes, antibiotics, and siderophores. These antagonistic bacteria specifically disintegrate the cells of pathogens by producing lytic enzymes, antibiotics, and bacteriocins. Similarly, antagonistic bacteria deprive the pathogen from iron by producing siderophores tochelate it, ultimately excluding the pathogen from niche. Eubacterial genera including *Bacillus*, *Burkholderia*, *Enterobacter*, *Herbaspirillum*, *Ochrobactrum*, *Pseudomonas*, *Serratia*, *Staphylococcus*, and *Stenotrophomonas* are well-known antagonistic bacteria.

13.5.1.5 Formation of Siderophores

Iron is among the bulk minerals present on the surface of the earth, yet it is unavailable in the soil for plants. Iron can occur in either the divalent (ferrous or Fe^{2+}) or trivalent (ferric or Fe^{3+}) states, which is determined by the pH and Eh (redox potential) of the soil and the availability of other minerals (e.g., sulfur is required to produce FeS_2 or pyrite). Iron has an important role to play in the plant photosynthetic system as it is an integral part of the light-absorbing chlorophyll and is also involved in a wide range of different biosynthetic mechanisms. Industrial waste contains iron and other metals as insoluble hydroxides and oxyhydroxides, which are not accessible to both plants and microbes, which are only slightly soluble and cannot be readily conveyed into the cells. Generally, bacteria have the ability to synthesis low-molecular-weight compounds termed as siderophores (~400–1,000 Da) capable of sequestering Fe^{3+}. Siderophores are small peptide molecules that have side chains and functional groups to which ferric ions can bind. Siderophores produced by PGPR bind to Fe^{3+} with an exceptionally high affinity (i.e., $Kd = 10^{-20}–10^{-50}$). They are iron chelators that serve as iron carriers and have a high affinity for some ligands. Quite a large number of them have been screened and used from microbes, and they can also be species-specific (Sandy and Butler, 2009). These siderophores are known to have high affinity for Fe^{3+} and thus make the iron available for plants. The siderophores are water-soluble and are of two types, viz., extracellular and intracellular. Fe^{3+} ions are reduced to Fe^{2+} and released into the cells by gram-positive and gram-negative rhizobacteria. This reduction results in destruction/recycling of siderophores. Siderophores are the low-molecular-weight substances that chelate mainly iron and other metals. Microorganisms encounter the nutritional requirements for iron using sidero-phores. In surrounding environment, the iron is solubilized when siderophores are released and a ferric-siderophore complex is formed, and this substance moves through diffusion process and reaches to cell surface. Membrane receptors of gram-positive and gram-negative bacteria recognize ferric-siderophore complex and start the active transport. Thus, the siderophore-producing bacteria can relieve plants from heavy metal stress and assist in iron uptake. Schematic representation of mechanism of siderophore for iron transport is shown in Figure 13.8. Rhizobial species, such as *R. meliloti, Rhizobium tropici, R. leguminosarum* bv. viciae, *R. leguminosarum* bv. trifolii, *R. leguminosarum bv. phaseoli, Sinorhizobium Meliloti*, and *Bradyrhizobium* sp., are known to produce siderophores. Siderophores have great affinity to form complex with ferric ion, improve its solubilization, and enable its removal from natural complexes or from minerals. Low ferric ions availability in the environment results in the reduced growth of pathogens, which ultimately exclude pathogen from niche. Siderophores are peptide molecules that include the functional groups along with side chains that enhance the regulation of ferric ions by forming high-affinity set of ligands.

Once bound, the now soluble iron-side rophore complex is taken up by specific receptors on the surfaces of bacteria or plants and internalized, and then following either reduction to the ferrous state (Fe^{2+}) or cleavage of the siderophore molecule, the iron is released from the siderophore. Bacterial siderophores are classified into four major classes based on the types of ligand and basic features of functional groups that form coordinate with iron. Main classes include phenol catecholates, carboxylate, pyoverdines, and hydroxamates. Hydroxamate-type siderophores are common with fungi, whereas catecholates, which bind iron more tightly than hydroxamates, are common in bacterial siderophores. Linear hydroxy- and amino-substituted aminocarboxylic acids, e.g., mugineic acid

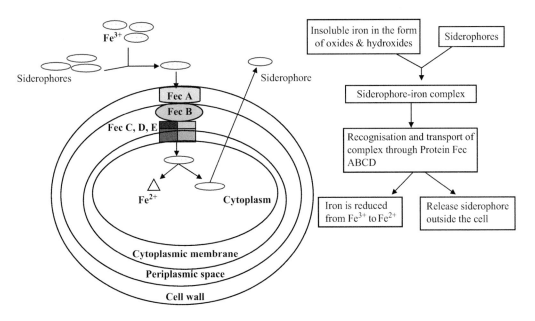

FIGURE 13.8 Schematic representation of mechanism of siderophore for iron transport.

and avenic acid are plant siderophores; they tend to bind iron more efficiently than bacterial sidero-phores. Other negatively charged molecules have lower affinity for iron than bacterial siderophores. In addition, some other trivalent and divalent metal ions also bind bacterial siderophores, though with a much lower affinity.

13.5.1.6   Production of Volatile Organic Compounds

Volatile organic compounds (VOCs) produced by PGPR are heavily involved in improving plant growth and ISR toward pathogens. ISR is a resistance mechanism in plants that is activated by infection. Its mode of action does not depend on direct killing or inhibition of the invading pathogen but rather on increasing physical or chemical barrier of the host plant. Like the systemic acquired resistance (SAR), a plant can develop defenses against an invader such as a pathogen or parasite if an infection takes place. In contrast to SAR, which is triggered by accumulation of pathogenesis-related proteins or salicylic acid, ISR instead relies on signal transduction pathways activated by jasmo-nate and ethylene. Several bacterial species, from diverse genera including *Bacillus*, *Pseudomonas*, *Serratia*, *Arthrobacter*, and *Stenotrophomonas*, produce VOCs that influence plant growth. Acetoin and 2,3-butanedIol synthesized by *Bacillus* are the best known of these compounds and are respon-sible for significant improvements in plant growth. Some other PGPR strains emit VOCs that can directly and/or indirectly mediate increases in plant biomass, disease resistance, and biotic and abiotic stress tolerance. VOC emission is a common property of PGPR. The identity and quantity of volatile compounds emitted are variable and depend on the species.

13.5.2   INDIRECT MECHANISMS OF PGPR FOR PLANT GROWTH PROMOTIONS

PGPR have the ability to produce many active principles for biocontrol of various phytopatho-gens with antibiosis production as indirect mechanism for plant growth during treatment of industrial and agricultural waste. Such mechanism includes (i) production of, (ii) production of exopolysaccharides,(iii) lipo chitooligosaccharides, (iv) bacteriocins,(v) lytic enzymes, i.e., chitin-ase and glucanase, (vi) induction of plant resistance by any of the metabolites mentioned above or by inducting the production of phenyl alanine lyase, antioxidant enzymes such as peroxidase, polyphenol oxidase, superoxide dismutase, catalase, lipoxygenase, and ascorbate peroxidase and also by phytoalexins and phenolic compounds in plant cells, (vii) HCN, phenazines, pyrrolnitrin, viscoinamide, and tensin by rhizobia are also reported as other mechanisms production, which inhibits electron transport and hence disruption of energy supply to the cells, (viii) competition for nutrients against phytopathogens and thereby occupies the colonizing site on root surface and other plant parts, and (ix) quorum quenching.

13.5.2.1   Production of Antibiotics

The production of antibiotics is assumed as most effective treatment and has antagonistic activity to suppress the phytopathogens. Antibiotics are organic compounds of low molecular weight that are involved in the inhibition of growth and metabolic activities of various microbes. The production of antibiotics is the most effective antagonistic activity to suppress the growth of phytopathogens. Thus, antibiotics play an important role in disease management, i.e., they can be used as biocon-trol agents. Antibiotics produced by PGPR include kanosamine, 2,4 diacetylphloroglucinol (2,4-DAPG), Martinez-Viveros oligomycin A, butyrolactones, xanthobaccin phenazine-1-carboxylic acid, pyrrolnitrin, zwittermycin A, and viscosinamide. Among them, DAPG is important since it has a broad-spectrum antibacterial, antifungal, and antihelminthic activity. The bacterial strains of *Pseudomonas fluorescens* BL915 involved in the production of antibiotic known as pyrrolnitrin have the ability to inhibit deterioration of *Rhizoctonia solani*. 2,4-DAPG is an extensively studied antibiotic involved in the membrane destruction of *Pythium* spp.. *Pseudomonas* spp. also synthe-size phenazine that contains theantagonistic activity against *Fusarium oxysporum*. Many *Bacillus* ssp. produced antibiotics such ascirculin, polymyxin, and colistin that are actively involved in the growth inhibition of pathogenic fungi as well as gram-negative and gram-positive bacteria. *B. subtilis* produce antibiotics such as fengycin and iturins and inhibit the growth of a fungus named *Podosphaera fusca*. Antibiotics play an important role in disease management, used as biocontrol agent and faced challenge due to limitations because antibiotics are prepared under natural circumstances. Ecological and other components that affect the antimicrobial action of antibiotics were examined to utilize the potential of antibiotics that are produced by PGPR in crop protection.

13.5.2.2 Production of Exopolysaccharides

Geesey in 1982 defined EPS as extracellular polymeric substances of biological origin, which involve in the formation of microbial aggregates. All polymers outside the cell not directly anchored in the outer membrane should be considered as EPS. EPS works as the house of biofilms because it shelters bacteria from physical, chemical, and biological challenges. EPS is made up with carbohydrates, proteins, humic substances, lipids, nucleic acids, uronic acids metabolites, and extracellular DNA (eDNA) and some inorganic components. The EPS is a mixture of polymers secreted from microorganisms and produced by cellular lysis and hydrolysis of macromolecules. In general, the most important role of EPS is the fundamental construction of EPS matrix formation. Polysaccharides and lectin-like proteins of EPS play very important roles in the formation of 3D networks of the EPS matrix by directly forming protein-polysaccharide cross-links, polysaccharide chains, and indirectly through multivalent cation bridges. Hence, EPS have a significant effect on the physicochemical properties of microbial aggregates, such as structure, flocculation, settling, adsorption, dewatering, and biodegradation. The EPS are divided into two categories. First category is bound EPS (capsular polymers, sheaths, loosely bound polymers, attached organic materials, and condensed gels), and second category is soluble EPS (colloids, soluble macromolecules, and slimes). Bound EPS can be further divided into tightly bound EPS and loosely bound EPS. Tightly bound EPS are closely adhered to the cell surface, which has a certain shape and is bound tightly and stably with the cell surface, whereas loosely bound EPS are the outer layers of EPS that are loosely bound to the cells without certain edge. The content of the loosely bound EPS in microbial aggregates is always less than that of the tightly bound EPS and thus may have some influence on the characteristics of microbial aggregates. Soluble EPS are soluble cellular components of EPS which are dissolved and release into surrounding liquor. This showed that interaction between soluble EPS and cells is very weak. Bacteria can use EPS as carbon and energy sources. Centrifugation is the technique by which we can separate out the bound EPS and soluble EPS. The polymers remained in the supernatant after centrifugation is known as soluble EPS, and pellets are considered as bound EPS. EPS is physically and chemically different from the bacterial capsule. EPS is highly hydrated because it binds with water through hydrogen bonding; hence it prevents the cells from desiccation, whereas chemical composition of EPS is slightly different in different species.

EPS may also contribute to the antimicrobial resistance properties of biofilms by reducing the transport of antibiotics through the biofilm, probably due to direct binding of EPS to antibiotic instead of bacterial cells. Most of the microorganisms of biological wastewater treatment systems are present in the form of microbial aggregates, such as biofilms, sludge flocs, and granules. Humic substances may also be a key component of the EPS (20% of the total EPS) of biofilms present in biological wastewater treatment system, whereas carbohydrates and proteins are major components of EPS. In addition, nucleic acids, lipids, uronic acids, and some inorganic components have been found in EPS. The EPS in microbial aggregates have many sites for the adsorption of metals and organic matters, such as aromatics, aliphatics in proteins, and hydrophobic regions in carbohydrates. This reveals the potential roles of EPS in heavy metal sorption to bacterial cells and transporting in environments. The EPS in microbial aggregates also have many charged groups (e.g., carboxyl, phosphoric, sulfhydryl, phenolic, and hydroxyl groups) and a polar groups (e.g., aromatics, aliphatics in proteins, and hydrophobic regions in carbohydrates) as hydrophilic areas. The formation of hydrophobic areas in EPS would be beneficial for organic pollutant adsorption. The presence of hydrophilic and hydrophobic groups in EPS molecules indicates that EPS are amphoteric. The presence of many negative charge in EPS, such as carboxyl, phosphoric, sulfhydryl, phenolic, and hydroxyl groups, can complex with heavy metals and be helpful for their remediation. On the basis of numbers of the available carboxyl and hydroxyl groups, the EPS are regarded to have a very high binding capacity. Proteins, carbohydrates, and nucleic acids in EPS have abilities to form complex with heavy metals. Furthermore, the soluble EPS might have a greater adsorptive ability for heavy metals than the bound EPS from sludge. The binding between EPS and divalent cations, such as Ca^{2+} and Mg^{2+}, is one of the main intermolecular interactions in maintaining the microbial aggregate structure. In the adsorption of heavy metal onto activated sludge, Ca^{2+} and Mg^{2+} were found to release into the solution simultaneously, indicating that the ion exchange mechanism was involved (Yuncu et al., 2006). EPS can also adsorb organic pollutants, such as phenanthrene, benzene, humic acids, and dye (Esparza-Soto and Westerhoff, 2003; Liu et al., 2000; Sheng et al., 2008). It might be due to presence of some hydrophobic regions in EPS. The EPS are always negatively charged and bind with the positively charged organic pollutants through electrostatic interaction. Moreover, proteins have a high binding property than humic substances. In soluble EPS, fraction of proteins is present in high amount than the bound EPS. Thus, soluble EPS have a greater binding capacity than the bound EPS. eDNA is common component of EPS different from chromosomal DNA due to its primary sequence. eDNA is released in EPS through lysis of cells in bacterial populations through QS-independent and QS-dependent mechanisms. QS-independent and QS-dependent mechanisms get activated during early and late exponential and early stationary growth phase, respectively. eDNA release in the matrix of biofilms is QS-independent mechanisms, which occurs via prophage-induced cell lysis controlled through flagella and type IV pili. But cell lysis is QS-dependent mechanisms, which concurrently generate elevated amounts of eDNA. QS molecules, such as AHLs and Pseudomonas quinolone signal (PQS), control the production of prophage and phenazine that are cell lysis factors that induce cell lysis, which leads to eDNA release. Microorganisms survive in harsh condition by different strategies such as fluctuations of some regulators, induction of SOS repair system, and increased eDNA uptake to improve genomic adaptability of the community. But the exact role of eDNA is not clear so far. Most of the EPS in anaerobic granular sludge were distributed in the outer layer, whereas the remainder was distributed through the rest of the granules. EPS of biofilms in remediation of pollutants were distributed in layers through the biofilm depth, and their yield varied along the biofilm depth. EPS are responsible for attachment often along with other bacteria to soil particles and root surfaces. EPS bind soil particles to aggregates, stabilizing soil structures and increasing water holding capacity and cation exchange capacity. EPS usually form an enclosed matrix of microcolonies, which confer protection against environmental stresses, water and nutrient retention, and epiphytic colonization. They are also indispensable for mature biofilm formation and functional nodules in legume-rhizobia symbiosis. EPS production and composition improve bacterial resistance to abiotic stress.

Exopolysaccharides (EPSs) are high-molecular-weight, biodegradable polymers that are formed of monosaccharide residues and their derivatives and biosynthesized by a wide range of bacteria, algae, and plants. EPSs play a central role maintaining water potential, aggregating soil particles, ensuring obligate contact between plant roots and rhizobacteria, sustaining the host under conditions of stress (saline soil, dry weather, or water logging) or pathogenesis and thus are directly responsible for plant growth and crop production. EPS producing PGPR, such as *R. leguminosarum, Azotobacter vinelandii, Bacillus drentensis, Enterobacter cloacae, Agrobacterium* sp., *Xanthomonas* sp., *and Rhizobium* sp., have an important role increasing soil fertility and contributing to sustainable agriculture.

13.5.2.3 Lipochitooligosaccharides

Lipochitooligosaccharides (LCOs) are secreted by rhizobia as nod-factors (NFs) in response to flavonoids present in root exudates and initiate nodule formation. LCOs are conserved at the core but diverge in the N-acetyl chain length, degree of saturation, and substitutions (glycosylation or sulfation), which are crucial in host specificity. Nod-factors also act as stress response signals in legumes, and NF synthesis is modulated by other PGPR and abiotic stresses.

13.5.2.4   Bacteriocins

Bacteriocins are protein aceous toxins that are secreted by bacteria that live in competitive micro-bial environment. They destroy the neighboring bacterial species by damaging the bacteriocino-genic cells. Bacteriocins are very effective in reducing or inhibiting the growth of phytopathogens. Bacteriocins have narrow killing spectrum as compared with conventional antibiotics, and these have damaging effect on the bacteria that are closely relative of bacteriocin-producing bacteria. Colicins are most prominent bacteriocins synthesized by *Escherichia coli*. Similarly, megacins are produced by *B. megaterium*; marcescins from *Serratia marcescens*; cloacins from *Enterobacter cloacae*; and pyocins from *Pseudomonas aeruginosa*. Bacteriocins that are produced by *Bacillus* spp. remarkably gain importance due to broad range of inhibition offungal, yeast, gram-positive, and gram-negative species that may have some pathogenic effects on animals and human beings.

13.5.2.5   Cell Wall–Degrading Lytic Enzymes

Many polymeric compounds such as cellulose, hemicellulose, chitin, and protein can be hydrolyzed by the lytic enzymes produced by PGPR. Microbes can directly suppress the growth and activi-ties of pathogens by secreting lyticenzymes. Hydrolytic enzymes including glucanases, proteases, chitinases, and lipases are involved in the lysis of fungal cell wall. The mechanism would involve the production of hydrolytic enzymes. Major fungal cell wall components are made up of chitin and beta-glucan; thus, chitinases and beta-glucanases producing bacteria would inhibit fungal growth. It is one of the important mechanisms for environmentally friendly control of soil-borne patho-gen. PGPR promotes plant growth through the control of phytopathogenic agents, primarily for the production of metabolites contributing to the antibiosis and antifungal properties used as defense systems. These enzymes also decompose nonliving organic matter and plant residues to obtain car-bon nutrition. Lytic enzymes produced by Myxobacteria are effective in the suppression of fungal plant pathogens. Antagonistic bacteria *Serratia marcescens* reduce mycelial network of *Sclerotium rolfsii* by expressing chitinase. Lysobacter is capable of producing glucanase that is involved in the control of diseases caused by *Bipolaris* and *Pythium* sp. Hydrolytic enzymes directly contribute in the parasitization of phytopathogens and rescue plant from biotic stresses. Apart from exhibiting the production of chitinase and beta-glucanases, *Pseudomonas* spp. Inhibit *R. solani* and *Phytophthora capsici*, two of the most destructive crop pathogens in the world.

13.5.2.6   Induction of Plant Resistance

PGPR are reported to trigger the resistance of plants against pathogens. This phenomenon is referred as ISR. In this process, a signal is generated involving jasmonate or ethylene pathway and thus inducing the host plant's defense response. PGPR are able to control the number of pathogenic bacteria through microbial antagonism, which is achieved by competing with the pathogens for nutrients, producing antibiotics, and the production of antifungal metabolites. Besides antagonism, certain bacteria-plant interactions can induce mechanisms in which the plant can better defend itself against environmental stress, pathogenic bacteria, fungi, and viruses. This is known as ISR and was first discovered in 1991 by Van Peer et al. (1991). The inducing rhizobacteria trigger a reaction in the roots that creates a signal that spreads throughout the plant, which results in the activation of defense mechanisms, such ascystein, ascorbic acid production, reinforcement of plant cell wall, and production of antimicrobial phytoalexins. Components of bacteria that can activate ISR include lipopolysaccharides (LPS), flagella, salicylic acid, and sideophores.

13.5.2.7   HCN Production

A number of biocontrol PGPR have the ability to synthesize HCN. If the HCN produced by these bacteria was the only biocontrol mechanism being used in most instances, the low level of HCN would not be particularly effective at preventing the proliferation of most fungal phytopathogens. However, it is often the case that biocontrol PGPR that can produce HCN also synthesize some antibioticsor cell wall–degrading enzymes. Moreover, it has been observed that the low level of HCN synthesized by the bacterium improves the effectiveness of antifungal directed against fungal pathogens, thereby ensuring that the fungi do not develop resistance to the particular antifungal in question. Thus, HCN synthesized by PGPR appears to act synergistically with other methods of biocontrol employed by the same bacterium. HCN toxicity is affected in its ability to inhibit cytochrome c oxidase as well as other important metalloenzymes. Many bacterial genera such as *Rhizobium*, *Pseudomonas*, *Alcaligenes*, *Bacillus*, and *Aeromonas* have shown to be HCN producers.

13.5.2.8 Competition

In addition to these mechanisms for biocontrol, PGPR produces substances that are inhibitory to phytopathogens; it is possible for some biocontrol PGPR to out compete the phytopathogens, either for nutrients or for binding sites on the plant root. Such competition can act to limit the binding of the phytopathogen to the plant, thereby making it difficult for it to proliferate. However, since it is not always possible to create mutants of PGPR that are either more or less competitive for binding to the plant surface, there are a relatively limited number of unequivocal demonstrations of the ability of biocontrol PGPR to outcompete phytopathogens and thereby prevent their functioning. In fact, it is generally thought that PGPR competitiveness works together with other biocontrol mechanisms to failure the functioning of phytopathogens.

13.5.2.9 Quorum Quenching

In the environment, bacterial cells use the mechanism of QS to detect the presence of similar as well as different types of bacteria to cope the stress conditions. With growing bacterial cells, once they have attained a certain critical cell density, the bacteria "sense" the cell density (through the production of chemical signals) and start to alter their metabolism by turning on different sets of genes, so that similar bacteria that are proximal to one another may begin acting in a coordinate manner. In most systems, bacteria synthesize low-molecular-weight chemicals called autoinducers that are typically secreted outside of the bacterial cells. When the bacterial ells population increases, the extracellular level of the autoinducers also increases until it exceeds some threshold level, binds to bacterial cellular receptors, and triggers a signal transduction cascade, thereby causing population-wide changes in bacterial gene expression with the actions of a unified group of cells, e.g., at a certain cell density, a bacterial plant pathogen may begin to become more virulent. Disrupting this QS (i.e., signaling among pathogens) can inhibit the pathogen from becoming increasingly virulent and prevent it from inhibiting plant growth. There are a number of biological means of quenching the phenomenon of QS. One way is to utilize a PGPR that produces an enzyme called a lactonase that degrades the pathogen-produced autoinducer and prevents infection.

13.6 FACTORS AFFECTING PGPR BIOREMEDIATION

The most important factors that influence the microbial flora of the rhizosphere effect are soil type and its moisture, soil amendments, soil pH, proximity of root with soil, plant species, age of plant, and root exudates.

a. Soil type and its moisture: In general, microbial activity and population is high in the rhizosphere region of the plants grown in sandy soils and least in the high humus soils, and rhizosphere organisms are more when the soil moisture is low. Nature of soil and moisture content greatly influence the PGPR growth; hence, it also affects the bioremediation potential of PGPR.

b. Soil amendments: Crop residues, animal manure, and chemical fertilizers applied to the soil cause no appreciable effect on the quantitative or qualitative differences in the microflora of rhizosphere. Hence, soil amendment also does not affect the remediation of waste by PGPR.

c. Soil pH/rhizosphere pH: Respiration by the rhizosphere microflora may lead to the change in soil rhizosphere pH. If the activity and population of the rhizosphere microflora is more, then the pH of rhizosphere region is lower than that of surrounding soil or nonrhizosphere soil. Rhizosphere effect for bacteria and protozoa is more in slightly alkaline soil and that of fungi is more in acidic soils.

d. Proximity of root with soil: Soil samples taken progressively closer to the root system have increasingly greater population of bacteria and actinomycetes and decrease with the distance and depth from the root system. Rhizosphere effect declines sharply with increasing distance between plant root and soil.

e. Plant species: Different plant species inhabit often somewhat variable microflora in the rhizosphere region. The qualitative and quantitative differences are attributed to variations in the rooting habits, tissue composition, and excretion products. In general, legumes show/produce a more pronounced rhizosphere effect than grasses or cereals. Biennials, due to their long growth period, exert more prolonged stimulation on rhizosphere effect than annuals. Accumulation of heavy metals by different plants might be due to the variability of PGPR with various plants species.

f. Age of plant: The age of plant also alters the rhizosphere microflora, and the stage of plant maturity controls the magnitude of rhizosphere effect and degree of response to specific microorganisms. The rhizosphere microflora increases in number with the age of the plant and reaches at peak during flowering, which is the most active period of plant growth and metabolism. Hence, the rhizosphere effect was found to be more at the time of flowering than in the seedling or full maturity stage of the plants. The fungal flora (especially, *cellulolytic* and *amylolytic*) of the rhizosphere usually increases even after fruiting and the onset of senescence due to accumulation of moribund tissue and sloughed off root parts/tissues, whereas bacterial flora of the rhizosphere decreases after the flowering period and fruit setting.

g. Root exudates: One of the most important factors responsible for rhizosphere effect is the availability of a great variety of organic substances at the root region by way of root exudates/excretions. The quantitative and qualitative differences in the microflora of the rhizosphere from that of general soil are mainly due to influences of root exudates. The spectrum of chemical composition root exudates varies widely, and hence, their influence on the microflora also varies widely.

13.7 GENETICALLY MODIFIED APPROACH IN PGPR FOR BIOREMEDIATION

The rhizosphere seems to be a promising environment for the bioremediation of contaminated soils, but as described above, many bacteria capable of degrading certain kinds of organic pollutants cannot survive and achieve bioremediation in the soil environment, because they are not competitive enough compared with other indigenous organisms. Meanwhile, many bacteria that are robust in the rhizosphere do not show or show only limited ability in degrading organic pollutants. With the development of molecular biology, the genetically engineered rhizobacteria with the contaminant-degrading gene are constructed to conduct the bioremediation in rhizosphere. For some pollutants such as trichloroethylene (TCE) and PCBs, the molecular mechanisms of degradation have been clearly studied. Another crucial problem to be solved is to select a suitable strain for gene recombination and inoculation into the rhizosphere. The following criteria should be considered: (i) the strain should be stable after cloning, and the target gene should have a high expression; (ii) the strain should be tolerant or insensitive to the contaminant; and (iii) some strains can survive only in several specific plant rhizospheres. Other methods are also considered besides the strain selection. For example, Villacieros et al. (2005) reported that the expression level of the bph genes in *Pseudomonas fluorescens* F113 was lower than that in the parental strain, which limited the ability of F113 to grow on biphenyl and therefore limited its ability to degrade PCBs. They found a way of increasing the biphenyl-degrading activity by increasing the transcription rate of the genes by changing the promoter regions. The heterologous rhizobial nodulation promoters (nod boxes) from *S. meliloti* and its regulatory systems were thus tested to drive the expression of the bph operon in *P. fluorescens* F113 derivatives. Barac et al. (2004) constructed the engineered endophytic bacteria to improve the phytoremediation of water-soluble, volatile, organic pollutants. The genetically modified endophytic strain showed the improved degradation and reduced the evapotranspiration of toluene, a moderately hydrophobic volatile compound. They hypothesized that the endophytic bacteria, possessing the genetic information required to efficiently degrade the organic contaminant, promoted its breakdown as it moved through the plant's vascular system. Due to the long transportation time of contaminant in the system, there was a sufficient time for the efficient degradation by endophytic bacteria in xylem.

Besides transition of gene between bacteria, transgenic plants have been constructed for higher remediation efficiency (Stearns et al., 2005). The expression of ACC deaminase in the plant exhibits several advantages against the bacteria: (i) during the initial stages of seed germination, the bacterial ACC deaminase activity is likely to be much lower than the activity in transgenic plants; (ii) it can constantly stimulate plant growth, which leads to a higher metal accumulation; (iii) in some cases, an increase in the shoot/root ratio; (iv) prompting metal uptake of certain fast-growing plants for the substitution of slow-growing hyperaccumulators.

13.8   BENEFICIAL AND HARMFUL ASPECTS OF PGPR

It is undisputed that rhizobacteria play a crucial role in maintaining soil fertility, phytoremediation of organic and inorganic pollutants, and upgradingplant growth and development. This growth betterment takes place with the help of several mechanisms as mentioned earlier, although the reverse is true in some other studies. For example, the production of cyanide is known to be a characteristic of certain *Pseudomonas* species. Here, cyanide production by the bacteria is considered as a growth promotion as well as a growth inhibition characteristic. Moreover, cyanide acts as a biocontrol agent against certain plant pathogen; on the other hand, it can also cause adverse effects on plant growth. The auxin production by PGPR can also cause positive as well as negative effects on plant growth. It is important to note that the effectiveness of auxin relies upon its concentration. For instance, at low concentrations, it enhances plant growth, whereas at a high level, it inhibits root growth. Furthermore, rhizobitoxine produced by *Bradyrhizobium elkanii* also has a dual effect. Since it is an inhibitor of ethylene synthesis, it can alleviate the negative effect of stress-induced ethylene production on nodulation. On the other hand, rhizobitoxine is also considered a plant toxin because it induces foliar chlorosis in soybeans. So far, the above discussion has proven that although PGPR are very effective at promoting plant growth and development, a select few bacterial species may inhibit growth. However, this negative impact may only occur under certain specific conditions and also by some particular traits. Thus, the selection of a particular strain is of the utmost importance in obtaining maximum benefits in terms of improved bioremediation, plant growth, and development.

13.9   CHALLENGES AND FUTURE PROSPECTS

The need of today's world is development of ecofriendly, cost-effective technology for treatment of industrial waste and sustainable agriculture. Hence, PGPR is now considered as safe means bioremediation of industrial waste, which may cause pollution and toxicity in the environment. Increasing yield in agriculture due to PGPR is promising solution for the environment. The most important function of PGPR is to protect plants from pollutants discharged from industries and chemicals that are used to kill pests and also cause harmful impact on the ecosystem. PGPR seems to have beneficial effect on laboratory as well as greenhouse experiment. An emerging field to improve and explore the PGPR strain is by genetic engineering, which enables to overexpress the traits so that strains with required characters are obtained. Besides being beneficial, there are several challenges faced by PGPR. The natural variation is a very important issue because it is difficult to predict how bacteria will act in laboratory and when placed in field. Another challenge is that under field conditions, PGPR need to be propagated to regain their viability and biological activity. This propagation can be according to seasonal and regional selectivity of host plants cultivation. Furthermore, another limited application is due to variable bioavailability of target pollutants and due to plant-mediated adsorption and transportation. Research on nitrogen fixation and phosphate solubilization by PGPR is progressed on, but little research can be done on potassium solubilization, which is third major essential macronutrient for plant growth. This will not only increase the field of the inoculants but also helpful for remediation of pollutant due to better plant growth. In addition, future marketing of bioinoculant of PGPR and release of transgenics into the environment as eco-friendly alternations to agrochemicals and for pollutant removal will depend on the generation of biosafety data. The major challenge in this area of research lies in the fact that, along with the identification of various strains of PGPRs and their properties, it is essential to dissect the actual mechanism of functioning of PGPRs for their efficacy toward pollutants causing harmful effects on ecosystem.

13.10 CONCLUSION

Environmental stresses due to discharge of organic and inorganic pollutants, which are persistent in nature, are becoming a major problem for environment. Our dependence on many industries produces different types of life-threatening pollutants, which are not only hazardous for human consumption but can also disturb the ecological balance. Chemical pesticides suppress phytopathogens for improved plant growth and health; nevertheless, they damage nontargeted beneficial microorganisms of soil and pollute soil environment. Use of rhizospheric microorganisms mainly PGPR in sustainable agriculture and environment management offers countless benefits. Their capability to survive in harsh environmental conditions makes them efficient candidates in different types of stress management, whereas their catabolic diversity can be used in the removal of recalcitrant pollutants. Secretion of specific lytic enzymes for degradation of recalcitrant compounds is also a suitable mechanism of PGPR. PGPR in rhizosphere soil are highly dynamic, more versatile in transforming, mobilizing, and solubilizing the nutrients. Therefore, the rhizobacteria are the dominant deriving forces in recycling the soil nutrients, and consequently, they are crucial for soil fertility. They may be extensively used in plant growth promotion as it acts as a plant nourishment and enrichment source, which would replenish the nutrient cycle between the soil and plant roots, exhibits detoxifying potential, controls phytopathogens, thereby exerts a positive influence on crop productivity and ecosystem functioning, and hence can be implemented in agriculture. PGPR can act as biofertilizers, biocontrol agents, and soil improvers. PGPR-containing inoculum may replace synthetic chemicals, which result in environmental hazards and pose a serious toxicological threat to the ecosystem. Beneficial microbes used in biocontrol tend to have high-target specificity and are environmentally friendly. Their utilization in the form of biofertilizers and biopesticides is becoming popular and providing substantial aid to the agro-ecosystems. Our understanding of the PGPR response in agroecosystems is increasing, and their potentially significant effects on environment restoration are also strengthening and collectively helping to obtain the goal of sustainable development. Impact of climate change particularly in the agriculture sector may also be minimized by the application of PGPR. However, with climate change perspective, effective use of PGPR requires more investigations and it has been realized that elucidation of mechanisms involved in their interaction with plants in extreme conditions may provide better chance of their vast application in environmental sustainability. Future research in rhizosphere biology will rely on the development of molecular and biotechnological approaches to increase our knowledge of rhizosphere biology and to achieve an integrated management of soil microbial populations. The application of multistrain bacterial consortium could be an effective approach for reducing the harmful impact of pollutants on plant growth. Biofertilizers can help solve the problem of feeding an increasing global population at a time when agriculture is facing various environmental stresses. It is important to realize the useful aspects of biofertilizers and implement its application to modern agricultural practices. The new technology developed using the powerful tool of molecular biotechnology can enhance the biological pathways of production of phytohormones. If identified and transferred to the useful PGPRs, these technologies can help provide relief from environmental stresses. However, the lack of awareness regarding improved protocols of biofertilizer applications to the field is one of the few reasons why many useful PGPRs are still beyond the knowledge of ecologists and agriculturists. However, the extent of success in realizing the benefits of PGPR tends to diminish as it moves from laboratory to greenhouse and to fields, which reflects the scarcity of research on the beneficial effects of PGPR under field conditions. Therefore, generation of comprehensive knowledge on screening strategies and intense selection of best rhizobacterial strain for rhizosphere competence and survival is the current need to enhance the field-level successes. Identification of such potential rhizobial strains and developing a robust technology for bioremediation of pollutants is still in its infancy. Thus, additional comprehensive research to exploit the potential of PGPR would provide for expansion of this research area and commercialization and improve sustainability and safe environment. The role of PGPR in environmental sustainability can be expanded if we get success in finding some unrevealed concepts related to their ecology, population dynamics, and functionality over a range of environments.

REFERENCES

Abd-Alla, M.H., 1994. Solubilization of rock phosphates by Rhizobium and Bradyrhizobium. *Folia Microbiol.* 39, 53–56.

Ahemad, M., Khan, M.S., 2010a. Influence of selective herbicides on plant growth promoting traits of phosphate solubilizing Enterobacter asburiae strain PS2. *Res. J. Microbiol.* 5, 849–857.

Ahemad, M., Khan, M.S., 2010b. Plant growth promoting activities of phosphate-solubilizing Enterobacter asburiae as influenced by fungicides. *Eurasia. J. Biosci.* 4, 88–95.

Ahemad, M., Khan, M.S., 2010c. Comparative toxicity of selected insecticides to pea plants and growth promotion in response to insecticide-tolerant and plant growth promoting Rhizobium leguminosarum. *Crop Prot.* 29(4), 325–329.

Ahemad, M., Khan, M.S., 2010d. Phosphate-solubilizing and plant-growth-promoting Pseudomonas aeruginosa PS1 improves greengram performance in quizalafop-p-ethyl and clodinafop amended soil. *Arch. Environ. Contam. Toxicol.* 58(2), 361–372.

Ahemad, M., Khan, M.S., 2011a. Toxicological assessment of selective pesticides towards plant growth promoting activities of phosphate solubilizing Pseudomonas aeruginosa. *Acta Microbiol. Immunol. Hung.* 58(3), 169–187.

Ahemad, M., Khan, M.S., 2011b. Assessment of plant growth promoting activities of rhizobacterium Pseudomonas putida under insecticide-stress. *Microbiol. J.* 1(2), 54–64.

Ahemad, M., Khan, M.S., 2011c. Effect of tebuconazole-tolerant and plant growth promoting Rhizobium isolate MRP1 on pea–Rhizobium symbiosis. *Scie. Hort.* 129(2), 266–272.

Ahemad, M., Khan, M.S., 2011d. Pseudomonasaeruginosa strain PS1 enhances growth parameters of greengram [Vignaradiata (L.) Wilczek] in insecticide-stressed soils. *J. Pest Sci.* 84(1), 123–131.

Ahemad, M., Khan, M.S., 2012a. Effect of fungicides on plant growth promoting activities of phosphate solubilizing Pseudomonas putida isolated from mustard (Brassica compestris) rhizosphere. *Chemosphere* 86, 945–950.

Ahemad, M., Khan, M.S., 2012b. Ecological assessment of biotoxicity of pesticides towards plant growth promoting activities of pea (Pisum sativum)-specific Rhizobium sp. strain MRP1. *Emirates J. Food Agric.* 24, 334–343.

Ahemad, M., Khan, M.S., 2012c. Alleviation of fungicide-induced phytotoxicity in greengram [Vigna radiata (L.) Wilczek] using fungicide-tolerant and plant growth promoting Pseudomonas strain. *Saudi J. Biol. Sci.* 19, 451–459.

Ahemad, M., Khan, M. S., 2012d. Alleviation of fungicide-induced phytotoxicity in greengram [Vigna radiata (L.) Wilczek] using fungicide-tolerant and plant growth promoting Pseudomonas strain. *Saudi J. Biol. Sci.* 19(4), 451–459.

Ahmad, F., Ahmad, I., Khan, M.S., 2008. Screening of free-living rhizospheric bacteria for their multiple plant growth promoting activities. *Microbiol. Res.* 163, 173–181.

Barac, T., Taghavi, S., Borremans, B., Provoost, A., Oeyen, L., Colpaert, J.V., 2004. Engineered endophytic bacteria improve phytoremediation of water-soluble, volatile, organic pollutants. *Nat. Biotechnol.* 22, 583–588.

Belimov, A.A., Hontzeas, N., Safronova, V.I., Demchinskaya, S.V., Piluzza, G., Bullitta, S., Glick, B.R., 2005. Cadmium-tolerant plant growth promoting rhizobacteria associated with the roots of Indian mustard (Brassica juncea L. Czern.). *Soil Biol. Biochem.* 37, 241–250.

Berraho, E.L., Lesueur, D., Diem, H.G., Sasson, A., 1997. Iron requirement and siderophore production in Rhizobium ciceri during growth on an iron-deficient medium. *World J. Microbiol. Biotechnol.* 13, 501–510.

Burd, G.I., Dixon, D.G., Glick, B.R., 2000. Plant growth promoting bacteria that decrease heavy metal toxicity in plants. *Can. J. Microbiol.* 46, 237–245.

Calheiros, C.S.C., Rangel, A.O.S.S., Castro, P.M.L., 2009. Treatment of industrial wastewater with twostage constructed wetlands planted with Typha latifolia and Phragmites australis. *Bioresour. Technol.* 100(13), 3205–3313.

Canbolat, M.Y., Bilen, S., Çakmakç, R., Sahin, F., Aydin, A., 2006. Effect of plant growth-promoting bacteria and soil compaction on barley seedling growth, nutrient uptake, soil properties and rhizosphere microflora. *Biol. Fertil. Soils* 42, 350–357.

Chandra, R., Bhargava, R.N., Rai, V., 2008.Melanoidin as major colorant in sugarcane molasses based distillery effluent and its degradation. *Biresour. Technol.* 99, 4648–4660.

Chandra, R., Chaturvedi, S., 2002. Isolation and characterization of bacterial population from rhizosphere of Cyperus papyrus growing in distillery effluent contaminated site. Abstract no. SIII/P-5. In: 2nd International Conference on Plants and Environmental Pollution held on 2002 at NBRI, Lucknow, 4–9 February, p. 39.

Chandra, R., Kumar, V., 2015. Mechanism of wetland plant rhizosphere bacteria for bioremediation of pollutants in an aquatic ecosystem. In: R. Chandra (Ed.), *Advances in Biodegradation and Bioremediation of Industrial Waste*. CRC Press, Boca Raton, FL, pp. 329–379.

Chandra, R., Kumar, V., 2017a. Detection of androgenic-mutagenic compounds and potential autochthonous bacterial communities during in situ bioremediation of post methanated distillery sludge. *Front. Microbiol.* 8, 887.

Chandra, R., Kumar, V., 2017b. Detection of Bacillus and Stenotrophomonas species growing in an organic acid and endocrine-disrupting chemicals rich environment of distillery spent wash and its phytotoxicity. *Environ. Monit. Assess.* 189, 26.

Chandra, R., Kumar, V., 2017c. Phytoextraction of heavy metals by potential native plants and their microscopic observation of root growing on stabilized distillery sludge as a prospective tool for in-situ phytoremediation of industrial waste. *Environ. Sci. Pollut. Res.* 24, 2605–2619.

Chandra, R., Kumar, V., 2018. Phytoremediation: A green sustainable technology for industrial waste management. In: R. Chandra, N.K. Dubey, V. Kumar (Eds.), *Phytoremediation of Environmental Pollutants*. CRC Press, Boca Raton, FL, pp. 1–50.

Chandra, R., Kumar, V., Singh, K., 2018a. Hyperaccumulator versus nonhyperaccumulator plants for environmental waste management. In: R. Chandra, N.K. Dubey, V. Kumar (Eds.), *Phytoremediation of Environmental Pollutants*. CRC Press, Boca Raton, FL, pp. 43–80.

Chandra, R., Kumar, V., Tripathi, S., 2018b. Evaluation of molasses-melanoidins decolourisation by potential bacterial consortium discharged in distillery effluent. *3 Biotech* 8, 187. DOI.10.1007/s13205-018-1205-3.

Chandra, R., Kumar, V., Tripathi, S., Sharma, P., 2017. Heavy metal phytoextraction potential of native weeds and grasses from endocrine-disrupting chemicals rich complex distillery sludge and their histological observations during in-situ phytoremediation. *Ecol. Eng.* 111, 143–156. DOI:10.1016/j.ecoleng.2017.12.007.

Chandra, R., Kumar, V., Tripathi, S., Sharma, P., 2018c. Phytoremediation of industrial pollutants and life cycle assessment. In: R. Chandra, N.K. Dubey, V. Kumar (Eds.), *Phytoremediation of Environmental Pollutants*. CRC Press, Boca Raton, FL, pp. 441–470.

Chandra, R., Saxena, G., Kumar, V., 2015. Phytoremediation of environmental pollutants: An eco-sustainable green technology to environmental management. In: R. Chandra (Ed.), *Advances in Biodegradation and Bioremediation of Industrial Waste*. CRC Press, Boca Raton, FL, pp. 1–29.

Deshwal, V.K., Pandey, P., Kang, S.C., Maheshwari, D.K., 2003. Rhizobia as a biological control agent against soil borne plant pathogenic fungi. *Indian J. Exp. Biol.* 41, 1160–1164.

Dey, R., Pal, K.K., Bhatt, D.M., Chauhan, S.M., 2004. Growth promotion and yield enhancement of peanut (Arachis hypogaea L.) by application of plant growth-promoting rhizobacteria. *Microbiol. Res.* 159, 371–394.

Esparza-Soto, M., Westerhoff, P., 2003. Biosorption of humic and fulvic acids to live activated sludge biomass. *Water Res.* 37(10), 2301–2310.

Gray, E.J., Smith, D.L., 2005. Intracellular and extracellular PGPR: Commonalities and distinctions in the plant–bacterium signaling processes. *Soil Biol. Biochem.* 37, 395–412.

Jha, P.N., Kumar, A., 2007. Endophytic colonization of Typha australis by a plant growth-promoting bacterium Klebsiella oxytoca strain GR-3. *J. Appl. Microbiol.* 103, 1311–1320.

Jiang, C., Sheng, X., Qian, M., Wang, Q., 2008. Isolation and characterization of a heavy metal-resistant Burkholderia sp. from heavy metal-contaminated paddy field soil and its potential in promoting plant growth and heavy metal accumulation in metalpolluted soil. *Chemosphere* 72, 157–164.

Khan, M.S., Zaidi, A., Aamil, M., 2002. Biocontrol of fungal pathogens by the use of plant growth promoting rhizobacteria and nitrogen fixing microorganisms. *Ind. J. Bot. Soc.* 81, 255–263.

Kloepper, J.W., 2003. A review of mechanisms for plant growth promotion by PGPR. In: Reddy, M.S., Anandaraj, M., Eapen, S.J., Sarma, Y.R., Kloepper, J.W. (Eds.), Abstracts and Short Papers. *Mechanisms and Applications of Plant Growth Promoting Rhizobacteria: Current Perspective. 6th International PGPR Workshop*, 5–10 October 2003, Indian Institute of Spices Research, Calicut, India, pp. 81–92.

Kloepper, J.W., Schroth, M.N., 1978. Plant growth-promoting rhizobacteria on radishes. In: C. Keel, B. Koller, G. Défago (Eds.), *Proceedings of the 4th International Conference on Plant Pathogenic Bacteria*, vol. 2. Station de Pathologie Vegetale et Phytobacteriologie, INRA, Angers, pp. 879–882.

Kumar, V., Behl, R.K., Narula, N., 2001. Establishment of phosphate solubilizing strains of Azotobacter chroococcum in the rhizosphere and their effect on wheat cultivars under greenhouse conditions. *Microbiol. Res.* 156, 87–93.

Kumar, V., Chandra, R., 2018. Bacteria-assisted phytoremediation of industrial waste pollutants and ecorestoration. In: R. Chandra, N.K. Dubey, V. Kumar (Eds.), *Phytoremediation of Environmental Pollutants*. CRC Press, Boca Raton, FL, pp. 159–200.

Kumar, V., Chandra, R., 2019. Bioremediation of melanoidins containing distillery waste for environmental safety. In: R.N. Bharagava, G. Saxena (Eds.), *Bioremediation of Industrial Waste for Environmental Safety*. Vol II—Microbes and Methods for Industrial Waste Management. Springer, Singapore, pp. 495–529.

Liu, D., Coloe, S., Baird, R., Pedersen, J., 2000. Rapid mini-preparation of fungal DNA for PCR. *J. Clin. Microbiol.* 38(1), 471.

Ma, Y., Rajkumar, M., Luo, Y., Freitas, H., 2011a. Inoculation of endophytic bacteria on host and non-host plants-effects on plant growth and Ni uptake. *J. Hazard. Mater.* 195, 230–237.

Ma, Y., Rajkumar, M., Vicente, J.A., Freitas, H., 2011b. Inoculation of Ni-resistant plant growth promoting bacterium Psychrobacter sp. strain SRS8 for the improvement of nickel phytoextraction by energy crops. *Int. J. Phytoremediation* 13, 126–139.

Noel, T.C., Sheng, C., Yost, C.K., Pharis, R.P., Hynes, M.F., 1996. Rhizobium leguminosarum as a plant growth promoting Rhizobacterium: Direct growth promotion of canola and lettuce. *Can. J. Microbiol.* 42, 279–283.

Poonguzhali, S., Madhaiyan, M., Sa, T., 2008. Isolation and identification of phosphate solubilizing bacteria from Chinese cabbage and their effect on growth and phosphorus utilization of plants. *J. Microbiol. Biotechnol.* 18, 773–777.

Rajkumar, M., Freitas, H., 2008. Effects of inoculation of plant growth promoting bacteria on Ni uptake by Indian mustard. *Bioresour. Technol.* 99, 3491–3498.

Rodrigues, E.P., Rodrigues, L.S., de Oliveira, A.L.M., Baldani, V.L.D., Teixeira, K.R.S., Urquiaga, S., Reis, V.M., 2008. Azospirillum amazonense inoculation: Effects on growth, yield and N2 fixation of rice (Oryza sativa L.). *Plant Soil* 302, 249–261.

Saravanakumara, D., Vijayakumarc, C., Kumarb, N., Samiyappan, R., 2007. PGPR-induced defense responses in the tea plant against blister blight disease. *Crop Prot.* 26, 556–565.

Sheng, M., Tang, M., Chen, H., Yang, B., Zhang, F., Huang, Y., 2008. Influence of arbuscular mycorrhizae on photosynthesis and water status of maize plants under salt stress. *Mycorrhiza* 18(6–7), 287–296.

Somers, E., Vanderleyden, J., Srinivasan, M., 2004. Rhizosphere bacterial signalling: A love parade beneath our feet. *Crit. Rev. Microbiol.* 30, 205–240.

Stearns, J.C., Shah, S., Dixon, D.G., Greenberg, B.M., Glick, B.R., 2005. Tolerance of transgenic canola expressing 1-aminocyclopropane-carboxylic acid deam inase to growth inhibition by nickel. *Plant Physiol. Biochem.* 43, 701–708.

Tak, H.I., Ahmad, F., Babalola, O.O., 2013. Advances in the application of plant growth promoting rhizobacteria in phytoremediation of heavy metals. In: D.M. Whitacre (Ed.), *Vol. 223, Reviews of Environmental Contamination and Toxicology*. Springer Science Business Media, New York, pp. 33–52.

Tripathi, M., Munot, H.P., Shouch, Y., Meyer, J.M., Goel, R., 2005. Isolation and functionalcharacterization of siderophore-producing lead- and cadmium-resistant Pseudomonas putida KNP9. *Curr. Microbiol.* 5, 233–237.

Van Peer, R., Niemann, G.J., Schippers, B., 1991. Induced resistance and phytoalexin accumulation in biological control of Fusarium wilt of carnation by Pseudomonas sp. strain WCS 417 r. *Phytopathology*, 81(7), 728–734.

Vessey, J.K., 2003. Plant growth promoting rhizobacteria as biofertilizers. *Plant Soil* 255, 571–586.

Villacieros, M., Whelan, C., Mackova, M., Molgaard, J., Sánchez-Contreras, M., Lloret, J., ... & Karlson, U., 2005. Polychlorinated biphenyl rhizoremediation by Pseudomonas fluorescens F113 derivatives, using a Sinorhizobium meliloti nod system to drive bph gene expression. *Appl. Environ. Microbiol.* 71(5), 2687–2694.

Vivas, A., Biro, B., Ruiz-Lozano, J.M., Barea, J.M., Azcon, R., 2006. Two bacterial strains isolated from a Zn-polluted soil enhance plant growth and mycorrhizal efficiency under Zn toxicity. *Chemosphere* 52, 1523–1533.

Wani, P.A., Khan, M.S., Zaidi, A., 2007a. Effect of metal tolerant plant growth promoting Bradyrhizobium sp. (vigna) on growth, symbiosis, seed yield and metal uptake by greengram plants. *Chemosphere* 70, 36–45.

Wani, P.A., Khan, M.S., Zaidi, A., 2007b. Synergistic effects of the inoculation with nitrogen fixing and phosphate solubilizing rhizobacteria on the performance of field grown chickpea. *J. Plant Nutr. Soil Sci.* 170, 283–287.

Yuncu, B., Sanin, F. D., Yetis, U., 2006. An investigation of heavy metal biosorption in relation to C/N ratio of activated sludge. *J. Hazard. Mater.* 137(2), 990–997.

14 Fungi Treatment of Synthetic Dyes by Using Agro-industrial Waste

Yago Araújo Vieira, Débora Vilar, Ianny Andrade Cruz, Diego Batista Menezes, Clara Dourado Fernandes, Nádia Hortense Torres, Silvia Maria Egues, Ranyere Lucena de Sousa, and Luiz Fernando Romanholo Ferreira
Tiradentes University and Institute of Technology and Research (ITP)

Ram Naresh Bharagava
Babasaheb Bhimrao Ambedkar University (A Central University)

CONTENTS

14.1 INTRODUCTION

Environmental legislation around the world increasingly demands the quality of the industrial effluents, since the damages caused by these wastes on the environment are even more intense due to the rising consumption and the increasing population in the globalized world (Kumar and Chandra 2018; Chandra et al. 2018; Chandra and Kumar 2015). In this scenario, it is necessary to highlight the generation of two extremely problematic wastes. On the one hand, there are agro-industrial residues, such as cassava bagasse, wheat bran, cob, sugarcane bagasse, rice bran, brewery beans, and carob pod, among others (Ergun and Urek 2017). On the other hand, effluents from synthetic dyes used in industries such as textile, graphic, and plastic are recalcitrant and pollute water bodies by the high staining and toxic substances formation or carcinogenic agents, for example, the aromatic amines of azo dyes (Upadhyay et al. 2016).

The agro-industrial residues generally accumulated in the environment have high polluting potential (Asgher et al. 2016), as well as the synthetic dyes applied in diverse industries, since both types of waste can present in common characteristics such as high rates of total organic carbon (TOC), biochemical oxygen demand (BOD), chemical oxygen demand (COD), acid and/or alkaline pH, and the presence of recalcitrant compounds (Singh 2017). The use of physical and chemical methods to treat these pollutants besides being expensive can lead to the negative consequence of concentrated sludge and by-products generation (Yesilada et al. 2018).

In this sense, mycoremediation is the fungi-based technology used for the organic contaminants' removal from waste and effluents in order to make them less toxic and not dangerous (Barrech et al. 2018). This bioremediation form has been widely estimated as a promising alternative; therefore, the capacity presented by fungi in the toxicity degradation and reduction occurs through the production of enzymes capable of degrading lignin. Within this enzymatic complex are manganese peroxidase (MnP, E.C 1.11.1.13), lignin peroxidase (LiP, E.C. 1.11.1.14), and laccase (Lac, E.C. 1.10.3.2) (Bilal et al. 2017).

Due to the low-substrate specificity of ligninolytic enzymes produced and expelled by fungi, it is possible to oxidize a wide range of compounds with structural similarities to lignin (Peralta et al. 2017), so mycoremediation plays an important role in the treatment of various toxic compounds in soils and wastewater. This system is highly versatile in nature and can be widely applied in biotechnology. The use in phenolic contaminants degradation, components oxidation, and consequent dyes decolorization has led to a drastic increase in the demand for these enzymes (Voběrková et al. 2018).

In recent years, several bioprocesses have been developed to take advantage of agro-industrial residues in techniques adoption that reduce environmental impacts, especially in order to decolorize dyes. There are two main ways of treating this effluent type (Figure 14.1): the first by direct application of the extracellular enzyme produced from agro-industrial effluent to the dye, so that the chromophore center cleavage process made by the action of ligninolytic enzymes promotes decolorization (Ali et al. 2010); this method is known as enzymatic degradation. Another way is through biosorption, where the mechanism proposed is based on the molecular linkages between the functional groups in the cell wall of the fungus adhered to the agro-industrial residue and the molecules of the dye, as well as in the kinetic models, equilibrium between the liquid phase and biomass, and thermodynamic behavior of biosorption and desorption kinetics (Guerrero-Coronilla et al. 2015).

Thus, agro-industrial waste acts as an efficient strategy to reduce the costs involved in the synthetic dye removal process (Wehaidy et al. 2018). In these circumstance, in order to meet the environmental demands for proper waste and effluent disposal, thereby guaranteeing sustainability, this chapter presents a review of the use of agro-industrial waste as a substrate for fungal production and its use in the discoloration of synthetic dyes.

14.2 SYNTHETIC DYES

Dyes are synthesized molecules that usually have a heterocyclic or aromatic nature and can be soluble in acidic, neutral, or basic medium (Shindy 2016). They are electrically unstable because they have a complex and unsaturated molecular structure. This complexity makes them difficult to

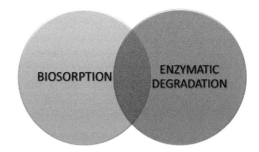

FIGURE 14.1 Main methods of mycoremediation of dyes with agro-industrial waste use.

biodegrade. Currently, there are more than 10,000 different dye types used in the cosmetic, plastic, and textile industries, commercially used in the world (Naraian et al. 2018). It is estimated that 700,000 tons of synthetic dyes are produced worldwide per year (Gürses et al. 2016). Those dyes can be classified in two different ways: according to the application method or depending on the functional groups in their chemical structure. According to Burkinshaw and Salihu (2013), a useful way of classifying the dyes is by the method of dyes application, since they can be acidic, mordant, metalized, direct, reactive, vat, dispersed, or basic.

For Yesilada et al. (2018), the chemical structure determines the properties of the dye and use, as well as provides the only rational basis for the classification of these compounds. This classification depends mainly on the chromophore group, which is in charge of giving each dye its characteristic color. The main chromophore groups are azo, indigo, anthraquinone, nitro, triphenylmethyl, and phthalene (Benkhaya et al. 2018). There is also the CI (Color Index International), which is the global reference classification form and nomenclature, and it has a bias for commercial purposes. This index is considered a dyes and pigments database developed by SDS (Society of Dyers and Colorists—UK) and AATCC (American Association of Textile Chemists and Colorists).

Within classification by the functional groups, azo dyes are the most used in the textile industry. They are characterized by the azo-type functional group appearance ($-N = N-$) (Dilarri et al. 2016), which can be found in different proportions (polyazo, triazo, diazo, or monoazo). The large-scale use of azo dyes in industries is due to four main factors: high molar extinction, high adaptability to application needs, structural variations possibility, and generally simple coupling reaction (Benkhaya et al. 2018).

In order to increase the color, and hence, the binding strength of the dyes to the fabric, the technology currently employed can further add metals to their composition. In this sense, a good part of the applied methods for effluent treatment becomes inefficient in the remediation of those that are produced by the textile industry, due to the generation of even more harmful by-products, once they are chemically modified for toxic or carcinogenic compounds (Kunjadia et al. 2016). Traditional technologies for degrading synthetic dyes (physical or physicochemical) are not efficient enough. In these technologies, the dye degradation products become more toxic than the original dyes (Plácido et al. 2016). Figure 14.2 shows the chemical structure of some dyes listed in this work.

14.3 MYCOREMEDIATION OF SYNTHETIC DYES

Among the main problems related to the wastewater production by the textile industry, it is possible to point out the correlation between the low efficiency of fiber dye aggregation, mainly reactive and dispersive and acid dyes, and the highly colored wastewater with high organic load generation, besides the large amount of water consumed by the sector. It is estimated that 30% of the initial dye concentration remains in the hydrolyzed form during the dyeing process. The affinity lack of the dye to the fabric results in the color accumulation to the effluent. It is imperative to remember that even in minimal amounts, these dyes can slightly alter the color of water bodies, which results in photosynthetic activity inhibition and dissolved oxygen concentration reduction, thus creating an anaerobic condition (Yesilada et al. 2018).

The reactive dyes have an electrophilic group capable of forming a covalent bond with the hydroxyl groups of cellulosic fibers, with amino, hydroxyl, and thiol groups of protein fibers and amino groups of polyamides (Epolito et al. 2005). In addition, they are water-soluble and recalcitrant under aerobic conditions found in conventional biological treatment systems. In contrast, in recent years, research has been intensifying several classes of fungi used in textile effluent treatment. The mycoremediation processes are accepted due to the fact that fungi, especially white-rot fungi and brown-rot fungi, present a nonspecific extracellular enzymatic system (Zahmatkesh et al. 2018).

The dye decolorization efficiency by ligninolytic enzymes depends on the enzyme type and the dye structure. There are several studies on dyes decolorization by laccases and several factors

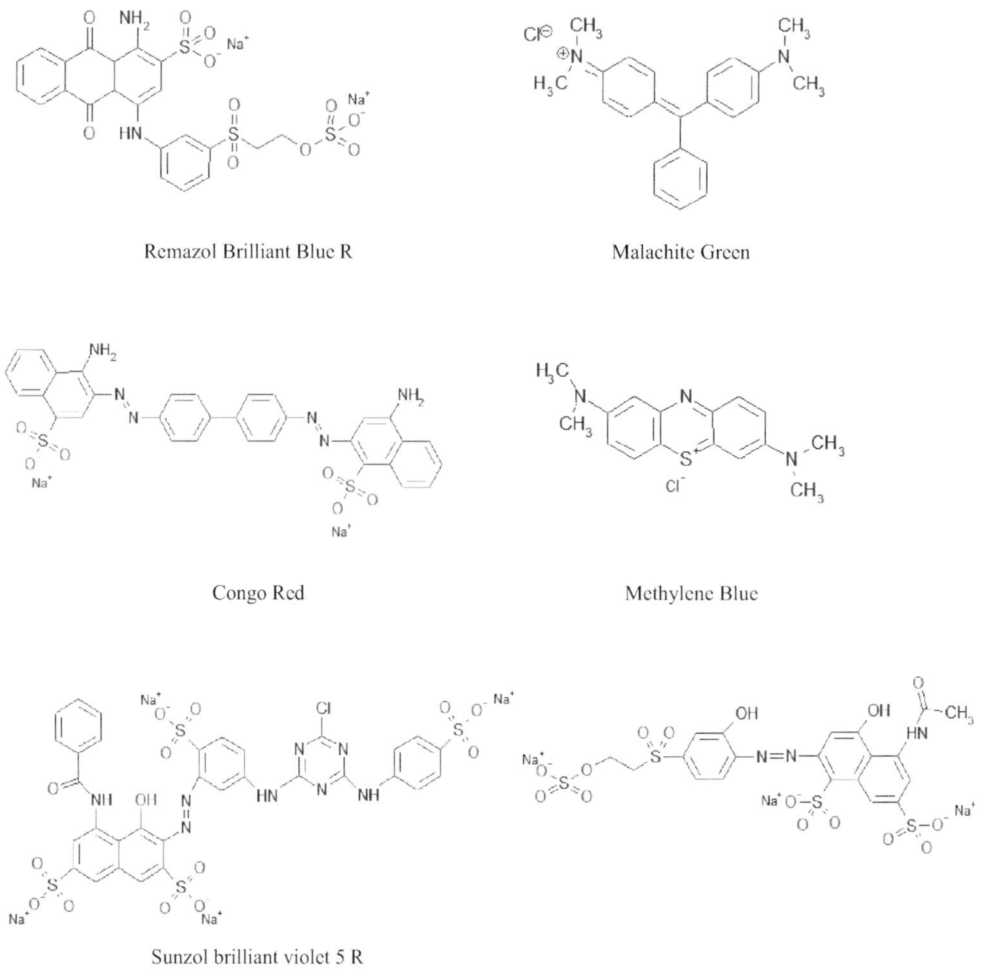

Remazol Brilliant Blue R

Malachite Green

Congo Red

Methylene Blue

Sunzol brilliant violet 5 R

FIGURE 14.2 Structure of some chemically different dyes.

influence the decolorization process, such as pH, reaction temperature, ionic strength, presence of redox mediators, enzymes type and concentration, and dyes (Cristóvão et al. 2008). Teerapatsakul et al. (2017) used laccase from *Ganoderma* sp. immobilized on copper alginate beads to degrade various synthetic dyes, especially Indigo Carmine; they obtained a high degradation efficiency with potential for industrial-scale implementation. In the work by Zhuo et al. (2016), *Pleurotus ostreatus* laccase was induced by various metal ions and aromatic structures and showed a strong ability of synthetic dyes decolorization. Jaiswal et al. (2016) immobilized the laccase on chitosan beads and used it for the Indigo Carmine decolorization, the color being completely removed in only 8 h, whereas with the free laccase, the decolorization was 56%. A new *Leucoagaricus naucinus* laccase was isolated and purified and showed efficiency in decolorizing activity in relation to azo, hetero-cyclic, and aromatic dyes, including Bromothymol Blue, Eriochrome T Black, Evans Blue, Basic Fuchsin, and Bright Blue Remazol R (Ning et al. 2016).

Treatment of textile effluent with fungi may be used in combination with other treatments. The ideal is to consider the cost, efficiency, risks to the environment, and water reuse in the industrial textile process. Table 14.1 shows different fungi application under different conditions for dyes decolorization.

TABLE 14.1

Different Fungi Application for Decolorization of Dyes and Removal Percentage

Dye	Fungi	Conditions	Decolorization (%)	Reference
Acid Blue 161	*Aspergillus terréus*	30°C, 14 days	84	Almeida and Corso (2018)
Acid Red 151	*Alternaria* spp.	30°C, 8 days, shaking	98,47	Ali et al. (2010)
Bromophenol Blue	*Paraconiothyrium variabile*	35°C, 3 h, shaking	100	Forootanfar et al. (2012)
Congo Red	*Cerrena unicolor* BBP6	30°C, 12 h.	53,9	Zhang et al. (2018)
Crystal Violet	*Cerrena unicolor* BBP6	30°C, 12 h.	80,9	Zhang et al. (2018)
Methylene Blue	*Aspergillus fumigatus*	Room temperature, 2 h	93,5	Kabbout and Taha (2014)
Orange II	*Alternaria* spp.	30°C, 8 days, shaking	58,66	Ali et al. (2010)
Procion Red MX-5B	*Aspergillus terréus*;	30°C, 14 days	92	Almeida and Corso (2018)
Reactive Black 5	*Cerrena* sp. WICC F39	30°C, 8 days	86	Hanapi et al. (2018)
Reactive Red 11	*Fusarium thapsinum* DB-42	25°C, 7 days	92,81	Yang et al. (2016a)
Remazol Brilliant Blue R	*Marasmius cladophyllus* UMAS MS8	Room temperature, 15 days	76	Sing et al. (2017)
Remazol Brilliant Violet 5R	*Irpex lacteus* CD2	Room temperature, 3 days	94,95	Qin et al. (2014)
Vat Brown 5	*Penicillium* sp.	30°C, 5 days, shaking	75	Ayla et al. (2018)

14.4 USE OF AGRO-INDUSTRIAL WASTE

There is a large amount of waste that is used as a nutrients source by fungi, including residues found in the nature of plant origin. Fungi are endowed with an enzymatic mechanism that transforms complex molecules into simple compounds through selective delignification, which is a removing process of lignin from wood or other cellulosic materials. When accumulated in the environment, agro-industrial waste can cause deterioration and promote the loss of natural resources. Many agricultural residues act as a substrate for fungal growth, and some of these residues are also described as inducing increased enzymatic activity, such as corn bran, rice, wheat, orange peel, grapefruit, tamarind, bagasse, and straw wheat (Manavalan et al. 2013).

Among the factors that can help in the development of a system whose enzymatic activity production is optimized, is the introduction of agro-industrial waste as a substrate in order to meet the nutrient needs of fungi and thus reduce production costs. Mendoza et al. (2014) used orange peels and sugarcane bagasse to grow *Galerina* sp. HC1, and in the enzyme subsequent application in the dye decolorization, they determined that the concentration of 20 g/L residue equals to the COD of 4 g/L glucose. With this residue addition, the enzymatic activity increased fourfold.

14.5 FUNGI TREATMENT OF DYES

Currently, the agro-industrial residues use has been shown to be efficient, inexpensive and highly applicable in several fungal strains' fermentation and subsequent application in textile dyes degradation. These residues, whether in the solid-state or as an effluent, can generally have similar characteristics as regards the pollutant potential, such as high BOD and COD, high organic matter content, mineral elements (Ca, S, Mg, N, and K), as well as high pigmentation. Some of these conditions, depending on the concentration, can influence the fungi development, extracellular enzymes, and biomass production.

Chicatto et al. (2018), in order to improve the textile dye treatment, cultivated white-rot fungi strains, such as *Ganoderma lucidum* EF 31, *G. lucidum* DSM 9621, and *Trametes versicolor* CECT 20817 in peach residue. They used three different fermentation techniques: submerged, solid-state, and adsorption. As a result, in the solid-state process, the treatment increased the textile effluent concentration after preliminary physical-chemical treatment. The highest decolorization obtained was 80% on the tenth day of *G. lucidum* DSM 9621 fermentation. Most of the studies cited in this chapter prove that the enzyme laccase is the major responsible for biodecolorization.

Wehaidy et al. (2018) covered the microbial laccase production cultivated in agro-industrial residues (orange peel, mandarin, and banana), as well as wheat bran, rice straw, and saw dust for the application of crude or partially purified enzyme in dyes such as Sunzol Brilliant Violet 5 R (reactive violet 5 or RV 5), Remazol Brilliant Blue R (reactive blue 19 or RB 19), and Dystar Supralan Blue 2 R (acid blue 225 or AB 225). *P. ostreatus* (NRRL 3501), *Penicillium pinophilum* (NRRL 1142), *Aspergillus niger*, *Polyporus durus* (ATCC 26726), *Agaricus subrufescens* (DSM 23611) and *Penicillium melini* (NRRL 848) were used for the search. The best results were obtained by using the wheat bran culture medium for fungal growth of submerged *Polyporus durus* with 1 mmol/L copper addition on the fifth day of fermentation. Laccase activity peaked at 2,297 IU/mL after 7 days of incubation. The laccase was then partially purified by acetone at 40% concentration, thereby achieving a purification factor of 8,1-fold. When crude and partially purified enzymes were applied to the Dystar Supralan blue 2 R, Remazol Brilliant Blue R and Sunzol Brilliant Violet 5 R dye solutions, they had 96.4% and 100% decolorization coefficients, respectively.

It is important to note that some of the agro-industrial effluents used in the research by Wehaidy et al. (2018) did not present satisfactory results regarding the fungi growth tested, as the saw dust and the orange and tangerine peels. In addition, remarkable attention is paid to the Cooper effect on process optimization; since laccase is also called "multicopper" oxidase, it directly uses the redox capacity of cupric ions, catalyzing various aromatic compounds oxidation and forming, at the end, water and oxygen molecules (Du et al. 2018).

This does not imply, however, that manganese peroxidase does not have a direct effect on dyes mycoremediation. Other studies have been done in the past years defending the best conditions for this enzyme production with agro-industrial effluents use; example is the research by Da Silva et al. (2017), which utilized *Pleurotus pulmonarius* CCB-19, cultivated on pineapple peel substrate and later application in Brilliant Blue R (RBBR, CI 61200) and Congo Red (CI 22120). The best response was obtained after the 14th day under solid-state conditions, where the enzyme activity reached a peak of 135.52 IU/mg. The MnP was then purified and reached a purification factor of 12.1 times, with yield of 28.4%. When the purified enzyme extract was applied to the liquid medium containing each dye + malonate buffer (manganese sulfate and H_2O_2, pH 4.5), under stirring at 100 rpm, 40°C for 2 h, 87% of the Brilliant Blue R dye decolorization and 63% of Congo Red.

Das et al. (2016) used *P. ostreatus* MTCC 142 grown on rice husk and corn husk cosubstrates. They were able to decolorize the Congo Red dye by extracting the crude laccase from the substrate and adding it to the dye, and 9.7 mL of citrate buffer. After 20 h incubation at 35°C, the Congo Red was decolorized in 36.84%. The optimal laccase activity condition was at pH 3.0, 35°C and Cu^{+2} addition to the cosubstrate. This result did not appear to be as high as those described above due to two factors: the first one was that the crude enzyme extract was used, which could interfere with the activity of other extracellular ligninolytic enzymes, and the second was the relatively short incubation time. Nonetheless, the work explores in a unique way the productive and efficient potential of laccase for azo dyes decolorization such as Congo Red.

Bankole et al. (2018) suggested the Cibracon Brilliant Red 3B-A decolorization using the white rot fungi *Daldinia concentrica* and *Daldinia Xylaria* grown separately and together in medium containing bean rind using solid-state fermentation under static and stirring conditions. In this study, laccase, manganese peroxidase, and lignin peroxidase activity were observed. The results revealed that under stirring the *D. concentrica* fungus, separately, was able to remove up to 95% of textile effluent.

When the dye was treated by *D. Xylaria*, decolorization occurred up to 97%, also under agitation conditions, while the two fungal lines consortium showed a 99% removal of Cibracon Brilliant Red 3B-A.

Bagewadi et al. (2017) purified the laccase from *Trichoderma harzianum,* previously cultured on wheat bran under solid state and applied it to decolorize the Synthetic Methylene Blue, Malachite Green, and Congo Red dyes. The laccase enzyme was purified and subsequently immobilized on calcium alginate-chitosan granule traps, sol-gel, copper alginate, and calcium alginate. After application of the sol-gel-immobilized laccase in the synthetic dyes in question, a 60% decolorization of Congo Red dye, 90% of Methylene Blue, and almost 100% of Malachite Green were achieved progressively, at the initial concentration of 200 mg/L, at 20, 18, and 16 h, respectively.

Yang et al. (2016b) made use of white rot fungus *Irpex lacteus* F17, cultivated in the laboratory from a wood chips pile, by solid-state fermentation. The manganese peroxidase presented activity of 66.32 IU/L and was applied to Malachite Green dye in its crude and purified states. In the optimized condition, which consisted of the crude enzyme extract addition, 2.0 mM $MnSO_4$, 0,1 mM H_2O_2, and 50 mM sodium malonate buffer to the dye (200 mg/L), there was an approximately 96% decolorization. It has been proven that the crude enzyme extract use has advantage over the purified enzyme, since the decolorization results being close, but the fact of purifying the enzyme makes the process expensive.

14.6 ENZYME SYSTEM

Although the literature is of various types, the most popular enzymes within the enzymes system formed by the fungi, whether of white, brown, or soft rot, are in order of importance with regard to the studies number, laccase (Lac), manganese peroxidase (MnP), and lignin peroxidase (LiP). These enzymes usually attack the complex structure of dyes and agro-industrial residues by extracellular agents, as well as release compounds with low molecular weight that function as enzymatic mediators. In this sense, the studies contained in this chapter show the decolorization rate that depends mainly on types of strains and concentration of ligninolytic enzymes (Hanapi et al. 2018).

14.7 INFLUENCE OF GROWTH FACTORS ON ENZYME ACTIVITY AND DYE DEGRADATION

The fungus growth is affected by several chemical and physical factors, such as temperature, humidity, oxygen concentration, pH, micronutrients, and carbon and nitrogen sources, among others. In this regard, the factors of greater influence for decolorization of dyes and mycoremediation of agro-industrial residues are presented here, considering the strains used in the literature (Figure 14.3).

14.7.1 pH EFFECT

The ligninolytic enzymes present maximum activity at pH where their active sites may have the highest possible interaction with the substrate. An energetic change in the medium pH can lead to the enzyme denaturation, which can lead to a significant activity loss (Das et al. 2016). Bagewadi et al. (2017) explain that the decrease in laccase activity occurs due to the increase in pH, and this circumstance may be linked to two factors: the variation in the enzyme's three-dimensional structure folding and the change in the active site ionic form.

Bankole et al. (2018) reported the ideal pH range between 4.5 and 6.5 for the Cibacron Brilliant Red 3B-A decolorization; according to them, it was observed that the lowest decolorization rate may be due to the presence of very basic and/or very acidic metabolites formed during the dye removal process. Desiring to find the best pH conditions for decolorization of several dyes by the *Yarrowia lipolytica* Lac, Darvishi et al. (2018) reported the best results within the range of pH 4.5 and 6.0.

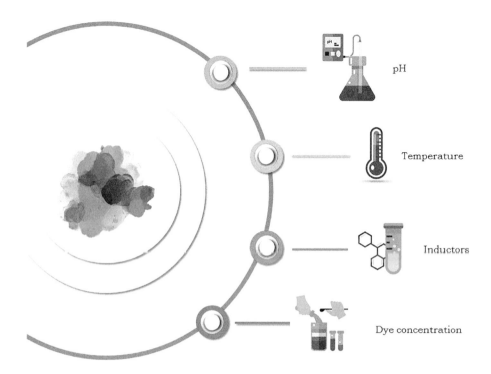

FIGURE 14.3   Influence factors for mycoremediation of synthetic dyes and agro-industrial waste.

Campos et al. (2016) reported that pH above 6.0 reduces the basidiomycete *Trametes trogii* Lac activity, and good results are obtained at pH 4.5 and 6.0, so that it was possible to decolorize efficiently Indigo Carmine, Malachite Green, Remazol Brilliant Blue R, Xylidine Ponceau, and Azure B.

Studies show that the major activities for the white rot fungi lignin peroxidase enzyme are obtained between pH 2 and 5 (Oliveira et al. 2016). Kunjadia et al. (2016) indicate that the interaction between the enzyme active sites and the dyes is affected because they are organic compounds of complex aromatic chains and thus have distinct ionization potentials at different pH values. The *Bjerkandera adusta* strain CX-9 LiP maintained half of its initial activity at pH 5 and 3, after incubation for 6 and 10 h, respectively, reaching the 89% Remazol Brilliant Blue R decolorization mark (Bouacem et al. 2018). According to the same authors, peroxidases with high stability in acidic environments have been much sought after the applications series. Lauber et al. (2017) using *Pleurotus sapidus* to decolorize Reactive Blue 5, with the phenolic substrates and ABTS addition, found that the ideal range for oxidation by peroxidase enzymes is between pH 3.5 and 4.5.

With regard to the enzyme MnP, da Silva et al. (2017) determined an ideal range for activity between pH 5 and 6. In reactive dye decolorization studies from *G. lucidum* IBL-05 MnP, the optimal condition for the free enzyme was at pH 5.0 (Bilal and Asgher 2015). The free fungal MnP presented the best enzymatic activity in the textile effluent decolorization at pH 5.0, and when immobilized on chitosan beads, a better result was found at pH 4.0 (Asgher et al. 2016).

According to the results of the cited studies, Figure 14.4 shows the pH conditions considered ideal for the activity of the main enzymes responsible for biodecolorization.

14.7.2   TEMPERATURE EFFECT

The temperature has a strong influence on the extracellular enzymatic activity production and, consequently, on decolorization of certain synthetic dyes. Yao et al. (2013), when fermenting *Schizophyllum* sp. in agro-industrial residues for Orange G, Orange IV, and Congo Red dyes, found

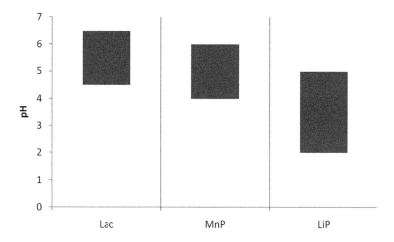

FIGURE 14.4   Ranges of ideal pH for the ligninolytic enzymes activity.

that the temperature effect on decolorization was associated with the crude MnP thermal stability, which was later investigated in the enzyme incubation at 25, 35, 45, and 65°C. MnP kept the activity high within the temperature range of 25–45°C. Moreover, the enzyme sustained about 80% of the original activity for 8 h. Jamil et al. (2018) found close value when using the *P. ostreatus* IBL-02 to decolorize the *Sandal-fix Turquoise blue GWWF, Sandal-fix Red C4BLN, Sandal-fix Black CKF, Sandal-fix Golden yellow CRL, and Reactive T blue GWF* dyes. The work showed that the free laccase showed maximum activity at 35°C, and the enzyme immobilization makes the temperature effect less influent; since in immobilizing the laccase in chitosan networks, it was possible to guarantee enzymatic activity at high temperature (60°C).

14.7.3   EFFECT OF INDUCERS' PRESENCE

Enzyme production may be increased by the addition of supplements to the substrates, which are known as inducers because they induce a higher enzyme production that may be the substrate of the enzyme itself or an analogous structural component. The addition of xenobiotic compounds such as xylidine, lignin, and veratrylic alcohol increases and induces laccase activity (Kumar et al. 2014). Manganese sulfate has been shown as a good inducer of *Phanerochaete chrysosporium* Manganese activity, its influence has been reported by Senthilkumar et al. (2011). Copper has also been shown to exhibit the induction effect on laccase activity (Passarini et al. 2015; Fonseca et al. 2016). Another inducer that can be cited is ethanol, which induced the production of Ganoderma lucidum laccase (Manavalan et al. 2013). Table 14.2 shows some researched inducers for various types of enzymes produced by fungi.

14.7.4   EFFECT OF INITIAL DYE CONCENTRATION

The initial concentration of dyes also exerts a strong influence on the biodecolorization, once the toxicity to the fungus is noted (Sen et al. 2016). According to the same authors, high concentration of dyes reduces fungal growth and consequently enzymatic production, which leads to the less efficiency of biodecolorization. The dye concentration factor reduces the decolorization rate, since it is necessary that the enzymes generate more hydroxyl radicals so that more organic molecules are oxidized (Neoh et al. 2015). The dye concentration was also observed by Medina-Moreno et al. (2012), which used a bioreactor of joint action between biosorption and enzymatic biodecolorization determining the best conditions for the system use. They found that at low initial concentration (from 50 to 200 mg/L) the *Trametes subectypus* B32 fungus has the ability to completely

TABLE 14.2
Inducers Used in the Enzyme Production

Inducer	Fungi	Enzyme	Reference
ABTS	*Trametes versicolor*	Lac	Amutha and Abhijit (2015)
Ammonium dihydrogen phosphate and ammonium chloride	*Phanerochaete chrysosporium*	MnP	Senthilkumar et al. (2011)
Cadmium	*Pleurotus ostreatus; Ganoderma lucidum.*	Lac	Kuhar and Papinutti (2014)
Copper	*Ganoderma lucidum; Pleurotus sajor-caju; Dichomitus squalens; Pleurotus florida; Trametes sp.*	Lac	Kuhar and Papinutti (2014), Fonseca et al. (2016)
Ethanol	*Ganoderma lucidum; Pycnoporus cinnabarinus.*	Lac	Manavalan et al. (2013)
Ferulic acid	*Pleurotus sajor-caju; Ganoderma lucidum.*	Lac	Kuhar and Papinutti (2014)
Gallic acid	*Ganoderma lucidum*	Lac	Manavalan et al. (2013)
Glucose	*Phanerochaete chrysosporium*	MnP, Lac and LiP	Senthilkumar et al. (2011)
Guaiacol	*Ganoderma lucidum; Trametes versicolor.*	Lac	Amutha and Abhijit (2015)
Manganese sulfate	*Irpex lacteus* F17	MnP	Yang et al. (2016b)
Syringaldazine	*Trametes versicolor.*	Lac	Amutha and Abhijit (2015)
Tannic acid	*Trametes versicolor; Pleurotus sp.*	Lac	Amutha and Abhijit (2015)
Vanillic acid	*Pleurotus eryngii; Ganoderma lucidum.*	Lac	Kuhar and Papinutti (2014)
Veratryl alcohol	*Irpex lacteus* F17	LiP	Yang et al. (2016b)

remove the Reactive Blue 4 dye; however, at concentrations between 250 and 800 mg/L, the color removal was significantly decreased. Neoh et al. (2015) used *Curvularia clavata* NZ2 to decolorize the dyes Acid Orange 7, Congo Red, and Reactive Black 5, varying concentrations of 100 and 1,000 ppm. According to them, the decolorization rate was reduced with increasing the initial concentration of dyes.

14.8 CONCLUSION

Industrial progress is made possible, in particular, by scientific advances, so that science is a key for sustainable development. In this context, agro-industrial wastes as well as effluents from synthetic dyes present chemical characteristics and structures with high polluting potential. The alternative that presents better results and less detachment for remediation of these pollutants is the mycoremediation by white or brown rot fungi. These fungi present a nonspecific extracellular enzyme complex that easily degrades compounds with lignin-like structures. Therefore, it was possible to present some techniques and studies done in recent years concerning the bioremediation of agro-industrial residues and synthetic dyes. So, there are many efforts on the researchers' part, especially in the biotechnology area and in the elaboration of techniques that can be applied in an industrial scale. With regard to the costs required to bring these techniques to the real plan, there are still few papers reported in the literature that bring an economic feasibility study, and this is usually a suggestion for future work.

REFERENCES

Ali N, Hameed A, Ahmed S (2010) Role of brown-rot fungi in the bioremoval of azo dyes under different conditions. *Brazilian J Microbiol* 41:907–915. doi:10.1590/S1517-83822010000400009.

Almeida EJR, Corso CR (2018) Decolorization and removal of toxicity of textile azo dyes using fungal biomass pelletized. *Int J Environ Sci Technol* 16:1–10. doi:10.1007/s13762-018-1728-5.

Amutha C, Abhijit M (2015) Screening and isolation of laccase producers, determination of optimal condition for growth, Laccase production and choose the best strain. *Bioremediat Biodegrad* 6:4–11. doi:10.4172/2155-6199.1000298.

Asgher M, Wahab A, Bilal M, Iqbal N (2016) Lignocellulose degradation and production of lignin modifying enzymes by *Schizophyllum commune* IBL-06 in solid-state fermentation. *Biocatal Agric Biotechnol* 6:195–201. doi:10.1016/j.bcab.2016.04.003.

Ayla S, Golla N, Pallipati S (2018) Production of ligninolytic enzymes from Penicillium Sp. and its efficiency to decolourise textile dyes. *Open Biotechnol J* 12:112–122. doi:10.2174/1874070701812010112.

Bagewadi ZK, Mulla SI, Ninnekar HZ (2017) Purification and immobilization of laccase from Trichoderma harzianum strain HZN10 and its application in dye decolorization. *J Genet Eng Biotechnol* 15:139–150. doi:10.1016/j.jgeb.2017.01.007.

Bankole PO, Adekunle AA, Govindwar SP (2018) Biodegradation of a monochlorotriazine dye, cibacron brilliant red 3B-A in solid state fermentation by wood-rot fungal consortium, Daldinia concentrica and Xylaria polymorpha: Co-biomass decolorization of cibacron brilliant red 3B-A dye. *Int J Biol Macromol* 120:19–27. doi:10.1016/j.ijbiomac.2018.08.068.

Barrech D, Ali I, Tareen M (2018) A review on mycoremediation—The fungal bioremediation. *Pure Appl Biol* 7:343–348.

Benkhaya S, Harfi S El, Harfi A El (2018) Classifications, properties and applications of textile dyes: A review. *Appl J Envir Eng* 3:311–320.

Bilal M, Asgher M (2015) Dye decolorization and detoxification potential of Ca-alginate beads immobilized manganese peroxidase. *BMC Biotechnol* 15:1–14. doi:10.1186/s12896-015-0227-8.

Bilal M, Asgher M, Parra-saldivar R, et al. (2017) Immobilized ligninolytic enzymes: An innovative and environmental responsive technology to tackle dye-based industrial pollutants—A review. *Sci Total Environ* 576:646–659. doi:10.1016/j.scitotenv.2016.10.137.

Bouacem K, Rekik H, Jaouadi NZ, et al. (2018) Purification and characterization of two novel peroxidases from the dye-decolorizing fungus *Bjerkandera adusta* strain CX-9. *Int J Biol Macromol* 106:636–646. doi:10.1016/j.ijbiomac.2017.08.061.

Burkinshaw SM, Salihu G (2013) The wash-off of dyeings using interstitial water. Part 4 : Disperse and reactive dyes on polyester/cotton fabric. *Dye Pigment* 99:548–560. doi:10.1016/j.dyepig.2013.06.006.

Campos PA, Levin LN, Wirth SA (2016) Heterologous production, characterization and dye decolorization ability of a novel thermostable laccase isoenzyme from *Trametes trogii* BAFC 463. *Process Biochem.* doi:10.1016/j.procbio.2016.03.015.

Chandra R, Kumar V (2015) Biotransformation and biodegradation of organophosphates and organohalides. In: R Chandra (Ed.), *Environmental Waste Management*. CRC Press, Boca Raton, FL, pp. 475–524.

Chandra R, Kumar V, Tripathi S (2018) Evaluation of molasses-melanoidins decolourisation by potential bacterial consortium discharged in distillery effluent. *3 Biotech* 8:187. doi:10.1007/s13205-018-1205-3.

Chicatto JA, Carolini H, Nunes A, et al. (2018) Strategies for decolorization of textile industry effluents by white- rot-fungi with peach palm residue. *Acta Sci* 40:1–9. doi:10.4025/actascitechnol.v40i1.35610.

Cristóvão RO, Tavares APM, Ribeiro AS, et al. (2008) Kinetic modelling and simulation of laccase catalyzed degradation of reactive textile dyes. *Bioresour Technol* 99:4768–4774. doi:10.1016/j.biortech.2007.09.059.

da Silva BP, Gomes Correa RC, Kato CG, et al. (2017) Characterization of a solvent-tolerant manganese peroxidase from *Pleurotus pulmonarius* and its application in dye decolorization. *Curr Biotechnol* 6:318–324. doi:10.2174/2211550105666160224004258.

Darvishi F, Moradi M, Jolivalt C, Madzak C (2018) Laccase production from sucrose by recombinant *Yarrowia lipolytica* and its application to decolorization of environmental pollutant dyes. *Ecotoxicol Environ Saf* 165:278–283. doi:10.1016/j.ecoenv.2018.09.026.

Das A, Bhattacharya S, Panchanan G, et al. (2016) Production, characterization and Congo red dye decolourizing efficiency of a laccase from *Pleurotus ostreatus* MTCC 142 cultivated on co-substrates of paddy straw and corn husk. *J Genet Eng Biotechnol* 14:281–288. doi:10.1016/j.jgeb.2016.09.007.

Dilarri G, Janaina É, Almeida R De, et al. (2016) Removal of dye toxicity from an aqueous solution using an industrial strain of Saccharomyces Cerevisiae (Meyen). *Water Air Soil Pollut* 227:269. doi:10.1007/s11270-016-2973-1.

Du W, Sun C, Wang J, et al. (2018) Isolation, identification of a laccase—Producing fungal strain and enzymatic properties of the laccase. *3 Biotech* 8:137. doi:10.1007/s13205-018-1149-7.

Epolito WJ, Lee YH, Bottomley LA, Pavlostathis SG (2005) Characterization of the textile anthraquinone dye reactive blue 4. *Dye Pigment* 67:35–46. doi:10.1016/j.dyepig.2004.10.006.

Ergun SO, Urek RO (2017) Production of ligninolytic enzymes by solid state fermentation using *Pleurotus ostreatus* Seyma. *Ann Agrar Sci* 43:582–597. doi:10.1016/j.aasci.2017.04.003.

Fonseca MI, Tejerina MR, Sawostjanik-afanasiuk SS, et al. (2016) Preliminary studies of new strains of Trametes sp. from Argentina for laccase production ability. *Braz J Microbiol* 47:287–297. doi:10.1016/j.bjm.2016.01.002.

Forootanfar H, Moezzi A, Aghaie-khozani M, et al. (2012) Synthetic dye decolorization by three sources of fungal laccase. *Environ Health* 9:1–10.

Guerrero-coronilla I, Morales-barrera L, Cristiani-urbina E (2015) Kinetic, isotherm and thermodynamic studies of amaranth dye biosorption from aqueous solution onto water hyacinth leaves. *J Environ Manage* 152:99–108. doi:10.1016/j.jenvman.2015.01.026.

Gürses A, Açıkyıldız M, Güneş K, Gürses MS (2016) Dyes and pigments: their structure and properties, 1st ed. In: *Dyes and Pigments*. Springer, Cham, pp. 13–29. doi:10.1007/978-3-319-33892-7-2.

Hanapi SZ., El nshasy H., Galil SA (2018) Isolation of a new efficient dye decolorizing white rot fungus *Cerrena* Sp. WICC F39. *J Sci Ind Res (India)* 77:399–404.

Jaiswal N, Pandey VP, Dwivedi UN (2016) Immobilization of papaya laccase in chitosan led to improved multipronged stability and dye discoloration. *Int J Biol Macromol* 86:288–295. doi:10.1016/j.ijbiomac.2016.01.079.

Jamil F, Asgher M, Hussain F, Bhatti HN (2018) Biodegradation of synthetic textile dyes by chitosan beads cross-linked laccase from *Pleurotus ostreatus* IBL-02. *J Anim Plant Sci* 28:231–243.

Kabbout R, Taha S (2014) Biodecolorization of textile dye effluent by biosorption on fungal biomass materials. *Phys Procedia* 55:437–444. doi:10.1016/j.phpro.2014.07.063.

Kuhar F, Papinutti L (2014) Optimization of 1 accase production by two strains of Ganoderma lucidum using phenolic and metallic inducers. *Rev Argent Microbiol* 46:144–149. doi:10.1016/S0325-7541(14)70063-X.

Kumar V, Chandra R (2018) Characterisation of manganese peroxidase and laccase producing bacteria capable for degradation of sucrose glutamic acid-Maillard products at different nutritional and environmental conditions. *World J Microbiol Biotechnol* 34:82.

Kumar V V, Sivanesan S, Cabana H (2014) Magnetic cross-linked laccase aggregates—Bioremediation tool for decolorization of distinct classes of recalcitrant dyes. *Sci Total Environ* 487:830–839. doi:10.1016/j.scitotenv.2014.04.009.

Kunjadia PD, Sanghvi G V., Kunjadia AP, et al. (2016) Role of ligninolytic enzymes of white rot fungi (Pleurotus spp.) grown with azo dyes. *Springerplus* 5:1487. doi:10.1186/s40064-016-3156-7.

Lauber C, Schwarz T, Nguyen QK, et al. (2017) Identification, heterologous expression and characterization of a dye-decolorizing peroxidase of Pleurotus sapidus. *AMB Express* 7:164. doi:10.1186/s13568-017-0463-5.

Manavalan T, Manavalan A, Thangavelu KP, Heese K (2013) Characterization of optimized production, purification and application of laccase from *Ganoderma lucidum*. *Biochem Eng J* 70:106–114. doi:10.1016/j.bej.2012.10.007.

Medina-moreno SA, Pérez-cadena R, Jiménez-gonzález A, Téllez-jurado A (2012) Modeling wastewater biodecolorization with reactive blue 4 in fixed bed bioreactor by *Trametes subectypus*: Biokinetic, biosorption and transport. *Bioresour Technol* 123:452–462. doi:10.1016/j.biortech.2012.06.097.

Mendoza L, Ibrahim V, Álvarez MT, Hatti-kaul R (2014) Laccase production by *Galerina* sp. and its application in dye decolorization. *J Yeast Fungal Res* 5:13–22. doi:10.5897/JYFR12.025.

Naraian R, Kumari S, Gautam RL (2018) Biodecolorization of brilliant green carpet industry dye using three distinct *Pleurotus* spp. *Environ Sustain* doi:10.1007/s42398-018-0012-4.

Neoh CH, Lam CY, Lim CK, Yahya A (2015) Biodecolorization of recalcitrant dye as the sole source of nutrition using *Curvularia clavata* NZ2 and decolorization ability of its crude enzymes. *Environ Sci Pollut Res Int* 22:11669–11678. doi:10.1007/s11356-015-4436-4.

Ning Y-J, Wang S-S, Chen Q-J, et al. (2016) An extracellular yellow laccase with potent dye decolorizing ability from the fungus *Leucoagaricus naucinus* LAC-04. *Int J Biol Macromol.* 93:837–842. doi:10.1016/j.ijbiomac.2016.09.046.

Oliveira SF, da Luz JMR, Kasuya MCM, et al. (2016) Enzymatic extract containing lignin peroxidase immobilized on carbon nanotubes: Potential biocatalyst in dye decolourization. *Saudi J Biol Sci* 25:651–659. doi:10.1016/j.sjbs.2016.02.018.

Passarini MRZ, Ottoni CA, Santos C, et al. (2015) Induction, expression and characterisation of laccase genes from the marine-derived fungal strains Nigrospora sp. CBMAI 1328 and *Arthopyrenia* sp. CBMAI 1330. *AMB Express* 5:19. doi:10.1186/s13568-015-0106-7.

Peralta RM, da Silva BP, Corrêa RCG, et al. (2017) Enzymes from basidiomycetes—Peculiar and efficient tools for biotechnology. In G Brahmachari (Ed.) *Biotechnology of Microbial Enzymes*. Academic Press, pp. 119–149.

Plácido J, Chanagá X, Monsalve SO, et al. (2016) Degradation and detoxification of synthetic dyes and textile industry effluents by newly isolated *Leptosphaerulina* sp. from Colombia. *Bioresour Bioprocess* 3. doi:10.1186/s40643-016-0084-x.

Qin X, Zhang J, Zhang X, Yang Y (2014) Induction, purification and characterization of a novel manganese peroxidase from irpex lacteus CD2 and its application in the decolorization of different types of dye. *PLoS ONE* 9:e113282. doi:10.1371/journal.pone.0113282.

Sen SK, Raut S, Bandyapadhyay P, Raut S (2016) Fungal decolouration and degradation of azo dyes: A review. *Fungal Biol Rev* 30:112–133. doi:10.1016/j.fbr.2016.06.003.

Senthilkumar S, Perumalsamy M, Prabhu HJ (2011) Decolourization potential of white-rot fungus *Phanerochaete chrysosporium* on synthetic dye bath effluent containing Amido black 10B. *J Saudi Chem Soc* 18:845–853. doi:10.1016/j.jscs.2011.10.010.

Shindy HA (2016) Basics in colors, dyes and pigments chemistry: A review. *Chem Int* 2:29–36.

Sing NN, Husaini A, Zulkharnain A, Roslan HA (2017) Decolourisation capabilities of ligninolytic enzymes produced by *Marasmius cladophyllus* UMAS MS8 on remazol brilliant blue R and other azo dyes. *Biomed Res Int* 2017. doi:10.1155/2017/1325754.

Singh L (2017) Biodegradation of synthetic dyes: A mycoremediation approach for degradation/decolourization of textile dyes and effluents. *J Appl Biotechnol Bioeng* 3:430–435. doi:10.15406/jabb.2017.03.00081

Teerapatsakul C, Parra R, Keshavarz T, Chitradon L (2017) Repeated batch for dye degradation in an airlift bioreactor by laccase entrapped in copper alginate. *Int Biodeterior Biodegrad* 120:52–57. doi:10.1016/j.ibiod.2017.02.001.

Upadhyay P, Shrivastava R, Agrawal PK (2016) Bioprospecting and biotechnological applications of fungal laccase. *3 Biotech* 6:1–12. doi:10.1007/s13205-015-0316-3.

Voběrková S, Solčány V, Vršanská M, Adam V (2018) Immobilization of ligninolytic enzymes from white-rot fungi in cross-linked aggregates. *Chemosphere* 202:694–707. doi:10.1016/j.chemosphere.2018.03.088.

Wehaidy HR, El-Hennawi HM, Ahmed SA, Abdel-naby MA (2018) Comparative study on crude and partially purified laccase from *Polyporus durus* ATCC 26726 in the decolorization of textile dyes and wastewater treatment. *Egypt Pharmaceut J* 17:94–103. doi:10.4103/epj.epj.

Yang P, Shi W, Wang H, Liu H (2016a) Screening of freshwater fungi for decolorizing multiple synthetic dyes. *Braz J Microbiol* 47:828–834. doi:10.1016/j.bjm.2016.06.010.

Yang X, Zheng J, Lu Y, Jia R (2016b) Degradation and detoxification of the triphenylmethane dye malachite green catalyzed by crude manganese peroxidase from *Irpex lacteus* F17. *Environ Sci Pollut Res Int* 23:9585–9597. doi:10.1007/s11356-016-6164-9.

Yao J, Jia R, Zheng L, Wang B (2013) Rapid decolorization of azo dyes by crude manganese peroxidase from Schizophyllum sp. F17 in solid-state fermentation. *Biotechnol Bioprocess Eng* 877:868–877. doi:10.1007/s12257-013-0357-6.

Yesilada O, Birhanli E, Geckil H (2018) Bioremediation and decolorization of textile dyes by white rot fungi and laccase enzymes. In R Prasad (Ed.) *Mycoremediation and Environmental Sustainability*. Springer International Publishing, Cham.

Zahmatkesh M, Spanjers H, Lier JB Van, et al. (2018) A novel approach for application of white rot fungi in wastewater treatment under non-sterile conditions: Immobilization of fungi on sorghum. *Environ Technol* 39:3330. doi:10.1080/09593330.2017.1347718.

Zhang H, Zhang J, Zhang X, Geng A (2018) Purification and characterization of a novel manganese peroxidase from white-rot fungus *Cerrena unicolor* BBP6 and its application in dye decolorization and denim bleaching. *Process Biochem* 66:222–229. doi:10.1016/j.procbio.2017.12.011.

Zhuo R, Yuan P, Yang Y, et al. (2016) Induction of laccase by metal ions and aromatic compounds in *Pleurotus ostreatus* HAUCC 162 and decolorization of different synthetic dyes by the extracellular laccase. *Biochem Eng J* 117:67–72. doi:10.1016/j.bej.2016.09.016.

15 Beneficial Microbes for Sustainable Agriculture

Aneesh Kumar Chandel, Hao Chen,
Hem Chandra Sharma, and Kaushik Adhikari
University of Arkansas at Pine Bluff

Bin Gao
University of Florida

CONTENTS

15.1 INTRODUCTION

Intensified agriculture is required to meet the food and textile demand of the growing population. For agriculture intensification, large quantities of agrochemicals are needed for plant nutrients supplement and disease control. Agrochemical input has been increased in both developed and developing countries for decades for the maximum yield of crops (Carvalho, 2017; Clark and Tilman, 2017). However, there is a growing concern worldwide over agrochemicals used that led to environment degradation (Clark and Tilman, 2017; Stehle and Schulz, 2015; Zhang et al., 2015). The natural roles of microorganisms in promoting plant growth maintaining soil fertility and biocontrol of plant pathogens represent a promising sustainable solution to improve agricultural production. Rhizosphere, the soil zone under the influence of plant roots, is characterized with a wide variety of bacteria species and highly dynamic plant-microbe-soil interactions (Lambers et al., 2009). Plant root exudates are easily degradable carbon compounds. They act as nutrients for macro- and microorganisms in the rhizosphere, including bacteria, fungi, virus, protozoa, algae, nematodes, and microarthropods. Among the different microorganisms found in the rhizosphere, current research is mostly focused on the plant beneficial microorganisms such as nitrogen-fixing bacteria and endo- and ectomycorrhizal fungi. Advancement in the analysis tools such as polymerase chain reaction, microarray method, and proteomic techniques has enabled scientists to identify the bacterial strains and to understand the bacterial diversity in molecular level.

15.2 BENEFICIAL MICROBES IN AGRICULTURE

15.2.1 BENEFICIAL FUNGI

Mycorrhizal fungi are the group of taxa that interact with the plants through different mechanisms from mutualistic symbiosis to parasitism (Behie and Bidochka, 2014). Mutualism-parasitism continuum is the best concept to describe the interaction between plants and fungi (Bonfante and Genre, 2010). Through this interaction, fungi develop hyphal conduits with the plant to transport nutrient and water and thus promote plant nutrient uptake and water use efficiency, and biotic and abiotic stress tolerance (Abeer et al., 2015; Volpe et al., 2018). Hyphal conduits are also responsible for interplant signaling (mechanism in which one plant protects other from insects or pest attack through providing signals) via the network of a vast association of hyphae called mycelium (common mycelial network) (Johnson and Gilbert, 2015). With highly evolved mutualistic relationship, arbuscular mycorrhiza fungi can penetrate the plant roots to form a branch structure to facilitate the nutrient transfer from the soil to the plants (Hodge and Storer, 2015; Miransari, 2010). Fungal rhizosphere diversity is much less studied because molecular tools for isolation and characterization of fungi have been developed much later than bacteria. Fungi endophytes have been considered as plant mutualists mainly by reducing herbivory via production of mycotoxins such as alkaloids (Saikkonen et al., 1998). In recent decades, fungi-based biopesticides have been extensively used as an alternative to traditional chemical-based pesticides because they are environmentally friendly, economical, and effective to some extent (Chandler et al., 2011).

15.2.2 BENEFICIAL BACTERIA

Bacteria are the most abundant microorganisms in the rhizosphere. Bacteria may cover up to 15% of the total root surface (Van Loon, 2007). Beneficial bacteria residing in the rhizosphere are called plant growth–promoting rhizobacteria (PGPR) consisting of genus such as *Pseudomonas*, *Bacillus*, *Arthrobacter*, *Rhizobia*, *Agrobacterium*, *Alcaligenes*, *Azotobacter*, *Mycobacterium*, *Flavobacter*, *Cellulomonas*, and *Micrococcus* (Tsavkelova et al., 2005).

It is reported that compare to gram-positive bacteria, gram-negative bacteria are predominant in the rhizosphere (Steer and Harris, 2000). *Pseudomonas* species have been reported as the most abundant gram-negative bacteria in soil (Teixeira et al., 2010). Moreover, recent studies reported that beneficial gram-positive bacteria of genus *Bacillus* associate with crops such as strawberry, oilseed rape, potato, wheat, and rice (Joshi and Bhatt, 2011; Rawat et al., 2011). The *Bacillus* can form endospores and produce antimicrobial substances that inhibit other bacteria in the rhizosphere. Biotic and abiotic are two major factors affecting rhizosphere bacterial diversity. The biotic factors include plant species, cultivars, age, development stage, and root characteristics (Smalla et al., 2001). Since specific bacterial strain responds to specific root exudates such as sugars, organic acids, and amino acids, plant species and their root exudates composition can be a controlling factor of the rhizosphere bacterial diversity (Kowalchuk et al., 2002). After microbial colonization around plant roots, different plant-microbial relationships may develop such as associative, symbiotic, and parasitic depending upon factors such as the nutrient status of the soil, soil environment, and plant defense mechanism (Parmar and Dufresne, 2011). Abiotic factors such as soil physiochemical conditions such as soil pH, soil salinity, soil organic matter content, and the presence of nutrient elements all affect the rhizosphere bacterial diversity (Garbeva et al., 2004).

15.3 FUNCTIONS OF BENEFICIAL MICROBES IN AGRICULTURE

15.3.1 Biofertilizer

15.3.1.1 Nitrogen fixation

N-based fertilizer is a major agrochemical input to promote crop growth. Nitrogen fixation microbial is able to use atmospheric N_2 to NH_3, a form that can be used by plants. Typically in rhizosphere enzymatic reaction, catalyzed nitrogenase system requires substantial quantities of energy in the specific form of ATP hydrolysis (Alberty, 1994).Symbiotic nitrogen-fixing bacteria include the cyanobacteria of the genera *Rhizobium, Bradyrhizobium, Azorhizobium, Allorhizobium, Sinorhizobium,* and *Mesorhizobium* (Tighe et al., 2000). Various types of of interactions occur between nitrogen-fixing bacteria and their host plants (Franche et al., 2009). Formation of root nodules on legumes is considered as the most efficient process for nitrogen fixation. Progress in the knowledge of the basic mechanisms underlying symbiotic and endophytic associations in legumes has been well studied (Zahran, 1999). A number of nonlegume plants have been studied to utilize nitrogen-fixing bacteria reducing deficiency (Franche et al., 2009). In recent years, a major break-through has been the demonstration of common genetic basis for plant root microbes, such as fungi, rhizobia, and *Frankia* bacteria, in both legumes and nonlegumes (Santi et al., 2013). The creation of artificial associations between nitrogen-fixing microorganisms and major nonlegume agricultural plants is a primary goal in sustainable agriculture to reduce the demand of chemical nitrogen fertilizers (Geddes et al., 2015). Recent advances in the understanding of nitrogen fixation with nonlegume plants may provide a new sustainable alternative for nonlegume crops.

15.3.1.2 Phosphate solubilization

Soil phosphate can be solubilized by rhizosphere microbes such as bacteria and fungi and become available to plant through various mechanisms. The most significant mechanism is the production of organic or inorganic acid and then lowering pH in the rhizosphere (Rodríguez and Fraga, 1999). The hydroxyl and carboxyl groups of these acids can chelate the cations (Al, Fe, Ca) associated with the phosphate, thus making phosphate soluble and available for plant uptake. It has been reported that the phosphate-solubilizing bacterial (PSB) strains have potential to solubilize 25–42 µg/mL phosphate and mineralize 8–18 µg/mL inorganic and organic phosphates, respectively (Guang-Can et al., 2008). Incorporation of PSB with single superphosphate and rock phosphate reduces the phosphate demand by 25% and 50%, respectively (Sundara et al., 2002). Research conducted by Zaidi et al. (2003) has shown that the application of PSB that converts fixed phosphate to solubilized form can increase crop yield up to 70%. Yazdani et al. (2009) also reported using coinoculation of PSB can reduce phosphate fertilizer application by 50% without affecting the corn yield. Ca phosphates can be solubilized by complexation effect, i.e., chelating with organic acids produced by soil bacteria (Yadav and Verma, 2012). However, the buffering capacity of the medium reduces the effectiveness of PSB in releasing calcium-bounded phosphate. Besides, acidification of the soil environment caused by organic acid produced by microbial can also help phosphate release from apatite by proton substitution or release of Ca^{2+}. In general, carboxylic anions with high affinity to calcium solubilize more phosphorus than acidification alone (Mohammadi, 2012). Carboxylic acids also play an important role in solubilizing Al phosphate and Fe phosphate. Carboxylic acids solubilize Al/Fe-bound phosphate through the direct dissolution of mineral phosphate due to anion exchange of phosphate by acid anion or by chelation of Fe and Al ions associated with phosphate, liberating phosphate available for plant uptake (Jones and Oburger, 2011).For organic ligands, their type, position, and acid strength govern their effectiveness in the phosphate solubilization. Phosphorus desorption potential of different carboxylic anions decreases with decrease in stability constants of Fe or Al organic acid complexes in the order: citrate > oxalate > malonate/malate > tartrate > lactate > gluconate > acetate > formate (Jones and Oburger, 2011).

PSB can also solubilize organic-bounded phosphate and play an important role in phosphate cycling in the farming system (Richardson and Simpson, 2011). Soil organic phosphate can be mineralized by microbial enzymes including acid phosphatases, alkaline phosphatases, and phytases (Sharma et al., 2013). It has been reported that most of the extracellular soil phosphatases are obtained from degradation of organic matter microbes (Sinsabaugh, 1994). Kim et al. (1998) reported *Enterobacter agglomerans* can solubilize phosphate hydroxyapatite and hydrolyze. In general, a mixed culture of PSBs (*Bacillus, Streptomyces, Pseudomonas*, etc.) is often more effective in mineralizing organic phosphates (Mohammadi, 2012).

15.3.2 BIOPESTICIDES

Microbial-based biopesticides typically contain beneficial microorganisms such as bacteria, fungi, viruses, or their metabolites. Naturally occurring microorganisms pathogenic to pest are identified, extracted, patented, mass-produced, and marketed as a microbial biopesticide. Based on the target pest group, biopesticides are categorized into biofungicides, bioinsecticides, bioherbicides, and bionematicides. Bioinsecticides are used against harmful insect pests such as aphids, grubs, and armyworms. One of the most common bioinsecticides is the bacterium *Bacillus thuringiensis* and its different subspecies (Bravo et al., 2011). Biofungicides are the microbial formulation that inhibits plant pathogenic fungi through antibiosis, parasitism, or rhizosphere competence. Application of biofungicide has become a successful strategy as a preventive method for growth media and seed treatment. Commercially available biofungicide product mostly contains *Trichoderma harzianum* and *Bacillus subtillis*, which controls plant pathogenic fungi of genus *Cylindrocladium, Fusarium, Rhizoctonia, Pythium, and Thielaviopsis* (Woo et al., 2014). Commercially available bioherbicides for weed control often contain three genera of fungi, i.e., *Colletotrichum* spp., *Chondrostereum* spp., and *Fusarium* spp. (Aneja, 2014; Bailey, 2014). Bionematicides such as *Bacillus megaterium, Trichoderma album, Trichoderma harzianum*, and *Ascophyllum nodosum* were used to control nematode (Radwan et al., 2012).

There is growing interest in biopesticides due to the adverse effect of synthetic chemical pesticides such as the development of pest resistance, an outbreak of secondary pest, soil and water pollution, and bioaccumulation in human and animal (Al-Zaidi et al., 2011). In general, biopesticides are host specific and nonpathogenic or nontoxic to nontargeted host. Thus, microbial pesticides can be sustainable alternatives to chemical pesticides.

15.3.3 BIOREMEDIATION

Bioremediation as promising strategies have gained increasing popularity in the past few decades for contaminants removal. Microorganisms associated with roots can facilitate the degradation of organic pollutants and transformation of toxic metals.

Degradation of organic contaminants in the rhizosphere using microbes involves the complex interactions of roots, root exudates, rhizosphere soil, and microbes that facilitate degradation of organics to nontoxic or less-toxic compounds (Kuiper et al., 2004). The success of rhizoremediation mainly depends on the colonizing efficiency and degradation capacity of microbes. Notably, rhizoremediation occurs naturally because roots secrete complex aromatic compounds such as flavonoids and coumarins; microflora consuming these aromatic compounds can also degrade aromatic contaminants with a similar structure such as PCBs or PAHs in the rhizosphere (Johnson et al., 2005). Rhizoremediation can be optimized by deliberate manipulation of the rhizosphere using suitable plant-microbe pairs. For example, a grass species incorporated with a naphthalene-degrading microbe that protects the grass seed from the toxic effects of naphthalene, and the growing roots propelled the naphthalene-degrading bacteria further into the soil which would not reach to that depth in the absence of roots (Germaine et al., 2009). In general, in situ rhizoremediation is more attractive than ex situ methods since it is easy to implement and often more economical than ex situ methods (excavation and incineration, off-site storage, soil washing) (Rohrbacher and St-Arnaud, 2016).

A recent study has shown that microbes in rhizosphere can form microbe-plant symbiosis and promote organic pollution phytoremediation efficiency. The rhizo-microflora secretes plant growth–promoting substances, siderophores, phytochelatins to alleviate metal toxicity and enhance the bioavailability of metals (phytoremediation) or complexation of metals (phytostabilization) (Kamaludeen and Ramasamy, 2008). Selection of right bacteria/consortia and inoculation to seed/ roots of suitable plant species will widen the perspectives of rhizoremediation. Wu et al. (2006) demonstrated that engineered rhizobacterium symbiosis with plant roots can significantly decrease cadmium phytotoxicity and increase cadmium accumulation in the plant root.

15.4 MOLECULAR TECHNIQUES IN THE STUDY OF BACTERIA IN THE RHIZOSPHERE

During the past few decades, a wide variety of molecular techniques have been developed and used as valuable tools for the study of diversity and function of rhizosphere bacteria (Lagos et al., 2015). However, the characteristics of each molecular technique must be considered and evaluated to better apply these techniques for rhizosphere bacteria study.

15.4.1 FUNCTIONAL GENE APPROACH

PCR (polymerase chain reaction) is a molecular technique for the analysis of DNA sequences by the amplification of selected DNA fragments; quantitative PCR (qPCR) with fluorescent detection technologies can record the accumulation of amplicons in "real time" during each cycle of the PCR amplification that allows the detection and quantification of those fragments and their gene expression. These techniques can provide important insight into the distribution of specific functional genes involved in relevant processes in the rhizosphere, such as nutrient cycling or pollution degradation. The accuracy of qPCR method is affected by several factors such as nucleic acid extraction efficiencies, amplification efficiencies, the concentration of RNA in rhizosphere sample, and the presence of PCR inhibitors (Schreiter et al., 2015). Besides, the lack of characterization of genes of interest or unavailability of the target sequence also limits the application of PCR-based technique.

15.4.2 MICROARRAY

Based on the nature of the probe used and target molecules used, three types of microarrays have been defined: community genome arrays (CGAs), rRNA-based oligonucleotide arrays (PhyloChips), and functional gene arrays (FGAs) (McGrath et al., 2010). CGAs analyze genomic DNA from environmental microbial communities (EMCs) and identify known microbes to species level; characterization of population changes of known microbial species is most suitable for this method. PhyloChips are designed with oligonucleotides that match rRNA sequences of specific microorganism taxa and reveal greater taxonomic diversity in environmental samples. Günther et al. (2006) designed a microarray containing 29 specific oligonucleotides that was applied to detect genus *Kitasatospora* belonging to actinomycetes in soil. FGAs are designed with probes targeting known genes and gene products with determined functions, for example, genes involved in nitrogen cycling, methane oxidation, sulfate reduction, and the degradation of pollutants (Tu et al., 2014). However, analysis of FGAs may be biased since it relies on known bacterial species genes, whereas unknown genes are not detectable (McGrath et al., 2010). A new method has been developed in which FGAs are constructed from randomly selected clones of cDNA libraries (Goel et al., 2017). The advantages of randomly selected cDNA library clones are that prior knowledge about functional genes, genomes, or species present in the environmental sample is not required, and FGA may provide less biased, culture-independent expression profiling of an entire microbial community (Fang et al., 2014).

15.4.3 Proteomics and Metaproteomics

Proteomic tools could be useful in gaining information about microbial community activity and to better understand the real interactions between roots and soil. Proteomics method is developed to analyze the entire protein complement expressed by a genome or by a cell, which can be used for evaluating expression and localization of proteins, as well as for analysis of posttranslational modifications (Wilkins et al., 1996). The typical proteomic analysis includes protein extraction and purification, tryptic digestion into peptides, protein or peptide separation, tandem mass spectrometry analysis, and comprehensive protein identification (Chandramouli and Qian, 2009). However, there are several major limitations of proteomic: the extraction of intracellular proteins from soil is a methodological challenge, due to the stability of proteins (protected against proteolysis) and the fact that they may be strongly adsorbed onto soil minerals or copurify with humic acids or soil colloids that will interfere with analysis of the complexity of biological structures and physiological processes. The quantity of data that are acquired with new techniques places new challenges on data processing and analysis. Metaproteomics provides a direct measure of proteins present in an environmental sample such as soil, offering information about the functional roles of soil microorganism, such as biogeochemical processes, degradation, or bioremediation processes (Lagos et al., 2015).

15.5 CONCLUSION

Beneficial bacteria are crucial in biogeochemical cycles and have been using sustainable agriculture management for decades. Rhizosphere beneficial bacteria are capable of promoting plant growth by colonizing the plant root and then reducing agrochemical input. Plant-bacterial interactions in the rhizosphere are the determinants for plant productivity and nutrient cycle in the agriculture system. Generally, bacteria play different roles to benefit plant growth: biofertilizer-promoting nutrients supply for plants, biopesticides lessening or preventing the plants from diseases, and bioremediates decrease soil contaminations. The advances in molecular and genomic techniques are crucial for the study of compositions and activities of soil microbial in the rhizosphere. However, there are still limitations to apply the molecular and genomic techniques for soil study, such as difficulties of extractions and purification of nucleic acids and proteins from complex environmental samples (Lagos et al., 2015). Advances in biotechniques such as expanding libraries DNA or database will continue to enable us to uncovering more genomes and functions of rhizosphere microbiota. Integrating beneficial bacteria into the agriculture system is needed for future sustainable developments, allowing the establishment of efficient nutrient management, benefitting the yield, and reducing agrochemical input.

REFERENCES

Abeer, H., Abd_Allah, E., Alqarawi, A., and Egamberdieva, D. (2015). Induction of salt stress tolerance in cowpea [Vigna unguiculata (L.) Walp.] by arbuscular mycorrhizal fungi. *Legume Research: An International Journal* **38**, 579–588.

Al-Zaidi, A., Elhag, E., Al-Otaibi, S., and Baig, M. (2011). Negative effects of pesticides on the environment and the farmers awareness in Saudi Arabia: a case study. *Journal of Animal and Plant Sciences* **21**, 605–611.

Alberty, R. (1994). Thermodynamics of the nitrogenase reactions. *Journal of Biological Chemistry* **269**, 7099–7102.

Aneja, K. R. (2014). Exploitation of phytpathogenic fungal diversity for the development of bioherbicides. *Kavaka* **42**, 7–15.

Bailey, K. L. (2014). The bioherbicide approach to weed control using plant pathogens. In D. P. Abrol (Ed.), *Integrated Pest Management: Current Concepts and Ecological Perspective*, pp. 245–266. Elsevier, San Diego, CA.

Behie, S. W., and Bidochka, M. J. (2014). Nutrient transfer in plant-fungal symbioses. *Trends in Plant Science* **19**, 734–40.

Bonfante, P., and Genre, A. (2010). Mechanisms underlying beneficial plant-fungus interactions in mycorrhizal symbiosis. *Nature Communications* **1**, 48.

Bravo, A., Likitvivatanavong, S., Gill, S. S., and Soberón, M. (2011). Bacillus thuringiensis: A story of a successful bioinsecticide. *Insect Biochemistry and Molecular Biology* **41**, 423–431.

Carvalho, F. P. (2017). Pesticides, environment, and food safety. *Food and Energy Security* **6**, 48–60.

Chandler, D., Bailey, A. S., Tatchell, G. M., Davidson, G., Greaves, J., and Grant, W. P. (2011). The development, regulation and use of biopesticides for integrated pest management. *Philosophical Transactions of the Royal Society B: Biological Sciences* **366**, 1987–1998.

Chandramouli, K., and Qian, P.-Y. (2009). Proteomics: Challenges, techniques and possibilities to overcome biological sample complexity. *Human Genomics and Proteomics: HGP* **2009**. 239204. doi:10.4061/2009/239204.

Clark, M., and Tilman, D. (2017). Comparative analysis of environmental impacts of agricultural production systems, agricultural input efficiency, and food choice. *Environmental Research Letters* **12**, 064016.

Fang, C., Xu, T., Ye, C., Huang, L., Wang, Q., and Lin, W. (2014). Method for RNA extraction and cDNA library construction from microbes in crop rhizosphere soil. *World Journal of Microbiology and Biotechnology* **30**, 783–789.

Franche, C., Lindström, K., and Elmerich, C. (2009). Nitrogen-fixing bacteria associated with leguminous and non-leguminous plants. *Plant and Soil* **321**, 35–59.

Garbeva, P. V., Van Veen, J., and Van Elsas, J. (2004). Microbial diversity in soil: selection of microbial populations by plant and soil type and implications for disease suppressiveness. *Annual Review of Phytopathology*. **42**, 243–270.

Geddes, B. A., Ryu, M.-H., Mus, F., Costas, A. G., Peters, J. W., Voigt, C. A., and Poole, P. (2015). Use of plant colonizing bacteria as chassis for transfer of N2-fixation to cereals. *Current Opinion in Biotechnology* **32**, 216–222.

Germaine, K. J., Keogh, E., Ryan, D., and Dowling, D. N. (2009). Bacterial endophyte-mediated naphthalene phytoprotection and phytoremediation. *FEMS Microbiology Letters* **296**, 226–234.

Goel, R., Suyal, D. C., Dash, B., and Soni, R. (2017). Soil metagenomics: A tool for sustainable agriculture. In V. C. Kalia, Y. Shouche, H. J. Purohit, P. Rahi (Eds.), *Mining of Microbial Wealth and MetaGenomics*, pp. 217–225. Springer, Singapore.

Guang-Can, T., Shu-Jun, T., Miao-Ying, C., and Guang-Hui, X. (2008). Phosphate-Solubilizing and-Mineralizing Abilities of Bacteria Isolated from Soils1. *Pedosphere* **18**, 515–523.

Günther, S., Groth, I., Grabley, S., and Munder, T. (2006). Design and evaluation of an oligonucleotide-microarray for the detection of different species of the genus Kitasatospora. *Journal of Microbiological Methods* **65**, 226–236.

Hodge, A., and Storer, K. (2015). Arbuscular mycorrhiza and nitrogen: implications for individual plants through to ecosystems. *Plant and Soil* **386**, 1–19.

Johnson, D., Anderson, D., and McGrath, S. (2005). Soil microbial response during the phytoremediation of a PAH contaminated soil. *Soil Biology and Biochemistry* **37**, 2334–2336.

Johnson, D., and Gilbert, L. (2015). Interplant signalling through hyphal networks. *New Phytologist* **205**, 1448–1453.

Jones, D. L., and Oburger, E. (2011). Solubilization of phosphorus by soil microorganisms. In E. K. Bünemann, A. Oberson, E. Frossard (Eds.), *Phosphorus in Action: Biological Processes in Soil Phosphorus Cycling*, pp. 169–198. Springer, Berlin and Heidelberg.

Joshi, P., and Bhatt, A. (2011). Diversity and function of plant growth promoting rhizobacteria associated with wheat rhizosphere in North Himalayan region. *International Journal of Environmental Sciences* **1**, 1135.

Kamaludeen, S., and Ramasamy, K. (2008). Rhizoremediation of metals: harnessing microbial communities. *Indian Journal of Microbiology* **48**, 80–88.

Kim, K. Y., Jordan, D., and McDonald, G. (1998). Enterobacter agglomerans, phosphate solubilizing bacteria, and microbial activity in soil: effect of carbon sources. *Soil Biology and Biochemistry* **30**, 995–1003.

Kowalchuk, G. A., Buma, D. S., de Boer, W., Klinkhamer, P. G., and van Veen, J. A. (2002). Effects of above-ground plant species composition and diversity on the diversity of soil-borne microorganisms. *Antonie Van Leeuwenhoek* **81**, 509.

Kuiper, I., Lagendijk, E. L., Bloemberg, G. V., and Lugtenberg, B. J. (2004). Rhizoremediation: a beneficial plant-microbe interaction. *Molecular Plant-microbe Interactions* **17**, 6–15.

Lagos, L., Maruyama, F., Nannipieri, P., Mora, M., Ogram, A., and Jorquera, M. (2015). Current overview on the study of bacteria in the rhizosphere by modern molecular techniques: A mini–review. *Journal of Soil Science and Plant Nutrition* **15**, 504–523.

Lambers, H., Mougel, C., Jaillard, B., and Hinsinger, P. (2009). Plant-microbe-soil interactions in the rhizosphere: An evolutionary perspective. *Plant and Soil* **321**, 83–115.

McGrath, K. C., Mondav, R., Sintrajaya, R., Slattery, B., Schmidt, S., and Schenk, P. M. (2010). Development of an environmental functional gene microarray for soil microbial communities. *Applied and Environmental Microbiology* **76**, 7161–7170.

Miransari, M. (2010). Contribution of arbuscular mycorrhizal symbiosis to plant growth under different types of soil stress. *Plant Biology* **12**, 563–569.

Mohammadi, K. (2012). Phosphorus solubilizing bacteria: Occurrence, mechanisms and their role in crop production. *Resources and Environment* **2**, 80–85.

Parmar, N., and Dufresne, J. (2011). Beneficial interactions of plant growth promoting rhizosphere microorganisms. In *Bioaugmentation, Biostimulation and Biocontrol*, A. Singh, et. al, (Eds). pp. 27–42. Springer, Berlin/Heidelberg.

Radwan, M., Farrag, S., Abu-Elamayem, M., and Ahmed, N. (2012). Biological control of the root-knot nematode, Meloidogyne incognita on tomato using bioproducts of microbial origin. *Applied Soil Ecology* **56**, 58–62.

Rawat, S., Izhari, A., and Khan, A. (2011). Bacterial diversity in wheat rhizosphere and their characterization. *Advances in Applied Science Research* **2**, 351–356.

Richardson, A. E., and Simpson, R. J. (2011). Soil microorganisms mediating phosphorus availability update on microbial phosphorus. *Plant Physiology* **156**, 989–996.

Rodríguez, H., and Fraga, R. (1999). Phosphate solubilizing bacteria and their role in plant growth promotion. *Biotechnology Advances* **17**, 319–339.

Rohrbacher, F., and St-Arnaud, M. (2016). Root exudation: the ecological driver of hydrocarbon rhizoremediation. *Agronomy* **6**, 19.

Saikkonen, K., Faeth, S. H., Helander, M., and Sullivan, T. (1998). Fungal endophytes: a continuum of interactions with host plants. *Annual review of Ecology and Systematics* **29**, 319–343.

Santi, C., Bogusz, D., and Franche, C. (2013). Biological nitrogen fixation in non-legume plants. *Annals of Botany* **111**, 743–767.

Schreiter, S., Eltlbany, N., and Smalla, K. (2015). Microbial communities in the rhizosphere analyzed by cultivation-independent DNA-based methods. In *Principles of Plant-Microbe Interactions*, B. Lugtenberg, (Ed). pp. 289–298. Springer International Publishing, New York.

Sharma, S. B., Sayyed, R. Z., Trivedi, M. H., and Gobi, T. A. (2013). Phosphate solubilizing microbes: Sustainable approach for managing phosphorus deficiency in agricultural soils. *SpringerPlus* **2**, 587.

Sinsabaugh, R. (1994). Enzymic analysis of microbial pattern and process. *Biology and Fertility of Soils* **17**, 69–74.

Smalla, K., Wieland, G., Buchner, A., Zock, A., Parzy, J., Kaiser, S., Roskot, N., Heuer, H., and Berg, G. (2001). Bulk and rhizosphere soil bacterial communities studied by denaturing gradient gel electrophoresis: plant-dependent enrichment and seasonal shifts revealed. *Applied and Environmental Microbiology* **67**, 4742–4751.

Steer, J., and Harris, J. (2000). Shifts in the microbial community in rhizosphere and non-rhizosphere soils during the growth of Agrostis stolonifera. *Soil Biology and Biochemistry* **32**, 869–878.

Stehle, S., and Schulz, R. (2015). Agricultural insecticides threaten surface waters at the global scale. *Proceedings of the National Academy of Sciences* **112**, 5750–5755.

Sundara, B., Natarajan, V., and Hari, K. (2002). Influence of phosphorus solubilizing bacteria on the changes in soil available phosphorus and sugarcane and sugar yields. *Field Crops Research* **77**, 43–49.

Teixeira, L. C., Peixoto, R. S., Cury, J. C., Sul, W. J., Pellizari, V. H., Tiedje, J., and Rosado, A. S. (2010). Bacterial diversity in rhizosphere soil from Antarctic vascular plants of Admiralty Bay, maritime Antarctica. *The ISME Journal* **4**, 989.

Tighe, S., De Lajudie, P., Dipietro, K., Lindström, K., Nick, G., and Jarvis, B. (2000). Analysis of cellular fatty acids and phenotypic relationships of Agrobacterium, Bradyrhizobium, Mesorhizobium, Rhizobium and Sinorhizobium species using the Sherlock Microbial Identification System. *International Journal of Systematic and Evolutionary Microbiology* **50**, 787–801.

Tsavkelova, E., Cherdyntseva, T., and Netrusov, A. (2005). Auxin production by bacteria associated with orchid roots. *Microbiology* **74**, 46–53.

Tu, Q., Yu, H., He, Z., Deng, Y., Wu, L., Van Nostrand, J. D., Zhou, A., Voordeckers, J., Lee, Y. J., and Qin, Y. (2014). GeoChip 4: a functional gene-array-based high-throughput environmental technology for microbial community analysis. *Molecular Ecology Resources* **14**, 914–928.

Van Loon, L. (2007). Plant responses to plant growth-promoting rhizobacteria. In P. A. H. M. Bakker, J. M. Raaijmakers, G. Bloemberg, M. Höfte, P. Lemanceau, B. M. Cooke (Eds.), *New Perspectives and Approaches in Plant Growth-Promoting Rhizobacteria Research*, pp. 243–254. Springer, Dordrecht.

Volpe, V., Chitarra, W., Cascone, P., Volpe, M. G., Bartolini, P., Moneti, G., Pieraccini, G., Di Serio, C., Maserti, B., and Guerrieri, E. (2018). The Association With Two Different Arbuscular Mycorrhizal Fungi Differently Affects Water Stress Tolerance in Tomato. *Frontiers in Plant Science* **9**.

Wilkins, M. R., Sanchez, J.-C., Gooley, A. A., Appel, R. D., Humphery-Smith, I., Hochstrasser, D. F., and Williams, K. L. (1996). Progress with proteome projects: why all proteins expressed by a genome should be identified and how to do it. *Biotechnology and Genetic Engineering Reviews* **13**, 19–50.

Woo, S. L., Ruocco, M., Vinale, F., Nigro, M., Marra, R., Lombardi, N., Pascale, A., Lanzuise, S., Manganiello, G., and Lorito, M. (2014). Trichoderma-based products and their widespread use in agriculture. *The Open Mycology Journal* **8**, 71–126.

Wu, C. H., Wood, T. K., Mulchandani, A., and Chen, W. (2006). Engineering plant-microbe symbiosis for rhizoremediation of heavy metals. *Applied and Environmental Microbiology* **72**, 1129–1134.

Yadav, B., and Verma, A. (2012). Phosphate solubilization and mobilization in soil through microorganisms under arid ecosystems. pp. 93–108. In M. Ali (Ed.), *The Functioning of Ecosystems*. InTech, Rajasthan.

Yazdani, M., Bahmanyar, M. A., Pirdashti, H., and Esmaili, M. A. (2009). Effect of phosphate solubilization microorganisms (PSM) and plant growth promoting rhizobacteria (PGPR) on yield and yield components of corn (Zea mays L.). *World Academy of Science, Engineering and Technology* **49**, 90–92.

Zahran, H. H. (1999). Rhizobium-legume symbiosis and nitrogen fixation under severe conditions and in an arid climate. *Microbiology and Molecular Biology Reviews* **63**, 968–989.

Zaidi, A., Khan, M. S., and Amil, M. (2003). Interactive effect of rhizotrophic microorganisms on yield and nutrient uptake of chickpea (Cicer arietinum L.). *European Journal of Agronomy* **19**, 15–21.

Zhang, X., Davidson, E. A., Mauzerall, D. L., Searchinger, T. D., Dumas, P., and Shen, Y. (2015). Managing nitrogen for sustainable development. *Nature* **528**, 51.

16 Detoxification of Biomedical Waste

Bamidele Tolulope Odumosu
University of Lagos

Tajudeen Akanji Bamidele
Nigerian Institute of Medical Research

Olumuyiwa Samuel Alabi
University of Ibadan

Olanike Maria Buraimoh
University of Lagos

CONTENTS

16.1   INTRODUCTION

The economic viability of any nation depends on its active workforce, which is directly proportional to the presence of industrial, agricultural, and infrastructural inputs that drive it. One effect of these inputs in a viable nation is a steady increase in the population growth, which will also require the establishment of more healthcare facilities such as hospitals, maternity homes, clinics, and diagnostic centers to take care of their health demands and maintenance. Healthcare facilities such as hospitals play a significant role as centers for treatments, prevention, and promotion of improving health conditions for the general public. The increasing number of healthcare facilities raises environmental concerns due to massive generation of biomedical wastes (BMWs), which must be properly addressed because of the risk of infectious microorganisms, toxic chemical wastes, and other potentially hazardous materials that are detrimental to public health. Some countries across the globe with sufficient financial resources and manpower to manage BMWs have taken practical steps to address the risk by introducing advance healthcare waste management systems. However, other countries with financial constraints are still struggling with how to manage their BMWs with limited measures. In the United Kingdom, the primary aim of healthcare waste management is to ensure it is properly handled, treated, and disposed of safely in a cost-effective manner that does not impact negatively on the environment and people. They do this by enacting key legislative laws and introducing waste hierarchy and disposal technologies. The waste hierarchy system provides a scale that ranks waste management options based on what is good or bad for the environment and with the best option at the top of the scale, whereas the least preferred option is at the bottom of the scale. In the United State, as part of efforts to rid the environment of hazardous BMWs, a microwave disinfection facility has been developed in California to enable the conversion waste to energy option for clinical wastes. US health facilities discard over 2 million tons of waste per year, of which unused equipment and devices constitute a larger fraction. Many of these US hospitals and other developed countries are fond of recycling or donating the scraps to the developing countries. According to a report by Manasi (2017), over 50% of hospitals in the United States send their single-use items to developing countries through reprocessors, who in turn resell them at relatively low prices after sterilization. Other countries such as Cambodia and the Philippines have improved on their legislation, which is directed to BMW management. In Malaysia, the government has implemented new strategies such as centralization and privatization of BMW disposal services all in a bid to ensure safe disposal of BMWs.

Nigeria, an example of a developing country with a unique nature where many health issues compete unfavorably with limited resources, does not give much attention to the management of BMWs. According to a Nigerian study on BMW management, it was revealed that there is lack of management commitment, poor waste handling practice, inadequate training of personnel on safety disposal practice, and practically nonexistence of BMW segregation of medical wastes (Abahand and Ohimain, 2011). According to another Nigerian surveillance study carried out by Olukanni et al. (2014), it was gathered that waste handlers do not carry out pretreatments of medical waste prior to disposal because some of the workers felt there is no need for pretreatment, whereas others felt handicapped by the cost of setting up pretreatment facilities. Oyekale and Oyekale (2017) observed there was low compliance with standard BMW management because of poor staff training and lack of requisite equipment for proper treatment and management of BMWs. The overall observation for Nigerian approach to BMW management is lack of awareness and specific healthcare policies that ensure that some of these hazardous wastes are properly disposed of, and consequently, an unsustainable waste management practice will be practiced. This is contrary to other developing countries such as India, Malaysia, and China where there is high sensitization of the general public on dangers of improper disposal and management of BMWs and government policies already enacted to law to safeguard the environment and the people.

Unsustainable waste BMW management can result in health hazards in many different ways. Firstly, it has a direct impairment of human and animal health in the environment. Secondly, it results in damages to the ecosystem or living organism attached to the ecosystem that can lead to economic losses. Thirdly, it can lead to loss of biodiversity of organisms in an ecosystem. Details about this shall be discussed in full in Section 16.5.3 of this chapter. Detoxification of BMW especially by microbial hydrolytic enzymes offers great potential because detoxification of BMW is rarer compared with conventional bioremediation of environmental wastes and pollutions. The latter is based on bioremediation by hydrolytic enzymes that are produced by bacteria for more environmentally friendly products. Detoxification of BMW can be done by either breaking it down, absorbing, or incorporating it into a nontoxic molecule. Several other molecular methods based on direct isolation and analysis of nucleic acids, proteins as well as other metabolites from environmental wastes have been previously reported (Ali et al., 2017). Several organisms have been discovered or engineered with the ability of detoxifying specific hazardous wastes. Microorganisms such as *Pseudomonas aeruginosa*, *Rhodococcus chlorophenolicus*, *Phanerochaete chrysosporium*, *Achromobacter* sp., and *Alcaligenes denitriticans* have been involved or used for detoxifying culture by a selection process as similar to the process of site cleanup. However, these organisms have limitations of use because most of them can only metabolize toxic substances in an aqueous medium, but many BMWs are difficult to dissolve in water.

Some of the challenges experienced during these processes include a decrease in efficiency such that after prolonged exposure to toxic wastes, they were able to degrade less than 50% of the waste. Another challenge is the death of microbial cultures or deactivation of the organism. Some organisms that are used for these processes grow poorly or die in the presence of hazardous BMWs other than those that are capable of degrading. For instance, *Eichornla crassipes*, a water hyacinths known to tolerate acids and many organic compounds, is known to be poisoned by pesticides, whereas *P. chrysosporium* has the capacity to degrade wide varieties of toxic substances without being affected.

16.2   CLASSES OF BIOMEDICAL WASTE

BMWs that share a lot of similarities with healthcare, medical, clinical, biohazardous, regulated medical, and infectious waste are as diverse as the different arms of biomedicine. Specific waste is generated in each arm although there will be some common wastes to all. On a broader sense, these wastes are generated from places such as hospitals and other health facilities, laboratories and research centers, mortuary and autopsy centers, animal research and testing laboratories, blood banks and collection services, and nursing homes for the elderly (WHO/UNICEF, 2015). The specifics are highlighted as follows and in Table 16.1.

TABLE 16.1

Classes of Biomedical Waste and Composition

SN	Class of Waste	Composition
1	Pathological and Anatomical	Human tissues/organs, body parts, biopsies, hair, teeth, nails
2	Human blood and products, body fluids, other potentially infectious materials (OPIMs)	Human blood, serum, plasma, platelets, semen, vaginal secretions, synovial, cerebrospinal, pleural, pericardial, peritoneal, sputum, saliva, amniotic fluids, urine, faeces, menstrual blood, contaminated urinary catheters, sanitary napkins
3	Microbiological	Bacterial, viral, parasite cultures, stocks of infectious agents, culture dishes, inoculating loop, specimens, vaccines (live, attenuated), cotton swabs, antibiotics, pipette tips, hand gloves
4	Sharps	Hypodermic needles, glassware pipettes, capillary tubes, razor blades, slides, coverslips, scalpel blades, syringes, broken bottles/glassware, lancet, wire gauze
5	Isolation waste	Materials from highly communicable infectious diseases such as Ebola viruses, Lassa fever, Monkey pox, and Marburg
6	Animal	Carcasses, body parts, beddings, animal cage/house
7	Radioactive	Radiotherapy and laboratory research liquids
8	Pharmaceuticals	Unused/expired vaccines, drugs, antibiotics, injectable, pills
9	Chemical	Disinfectants, solvents, heavy metals, batteries
10	Genotoxic	Ethidium bromide, drugs for cancer treatment
11	Others	Stationeries, paper towel, equipment

16.2.1 PATHOLOGICAL AND ANATOMICAL WASTE

This includes those human parts, such as tissues, organs, body parts, biopsies, hair, teeth, and nailset cetera, which have been removed either during surgical procedures or for special diagnosis. The anatomical waste refers to those recognizable parts that form the whole body.

16.2.2 HUMAN BLOOD AND PRODUCTS, BODY FLUIDS, OTHER POTENTIALLY INFECTIOUS MATERIALS

All human bulk body fluids might be from the blood such as serum, plasma, platelets. The human seminal fluids, vaginal secretions, and other fluids such as cerebrospinal, synovial, pleural, pericardial, and amniotic fluids are in this category. Others are saliva, sputum, urine/feces, etc.

16.2.3 MICROBIOLOGICAL

This can have some of the human products as listed in all the classes, but specifically, all bacterial, viral, parasitic cultures, stocks of infectious agents, Petri dishes, inoculating loop, samples, vaccines whether living or attenuated, cotton swabs, antibiotics, pipette tips, and hand gloves all constituted wastes in this category.

16.2.4 SHARPS

These are all objects with pointed ends that can puncture or cut the skin or other materials (Practice Greenhealth 2018, MedPro 2018); the hypodermic needles, broken glassware, capillary tubes, razor blades, microscopic slides, coverslips, lancet, and wire gauze are categorized as sharps.

16.2.5 ISOLATION WASTE

In the events of certain outbreaks of highly communicable diseases, materials from the infected patients, for instance, who have been infected with Lassa fever, Marburg, Monkey pox, and Ebola viruses, are treated as isolation waste (Practice Greenhealth 2018; CDC, 2003).

16.2.6 ANIMAL

All materials of animal origin especially experimental are under this category. This includes carcasses, body parts, beddings, droppings/dungs/feces, cage, or motorized house.

16.2.7 PHARMACEUTICALS

These are unused or expired vaccines, drugs, antibiotics, injectable, and pills.

16.2.8 2 RADIOACTIVE AND GENOTOXIC WASTES

Radiotherapy and some laboratory research liquids are radioactive, whereas ethidium bromide and drugs for cancer treatment have genotoxic effects.

16.2.9 CHEMICALS

All disinfectants, solvents used in the laboratory or another unit within medical facilities, are in this category. Also included are heavy metals and batteries.

16.2.10 LIQUID BIOMEDICAL WASTE

The liquid from human sources, laboratory, surgical/operating rooms, clinics, etc. such as blood, spinal fluids, saliva, dialysis waste, amniotic fluid, other bodily secretions, laboratory cultures, medications, and containers of liquid waste, which when pressed will release contents, all constitute liquid/fluid wastes. Also are buffers used in running ethidium bromide–impregnated agarose gel electrophoresis, reagents, and reconstituted enzymes in the molecular laboratory.

A large portion of surgical waste is liquid in nature; this is usually collected in disposable plastic suction canisters (Figure 16.1), which, studies have shown, comprise 25%–40% of regulated or liquid medical waste in the hospital (Mathias, 2004). This type of waste is disposed of in either of two ways: first, the health worker manually opens the suction canisters and empties them into a drain, which goes off into municipal or public sewer system, or second, by addition of solidifier that, by its constituents (chlorine, glutaraldehyde), also constitutes a waste. The solidifier has been reported not to be entirely solid and that it can still splatter if the container is dropped, although it has been demonstrated that a-2-min solidification can be achieved (Barlow, 2004). Whether they solidify or not, glutaraldehyde, for instance, causes illnesses such as throat/lung irritation, headache, nausea, nose bleeding, burning eyes, wheezing, and asthma (CDC-NIOSH, 2001).

16.2.11 SOLID BIOMEDICAL WASTE

This refers to the tangible elements in BMWs that can be held and does not flow in its container. These are culture media, expired antibiotics, contaminated personal protective equipment, sharps, pipette tips, glass wares, etc. The treatment of solid waste is quite different from liquid, and this also is dependent on individual waste categories. The sharp containers and all the contents inside, for instance, cannot be removed by unauthorized persons for the safety of staff, patients, and largely the environment. They are rather collected by a licensed medical waste transporter or vendor who has the legal responsibility of decontaminating them by either incineration or steam sterilization.

FIGURE 16.1 A disposable suction canister fitted with aerostat filter for collection of liquid biomedical waste. The two outlets at the top are connected to a suction to extract liquid waste.

The methods of disposal differ depending on settings, but the factors generally considered are patients, healthcare worker, and environmental safety. Some medical facilities adopt processes such as the provision of readily available small and large red bags, at the operating room (OR) and soiled utility area where designated service provider picks up, sterilizes, and shreds before final disposal. At some other facilities, wastes are collected at the point-of-use site such as bedside, examination room, or suite, et cetera. The containers are removed when full, or at set intervals. These are taken into a soiled utility room and kept in a motorized container with a lid. The containers are thereafter taken at a predetermined time to the containers of a regulated BMW disposer.

16.3 TOXIC SUBSTANCES ASSOCIATED WITH BIOMEDICAL WASTE

Among the 10 categories of BMWs that have been stated, discarded medicines and cytotoxic drugs, as well as chemical wastes generated during production of biological and disinfection, are substances that could serve as potential toxic agents when disposed inappropriately into the environments. Toxic substances such as dioxins and furans may also be released into the environment through the fume emitted from incinerated BMWs particularly the waste sharps (used and unused syringes, cannulas, scalp vein needle, etc.) (WHO, 2016) and solid wastes (tubing, catheters, infusion/blood giving sets, etc.). Materials from incomplete combustion, dust particles, sulfur dioxide, nitrogen oxide, carbon monoxide, hydrogen chloride, and heavy metals such as mercury, cadmium, and lead are some other pollutants that may, through incineration of certain BMWs, find their way to the environment (WHO, 1999). Because of the various challenges being faced by physicians with regard to the failing of conventional chemotherapeutic agents used in the treatment of most infections and diseases, there has been a consistent increase in the use of newly developed potent chemotherapeutic agents, and this has consequently increased the level of pharmaceutical wastes. Inappropriate disposal of pharmaceutical wastes (used and unused medicines including cytotoxic drugs) by healthcare practitioners could have either directly or indirectly serious environmental health consequences on humans, animals, and even aquatic creatures. Management of pharmaceutical wastes is highly complex because of the presence of certain constituents of the pharmaceutical products. Healthcare practitioners usually during the cause of their training do not receive any formal training on the management of hazardous BMWs. The BMW managers, on the other hand, do not have the knowledge of the various constituents that make up the pharmaceutical wastes and thus mismanage the disposal of these drugs (Zile et al., 2001; Tarbish, 2005).

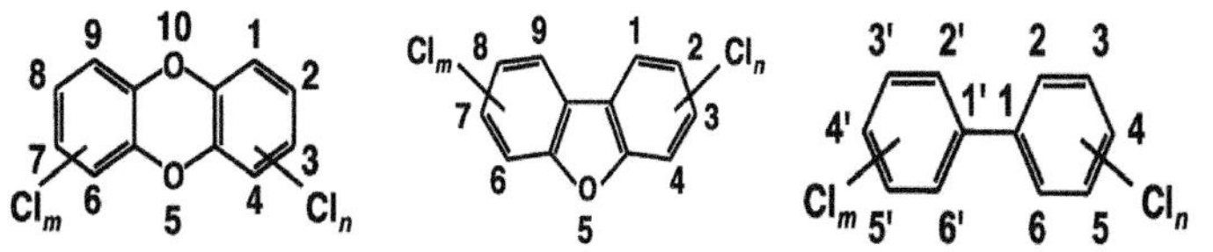

FIGURE 16.2 The parent structures of polychlorodibenzo-p-dioxin, polychlorodibenzo-furan, and polychlorobiphenyl. (https://openi.nlm.nih.gov/detailedresult.php?img=PMC4159838_ijms-15-14044-g001&req=4.)

Inui et al 2014. (Inui H., Itoh T., Yamamoto K., Ikushiro S.-I., Sakaki T. Mammalian cytochrome P-450-dependent metabolism of polychlorinated dibenzo-p-dioxins and coplanar polychlorinated biphenyls. International Journal of Molecular Sciences. 2014;15(8):14044–14057. doi: 10.3390/ijms150814044)

16.3.1 COMMON TOXIC SUBSTANCES FROM BIOMEDICAL WASTES

16.3.1.1 Cytotoxic Substances

These are substances used in the treatment of malignant and related health conditions. Some of these drugs include methotrexate, cytarabine, vinblastine, vincristine, etc. They act by destroying the rapidly growing malignant cells but themselves can cause cancer (carcinogenic), genetic mutation (mutagenic), and fetal anatomical or functional defects (teratogenic) in healthy subjects exposed to them either deliberately or accidentally (Gambrell and Moore, 2006; NSWH, 2008).

16.3.1.2 Other Pharmaceutical Substances

These are wastes generated through unused, leftover, unwanted, or expired medications and needed to be disposed of the hospitals or homes. Several factors, such as inappropriate donations, inadequacies in stock management and distribution at the dispensary, and erroneous purchase of substandard or misbranded drugs, contribute to the increase in pharmaceutical wastes particularly in hospitals. Table 16.2 shows the list of common drugs that are improperly discarded in the environment. This consists of mainly expired and leftover drugs that are discarded along with hospital waste and sometimes improperly discarded in the environment.

16.3.1.3 Dioxins

Dioxin is a term often used for the family of compounds that are structurally and chemically related to the polychlorinated dibenzo para dioxins (PCDDs) and the polychlorinated dibenzo-furans (PCDFs) including certain dioxin-like polychlorinated biphenyls (PCBs). Dioxins (tetra-chlorodibenzo para dioxin, heptachlor dibenzo-p-dioxin, etc.) belong to a group of environmental chemicals known as persistent organic pollutants (POPs) (WHO, 2016) (Figure 16.2).

Dioxins are usually by-products of industries, volcanic eruptions, and burnt forest residues. Some manufacturing processes such as smelting, herbicides and pesticides production and bleaching of paper pulp with chlorine contribute immensely to the release of dioxins to the environment. Also, the uncontrolled incineration of some solid and hospital wastes results in the production of dioxins due to

TABLE 16.2

Pharmaceutical Products (Drugs) Commonly Found among Disposed of Wastes

Type of Drug	Common Names and Examples	Places Found
Antibiotics	Tetracycline, ampicillin, amoxicillin, penicillin, streptomycins, metronidazole	Dumpsites, farmlands, hospital waste
Analgesic	Paracetamol, ibuprofen, aspirins, codeine	Dumpsites, hospital waste
Anti-malaria	Chloroquine, amodiaquine, lumefantrine	Dumpsites, hospital waste
Other drugs	This includes vitamins, antidepressant drugs, diabetics, and so on	Dumpsites and hospital wastes

incomplete burning of the wastes. Dioxins are chemically stable and thus have long half-life within the fatty tissues of the body where they are usually stored. They are known throughout the world to accumulate in the food chain through which they accumulate in the fatty tissues of humans and animals. They are highly toxic to most organs of the body and may cause serious alterations to the reproductive, immune, and hormonal systems. They could also compromise the developmental stages in humans and animals and are usually carcinogenic in nature when there is long-term exposure to them. Short-term exposure of high-level dioxins to humans may cause skin lesions and liver function alteration. Pregnant women are at high risk of alteration in fetal development when exposed to dioxins as well as newborn.

### 16.3.1.4	Heavy Metals

These are elements that have higher density compared with the other metals. Examples are lead, mercury, cadmium, iron, and arsenic. These metals are usually present in a small amount within the environment, food chain, water, etc. and are known to be beneficial to human and animal health when taken as a trace element in diet and drinking water. However, consumption of a large amount of these metals may result into acute or chronic toxicity such as damaged or reduced mental and central nervous function, damage to blood cells, lungs, kidneys, liver, and other vital organs of the body. Clinical presentations associated with long-term exposure to these metals are characterized by slow and progressive physical, muscular, and neurological degeneration, which could manifest to Alzheimer's and Parkinson's diseases, muscular dystrophy, multiple sclerosis, allergies, and cancer.

The increase in the demand for pharmaceuticals has an indirect effect on the level of pharmaceutical wastes being generated, most of which contain high quantity of these heavy metals that are being discharged into the environment usually as either untreated or poorly treated pharmaceutical effluents. The pharmaceutical effluents containing the different heavy metals are usually discharged into water bodies and undergrounds where they may come in contact with animals and humans and thus indirectly accumulate in the system through the food chain.

### 16.3.1.5	Inorganic Substances

Inorganic compounds such as sulfur dioxide, carbon monoxide, and hydrogen chloride are common emissions from the incineration of BMWs. Sulfur dioxide is a toxic gas with a characteristic sharp, choking smell like that of a burnt match, and it is produced from the burning of sulfur-containing materials. Sulfur dioxide gas could dissolve in rainwater to make it more acidic than normal (acid rain). Inhalation of the gas may be associated with increased respiratory disorders and diseases, difficulty in breathing, and eventually premature death. Carbon monoxide is a toxic gas with no characteristic smell generated from the combustion of carbon-based materials. The gas reduces the amount of oxygen carried by hemoglobin in the red blood cells and thus results in oxygen supply to the body, especially some vital organs such as the brain, heart, and the nervous system. The health hazard as a result of carbon monoxide inhalation includes headache, dizziness, nausea, vomiting, body weakness, blurred vision, and confusion. Exposure to the gas at moderate to high level for a long period of time has been linked with an increased risk of heart disorders. Inhalation at high concentration may result in loss of consciousness and eventual death without warning. The main source of hydrogen chloride in BMWs is plastic when combusted. Hydrogen chloride is a colorless, nonflammable, corrosive gas with a characteristic pungent odor. It is not absorbed through the skin but dissolves in moisture to form hydrochloric acid, a highly corrosive solution that can cause irritation and burns. The waste stream from X-ray units has chemical contamination of silver bromide (Faxon), glutaraldehyde, hydroquinone, and potassium hydroxide. The sterilization of dialysis units, operation theaters, and private wards contributes formaldehyde to the waste.

### 16.3.1.6	Radioactive Substances

The use of radioactive isotopes, in the diagnosis and therapy of various diseases, is on the increase, and this has eventually contributed to the increase in the effluent of radioactive wastes generated from health facilities. Some of the commonly used radioactive compounds are isotopes of iodine (I-131,

I-125, I-123), which are commonly used in the diagnosis of thyroid function and treatment of hyper-thyroidism; fluorine (F-18), which is a good positron-emitting radioisotope for the development of radiopharmaceuticals for positron emission tomography (PET), a powerful nuclear medicine imaging technique; and carbon (C-14), which is used as a tracer in medical test and to date organic material.

The radioactive wastes are usually liquid with some being solid or gaseous in nature. Medical devices such as needles, syringes, cotton swabs, absorbent materials, and hand gloves contaminated with these radioactive substances constitute the solid radioactive wastes, which are expected to be disposed of in a manner to ensure minimal exposure to the environment within the recommended safe limits (ICRP, 1995; Murthy, 2000).

16.4 BIOMEDICAL WASTE MANAGEMENT PRACTICE AND REMOVAL

Sustainable waste management practice became imperative in the face of hazardous pollution incidents and public concerns about inadequate disposals and legislation associated with incorrect waste management practices. New regulatory frameworks that deal with untenable waste management practice have since been introduced by many national and regulatory agencies (Pruss et al., 1999; Guisty, 2009). These are based on most environmentally friendly waste prevention/minimization, recycling, or composting methods. Landfilling, composting, incineration, and recycling are the most adopted practices in the United States, France, Japan, UK, and Italy. In clinical terms, BMW management requires extra attention because of its potential risk on public health. The World Health Organization (WHO) prescription on BMW disposal method emphasized on the cost-effectiveness, implementation, and the preservation of the environment (WHO, 2018). This has become of necessity because of the presence of a viable pathogenic microorganism in BMWs, which must be properly inactivated. Hence, the management of BMWs must ensure that the environment is free from infectious diseases coming from the disposed of wastes. Several technologies have been proposed and introduced for the purpose of BMW management but are unable to adequately ensure total inactivation of microorganisms; hence better, efficient methods and strategies of minimizing pathogenic microorganisms in BMWs are needed.

16.4.1 NONINCINERATOR METHOD OF BIOMEDICAL WASTE MANAGEMENT

Management of BMWs by the use of incinerator can cause significant health and environmental hazards. Incineration of medical wastes releases toxic substances such as dioxins, furans, heavy metals, fine particles, as well as gaseous impurities such as carbon monoxide, hydrogen chlorides, nitrogen oxide, and products of incomplete combustion into the environment. These compounds generated by incineration have serious and highly complicated health effects on humans, animals, and plants within the environment. While some of these compounds have been found to be carcinogenic to humans and animals, some can cause serious complications to the biological systems and organs of the body such as the hormonal, respiratory, and reproductive systems as well as several metabolic activities within the body.

The installation of certain devices to the incinerator in order to reduce the emission of these toxic gaseous substances could be effective in preventing the contamination of the atmosphere, but this will usually result in an increase of these pollutants in the solid waste phase, as the gaseous phase will be transformed to the solid phase. Furthermore, the high cost of installation of such devices and the level of efficiency of filters in capturing fine particles pose serious disadvantages to the use of incinerator. Ultrafine particles with a size below 1 μm are hardly captured by the filter. The ultrafine particles are usually highly reactive and could cause a serious health effect on humans and animals when inhaled. Hence the development of a nonincinerator method of disposing BMWs is of high advantage and safer compared with the incineration method.

The development of the nonincinerator method of BMW disposal encompasses a robust strategic framework that cut across various aspects of medical waste management with the aim of ensuring maximum environmental and occupational safety as well as economic benefits. Before now, all wastes generated within the different departments of the hospital (wards, pharmacy, operation room, kitchen, etc.) were usually not sorted out but were lumped together to be destroyed in the incinerator. However, the concern for the protection of the environment, public health issues, and cost reduction in the management of medical waste by the nonincineration techniques requires a new framework for handling the wastes being generated within the health facilities, particularly the medical wastes. The main components of this strategic framework are waste minimization and segregation. Waste minimization entails the reduction to the bearest minimum, of wastes that are potentially hazardous and must be disposed of properly. It is of great importance to minimize waste as much as possible, as this will save cost, reduce liability, and enhance regulatory compliance and occupational and environmental safety. Some waste minimization techniques include reduction at source, products that can result in hazardous waste, segregation of similar waste types into different collection units, treatment of wastes for reuse, and recycling of wastes.

Nonincineration methods of BMWs can be categorized in so many ways, but the most exclusive category is that based on the techniques for decontamination of the wastes. The different decontamination techniques for BMWs include thermal, chemical, radiation, and biological decontamination.

16.4.1.1 Thermal Decontamination of Biomedical Wastes

This process utilizes thermal energy for the decontamination processes and can be low, medium, or high heat thermal processes. However, medium and high heat thermal processes are known to cause chemical breakdown, support pyrolysis, and combustion of organic and inorganic materials, which may lead to the production of hazardous gaseous emission similar to that released during incineration process. Hence, the use of medium- and high-heat thermal processes is not appropriate for the decontamination of BMWs.

Low-heat thermal process involves the application of heat within the range of 90°C–180°C for the decontamination of selected medical wastes that may be a potential source of pathogenic contamination to the environment. In low-heat thermal process, the heat could be delivered through steam (wet heat) or through dry heat (hot air).

16.4.1.2 Wet Heat Thermal Process

Steam autoclave has been considered as one of the efficient BMW management processes and has been receiving considerable attention. Autoclave has been a method of sterilization of different medical equipment for more than a century. They are generally used for the sterilization of items contaminated with organic matters and waste generated from treatment wards and surgical rooms such as bandages, blood, gauze, and other nonchemical laboratory wastes. The autoclave of BMWs is more costly in comparison with the incineration method even though incineration has been dismissed by many developing countries because of the hazardous products that are released from its uses. Autoclaved waste often requires another treatment after the initial process; hence it has been considered a double treatment option of BMWs and cost. The process of decontamination of waste through steam by wet heat is carried out under specific temperature and time (115°C for 30 min or 121°C for 15 min), and at a specific pressure. The autoclave must be dedicated for this purpose solely. For gravity flow autoclave, temperature should be 121°C, 135°C, or 149°C, and pressure of 15, 31, or 52 lb/in.^{2} (psi) for a residence time of not less than 60, 45, or 30 min, respectively, is required. For a vacuum autoclave, the temperature should not be less than 121°C or 135°C, and a pressure of 15 or 31 psi per an autoclave residence time of not less than 45 or 30 min, respectively, is required.

Aside from the fact that it cannot handle large-size wastes because of their mass and other characteristics, autoclave use is restricted especially for certain wastes in the category of hazardous, e.g., radioactive wastes, chemotherapy treatment residues, volatile organic compounds, and the related. Autoclaving of liquid BMWs is more efficient for inactivation of infectious microbes that can be readily transmitted to humans by aerosols that are present in the waste generated in the hospital before final disposal.

16.4.1.3 Microwaving

Heating through steam can also be done in a microwave, which involves the use of direct delivery of heat energy to waste materials. This technology works best on semisolids and liquids in order to allow moisture heat to penetrate deeper and sterilize (Bélanger et al., 2008). Microwave treatment is not suitable for certain waste such as cytotoxic or radioactive wastes, animal carcasses that may have been contaminated, body parts, and other large metal items because they cannot be altered physically by these methods. To achieve better results, medical wastes are shredded into smaller bits and moisturized to achieve the desired effect during sterilization. Regarding the ability to destroy pathogens by its sterilization power, convincing evidence has shown that specially constructed microwave system is capable of inactivating bacteria and their spores through the action of moisture and low heat (Cha and Carlisle, 2001). Some of the major benefits of using microwave technology for waste management are the elimination of long heating periods, reduction of energy loss to the environment, and economic competitiveness. However, the methods are not appropriate for large-scale treatment of BMWs; they are not cost-effective and may not be affordable by most developing countries.

16.4.1.4 Dry Heat Thermal Process

The process of decontamination of medical waste by hot air is carried out in a hot air oven usually at a temperature between 130°C and 180°C for 1–2 h. In dry heat process, water is not added to the medical waste, and heat is generated through an electric heater surrounded by a good metal conductor of heat and the hot dry air circulated within the oven by a rolling fan. In some cases, heat is generated through infrared or any other radiation.

16.4.2 SUPERCRITICAL CARBON DIOXIDE (Sc-CO$_2$)

This is a nonthermal sterilization technique used extensively for the inactivation and removal of viable microorganism for commercial purpose, especially in food and pharmaceutical industries. The technique works on the principle of solubility and diffusion of pressure through the microbial cell membrane. It avoids the use of high temperature during extraction since many supercritical fluids are best kept at low temperature in order to preserve the integrity of such products under study. Supercritical carbon dioxide (Sc-CO$_2$) is environmentally friendly because it is nontoxic although inflammable (Zhang et al., 2008). This method is highly effective for the inactivation of infectious organisms and their spores (Banana, 2013); it is based on three principles in combination: carbon dioxide, temperature, and pressure that are all critical for the growth and development of many living cells. Sc-CO$_2$ has been demonstrated to be effective methods for remediation of high solvent loads in the environment, e.g., toluene, IPAc, MTBE, etOAc, and so on. The major advantage of this method is in the recovery of the CO$_2$ for reuse and further increasing the preservation of the environment and the green chemistry.

16.4.3 CHEMICAL DECONTAMINATION OF BIOMEDICAL WASTE

This method involves the use of chemical disinfectants such as sodium hypochlorite, peracetic acid, dissolved chlorine dioxide, hydrogen peroxide, sodium hydroxide solution, dry inorganic chemical, and ozone to disinfect the medical waste before disposal depending on the waste's characteristics. Catalytic oxidation and hydrolysis of tissues in heated stainless steel tanks using alkali are novel methods of decontaminating BMWs. For solid BMWs, it may require that the wastes are first shredded before mixing with the disinfectant so as to increase the area of contact.

16.4.4 IRRADIATION PROCESS OF DECONTAMINATION OF BIOMEDICAL WASTES

This involves the use of radiations of a particular wavelength and frequency to decontaminate BMWs. Radiations such as electron beams, ultraviolet, and ionizing radiation are usually employed in the decontamination of the BMWs in an enclosed chamber. The efficacies of the radiations depend on the density of the waste and the energy of the radiations. Because of the hazard associated with radiations, operators must be shielded from the radiation when in operation. Also because the BMWs are not altered physically by the radiation, they should be shredded before final disposal.

16.4.5 Biological Method of Decontamination of Biomedical Wastes

This method involves the use of biological enzymes in the treatment of BMWs. The enzymes reaction may not only destroy the pathogenic microorganisms but also destroy all the organic constituents leaving only the inert materials as residues.

16.4.6 Mechanical Destruction of Biomedical Waste

This involves an alteration in the physical characteristics of the solid BMWs by mechanical processes such as shredding, grinding, milling, mixing, and compaction. This method does not decontaminate the solid BMWs but renders it unrecognizable and thus usually is used to supplement other treatment processes. Mechanical destruction of needles and syringes helps to prevent accidental injury to the waste managers during collection and separation for disposal.

The use of mechanical processes such as shredding and mixing during thermal or chemical treatment of BMWs helps in the circulation of heat in the thermal process. It also increases the exposure of more surfaces of the wastes to the disinfection process of the chemical agents and thus enhances the efficiency of the decontamination processes.

Except in an enclosed system where mechanical process and any other decontamination process will take place simultaneously, the mechanical process should be used after the BMWs have been decontaminated to avoid unintentional or accidental transfer of the wastes' pathogens to the environment and the operators.

16.5 MICROBIOLOGY OF BIOMEDICAL WASTE

Until recently, hospital wastes were usually disposed of without any prior treatment to remove any infectious agents that may be present within the waste. They are often disposed of at locations where they get mixed up with municipal solid waste usually in uncontrolled or illegal landfills within a residential area. Some of these BMWs such as used needles, syringes, bandages, cotton wool, gauzes as well as amputated human parts, placenta, wastes from dialysis equipment, and reagents and materials for diagnosis in the microbiology laboratory may all be sources of infections, which are hazardous to the public health. A discarded medical product such as blood products serves as a significant template for the proliferation of hazardous diseases. Wastewater from sterilization of surgical instruments and those from the analysis in the laboratory usually harbor infectious substances and cells. Furthermore, unhygienic conditions in general ward toilets, coupled with the unprofessional activities of the health facility cleaners, create what is the virtually secondary route of infectious diseases within the hospital premises. Of major concern is the transmission of viral infections such as human immunodeficiency virus (HIV), enteroviruses, and hepatitis A, B, or C viruses as well as other bacterial (*Vibrio cholerae*, pathogenic *Escherichia coli*, *Salmonella typhi*, *Shigella* spp., *Clostridium tetani*, *Staphylococcus* spp., *Streptococcus* spp., *Pseudomonas* spp., *Borrelia* spp., etc.), fungal (dermatitis, etc.), and parasitic (*Giardia lamblia*, *Wucheraria bancrofti*, etc.) infections from uncontrolled and illegally disposed hospital waste to animals and humans. Whatever the amount of hospital waste being disposed of, it proves to be harmful to the community and thus needs immediate treatment and effective disposal to prevent outbreaks of infections (NACO, 1998).

16.6 IMPACT OF BIOMEDICAL WASTE ON HUMAN HEALTH

The effect of pollution as a result of the generation of BMWs is a potential threat to life and the environment because it promotes the proliferation of infectious and hazardous substances. In most developing countries, waste management is poor, and open dumpsites are common in some climes as a result of low or no budget for proper waste disposal and lack or insufficient numbers of trained personnel. Other waste disposal methods such as incinerator often give out hazardous emission, which has been reported to be detrimental to human health. The impact of these wastes is directly on man and animals, e.g., via drinking water, plants consumption, microbial diversity, and ecosystem, e.g., via surface water, soil, etc. Let us look at some important sources of waste with serious impact to human and the environment.

16.6.1 IMPACT FROM PHARMACEUTICAL WASTE

Pharmaceutical industries manufacturing antimicrobial drugs such as antibiotics contribute immensely to the human health hazard and that of the environment if they dispose of expired, unused drugs and pharmaceutical effluents indiscriminately to their immediate or remote environment without proper predisposal inactivation of the waste. Antibiotic usage for the treatment of infections caused by pathogenic bacteria has gained significant attention in both developed and developing countries; hence, production of this drug will continue for a long time to come. A major notable impact among others on humans is the proliferation of antibiotic-resistant genes (ARGs) among pathogenic bacteria, which can lead to treatment failures due to antibiotic resistance among such bacteria that find their way into the human system and cause different types of bacterial infections. Carbapenem-resistant *Enterobacteriaceae*, Methicillin-resistant *Staphylococcus aureus*, *Pseudomonas aeruginosa*, and vancomycin-resistant enterococcus to mention but a few among globally recognized pathogenic bacteria are capable of acquiring ARGs from the environment and wreak havoc in humans and animals. Longer hospital stay, high mortality and morbidity rates, and high economic burden are also major impacts of contamination of the environment by pharmaceutical wastes.

Although pharmaceutical wastes that pollute the environment are in most cases not directly from the manufacturers; in most cases, these pollutions are usually indirect because some advanced industries have a well-developed waste management system, thereby limiting indiscriminate disposal of their products in the environment (Bdour, 2004). There are indirect routes by which this sector contributes to pollution, for instance, antibiotics may be released directly into the environment by antibiotic consumers, e.g., patient who excretes nonmetabolized antibiotic compounds into wastewater bodies that also finds its way into larger urban wastewater treatment plants. Wastewater treatment normally goes through three stages: the primary, secondary, and tertiary treatment processes. Primary treatment is aimed at removal of solid contents by filtration and sedimentation process, whereas secondary treatment removes dissolved organic matters via aerobic decomposition, and tertiary treatment of wastewater aimed at purifying the water by chlorination, chemical oxidation, absorption, and filtration (Moreira et al., 2016). At the final stage is where a broad range of pollutants are removed in the treatment process, certain pharmaceuticals at a very low concentration still persist through the treatment processes and are consumed at various homes and utilities. Hence, long-term persistence of these antimicrobial components in water systems at subinhibitory concentrations often leads to selective pressures, resulting in the development of ARG in bacteria.

Improper agricultural practice also contributes to biomedical waste pollution in the environment. Nowadays, antibiotics are used not only for prevention and therapy of livestock but also in their breeding and fattening in order to make more profits especially in poultry and pig farming (Scoppetta et al., 2016). Extensive use of antibiotics is also found in crop farming to prevent bacterial infections and crop promotion. The use of antibiotics in bee-keeping and aquaculture for therapeutic and as a prophylactic agent is popular (Serrano, 2005). It is worthy of note that these antibiotics being used for all these various purposes are similar in structures and activities to those prescribed for human use. Therefore, the introduction of these agents to food-producing animals that are being consumed poses great dangers to the gut microbiota and pathogenic strains in the selection of resistance due to the low amount of antibiotic exposure and exchange of antibiotic resistance genes via horizontal gene transfer.

16.6.2 Impact from Clinical Waste

All solid and liquid wastes that emerge during diagnosis, treatments, surgery, and experimental research or during medical production of certain health materials are collectively referred to as clinical wastes. Hospitals and other related facilities such as veterinary and dental clinics generate infectious waste, which are sometimes poorly disposed of and managed, are threats to the public because they are capable of transmitting infectious diseases. For an infection to be established, the source must be capable of transmission, the inoculum size of the infectious agent must be sufficient enough to infect a susceptible individual that comes in contact with it, and there must be a portal of entry. Hence, different clinical waste with a potential hazard, such as sharps, used needles, anatomical parts, and blood harbor high microbial burden that can be transmitted easily from one handler to another due to errors during packaging of clinical wastes.

Hospital personnel are at high risk of hepatitis B and C, and HIV from used needles, syringes, and other sharp instruments through cuts and injury. Studies from around the world revealed that more than 50% of healthcare personnel especially nursing staff is liable to needle injury, whereas other supporting medical staff such as attendant and laboratory technicians come across such injury during their duty (Himmelreich et al., 2013). Clinical waste impacts on human health are often associated with direct contact with waste materials or in some cases indirect contact, e.g., water. The impacts of untreated liquid clinical waste are more serious and spread faster than solid BMWs. Improper handling and disposal of liquid BMWs has the ability to percolate watershed and contaminate ground and surface water, thereby leading to serious waterborne infection such as typhoid fever and cholera, etc. open dump of refuse containing medical waste and it also contributes to community-acquired infection especially from pathogens such as *Staphylococcus aureus* and others.

Lastly, certain clinical wastes are carcinogenic because they contain genotoxic materials that are hazardous and have harmful effects such as tissue damage and so on. The open burning or incineration of such genotoxic matters is often released into the environment; extreme harmful pollutants such as dioxins are known to cause cancer and other health hazards, i.e., birth defects, female infertility, low sperm counts, etc. (Figure 16.3).

16.6.3 Impact of Biomedical Waste on the Ecosystem

Biomedical wastes that contain infectious matter or potentially infectious substances such as blood that are not properly disposed of or decontaminated constitute a significant ecosystem imbalance in many ways. These BMWs end up in water bodies such as lakes and a terrestrial environment where abundant wildlife, birds, and fauna live. A scientific study has shown that wildlife is so curious about certain experiences that intoxicate and alter its state of consciousness (*Listverse, 2019*). In the same vein, this wildlife is curious about consumption of wastes, especially pharmaceuticals that are disposed of carelessly in the environments, which are injurious to them or can even kill them.

Biomedical wastes such as pharmaceuticals and their metabolites in aquatic environments are capable of undergoing biotic and abiotic transformations as well as accumulate in the tissues of aquatic organisms. Some are carcinogenic, whereas others are too toxic at lower concentrations. The ecology of the aquatic habitat becomes affected, and the physicochemical conditions become impaired. The aquatic food webs become affected directly by the bioaccumulation of

FIGURE 16.3 An open dump site of mixed waste beside a major road where several people buy and sell.

pharmaceuticals and metabolites in the food chain to toxic levels or indirectly as a result of the loss of important species that are particularly sensitive to the waste constituents.

Landfills and leachates significantly affect the groundwater quality and make it different from the recommended quality by the Word Health Organization. The study carried out in Egypt by Abd El-Salam and Abu-Zuid (2015) demonstrated groundwater quality to be significantly different from the recommended quality, which suggested a negative influence of the leachate and landfill. In the same vein, Koda et al. (2017) reported this influence of the leachate on the underground water. Good water quality is very important for human consumption as well as livestock. Poor quality may lead to an outbreak of diseases and infections, which may increase the mortality and morbidity rate.

16.7 MICROBIAL TOXIN METABOLISM OF BIOMEDICAL WASTE

Methods for toxic substance removal from wastes such as discussed in Section 16.4 of this chapter have been conventionally employed for biomedical wastes. However, there are instances of shortcomings and demerits associated with one or more of these methods. Hence, biotechnological tool for the removal of toxic waste has been developed. Microbial removal of toxic waste is environmentally friendly and cost-effective. We shall discuss below a few methods of microbial detoxification of BMWs and their potential uses.

16.7.1 Biosorption and Bioaccumulation

Biosorption is an efficient and economical method that makes use of sorbents, such as bacteria, fungi, algae, and industrial and agricultural waste for the removal of metalloids and particulates from a solution. Application of biosorption by microbial biomass is of great interest because of the ability to remove potentially hazardous substances, such as metals and metalloids in liquid wastes. The two main methods of transformation and immobilization of heavy metals by microorganisms are bioaccumulation and biosorption. The former is based on the incorporation of the toxic substance into living biomass, whereas the latter maintains the toxic substance on the cellular surface by other mechanisms, which may include complexation, physical adsorption, and movement across the cell membrane, ion exchange, and precipitation depending on the functional group on the cell wall. The two methods are different from each other in many ways, for instance, in terms of operational cost, the volume of biological waste to be treated and disposed of, the efficiency of the detoxification process, and nutritional requirement.

The efficiency of bioaccumulation by living cells depends on the conditions under which they grow, their physiological state, and their age. For example, Leonila et al. (2018) authors reported that a pretreatment of *Saccharomyces cerevisiae* cells with glucose (10–20 mmol/L) enhanced removal of between 30%–40% for Cd^{2+}, Cr^{3+}, Cr^{6+}, Cu^{2+}, Pb^{2+}, Ni^{2+}, and Zn^{2+} from galvanoplastic effluents. Hence, it was observed that the pretreatment of yeast cell with glucose was observed to be more efficient than directly adding glucose to the effluent.

16.7.2 Phytoremediation

The use of microalgae and macroalgae for the removal or biotransformation of toxic substance from wastewater is called phytoremediation. These methods have a better advantage among biological and nonbiological methods, which have been developed for detoxification of BMWs because of algae's ability to bind essential quantities of toxic substance, e.g., pollutant in a medium. The differences in the composition of the contaminant as well as microalgae species play an important role in the efficiency of this process. Water hyacinth has also been reported to be an important environmental utility for efficient accumulation and translocation of heavy metals from its environment to its organelles (stems and leaves).

16.7.3 Aerobic and Anaerobic Digestion of Biomedical Waste

Aerobic composting application in BMWs works on the principle of the organic breakdown of BMWs into carbon dioxide, water, and ammonia by aerobic bacteria in the presence of oxygen, whereas the same action by anaerobic organisms in the absence of oxygen usually leads to fermentation and the production of gas. In the anaerobic composting method, just like their aerobic counterparts, nitrogen, phosphorus, and other nutrients are used for the development of protoplasm, but their difference is in the reduction of certain elements such as nitrogen to inorganic acids and ammonia. Carbon-rich BMWs can provide a good source for essential by-product such as methane gas that is used for heating domestically and industrially. Anaerobic composting is carried out in biodigesters.

16.8   BIOMEDICAL WASTES AS VALUE-ADDED PRODUCTS

Both liquid and solid biomedical wastes especially organic type could be used as substrates for the growth of biomass and novel microorganisms. In addition, enzymes, exopolysaccharides of biotechnological importance, and various chemicals of medical, industrial, and biotechnological importance can also be obtained from aerobic or anaerobic fermentation processes. Biotechnology requires microbial enzymes that are capable of withstanding harsh environmental conditions for use in the food, medical cosmetics, and other industries as well as for environmental bioremediation purposes. Isolation could be achieved using conventional methods or using new strategies such as the application of systematics to the development of selective isolation techniques, metagenomics, proteomics, transcriptomics, and other techniques.

Also, biomedical wastes either liquid or solid could serve as substrates to produce value-added products. After the destruction of infectious agents using an autoclave, microwave, and other treatment methods, wastes, including animals used in research and waste originating from veterinary hospitals and animal houses, have a high potential for bioconversion into value-added products. Below are some biomedical wastes that have potentials for value-added products.

To produce value-added products as listed in Table 16.3, noninfectious solid/slurry can be used as a substrate for solid-state fermentation (SSF).

TABLE 16.3
Biomedical Wastes and Potential Value-Added Products

Waste	Product	Microorganisms
Skin, hides, flesh, blood, fluids, fats, horns, shavings, bones, intestines	Enzymes, biofertilizers, glues, surfactants, lubricants, organic acids, and animal feed stuff	Several bacterial species (such as *Streptomyces, Bacillus* and *Pseudomonas*) and fungal. (*Aspergillus and Trichoderma*)

16.9 IMMOBILIZATION OF TOXIC BIOMEDICAL WASTE

Immobilization of biomedical wastes can be achieved using diverse techniques depending on waste types. Containment techniques have been known to prevent or control leachate. This may include pumping, capping, draining techniques, and the installation of slurry walls. Another method known is the solidification method; BMWs have been known to be solidified into durable physical forms that are more compatible for reuse or landfill. It acts by creating barriers between the environment and waste components through reduction of permeability of the waste or by reduction of the surface area meant for diffusion with or without chemical reaction (solidification includes dewatering, mixing with adsorbents, and vegetative stabilization).

16.10 CONCLUSION

A clean environment reduces the risk of diseases and infections, and we are all entitled to live in a clean and safe environment. Constant generation of BMWs without proper plans for detoxification may jeopardize the chances of living in good health and clean environment. The government can also look toward the potential benefits of microbial removal and detoxification of BMWs for commercial purpose such as those enumerated in this chapter.

REFERENCES

Abahand, S.O., Ohimain, E.I. (2011) Healthcare waste management in Nigeria: a case study. *Journal of Public Health and Epidemiology* 3(3): 99–110.

Abd El-Salam, M.M., Abu-Zuid, G.I. (2015) Impact of landfill leachate on the groundwater quality: A case study in Egypt. *Journal of Advanced Research* 6(4): 579–586.

Ali, N., Rampazzo, R.C.P., Costa, A.D.T., Krieger, M.A. (2017) Current nucleic acid extraction methods and their implications to point-of-care diagnostics. *BioMed Research International* 2017: 9306564.

Banana, A.A.S. (2013) Inactivation of pathogenic bacteria in human body fluids by steam autoclave, microwave and supercritical carbon dioxide. Ph.D. Thesis, Environmental Technology Division, School of Industrial Technology, UniversitiSains Malaysia (USM), Penang, Malaysia.

Barlow, R.D. (2004) Proper liquid waste disposal mines solid gold bottom line. *Healthcare Purchasing News*. Accessed on October 18, 2018. Available at: http://www.hpnonline.com/inside/2004-06/liquid_waste_disposal.htm

Bdour, A. (2004) *Guideline for the Safe Management of Medical, Chemical, and Pharmaceutical Waste.* National Institute for Environmental Training, Riyadh.

Bélanger, J.M., Paré, J.R., Poon, O., Fairbridge, C., Ng, S., Mutyala, S., Hawkins, R. (2008) Remarks on various applications of microwave energy. *The Journal of Microwave Power and Electromagnetic Energy* 42: 24–44.

Center for Disease Control and Prevention (2003). Guidelines for Environmental Infection Control in Health-Care Facilities. Accessed on October 17, 2018. Available at: https://www.cdc.gov/infectioncontrol/guidelines/environmental/background/medical-waste.html#table27

Cha, C.Y., Carlisle, C.T. (2001) Microwave process for volatile organic compound abatement. *Journal of Air and Waste Management Association* 51: 1628–1641.

Gambrell, J., Moore, S. (2006) Assessing workplace compliance with handling of antineoplastic agents. *Clinical Journal of Oncolgoy Nursing* 10(4): 473–477.

Guisty, L. (2009) A review of waste management practices and their impact on human health. *Waste Management* 29: 2227–2239.

Himmelreich, H., Rabenau, H.F., Rindermann, M., Stephan, C., Bickel, M., Marzi, I., Wicker, S. (2013) The management of needlestick injuries. *Deutsches Ärzteblatt International* 110(5): 61–67.

ICRP (1995) Recommendations of the international commission on radiological protection. *British Institute of Radiology*, Suppl 6. 32. London. Access on September 25, 2019. Available at: https://journals.sagepub.com/doi/pdf/10.1016/S0074-27402880014-6

Koda, E., Miszkowska, A., Sieczka, A. (2017) Levels of organic pollution indicators in groundwater at the old landfill and waste management site. *Applied Sciences* 7(6): 638. DOI:10.3390/app7060638.

Leonila, M.L., Acioly, D.C., Marcos, C.L., José, C.V. Jnr., Rosileide, F.S.A, Thayse, A.L.S., Camilo, E.L.R., Galba, M.C.-T. (2018) Cadmium removal from aqueous solutions by strain of *Pantoea agglomerans* UCP1320 isolated from laundry effluent. *Open Microbiology Journal* 12: 297–307.

Listerverse (2019). 10 Unusual stories involving drunk animals. https://listverse.com/2018/12/18/10-unusual-stories-involving-drunk-animals/

Manasi, S. (2017) Challenges in biomedical waste management in cities: a ward level study of Bangalore. *Advances in Recycling & Waste Management* 2: 119. DOI:10.4172/2475-7675.1000119.

Moreira, F.C., Soler, J., Alpendurad, M.F., Boaventura, R.A.R., Brillas, E., Vilar, V.J.P. (2016) Tertiary treatment of a municipal wastewater toward pharmaceuticals removal by chemical and electrochemical advanced oxidation processes. *Water Research* 105: 251–263.

Murthy, B.K.S. (2000) Operational limits. Training workshop on radiation safety in nuclear medicine and RSO certification examination. BARC, Mumbai, 6.1–6.6.

National AIDS Control Organisation Manual of Hospital Infection Control. (1998) New Delhi, 50–66.

National Institute of Occupational Safety and Health, Centers for Disease Control and Prevention. (2001) Glutaraldehyde – occupational hazards in hospitals. NIOSH Publication No. 2001-115. Accessed on October 19, 2018. Available at: http://www.cdc.gov/niosh/docs/2001-115/#2

New South Wales Health. (October 2008) Cytotoxic drugs and related waste: safe handling in the NSW public health system. Department of Health, New South Wales Government.

Olukanni, D.O., Azuh, D.E., Toogun, T.O., Okorie, U.E. (2014) Waste management practices among selected health-care facilities in Nigeria: a case study. *Scientific Research and Essays* 9(10): 431–439.

Mathias, J.M. (2004) Safe options for suction canister waste. *OR Manager* 20(4): 16–18. Accessed on October 18, 2018. Available at: http://www.orbusmgt.com/catalog/ download/ORMVol20No4SuctionWaste.pdf

MedPro (2018). What is Medical Waste? Medical Waste definition, types, examples, and more. Accessed on October 20, 2018. Available at: https://www.medprodisposal.com/what-is-medical-waste-medical-waste-definition-types-examples-and-more.

Oyekale, A.S., Oyekale, T.O. (2017) Healthcare waste management practices and safety indicators in Nigeria. *BMC Public Health* 17(1): 740.

Practice Greenhealth (2018). Typical Categories of Medical Waste. Accessed on October 17, 2018. Available at: https://practicegreenhealth.org/topics/waste/waste-categories-types/regulated-medical-waste/typical-categories-medical-waste.

Pruss, A., Giroult, E., Rushbrook, P. (Eds.) (1999) *Safe Management of Wastes from Healthcare Activities*. World Health Organization, Geneva.

Scoppetta, F., Cenci, T., Valiani, A., Galarini, R., Capuccella, M. (2016) Qualitative survey on antibiotic use for mastitis and antibiotic residues in Umbrian dairy herds. *Large Animal Review* 22: 11–18.

Serrano, P.H. (2005) Responsible use of antibiotics in aquaculture. Fisheries Technical Paper 469, Food and Agriculture Organization of the United Nations (FAO), Rome.

Tabish, S.A. (2005) Ecohealth: management of biomedical waste. In: *Hospital Infection Control: Conceptual Framework*. Academa Publishers, pp. 139–145.

WHO/UNICEF (2015) Water, sanitation and hygiene in health care facilities: status in low- and middle-income countries. World Health Organization, Geneva.

World Health Organization (1999) Guidelines for safe disposal of unwanted pharmaceuticals in and after emergencies. Accessed on October 2018. Available at: http://www.who.int/water_sanitation_health/medicalwaste/unwantpharm.pdf

World Health Organization (2018) Health Care Waste. Available at: http://www.who.int/news-room/fact-sheets/detail/health-care-waste.

World Health Organisation Fact Sheet (2016). Dioxins and their effects on human health. Accessed on October 2018. Available at: http://www.who.int/news-room/fact-sheets/detail/dioxins-and-their-effects-on-human-health.

Zhang, Q.Y., Qian, J.Q., Guo, H., Yang, S.L. (2008) Supercritical CO2: a novel environmentally friendly mutagen. *Journal of Microbiological Methods* 75: 25–28.

Zile, S., Bhalwar, R., Jayaram, J., Tilak, V.W. (2001) An introduction to essentials of Bio-Medical Waste Management. *Medical Journal Armed Forces India* 57(2): 144–147.

17 Immobilized Enzyme-Based Biocatalytic Cues
An Effective Approach to Tackle Industrial Effluent Waste

Muhammad Bilal
Huaiyin Institute of Technology, China

Shahid Mehmood
Chinese Academy of Sciences, China

Hafiz M. N. Iqbal
Tecnologico de Monterrey, Mexico

CONTENTS

17.1 INTRODUCTION

World population has incredibly increased to 6 billion just before the 20th century, and during 21st century, the growth rate is approximated to be rising by 400% (Akan et al., 2009). This exponential increase in the population directly influences consumption of resources, energy, food, and other natural resources (Athira and Jaya, 2018). Similarly, increased growth rate also largely affects physical atmosphere, including air, land soil, water, and nonliving environment. For instance, human needs more land for cultivation, more natural resources, and more fertilizers such as pesticides and fungicides. The

TABLE 17.1

Major Industrial Pollutants and Their Effects on Human Health

Industry	Pollutants	Disease	References
Iron and steel	BOD, COD, oil, metals, acids, phenols, and Cyanide	Kidney dysfunction, infertility, CNS and PNS damage	Ogwuegbu and Muhanga (2005)
Textiles and leather	BOD, solids, sulfates, and chromium	Carcinogenic, liver, kidney, and skin ulcer	Martin and Griswold (2009)
Pulp and paper	BOD, COD, solids, chlorinated organic Compounds	NS effects, e.g., a headache, irritability dizziness, nausea, vomiting, tremors, excitation, convulsions, loss of consciousness, respiratory and CNS depression, possible death	World Health Organization (2010)
Petrochemicals and refineries	BOD, COD, mineral oils, phenols, and chromium	Carcinogen, skin irritation, stomach problem, nausea, and asthma	Haddis and Devi (2008)
Chemicals	COD, organic chemicals, heavy metals, SS, and cyanide	Carcinogen, infertility, CNS and PNS damage	Järup (2003)
Nonferrous metals	Fluorine and SS	Bone marrow depression, hemolysis, hepatomegaly, melanosis, polyneuropathy, and encephalopathy may be observed. Ingestion of inorganic arsenic may induce peripheral vascular disease	Hassaan and El Nemr (2017b)

BOD—biochemical oxygen demand; COD—chemical oxygen demand; SS—sediments and suspended solids; CNS—central nervous system; PNS—peripheral nervous system NS—nervous system.

excessive use of fertilizer leads to increased infertility of soil and is also responsible for the natural resources imbalance (Sarayu and Sandhya, 2012). Pollution is the biggest challenge of the modern era. Our ecosystem is in trouble due to the modern, technological, and evolutionary lifestyle (Soares et al., 2017). Modern lifestyle facilitates human life, but on the other hand, our ecosystem faced ample problems including air, water, and soil pollution. Ecosystem badly faced numerous contaminations, i.e., industrial hazardous chemical toxins pollute streams, rivers, and canals; smog contaminates air; and blows over the city skies come from industries (Anastasi et al., 2011).

Industrial waste effluents may also alter the soil chemistry by converting into muddy landscapes, altering pH and underwater passages. These industrial pollutants may lead to serious global alarming situations (Bilal et al., 2017a). Long-term exposure with these substances can cause life-threatening problems, including lung cancers, respiratory cancer, and infertility. According to the WWF report, about 1 billion people have no access to clean water, and about 2.4 billion are entangled with lethal diseases due to inadequate amenities. The mortality rate is up to 12–20 million per year out of 220 million due to inadequate management of industrial waste effluents (https://www.worldwildlife.org/threats/pollution) (Hassaan and El Nemr, 2017a) (Table 17.1).

17.2 HAZARDOUS DYES CONTAINING EFFLUENT WASTE: PAST AND PRESENT THREATS

The use of industrial dyes increases tremendously since 1856. It has been reported that more than 10,000 different kinds of dyes are available for commercial usage and approximately 7×10^5 metric tons dyes are produced every year (Aksu and Tezer, 2005; Ansari and Thakur, 2002).

Dyes are an integral part of the industries, including plastic, tannery, paper and pulp, and textile and electroplating. A large number of dyes are also used in the food and pharmaceutical industries (Harazono and Nakamura, 2005). The exponential growth in the manufacturing of dyes especially in textile and tanning industries has led to increased risk of wastewater expulsion to the environment. A major reason for the increase in complexity is that the dyes are directly dumped into the surface water without any treatment (Verma and Madamwar, 2003; Wesenberg et al., 2003). Previous research reported that 15%–20% dyes had been discarded while processing and about 50% dyes are dumped in water after processing (Shah, 2016). Industrial effluents especially containing synthetic dyes are hard to recycle or treat because of the complex aromatic chemical structure of the dyes. Therefore, these dyes are more stable and resist to degrade or difficult to remediate (Padmesh et al., 2005). However, these hazardous pollutants have adverse and life-threatening effects such as mutagenic and carcinogenic induction to living organisms, both flora and fauna, as well as on the environment (Figure 17.1). Most of the dyes are well known as a serial killer such as benzidine and DDT (Hao et al., 2000). Industrial effluent harboring dyes pose serious effects on an animal underwater and the earth such as nitro- and azo-containing dyes that are carcinogenic to aquatic animals. Similarly, these dyes also cause serious abnormalities in humans, such as nausea, hemorrhage, ulceration of skin and mucous membranes, kidney damage, reproductive system infertility, liver cirrhosis, brain retardation, and central nervous system damage. These trepidations led to implement new strategies for the regulation and treatment of dyes and also regulate the cleaner technology for the industrial area (Hai et al., 2007; Hao et al., 2000; Rott, 2003) (Table 17.2).

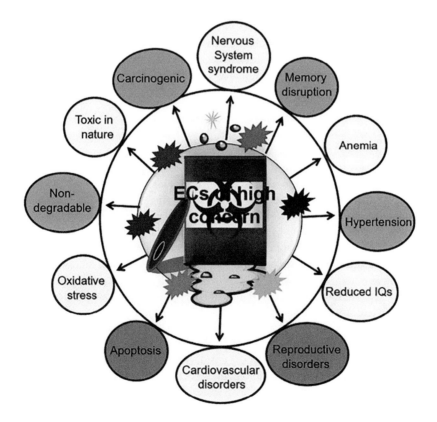

FIGURE 17.1 Major consequences and adverse effects of ECs of high concern on human's health and the environment. (Adapted from Rasheed et al. 2019, with permission from Elsevier. Copyright (2018) The Authors. Published by Elsevier Ltd.)

TABLE 17.2

Hazardous Dyes and Their Serious Impact on Human Health

Dyes Name	Impact on Human Health	References
Acid red (Azo dye)	Dermatitis, carcinogen especially liver cancer, intestinal cancer, allergic reaction	Hassaan and El Nemr (2017b)
Reactive black 5	Hypersensitivity, hypoxia, mutagen, carcinoma, hepatocellular, asthma	Zheng et al. (2016)
Reactive blue 4	Mutagenic, growth inhibitors of normal soil flora	Nga et al. (2017)
Methylene blue	Central nervous system toxicity, methemoglobinemia, vasoplegia	Gillman (2011)
Congo red	Carcinogenic and mutagenic	Bazrafshan et al. (2012)
Reactive red	Reduce intestinal microbial flora, skin irritation, and allergic reactions	Bazrafshan et al. (2012)
Orange G	No data available in human toxicity	Bazrafshan et al. (2012)

17.3 INDUSTRIES GENERATING EFFLUENT WASTE

During modern evolution, industries grow rapidly as human population exponentially increases. On the other hand, industrial revolution is the biggest source of disruption in ecosystem regulation, including balancing food chain and sustainability and self-regulation fitness of biosphere (Borchert and Libra, 2001). Industrial effluent waste (EW) has ample hazardous material, especially heavy metals, polyaromatic hydrocarbons (PAHs), petrochemicals, aromatic compounds, phenolic compounds, soil microbial, polychlorinated biphenyl, and dioxin. Industrial EW is dumped into water that is accumulated under water and causes serious and lethal effects on aquatic biota and change the aquatic environmental chemistry especially pH, color, and salt balancing (Kanu and Achi, 2011).

17.3.1 TEXTILE INDUSTRY

The textile industries are the most flourishing and prominent sectors in the modern industrial revolution since the 21st century. All textile industries consume a large amount of water to execute a large number of manufacturing processes such as bleaching, dyeing, and washing and resultantly eliminates a huge amount of effluents into surroundings. The chemicals present in the wastewater can harm both human health and the environment, by blocking the sunlight passage through the water, hamper photosynthesis, increase the biological oxygen demand and affect the aquatic life. The direct discharge of this wastewater into the environment affects its ecological status by causing various undesirable changes (Ghaly et al., 2014). Also, other chemicals are widely used for colorization and decolorization of fabrics; for example, dyeing and printing sometimes involve dangerous chemicals and substances such as arsenic, lead, and mercury. Approximately, 40% of globally used colorants contain organically bound chlorine, a known carcinogen. Chemicals evaporate into the air we breathe or are absorbed through our skin, which causes allergic reactions and may harm children even before birth. Due to this chemical pollution, the normal functioning of cells is disturbed, and this, in turn, may cause an alteration in the physiology and biochemical mechanisms of animals, resulting in impairment of important functions such as respiration, osmoregulation, reproduction, and even mortality. Heavy metals, present in the textile industry effluents, accumulate in the body organs, leading to various symptoms of diseases. For instance, even very low quantities of potentially toxic metals such as lead, cadmium, and mercury can cause a large number of diseases to human life, including nervous system syndrome, memory disruption, anemia, hypertension, reduced IQs, and reproductive and cardiovascular issues. Figure 17.2 illustrates various sources of lead, cadmium, and mercury. In the textile industry, both natural and synthetic dyes are randomly and widely used in final product manufacturing (Rahman, 2016). Manufacturing rayon, an artificial fabric prepared from wood pulp, has resulted in the loss of many old-growth forests. During the fabrication process, the pulp is treated with hazardous chemicals that eventually find their way into the environment. Synthetic or human-made fabrics such as nylon and polyester from petrochemicals and fossil fuels require lots of water and energy. Nylon manufacturing generates greenhouse gases that harm the air.

17.3.2 Pharmaceutical Industry

Pharmaceutical and personal care products (PPCPs) play an integral part in the socioeconomic development worldwide. Previous studies revealed some drug effluents in groundwater with significant undesirable effects (Izah et al., 2016). PPCPs exaggerated in the environment cause microbes inhibition, unpleasant odor, and reduction in the cell-aqueous phase transfer rates. The biomedical or pharmaceutical industries waste comprising infectious waste, anatomical waste, medical waste, genotoxic waste, chemical waste, heavy metal waste, and radioactive waste may cause ecotoxic damage and carcinogenic consequences. Many conventional biological technologies have been used for the eradication of waste material containing carbon and nitrogen compounds. However, these eradication technologies were found insufficient for a lot of chemical waste extermination. The breakthroughs in PPCP wastes are that removal efficiencies of every product are different and specific eradication technology is used for specific waste products (Kanu and Achi, 2011).

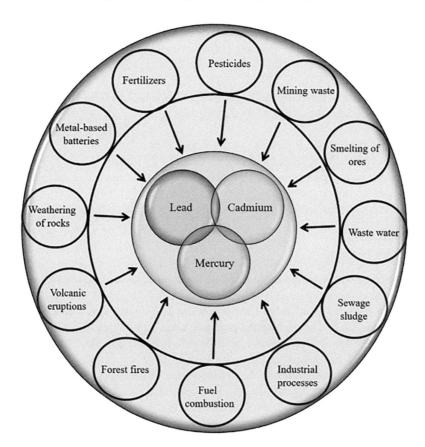

FIGURE 17.2 Various sources of lead, cadmium, and mercury. (Adapted from Rasheed et al. 2018, with permission from Elsevier. Copyright (2017) Elsevier.)

17.3.3 FOOD INDUSTRY

Food industries are one of the rapidly growing manufacturing units and recognized as the chief contributor to the world economy. Each step in the food industry from production to marketing has a great impact on the environment. Along with a pronounced positive impact, these industries leave marks on the environment in the form of ecological hazardous since they release harmful chemicals. Food industry largely relies on water supply and adequate water quality from primary food production to processing; thus, water pollution is one of the major pollution marks left by the food industry. As compared with other industries, the food industry necessitates a huge amount of water, for example, in cleaning, decontamination, freezing, and transportation (Kanu et al., 2006).

The principal environmental issues due to food industry include the following:

1. Wastewater bearing biochemical oxygen demand (BOD), total suspended solids (TSS), excessive nutrient loading (nitrogen and phosphorous compounds), pathogenic organisms (resulting from animal processing), and residual chlorine and pesticide level.
2. Solid wastes include both the organic and inorganic wastes, whereas the organic wastes comprise the rinds, seeds, and bones from raw materials and inorganic wastes include the packaging materials such as plastics, glass, and metal.

Chocolate industries act like a monster to cause water pollution as compared with other food industries. Similarly, the brewery industry produces enormous wastewater from liquor after the processing of yeast and grains. Wastewater eliminated by brewery industry has diverse properties, i.e., acidic pH and fermented odor. Brewery effluents are rich with organic waste products, including nitrogenous compound, carbohydrates, and other decolorizing reagents that change microfloral chemistry (Kanu et al., 2006). The possible pollution caused by food industry can lead to great loss of valuable biomass and nutrients if not recovered by appropriate methods and technologies for upgrading, bioconversion, or reutilization. Therefore, a number of mitigation measures must be carried out to ensure ecofriendly environmental conditions. The following measures could be considered to limit the adverse effects of food industries:

a. Utilization of best available technology for all pollution
b. Payment of optimal liability compensation to the nearby settlers
c. The manufacturing processes should be designed in a way to maximize recycling processes and minimize as much effluents/wastes as possible
d. New technologies that are low- and nonwaste generators can be imploded

17.3.4 LEATHER INDUSTRY

Leather or tannery industry is considered one of the most ancient industries all over the world. It plays a key role in the economic growth by offering employment and earning country income through exportation, but although it has the positive impact, it poses huge environmental pollution (Kanagaraj et al., 2006). The environmental pollution caused by the leather industry is categorized into three aspects: wastewater, solid waste, and air emissions. The leather industry is categorized as one of the biggest hazardous and life-threatening chemical-producing units, and thereby principal ecological damaging industry. According to the United Nations Industrial Development Organization (UNIDO) report, 3.5 million of various chemicals are used in the leather industry, and a considerate amount of these chemicals is discharged as effluents (Ludvik, 1996). The hazardous chemicals and toxins produced by leather industry include COD, BOD, a heavy amount of heavy metals, and genotoxic salts (sulfide, chromium, chloride, sodium) (Chaudry et al., 1998; Cooman et al., 2003). Several studies have shown that approximately 90% of water used during leather processing is discharged as waste. This industry also causes air pollution due to the expulsion of gaseous waste into the environment. According to reports, 40% of chromium salts are discharged following completion of chrome tanning process (Tariq et al., 2005; Leghouchi et al., 2009). Various studies have proved that chromium salt is group A carcinogenic agent. Acquaintance with chromium causes various health hazards, including dermatitis, ulcer nasal septum perforation, and lung cancer (Kornhauser et al., 2002; Chowdhury et al., 2013).

17.4 CONVENTIONAL VS. GREEN TECHNOLOGIES

Increased industrialization due to increasing urbanization poses a severe threat to ecosystem especially aquatic ecosystem including human. Due to the rapid increase in industrial growth, underground water is contaminated with hazardous effluents (Zhang and Balay, 2014). Globally, water consumption is mostly dependent on alternative resources, including recycling, treated water other than surface and underground water. The primary concern is to treat the existing water to make it useable and to save for life as well as to protect natural water (Zheng et al., 2015; Umamaheswari and Shanthakumar, 2016).

Nowadays, the major focus is being devoted to industrial effluents, particularly wastewater making it expedient for life. Worldwide, especially in underdeveloped countries such as Nigeria and Ghana, wastewater treatment is now shifting from centralized to more defensible decentralized wastewater treatment (DEWATS) because conventional methods are no more feasible (Adu-Ahyiah and Anku, 2010). This section describes conventional technologies in comparison with green technologies for waste management.

Conventional treatment approaches including chemical, physical, and biological methods have been widely attempted for the disposal of organic, inorganic, and solid waste. In conventional technology, waste treatment is designated at a different level, i.e., preliminary, primary, secondary, tertiary, and advanced waste treatment (Awaleh and Soubaneh, 2014). Preliminary treatment of waste removed solid and floating waste material containing papers, plastic debris, peels, wood shards, and heavy inorganic solids. Various treatment methods such as detritus, comminutors, chamber, grit chamber, Skimming, and tank screening have been used in preliminary treatment. In the primary method, waste material either organic or inorganic has been removed by sedimentation and advanced method of skimming. 25%–50% BOD, 50%–70% suspended solids (SS), and 65% of oil and grease waste have been eliminated through this method. Similarly, some other organic wastes such as nitrogen, organic phosphorus, and heavy metals attached with solid sediments were also removed (Liew et al., 2015).

The secondary treatment was used to remove the remittent effluents containing solids and organic residues. This method followed by primary treatment with some advancements that are aerobic biological methods was used for removal of colloidal organic and biodegradable materials. The secondary treatment plant was employed by using aerobic microorganisms specifically bacteria that catalyzed the waste materials into inorganic end products such as H_2O, NH_3, and CO_2 (Gupta et al., 2012). Several processes including activated sludge processes, oxidation ditches, biofilters, and rotating biological contactors (RBCs) were used for secondary treatment of waste effluents especially water. For municipal water treatment, a combinatorial method, that is, activated sludge combined with the biofilters process (make it for efficient), was used. Sometimes, it has more organic compounds than industrial waste (Eckenfelder and Cleary, 2013). Thus, tertiary treatment was used to accompany previous treatments for removal of lingering organic and inorganic residues, and dissolved and refractory substances. This treatment comprises oxidation, disinfection, optimizing chemical treatment, filtration, ion exchange, reverse osmosis, and activated carbon treatment (Koivunen et al., 2003).

Although conventional methods have been widely used for the elimination of industrial waste, these methods have several limitations imposing a hazardous impact on the ecosystem. Currently, researchers introduced advanced green technology (AGT) that is eco-friendlier and fewer precincts. The term "green technology" is defined as the "technology intended to mitigate or reverse the effects of human activity on the environment." The main purpose of AGT for industrial effluents elimination is to:

   i. Minimize the water usage and moderate the implication of nonrenewable sources
  ii. Reduce the contamination and mismanagement of water resources
 iii. Save the territory, biodiversity, and the environment

Presently, there is the hallmark of ecosystem matter to eliminate the toxic chemicals. These chemicals are undesirable to the environment. Advanced oxidation processes (AOPs) is another well-developed process for eliminating industrial effluents. AOPs are "defined as a set of chemical treatment processes designed to remove organic compounds in wastewater by several oxidation reactions." AOPs are based on reactive ions (radicals) especially hydroxyl ions that react with pollutants and convert into salt dome. These free radicals can be harvested by the oxidation process of agents such as H_2O_2, O_3, UV radiation, and catalyst (Goh et al., 2015). Membrane bioreactor (MBR) has become a more efficient and promising approach to eradicating the waste effluents from the water owing to its fascinating advantages such as reducing sludge fabrication and greater separation competency. (Neoh et al., 2016). MBR has a high tendency to eliminate the molecules with smaller molecular mass, which are called micropollutants. This process modified with ultrafiltration has increased the efficiency even to eradicate the nanopollutants like viruses (Rodríguez et al., 2011).

17.4.1 Immobilized Enzyme—Physiochemical and Biocatalytic Characteristics

In recent years, increasing focus has been put on biocatalysis to furnish chemical industries environmentally more safe, friendlier, and sustainable. Undeniably, a plethora of research investigations impressively corroborate the perspective of *White Biotechnology* to diminish resource consumption and thereby waste generation (Hollmann and Arends, 2012). Accelerating the rate of nearly all chemical reactions within a cell is one of the most significant functions of natural biocatalysts. However, the real-time practicality of enzyme is generally associated with several drawbacks, resulting from amenability to process conditions, marginal stability, or inhibiting tendency by an elevated level of reaction components. Most of the native enzymes are quite unstable, and their industrial-scale employment is often hindered by lacking durable operational stability, reusability, and the challenging recovery of the enzyme from the reaction mixture. These circumstances leave ample room for improvement. Among the different cost-reduction methods taken into practice, immobilization is one of the important approaches to realize enzyme application in biotechnological processes more favorable. Immobilized enzymes, enzymes physically confined in a certain defined region, retain their catalytic potentialities and can be employed recurrently in several successive cycles. Apart from the convenient handling and separation of the enzyme from the product, the reaction process can be controlled with a better quality product by immobilization. It also substantially diminishes the cost of the enzyme by enhancing its stability under both operational and storage conditions (Sheldon, 2001, 2007).

Immobilization/insolubilization of the biocatalyst to a supporting matrix imparts several advantages that make them more promising for numerous biotechnological applications such as follows:

1. Rapid termination of the reaction by eliminating enzyme from the reaction media
2. Improved enzyme tolerance against wide-working pH, temperature, and solvent impurities and other contaminants
3. Efficient retrieval and recycling ability of the enzymes allowing their use in a continuous-flow packed-bed bioreactor operation
4. Possibly better quality products with low or no contamination

17.4.2 ENZYME-IMMOBILIZED BIOCATALYTIC MEMBRANES TO TREAT EFFLUENT WASTE

Biocatalytic membranes with immobilized enzymes offer distinct functionalities of enzyme stabilization and reusability, alleviate product inhibition, and promote continuous bioprocessing. Consequently, the design and development of biocatalytic membranes with physically "anchored" enzymes have recently fascinated increasing research attention for widespread applications such as food, pharmaceutical, and water treatment industries (Chandra et al., 2015; Zhang et al., 2018). This is mainly attributed to easily tailored properties, which furnish them desired host platforms for numerous enzymes, such as that the huge specific surface area of these membranes enables effective attachment of the enzyme molecules (Handayani et al., 2012). Moreover, the good porosity, well-defined pore sizes, and structure allow the attachment of enzymes not only on the membrane surface but also in its pores. Importantly, the reaction mixture passing through the membrane has relatively easy accessibility to the catalytic sites of the enzyme, thus reducing diffusional restrictions. Membranes with tailor-made properties can be fabricated in a variety of different shapes and geometrical configurations (Gupta et al., 2013). Immobilization of enzymes is the most critical step for designing biocatalytic membranes. Notably, physical adsorption, entrapment, cross-linking, affinity, and covalent coupling have been regarded as the leading mechanisms for enzyme immobilization onto the porous membrane. Figure 17.3 illustrates the whole schematic process of laccase immobilization by covalent binding on ceramic support (Barrios-Estrada et al., 2018). Apart from membranes prepared from biopolymers, i.e., alginate, agarose, cellulose, polyacrylamide, and chitosan, many synthetic polymers including poly(vinylidene fluoride), poly(vinyl alcohol), and polyurethane have also been used to develop membranes for enzyme immobilization. In contrast to other carrier supports, lack of any additional separation and purification of the reaction media is the paramount advantage of polymer-based membranes (Morthensen et al., 2015). Polymer-based membranes play a crucial role in the case of enzymatic membrane reactors (EMRs) as support materials for enzymes (Table 17.3).

17.4.3 ENZYME-IMMOBILIZED BEADS/MICROSPHERES TO TREAT EFFLUENT WASTE

Several polymeric materials have been used in the form of beads, microspheres, or capsules as promising support materials to engineer the biocatalytic, thermal, storage, and recyclability characteristics of the enzymes. Natural biopolymers such as agarose, alginate, agar-agar, carrageenan, chitin, chitosan, cellulose, and keratins have been employed in the form of microspheres to immobilize different biomolecules. An exceptional set of properties, i.e., nontoxicity, biodegradability, biocompatibility, and increased binding affinity to proteins or enzymes makes biopolymers as ideal candidates for enzymes attachment. In addition, the presence of reactive functional groups in their structure mainly amine, carbonyl, and hydroxyl moieties allows direct reaction between the support and the biocatalyst as well as enables easier surface modification (Kurita, 2001). Literature survey revealed chitosan as the most frequently employed biopolymer as support material for immobilizing enzymes in different forms and geometrical shapes for environmental applications. For instance, Jamil et al. (2018) reported the immobilization of an indigenously produced laccase from *Pleurotus ostreatus* IBL-02 on chitosan beads using glutaraldehyde as a cross-linking agent and utilized for the degradation of toxic dye pollutants. Intestinally, the immobilized biocatalytic system efficiently catalyzed the decolorization of five different synthetic textile dyes such as S.F. Turq Blue FBN, S.F. Turq Blue GWWF, S.F. Red C4BLN, S.F. S.F. Black CKF, and Golden Yellow CRL with a maximum of 90% decolorization of S.F. Golden Yellow CRL under optimal reaction conditions. Similarly, glutaraldehyde-activated chitosan beads–entrapped lignin peroxidase also demonstrated some superior dye degradation properties than that its free counterpart (Sofia et al., 2016). In the presence of redox mediator, complete color removal of a textile dye, i.e., Sandal-fix Red C4BLN, was achieved from an aqueous solution by the laccase-bound chitosan microspheres prepared from 2.5% (w/v) chitosan concentration using 1.5% (v/v) glutaraldehyde as an activating agent after 4.0 h

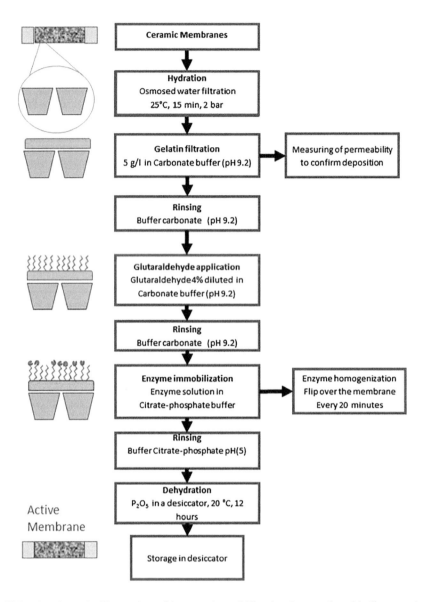

FIGURE 17.3 A schematic illustration of laccase immobilization by covalent binding on the ceramic support. (Adapted from Barrios-Estrada et al. 2018, with permission from Elsevier. Copyright (2017) Elsevier.)

of treatment. The immobilized biocatalyst also maintained 80.19% of its original activity after ten successive dye decolorization batches. Use of alginates is also a prevalent option, and significant attention has recently been given to alginates as a supporting matrix for enzyme immobilization because of a number of characteristics such as low cost, nontoxicity, biocompatibility, and gelation under an extremely mild environment by the addition of divalent cations such as Ca^{2+}. Single or numerous enzymes can be encapsulated or entrapped in the alginate-based beads. However, low mechanical stability and diffusional restrictions may hamper the utilization of alginate gels for large-scale enzyme immobilization. Alginates-entrapped tyrosinase enzyme exhibited much higher pH and temperature resistance in comparison with free enzyme. In addition, the resulting biocatalytic system showed a pronounced removal efficiency of bisphenol A (BPA) from the aqueous

TABLE 17.3

Enzyme-Immobilized Biocatalytic Membranes/Hydrogel/Fibers to Treat Effluent Waste from Different Industries

Enzyme Species	Immobilization Matrix	Immobilization Method	Environmental Application	Reference
Laccase	Multichannel ceramic membrane	Cross-linking	100% degradation of bisphenol A in less than 24 h	Barrios-Estrada et al. (2018)
Laccase	Monoaminoethyl-N-aminoethyl (MANAE–agarose)	Ionic adsorption	The immobilized biocatalysts showed potential efficiency for degradation of bisphenol A. It retained more than 90% of its original capacity to degrade bisphenol A after 15 repeated cycles	Brugnari et al. (2018)
Ginger peroxidase	Alginate-guar gum hydrogel	Entrapment	Decolorization of about 69% of the textile effluent even after 1 month of continuous operation at 30°C in a continuous packed bed reactor	Ali and Husain (2018)
Ginger peroxidase	Agarose guar gum hydrogel	Entrapment	Up to 80% color removal of the effluent after continuous operation of 30 days at room temperature in a packed bed reactor	Ali and Husain (2018)
Laccase	Ca-alginate, agarose-agar, agar-agar, alginate-gelatin mixed gel	Entrapment	High decolorization and degradation efficiency of the immobilized laccase for Congo Red dye at 50°C after 24 h of incubation	Reda et al. (2018)
Laccase	Graphene oxide membrane	Covalent attachment	Significant degradation potential of the biocatalytic membrane with immobilized laccase enzyme for Reactive Brilliant Blue, Reactive Brilliant Blue, Methyl Orange, and Crystal Violet	Xu et al. (2018)
Laccase	Bacterial cellulose/TiO_2-functionalized composite membranes	Covalent attachment	Laccase-incorporated biocatalytic membrane exhibited combined photocatalytic and biocatalytic degradation of textile dye	Li et al. (2017a)
Laccase	Fibrous polymer-grafted polypropylene chloride film	Adsorption	Effective degradation of three structurally different dyes including Procion Green H4G, Brilliant Blue G, and Crystal Violet in both batch and enzyme reactor system	Arica et al. (2017)

(Continued)

TABLE 17.3 (*Continued*)
Enzyme-Immobilized Biocatalytic Membranes/Hydrogel/Fibers to Treat Effluent Waste from Different Industries

Enzyme Species	Immobilization Matrix	Immobilization Method	Environmental Application	Reference
Manganese peroxidase	Gelatin hydrogel	Entrapment	Potential efficacy for the decolorization and degradation of Sandal-fix Red C4BLN dye in several consecutive batches A significant reduction in water quality parameters such as pH, total organic carbon, and chemical oxygen demand Cytotoxicity reduction of the treated dye sample as determined by brine shrimp and Daphnia Magna bioassays	Bilal et al. (2016a)
Manganese peroxidase	Agar-agar matrix	Entrapment	Decolorization of synthetic textile dyes including Reactive Blue 21 (84.7%), Reactive Red 195A (78.6%), and Reactive Yellow 145A (81.2%) after 12 h in the presence of $MnSO_4$ as a redox mediator A substantial reduction in the toxicity of tested dyes solutions after treatment	Bilal et al. (2016b)
Esterase SulE	Cross-linked poly (γ-glutamic acid)/gelatin hydrogel	Entrapment	High degradation efficiency of chlorimuron ethyl (76.7%–87.7%) of the immobilized enzyme in both soil and water system, particularly in an acidic environment	Yang et al. (2015)
Lignin peroxidase, manganese peroxidase, and laccase	Polyacrylamide hydrogel	Encapsulation	High degradation of bisphenol A (more than 90%) after a short time of 8 h	Gassara et al. (2013)
Laccase	Alginate-gelatin hybrid gel	Entrapment	Removal of eight different synthetic dyes including Amido Black 10B, Bromothymol Blue, Coomassie Blue G-250, Crystal Violet, Eosin, Malachite Green, and Methyl Green, and Methyl Red	Mogharabi et al. (2012)
Laccase	Green coconut fiber	Covalent attachment	Great efficiency in the continuous decolorization of reactive textile dyes such as Reactive Black 5, Reactive Blue 114, Reactive Yellow 15, Reactive Yellow 176, Reactive Red 239, Reactive Red 180	Cristóvão et al. (2012)

solution in a relatively shorter time. Nevertheless, a rapid decrease in the biocatalytic activities during the repeated catalytic cycles might be related to the enzyme leakage presumably due to weak binding (Kampmann et al., 2014). Sondhi et al. (2018) developed a continuous-flow packed-bed bioreactor by entrapping laccase-ABTS in Cu alginate beads under standardized conditions and attempted for decolorizing dyes-harboring textile effluent in a continuous system. As compared with free enzyme treatment, the prepared biocatalytic system resulted in a significant reduction in color (60%), biological oxygen demand, and chemical oxygen demand (COD) of the tested wastewater effluent. UV-VIS spectrum analysis revealed the appearance of new peaks or degradation products because of dye degradation by the catalytic action of immobilized laccase that demonstrates its usefulness for on-site industrial exploitability. In addition, agarose has also been widely used as biocompatible and eco-friendly support for environmental applications. In a recent study, Bilal et al. (2017b) entrapped a manganese peroxidase enzyme on agarose beads fabricated using an agarose concentration of 3.0% and exploited for the degradation of dye-based textile industry effluent. In addition to better tolerance pH and temperature variations, the agarose-entrapped MnP catalyzed the decolorization of textile industry effluents to various extents. The repeatability of the enzyme was recorded up to six continuous phases with notable color removing efficiency that reflects its bioremediation applicability (Table 17.4).

TABLE 17.4
Enzyme-Immobilized Beads/Microspheres/Microcapsules to Treat Effluent Waste from Different Industries

Enzyme Species	Immobilization Matrix	Immobilization Method	Environmental Application	Reference
Laccase	Calcium alginate beads	Entrapment	The biocatalyst degraded more than 99% bisphenol A (20 mg/L) from aqueous solution at optimized conditions of pH 5.0, 30°C within 2 h. Immobilized enzyme was capable to eliminate bisphenol A up to ten continuous batches retaining an initial efficiency of greater than 70%	Lassouane et al. (2019)
Laccase	Chitosan beads	Cross-linking	Immobilized laccase displayed potential efficiency for degradation of five different industrial dyes including Sandal-fix Turquoise blue GWWF, Sandal-fix Black CKF, Sandal-fix Red C4BLN, Sandal-fix Golden Yellow CRL, and Reactive T blue GWF	Jamil et al. (2018)
Laccase	Calcium alginate capsules	Encapsulation	Enhanced biodegradation of sediment-bound heavy hydrocarbon fractions, polycyclic aromatic hydrocarbons, total petroleum hydrocarbons, and fractions of saturates, aromatics, resins, and asphaltenes	Kucharzyk et al. (2018)
Laccase	Copper alginate bead	Entrapment	Degradation of different dyes to various extents. Up to 66% color removal of the textile effluent in a continuous-flow packed-bed bioreactor	Sondhi et al. (2018)

(Continued)

TABLE 17.4 (*Continued*)

Enzyme-Immobilized Beads/Microspheres/Microcapsules to Treat Effluent Waste from Different Industries

Enzyme Species	Immobilization Matrix	Immobilization Method	Environmental Application	Reference
Lignin peroxidase	Calcium alginate beads	Glutaraldehyde cross-linking	Encouraging biocatalytic degradation of textile reactive dyes in the range of 80%–93% Reduction in water quality parameters such as BOD (66.44%–98.22%), COD (81.34%–98.82%), and TOC (80.21%–97.77%) Cytotoxicity reduction of treated dye solutions up to 2.10%–5.06% and 5.43%–9.23% for hemolytic and brine shrimp lethality tests, respectively	Shaheen et al. (2017)
Laccase	Copper alginate beads	Entrapment	Fourteen successive batch runs of complete degradation of Indigo Carmine dye by means of only a single enzyme supplementation in an airlift bioreactor	Teerapatsakul et al. (2017)
Laccase	Chitosan microspheres	Entrapment	Immobilized laccase catalyzed the decolorization of five different textile reactive dyes with the highest of 100% decolorization after 4.0 h It preserved 80.19% of its original catalytic activity after 10-times repeated color removal cycles for Sandal-fix Red C4BLN	Asgher et al. (2017)
Horseradish peroxidase	Polyvinyl alcohol-alginate beads	Entrapment	Complete degradation of Methyl Orange dye in a batch mode by the immobilized horseradish peroxidase (HRP) Identification and analysis of biodegraded fragments by ultra-performance liquid chromatography coupled with mass spectrometry	Bilal et al. (2017c)
Manganese peroxidase	Agarose beads	Entrapment	Bioremediation of textile industry effluent. After six sequential cycles, the tested effluents were degraded to varying extents with a maximum of 98.4% decolorization	Bilal et al. (2017b)
Manganese peroxidase	Chitosan beads	Entrapment	The enhanced catalytic potential for decolorization and detoxification of textile effluent A considerable reduction in BOD, COD, TOC parameters along with toxicity removal	Bilal et al. (2016c)

(Continued)

TABLE 17.4 (*Continued*)
Enzyme-Immobilized Beads/Microspheres/Microcapsules to Treat Effluent Waste from Different Industries

Enzyme Species	Immobilization Matrix	Immobilization Method	Environmental Application	Reference
Lignin peroxidase	Chitosan beads	Glutaraldehyde cross-linking	Higher decolorization efficiency for six reactive dyes with a decolorization rate of S. F. Foron Blue E2BLN (89.71%), S. F. Red C4BLN (87.54%), S. F. Turq Blue GWWF (95.43%), S. F. Golden Yellow CRL (83.87%), S. F. Black CKF (63.76%), and Reactive T Blue GWF (69.86%) within 6 h A significant reduction in cytotoxicity of the treated dye sample as determined by hemolytic assay	Sofia et al. (2016)
Horseradish peroxidase	Calcium alginate beads	Covalent binding	Immobilized HRP exhibited great potentiality for the decolorization of commercial dyes, including Reactive Red 120, Reactive Blue 4, and Reactive Orange 16 in successive dye-decolorizing batch reactions in a packed-bed bioreactor	Bilal et al. (2016d)
Tyrosinase	Calcium alginate beads	Entrapment	Efficient removal of phenolic compounds from aqueous solution within few hours	Roy et al. (2014)
Tyrosinase	Sodium alginate capsules	Entrapment	Almost 100% conversion of endocrine disrupting bisphenol A for 11 days under continuous stirring and 50%–60% conversion after 20 days with no stirring in repeated batch experiments	Kampmann et al. (2014)

17.4.4 ENZYME-IMMOBILIZED BIOCOMPOSITES TO TREAT EFFLUENT WASTE

Considering the unique characteristics of different immobilization matrices, a great deal of research direction has been rekindled to integrating these supports for acquiring maximum benefits. The synthesis of the biocatalytic system by hybrid and reinforced materials ensures some sophisticated properties not detected in their individual counterparts and therefore could be subjected to wide-ranging environmental applications. Use of hybrid composite materials enables target enzyme more robust and stable under unfavorable reaction environments. In addition, provision of suitable and biocompatible conditions for biomolecules promotes the biocatalytic system maintaining enhanced catalytic potentiality and shields its structural organization during storage and reprocessing. Consequently, intriguing research attention is being paid to hybrid, blended, or composite support matrices for conjugating different biocatalysts for the decolorization and degradation of environmental pollutants. Chao et al. (2018) immobilized laccase onto polyvinyl alcohol/halloysite hybrid beads by an entrapment approach and used for the removal of a Reactive Blue dye. The resulting biocatalyst not only demonstrated to be efficient in removal of dye pollutants (93.41%) but also showed good thermal, storage, and reusability. In another study, covalently immobilized laccase on chitosan-functionalized supermagnetic halloysite nanotubes functioned as an environmentally friendly biocatalyst for degrading 87% of Direct Red 80 with significant retention of 33% decolorization efficiency up to 11 repeated cycles (Kadam et al., 2018).

Likewise, multiwalled carbon nanotubes/polysulfone-immobilized laccase membranes efficiently catalyzed the degradation of a mixture of phenolic compounds (Costa et al., 2019). Figure 17.4 illustrates the possible breakdown/degradation pathway for a phenol-containing compound, i.e., phenanthrene, under the MnP-assisted environment (Bilal et al., 2017a). A recent list of studies reporting the use of various kinds of novel hybrid materials as carrier support to develop a biocatalytic system for the treatment of different industrial wastes is portrayed in Table 17.5.

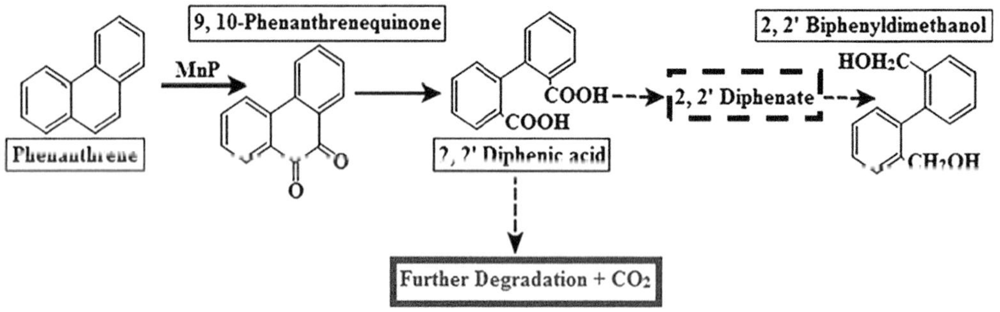

FIGURE 17.4 A hypothetical breakdown/degradation pathway for a phenol-containing compound, i.e., phenanthrene under the MnP-assisted environment. (Reproduced from Bilal et al. 2017a, with permission from Elsevier. Copyright (2016) Elsevier.)

TABLE 17.5
Enzyme-Immobilized Biocomposites or Blends to Treat Effluent Waste from Different Industries

Enzyme Species	Immobilization Matrix	Immobilization Method	Environmental Application	Reference
Ginger peroxidase	Polypyrrole-cellulose-graphene oxide nanocomposite	Adsorption	The support-bound biocatalyst showed higher than 99% decolorization efficiency for Reactive Blue 4 dye in 3 h stirred batch treatment It also caused a significant reduction of its genotoxicity	Ali et al. (2018)
Horseradish peroxidase (HRP)	ZnO nanowires/macroporous SiO_2 composites	Cross-linking	Immobilized HRP showed a high decolorization efficacy of 94.3% and 95.9% for Acid Violet 109 and Reactive Blue 19 within the first 30 min, respectively. It completely decolorized these dyes within 2–3 h	Sun et al. (2018)
Chloroperoxidase	ZnO nanowire/macroporous SiO2 composite	Cross-linking	Enhanced decolorization percentage of 95.4, 92.3, and 89.1%, for Acid Blue 113, Direct Black 38, and Acid Black 10 BX respectively, under optimal conditions Retention of 83.6% and 80.9% of initial activity for decolorizing Acid Blue 113 after incubation at 4°C for 2 months and 12 repeated cycles, respectively	Jin et al. (2018)

(Continued)

TABLE 17.5 (*Continued*)

Enzyme-Immobilized Biocomposites or Blends to Treat Effluent Waste from Different Industries

Enzyme Species	Immobilization Matrix	Immobilization Method	Environmental Application	Reference
Laccase	Porous polyvinyl alcohol/ halloysite hybrid beads	Entrapment	High removal rate (93.41%) of the immobilized enzyme for reactive blue in the presence of redox mediator resulting from simultaneous adsorption and degradation	Chao et al. (2018)
Laccase	Polyacrylonitrile-biochar composite nanofibrous membrane	Covalent binding	Degradation of chlortetracycline by the immobilized enzyme in continuous mode exhibiting 58.3%, 40.7%, and 22.6% removal efficiency at flux rates of 1, 2, and 3 mL/h cm^2	Taheran et al. (2017)
Laccase	Bacterial cellulose/TiO$_2$ functionalized composite membranes	Covalent binding	Collective bio- and photocatalytic potentially of the immobilized enzyme for industrial textile dye degradation	Li et al. (2017a)
Ginger peroxidase	Amino-functionalized silica-coated titanium dioxide nanocomposite	Adsorption	Highly efficient in removing acid yellow 42 dye in a stirred batch process More than 90% of the dye was decolorized within 90 min in contrast to only 69% by the free enzyme in the same period	Ali et al. (2017)
Laccase	Polyamide 6/ chitosan nanofibers	Covalent binding	Potential capability of the biocatalyst to remove a mixture containing two endocrine-disrupting chemicals: 17α-ethinylestradiol and bisphenol A	Maryšková et al. (2016)
Laccase	Poly(acrylamide-crotonic acid)/ Na Alginate	Covalent immobilization	A high decolorization of 73% Acid Orange 52 by the immobilized enzyme in the presence of a mediator	Koklukaya et al. (2016)
Laccase	Chitosan-grafted polyacrylamide hydrogel	Encapsulation	Continuous decolorization of two dye pollutants, i.e., Acid Orange 7, Malachite Green with much better durability	Sun et al. (2015)

17.4.5 ENZYME-IMMOBILIZED NANOPARTICLE/NANOFIBERS/NANOFLOWERS TO TREAT EFFLUENT WASTE

Current interest in nanoscience and nanotechnology has led to the development of a large variety of new nanoscaffolds exhibiting biocompatible surfaces for immobilizing different enzymes. Many exceptional features such as efficient enzyme loading, high specific surface area, and profound catalytic potential render nanosupport as perfect candidates for large-scale biocatalysis in aqueous/nonaqueous media. Application of nanostructured materials allows the construction of a biocompatible microenvironment surrounding the enzyme molecule for optimal reaction efficiencies. The fabrication, biocatalytic performance, stability, and application robustness of nanocatalysts are explicitly broader as compared with their conventional supports–based immobilized enzyme counterparts. Recycling ability up to several successive cycles along with easy separation of nanostructured carriers, particularly magnetic nanosupport by an external magnetic field, can significantly diminish the cost of enzyme-based industrial bioprocesses (Adeel et al., 2018). In recent years, an immense variety of functionalized nanostructured materials/<u>nanoscaffolds,</u> including nanofibers, nanospheres, nanotubes, mesoporous/nanoporous carriers, carbon nanotubes, magnetic or nonmagnetic nanoparticles, nanocomposites, nanocontainers, and polymers and zeolite-based gold nanoparticles, have been endeavored as support biocompatible carriers to develop the <u>nanobiocatalytic</u> system. The driving force to introduce and investigate newer nanocarriers with diverse structures and physicochemical properties arises from the accelerating biotechnological interest of nanocatalysts, and the employment of such advanced <u>nanoscaffold</u> variants might upgrade the desired enzymes properties. Owing to their greater robustness and tensile strengths, nanosized materials are often able to tolerate mechanical shear in the running bioreactor, and optimization of various operating parameters, such as the assortment of right immobilization technique and the type of enzyme, may result in the development of faster and more efficient enzyme reactor. Recently, poly(vinyl alcohol) and polyacrylamide bicomponent nanofibers–conjugated horseradish peroxidase showed almost 29.68% phenol removal efficiency in real wastewater samples at 3 h of treatment (Temoçin et al., 2018). Shao et al. (2019) reported that modified hollow mesoporous carbon spheres–immobilized laccase presented a pronounced removal efficiency for two antibiotics, namely, ciprofloxacin hydrochloride and tetracycline hydrochloride, in the presence of redox mediator indicating its potential in environmental remediation. As compared with free enzyme, the ginger peroxidase adsorbed on a novel polypyrrole-cellulose-graphene oxide nanocomposite resulted in 99% decolorization of Reactive Blue 4 under 3 h of stirring batch treatment. It also exhibited greater operational steadiness retaining about 72% of its starting decolorization efficacy even after 10 consecutive dye-degrading cycles. Genotoxic evaluation of immobilized enzyme-treated dye samples showed a substantial decrease in its genotoxic potential (Ali et al., 2018). Similarly, laccase/graphene oxide biocatalytic membrane showed a significantly improved degradation potential for Reactive Brilliant Blue, Methyl Orange, and Crystal Violet dyes than the membrane lacking immobilized laccase (Xu et al., 2018). Table 17.6 represents a detailed list of contemporary studies reporting the use of enzyme-immobilized nanoparticles/nanofibers/nanoflowers to treat EW from different industries.

TABLE 17.6
Enzyme-Immobilized Nanoparticle/Nanofibers/Nanoflowers to Treat Effluent Waste from Different Industries

Enzyme Species	Immobilization Matrix	Immobilization Method	Environmental Application	Reference
Laccase	Magnetic nanoflowers	Covalent binding	Immobilized biocatalyst exhibited 100% bisphenol A degradation within only 5 min under optimal conditions	Fu et al. (2019)
Horseradish peroxidase	Electrospun poly(vinyl alcohol)-polyacrylamide blend nanofiber membrane	Covalent binding	Conversion of phenol (29.68%) by the immobilized horseradish peroxidase HRP after 180 min in real wastewater	Temoçin et al. (2018)
Lignin peroxidase	Carboxylated carbon nanotubes	Covalent binding	Immobilized enzymatic extract exhibited a great promise for the decolorization of Remazol Brilliant Blue R dye along with an elevated catalytic activity, thermal stability, and reusing capacity in the continuous dye degradation processes	Oliveira et al. (2018)

(Continued)

TABLE 17.6 (*Continued*)
Enzyme-Immobilized Nanoparticle/Nanofibers/Nanoflowers to Treat Effluent Waste from Different Industries

Enzyme Species	Immobilization Matrix	Immobilization Method	Environmental Application	Reference
Laccase	Protein-metal hybrid nanoflower	Encapsulation	High oxidation potential (265% higher than that of the native enzyme) by the immobilized enzyme toward phenolic compounds Retention of up to 84.6% residual decolorization efficiency for synthetic dyes in repeated batch reactions of 10 cycles	Patel et al. (2018)
Horseradish peroxidase	Modified reduced graphene oxide nanosheets	Physical adsorption	Complete removal efficiency for an elevated concentration of phenol (2,500 mg/L) by the immobilized HRP	Vineh et al. (2018)
Laccase	Amino-functionalized magnetic Fe_3O_4 nanoparticles	Cross-linking	Efficient decolorization of acid fuchsin with the maximum of 77.41% decolorization rate	Gao et al. (2018)
Laccase	Hyperbranched polyethyleneimine/ polyethersulfone electrospun nanofibrous membranes	Covalent binding	The immobilized laccase displayed a high bisphenol A of 89.6% It retained the removal rate of up to 79% even after four filtration cycles	Koloti et al. (2018)
Laccase	Silica-coated magnetic nanoparticles	Covalent binding	Biodegradation and biotransformation of the endocrine-disrupting compound, i.e., bisphenol A Enzymatic biotransformation of xenobiotics	Moldes-Diz et al. (2018)
Laccase	Graphene oxide-enzyme hybrid nanoflowers	Covalent binding	Efficient removal of organic dye and micropollutants	Li et al. (2017b)
Laccase	Titania nanoparticles	Direct immobilization	Enhanced degradation capability of the biocatalytic system for micropollutants, namely, bisphenol-A and carbamazepine that are commonly detected in sewage	Ji et al. (2017)
Turkish black radish peroxidase	Inorganic hybrid nanoflowers	Covalent Immobilization	Excellent degradation efficiency for Victoria Blue dye with higher than 90% within 60 min Immobilized biocatalyst was repeatedly used for ten continuous cycles causing 77% dye decolorization efficiency	Altinkaynak et al. (2017)

17.5 CONCLUSIONS AND HORIZONS

Enzyme-assisted treatment approaches are increasingly preferred over the classical chemical routes in industrial bioprocesses. Exploitation of insolubilized enzymes as (bio)catalysts is continually drawing substantial attention from both industries and academia as a highly proficient, cost-efficient, and environmentally friendly technique. The manufacturing and engineering of novel, cheaper, and more robust matrices as support materials for enzyme immobilization can potentially trim down the cost of industrial processes. In addition, continuous studies focused on circumventing the existing downsides in immobilization approaches, accompanied by the development of a simple and durable enzyme immobilization technique, could significantly bring down the overall cost of carrier-bound enzymes-based industrial processes. Interdisciplinary integration of nanotechnology and biotechnology offers a novel paradigm for improving the efficacy of immobilized enzymes for environmental applications. Repetitive usage of robust and stable biocatalytic systems could impressively improve the economic viability of the enzymes exploitability in the upcoming years.

ACKNOWLEDGMENTS

All authors are grateful to their representative institutes for providing literature facilities.

CONFLICT OF INTEREST

Authors declare that they do not have a conflict of interest in any capacity including competing or financial.

REFERENCES

Adeel, M., Bilal, M., Rasheed, T., Sharma, A., & Iqbal, H. M. (2018). Graphene and graphene oxide: functionalization and nano-bio-catalytic system for enzyme immobilization and biotechnological perspective. *International Journal of Biological Macromolecules, 120*, 1430–1440.

Adu-Ahyiah, M., & Anku, R. E. (2010). Small scale wastewater treatment in Ghana (a Scenerio): Retrieved.

Akan, J., Abdulrahman, F., Ayodele, J., & Ogugbuaja, V. (2009). Impact of tannery and textile effluent on the chemical characteristics of Challawa River, Kano State, Nigeria. *Australian Journal of Basic and Applied Sciences, 3*(3), 1933–1947.

Aksu, Z., & Tezer, S. (2005). Biosorption of reactive dyes on the green alga Chlorella vulgaris. *Process Biochemistry, 40*(3–4), 1347–1361.

Ali, M., & Husain, Q. (2018). Guar gum blended alginate/agarose hydrogel as a promising support for the entrapment of peroxidase: Stability and reusability studies for the treatment of textile effluent. *International Journal of Biological Macromolecules, 116*, 463–471.

Ali, M., Husain, Q., Alam, N., & Ahmad, M. (2017). Enhanced catalytic activity and stability of ginger peroxidase immobilized on amino-functionalized silica-coated titanium dioxide nanocomposite: a cost-effective tool for bioremediation. *Water, Air, & Soil Pollution, 228*(1), 22.

Ali, M., Husain, Q., Sultana, S., & Ahmad, M. (2018). Immobilization of peroxidase on polypyrrole-cellulose-graphene oxide nanocomposite via non-covalent interactions for the degradation of Reactive Blue 4 dye. *Chemosphere, 202*, 198–207.

Altinkaynak, C., Tavlasoglu, S., Kalin, R., Sadeghian, N., Ozdemir, H., Ocsoy, I., & Özdemir, N. (2017). A hierarchical assembly of flower-like hybrid Turkish black radish peroxidase-Cu^{2+} nanobiocatalyst and its effective use in dye decolorization. *Chemosphere, 182*, 122–128.

Anastasi, A., Parato, B., Spina, F., Tigini, V., Prigione, V., & Varese, G. C. (2011). Decolourisation and detoxification in the fungal treatment of textile wastewaters from dyeing processes. *New Biotechnology, 29*(1), 38–45.

Ansari, A., & Thakur, B. (2002). Bio-chemical reactor for treatment of concentrated textile effluent. *Colourage, 49*(2), 27–30.

Arica, M. Y., Salih, B., Celikbicak, O., & Bayramoglu, G. (2017). Immobilization of laccase on the fibrous polymer-grafted film and study of textile dye degradation by MALDI–ToF-MS. *Chemical Engineering Research and Design, 128*, 107–119.

Asgher, M., Noreen, S., & Bilal, M. (2017). Enhancing catalytic functionality of Trametes versicolor IBL-04 laccase by immobilization on chitosan microspheres. *Chemical Engineering Research and Design, 119*, 1–11.

Athira, N., & Jaya, D. (2018). The use of fish biomarkers for assessing textile effluent contamination of aquatic ecosystems: a review. *Nature Environment and Pollution Technology, 17*(1), 25–34.

Awaleh, M. O., & Soubaneh, Y. D. (2014). Waste water treatment in chemical industries: the concept and current technologies. *Hydrology: Current Research, 5*(1), 1.

Barrios-Estrada, C., de Jesús Rostro-Alanis, M., Parra, A. L., Belleville, M. P., Sanchez-Marcano, J., Iqbal, H. M., & Parra-Saldívar, R. (2018). Potentialities of active membranes with immobilized laccase for Bisphenol A degradation. *International Journal of Biological Macromolecules, 108*, 837–844.

Bazrafshan, E., Mostafapour, F. K., Hosseini, A. R., Raksh Khorshid, A., & Mahvi, A. H. (2012). Decolorisation of reactive red 120 dye by using single-walled carbon nanotubes in aqueous solutions. *Journal of Chemistry, 2013*(938374), 8 pages, DOI:10.1155/2013/938374.

Bilal, M., Asgher, M., Hu, H., & Zhang, X. (2016a). Kinetic characterization, thermo-stability and Reactive Red 195A dye detoxifying properties of manganese peroxidase-coupled gelatin hydrogel. *Water Science and Technology, 74*(8), 1809–1820.

Bilal, M., Asgher, M., Iqbal, H. M., Hu, H., Wang, W., & Zhang, X. (2017b). Bio-catalytic performance and dye-based industrial pollutants degradation potential of agarose-immobilized MnP using a packed bed reactor system. *International Journal of Biological Macromolecules, 102*, 582–590.

Bilal, M., Asgher, M., Iqbal, M., Hu, H., & Zhang, X. (2016c). Chitosan beads immobilized manganese peroxidase catalytic potential for detoxification and decolorization of textile effluent. *International Journal of Biological Macromolecules, 89*, 181–189.

Bilal, M., Asgher, M., Parra-Saldivar, R., Hu, H., Wang, W., Zhang, X., & Iqbal, H. M. (2017a). Immobilized ligninolytic enzymes: an innovative and environmental responsive technology to tackle dye-based industrial pollutants–a review. *Science of the Total Environment, 576*, 646–659.

Bilal, M., Asgher, M., Shahid, M., & Bhatti, H. N. (2016b). Characteristic features and dye degrading capability of agar-agar gel immobilized manganese peroxidase. *International Journal of Biological Macromolecules, 86*, 728–740.

Bilal, M., Iqbal, H. M., Shah, S. Z. H., Hu, H., Wang, W., & Zhang, X. (2016d). Horseradish peroxidase-assisted approach to decolorize and detoxify dye pollutants in a packed bed bioreactor. *Journal of Environmental Management, 183*, 836–842.

Bilal, M., Rasheed, T., Iqbal, H. M., Hu, H., Wang, W., & Zhang, X. (2017c). Novel characteristics of horseradish peroxidase immobilized onto the polyvinyl alcohol-alginate beads and its methyl orange degradation potential. *International Journal of Biological Macromolecules, 105*, 328–335.

Borchert, M., & Libra, J. A. (2001). Decolorization of reactive dyes by the white rot fungus Trametes versicolor in sequencing batch reactors. *Biotechnology and Bioengineering, 75*(3), 313–321.

Brugnari, T., Pereira, M. G., Bubna, G. A., de Freitas, E. N., Contato, A. G., Corrêa, R. C. G., … & Peralta, R. M. (2018). A highly reusable MANAE-agarose-immobilized Pleurotus ostreatus laccase for degradation of bisphenol A. *Science of The Total Environment, 634*, 1346–1351.

Chandra R., Kumar V., Yadav S. (2015) Extremophilic ligninolytic enzymes. In: R. Sani, R. N. Krishnaraj (Eds.), *Extremophilic Enzymatic Processing of Lignocellulosic Feedstocks to Bioenergy*. Springer International Publishing AG. DOI:10.1007/978-3-319-54684-1_8.

Chao, C., Guan, H., Zhang, J., Liu, Y., Zhao, Y., & Zhang, B. (2018). Immobilization of laccase onto porous polyvinyl alcohol/halloysite hybrid beads for dye removal. *Water Science and Technology, 77*(3), 809–818.

Chaudry, M. A., Ahmad, S., & Malik, M. (1998). Supported liquid membrane technique applicability for removal of chromium from tannery wastes. *Waste Management, 17*(4), 211–218.

Chowdhury, M., Mostafa, M., Biswas, T. K., & Saha, A. K. (2013). Treatment of leather industrial effluents by filtration and coagulation processes. *Water Resources and Industry, 3*, 11–22.

Cooman, K., Gajardo, M., Nieto, J., Bornhardt, C., & Vidal, G. (2003). Tannery wastewater characterization and toxicity effects on Daphnia spp. *Environmental Toxicology: An International Journal, 18*(1), 45–51.

Costa, J. B., Lima, M. J., Sampaio, M. J., Neves, M. C., Faria, J. L., Morales-Torres, S., … Silva, C. G. (2019). Enhanced biocatalytic sustainability of laccase by immobilization on functionalized carbon nanotubes/polysulfone membranes. *Chemical Engineering Journal, 355*, 974–985.

Cristóvão, R. O., Silvério, S. C., Tavares, A. P., Brígida, A. I. S., Loureiro, J. M., Boaventura, R. A., … Coelho, M. A. Z. (2012). Green coconut fiber: a novel carrier for the immobilization of commercial laccase by covalent attachment for textile dyes decolourization. *World Journal of Microbiology and Biotechnology, 28*(9), 2827–2838.

Eckenfelder, W. W., & Cleary, J. G. (2013). *Activated Sludge Technologies for Treating Industrial Wastewaters.* DEStech Publications, Inc. Lancaster, PA.

Fu, M., Xing, J., & Ge, Z. (2019). Preparation of laccase-loaded magnetic nanoflowers and their recycling for efficient degradation of bisphenol A. *Science of the Total Environment, 651,* 2857–2865.

Gao, Z., Yi, Y., Zhao, J., Xia, Y., Jiang, M., Cao, F., … Yong, X. (2018). Co-immobilization of laccase and TEMPO onto amino-functionalized magnetic Fe 3 O 4 nanoparticles and its application in acid fuchsin decolorization. *Bioresources and Bioprocessing, 5*(1), 27.

Gassara, F., Brar, S. K., Verma, M., & Tyagi, R. D. (2013). Bisphenol A degradation in water by ligninolytic enzymes. *Chemosphere, 92*(10), 1356–1360.

Ghaly, A., Ananthashankar, R., Alhattab, M., & Ramakrishnan, V. (2014). Production, characterization and treatment of textile effluents: a critical review. *Chemical Engineering Processing: Process Intensification, 5*(1), 1–19.

Gillman, P. K. (2011). CNS toxicity involving methylene blue: the exemplar for understanding and predicting drug interactions that precipitate serotonin toxicity. *Journal of Psychopharmacology, 25*(3), 429–436.

Goh, S., Zhang, J., Liu, Y., & Fane, A. G. (2015). Membrane distillation bioreactor (MDBR)–A lower green-house-gas (GHG) option for industrial wastewater reclamation. *Chemosphere, 140,* 129–142.

Gupta, V. K., Ali, I., Saleh, T. A., Nayak, A., & Agarwal, S. (2012). Chemical treatment technologies for waste-water recycling—an overview. *Rsc Advances, 2*(16), 6380–6388.

Gupta, S., Bhattacharya, A., Murthy, C. N. (2013). Tune to immobilize lipases on polymer membranes: techniques, factors and prospects. *Biocatalysis and Agricultural Biotechnology, 2*(3), 171–190.

Haddis, A., & Devi, R. (2008). Effect of effluent generated from coffee processing plant on the water bodies and human health in its vicinity. *Journal of Hazardous Materials, 152*(1), 259–262.

Hai, F. I., Yamamoto, K., & Fukushi, K. (2007). Hybrid treatment systems for dye wastewater. *Critical Reviews in Environmental Science and Technology, 37*(4), 315–377.

Handayani, N., Loos, K., Wahyuningrum, D., & Zulfikar, M. A. (2012). Immobilization of mucor miehei lipase onto macroporous aminated polyethersulfone membrane for enzymatic reactions, *Membranes, 2*(2), 198–213.

Hao, O. J., Kim, H., & Chiang, P.-C. (2000). Decolorization of wastewater. *Critical Reviews in Environmental Science and Technology, 30*(4), 449–505.

Harazono, K., & Nakamura, K. (2005). Decolorization of mixtures of different reactive textile dyes by the white-rot basidiomycete Phanerochaete sordida and inhibitory effect of polyvinyl alcohol. *Chemosphere, 59*(1), 63–68.

Hassaan, M. A., & El Nemr, A. (2017a). Advanced oxidation processes for textile wastewater treatment. *International Journal of Photochemistry and Photobiology, 2*(3), 85–93.

Hassaan, M. A., & El Nemr, A. (2017b). Health and environmental impacts of dyes: mini review. *American Journal of Environmental Science and Engineering, 1*(3), 64–67.

Hollmann, F., & Arends, I. W. (2012). Enzyme initiated radical polymerizations. *Polymers, 4*(1), 759–793.

Izah, S. C., Chakrabarty, N., & Srivastav, A. L. (2016). A review on heavy metal concentration in potable water sources in Nigeria: human health effects and mitigating measures. *Exposure and Health, 8*(2), 285–304.

Jamil, F., Asgher, M., Hussain, F., & Bhatti, H. N. (2018). Biodegradation of synthetic textile dyes by chito-san beads cross-linked laccase from *Pleurotus ostreatus* IBL-02. *JAPS: Journal of Animal & Plant Sciences, 28*(1), 231–243.

Järup, L. (2003). Hazards of heavy metal contamination. *British Medical Bulletin, 68*(1), 167–182.

Ji, C., Nguyen, L. N., Hou, J., Hai, F. I., & Chen, V. (2017). Direct immobilization of laccase on titania nanoparticles from crude enzyme extracts of P. ostreatus culture for micro-pollutant degradation. *Separation and Purification Technology, 178,* 215–223.

Jin, X., Li, S., Long, N., & Zhang, R. (2018). Improved biodegradation of synthetic azo dye by anionic cross-linking of chloroperoxidase on ZnO/SiO_2 nanocomposite support. *Applied Biochemistry and Biotechnology, 184*(3), 1009–1023.

Kadam, A. A., Jang, J., Jee, S. C., Sung, J. S., & Lee, D. S. (2018). Chitosan-functionalized supermagnetic hal-loysite nanotubes for covalent laccase immobilization. *Carbohydrate Polymers, 194,* 208–216.

Kampmann, M., Boll, S., Kossuch, J., Bielecki, J., Uhl, S., Kleiner, B., & Wichmann, R. (2014). Efficient immobilization of mushroom tyrosinase utilizing whole cells from *Agaricus bisporus* and its application for degradation of bisphenol A. *Water Research, 57,* 295–303.

Kanagaraj, J., Velappan, K. C., Babu, N. K., & Sadulla, S. (2006). Solid wastes generation in the leather industry and its utilization for cleaner environment—a review. *Journal of Scientific & Industrial Research, 65,* 541–548.

Kanu, I., & Achi, O. (2011). Industrial effluents and their impact on water quality of receiving rivers in Nigeria. *Journal of Applied Technology in Environmental Sanitation, 1*(1), 75–86.

Kanu, I., Achi, O., Ezeronye, O., & Anyanwu, E. (2006). Seasonal variation in bacterial heavy metal biosorption in water samples from Eziama River near soap and brewery industries and the environmental health implications. *International Journal of Environmental Science & Technology, 3*(1), 95–102.

Koivunen, J., Siitonen, A., & Heinonen-Tanski, H. (2003). Elimination of enteric bacteria in biological–chemical wastewater treatment and tertiary filtration units. *Water Research, 37*(3), 690–698.

Koklukaya, S. Z., Sezer, S., Aksoy, S., & Hasirci, N. (2016). Polyacrylamide-based semi-interpenetrating networks for entrapment of laccase and their use in azo dye decolorization. *Biotechnology and Applied Biochemistry, 63*(5), 699–707.

Koloti, L. E., Gule, N. P., Arotiba, O. A., & Malinga, S. P. (2018). Laccase-immobilized dendritic nanofibrous membranes as a novel approach towards the removal of bisphenol A. *Environmental Technology, 39*(3), 392–404.

Kornhauser, C., Wrobel, K., Wrobel, K., Malacara, J. M., Nava, L. E., Gómez, L., & González, R. (2002). Possible adverse effect of chromium in occupational exposure of tannery workers. *Industrial Health, 40*(2), 207–213.

Kucharzyk, K. H., Benotti, M., Darlington, R., & Lalgudi, R. (2018). Enhanced biodegradation of sediment-bound heavily weathered crude oil with ligninolytic enzymes encapsulated in calcium-alginate beads. *Journal of Hazardous Materials, 357*, 498–505.

Kurita, K. (2001). Controlled functionalization of the polysaccharide chitin. *Progress in Polymer Science, 26*(9), 1921–1971.

Lassouane, F., Aït-Amar, H., Amrani, S., & Rodriguez-Couto, S. (2019). A promising laccase immobilization approach for Bisphenol A removal from aqueous solutions. *Bioresource Technology, 271*, 360–367.

Leghouchi, E., Laib, E., & Guerbet, M. (2009). Evaluation of chromium contamination in water, sediment and vegetation caused by the tannery of Jijel (Algeria): a case study. *Environmental Monitoring and Assessment, 153*(1–4), 111.

Li, H., Hou, J., Duan, L., Ji, C., Zhang, Y., & Chen, V. (2017b). Graphene oxide-enzyme hybrid nanoflowers for efficient water soluble dye removal. *Journal of Hazardous Materials, 338*, 93–101.

Li, G., Nandgaonkar, A. G., Wang, Q., Zhang, J., Krause, W. E., Wei, Q., & Lucia, L. A. (2017a). Laccase-immobilized bacterial cellulose/TiO2 functionalized composite membranes: Evaluation for photo-and bio-catalytic dye degradation. *Journal of Membrane Science, 525*, 89–98.

Liew, W. L., Kassim, M. A., Muda, K., Loh, S. K., & Affam, A. C. (2015). Conventional methods and emerging wastewater polishing technologies for palm oil mill effluent treatment: a review. *Journal of Environmental Management, 149*, 222–235.

Ludvik, J. (1996). Cleaner tanning technologies, UNIDO report, 18–25.

Martin, S., & Griswold, W. (2009). Human health effects of heavy metals. *Environmental Science and Technology Briefs for Citizens, 15*, 1–6.

Maryšková, M., Ardao, I., García-González, C. A., Martinová, L., Rotková, J., & Ševců, A. (2016). Polyamide 6/chitosan nanofibers as support for the immobilization of Trametes versicolor laccase for the elimination of endocrine disrupting chemicals. *Enzyme and Microbial Technology, 89*, 31–38.

Mogharabi, M., Nassiri-Koopaei, N., Bozorgi-Koushalshahi, M., Nafissi-Varcheh, N., Bagherzadeh, G., & Faramarzi, M. A. (2012). Immobilization of laccase in alginate-gelatin mixed gel and decolorization of synthetic dyes. *Bioinorganic Chemistry and Applications, 2012*(823830), 6 pages, DOI:10.1155/2012/823830

Moldes-Diz, Y., Gamallo, M., Eibes, G., Vargas-Osorio, Z., Vazquez-Vazquez, C., Feijoo, G., … Moreira, M. T. (2018). Development of a superparamagnetic laccase nanobiocatalyst for the enzymatic biotransformation of xenobiotics. *Journal of Environmental Engineering, 144*(3), 04018007.

Morthensen, S. T., Luo, J., Meyer, A. S., Jørgensen, H., Pinelo, M. (2015). High performance separation of xylose and glucose by enzyme assisted nanofiltration. *Journal of Membrane Science, 492*, 107–115.

Neoh, C. H., Noor, Z. Z., Mutamim, N. S. A., & Lim, C. K. (2016). Green technology in wastewater treatment technologies: integration of membrane bioreactor with various wastewater treatment systems. *Chemical Engineering Journal, 283*, 582–594.

Nga, N. K., Chinh, H. D., Hong, P. T. T., & Huy, T. Q. (2017). Facile preparation of chitosan films for high performance removal of reactive blue 19 dye from aqueous solution. *Journal of Polymers and the Environment, 25*(2), 146–155.

Ogwuegbu, M., & Muhanga, W. (2005). Investigation of lead concentration in the blood of people in the copper belt province of Zambia. *Journal of Environmental, 1*, 66–75.

Oliveira, S. F., da Luz, J. M. R., Kasuya, M. C. M., Ladeira, L. O., Junior, A. C. (2018). Enzymatic extract containing lignin peroxidase immobilized on carbon nanotubes: potential biocatalyst in dye decolourization. *Saudi Journal of Biological Sciences, 25*, 651–659.

World Health Organization. (2010). Persistent organic pollutants: impact on child health.

Padmesh, T., Vijayaraghavan, K., Sekaran, G., & Velan, M. (2005). Batch and column studies on biosorption of acid dyes on fresh water macro alga Azolla filiculoides. *Journal of Hazardous Materials, 125*(1–3), 121–129.

Patel, S. K., Otari, S. V., Li, J., Kim, D. R., Kim, S. C., Cho, B. K., … Lee, J. K. (2018). Synthesis of cross-linked protein-metal hybrid nanoflowers and its application in repeated batch decolorization of synthetic dyes. *Journal of Hazardous Materials, 347*, 442–450.

Rahman, F. (2016). The treatment of industrial effluents for the discharge of textile dyes using by techniques and adsorbents. *Journal Textile Science and Engineering, 6*, 242.

Rasheed, T., Bilal, M., Nabeel, F., Adeel, M., & Iqbal, H. M. (2019). Environmentally-related contaminants of high concern: Potential sources and analytical modalities for detection, quantification, and treatment. *Environment International, 122*, 52–66.

Rasheed, T., Bilal, M., Nabeel, F., Iqbal, H. M., Li, C., & Zhou, Y. (2018). Fluorescent sensor based models for the detection of environmentally-related toxic heavy metals. *Science of the Total Environment, 615*, 476–485.

Reda, F. M., Hassan, N. S., & El-Moghazy, A. N. (2018). Decolorization of synthetic dyes by free and immobilized laccases from newly isolated strain Brevibacterium halotolerans N11 (KY883983). *Biocatalysis and Agricultural Biotechnology, 15*, 138–145.

Rodríguez, F. A., Poyatos, J. M., Reboleiro-Rivas, P., Osorio, F., González-López, J., & Hontoria, E. (2011). Kinetic study and oxygen transfer efficiency evaluation using respirometric methods in a submerged membrane bioreactor using pure oxygen to supply the aerobic conditions. *Bioresource technology, 102*(10), 6013–6018.

Rott, U. (2003). Multiple use of water in industry—the textile industry case. *Journal of Environmental Science and Health, Part A, 38*(8), 1629–1639.

Roy, S., Das, I., Munjal, M., Karthik, L., Kumar, G., Kumar, S., & Rao, K. V. B. (2014). Isolation and characterization of tyrosinase produced by marine actinobacteria and its application in the removal of phenol from aqueous environment. *Frontiers in Biology, 9*(4), 306–316.

Sarayu, K., & Sandhya, S. (2012). Current technologies for biological treatment of textile wastewater–a review. *Applied Biochemistry and Biotechnology, 167*(3), 645–661.

Shah, M. (2016). Industrial wastewater treatment: a challenging task in the industrial. *Advances in Recycling & Waste and Management, 2*: 115.

Shaheen, R., Asgher, M., Hussain, F., & Bhatti, H. N. (2017). Immobilized lignin peroxidase from Ganoderma lucidum IBL-05 with improved dye decolorization and cytotoxicity reduction properties. *International Journal of Biological Macromolecules, 103*, 57–64.

Shao, B., Liu, Z., Zeng, G., Liu, Y., Yang, X., Zhou, C., & Yan, M. (2019). Immobilization of laccase on hollow mesoporous carbon nanospheres: noteworthy immobilization, excellent stability and efficacious for antibiotic contaminants removal. *Journal of Hazardous Materials, 362*, 318–326.

Sheldon, R. (2001). Catalytic reactions in ionic liquids. *Chemical Communications, (23)*, 2399–2407.

Sheldon, R. A. (2007). Cross-linked enzyme aggregates (CLEAs): stable and recyclable biocatalysts. *Biochemical Society, 35*(6), 1583–1587.

Soares, P. A., Souza, R., Soler, J., Silva, T. F., Souza, S. M. G. U., Boaventura, R. A., & Vilar, V. J. (2017). Remediation of a synthetic textile wastewater from polyester-cotton dyeing combining biological and photochemical oxidation processes. *Separation and Purification Technology, 172*, 450–462.

Sofia, P., Asgher, M., Shahid, M., & Randhawa, M. A. (2016). Chitosan beads immobilized *Schizophyllum commune* IBL-06 lignin peroxidase with novel thermo stability, catalytic and dye removal properties. *JAPS: Journal of Animal & Plant Sciences, 26*(5), 1451–1463.

Sondhi, S., Kaur, R., Kaur, S., & Kaur, P. S. (2018). Immobilization of laccase-ABTS system for the development of a continuous flow packed bed bioreactor for decolorization of textile effluent. *International Journal of Biological Macromolecules, 117*, 1093–1100.

Sun, H., Jin, X., Jiang, F., & Zhang, R. (2018). Immobilization of horseradish peroxidase on ZnO nanowires/macroporous SiO_2 composites for the complete decolorization of anthraquinone dyes. *Biotechnology and Applied Biochemistry, 65*(2), 220–229.

Sun, H., Yang, H., Huang, W., & Zhang, S. (2015). Immobilization of laccase in a sponge-like hydrogel for enhanced durability in enzymatic degradation of dye pollutants. *Journal of Colloid and Interface Science, 450*, 353–360.

Taheran, M., Naghdi, M., Brar, S. K., Knystautas, E. J., Verma, M., & Surampalli, R. Y. (2017). Degradation of chlortetracycline using immobilized laccase on Polyacrylonitrile-biochar composite nanofibrous membrane. *Science of the Total Environment, 605*, 315–321.

Tariq, S. R., Shah, M. H., Shaheen, N., Khalique, A., Manzoor, S., & Jaffar, M. (2005). Multivariate analysis of selected metals in tannery effluents and related soil. *Journal of Hazardous Materials, 122*(1–2), 17–22.

Teerapatsakul, C., Parra, R., Keshavarz, T., & Chitradon, L. (2017). Repeated batch for dye degradation in an airlift bioreactor by laccase entrapped in copper alginate. *International Biodeterioration & Biodegradation, 120*, 52–57.

Temoçin, Z., İnal, M., Gökgöz, M., & Yiğitoğlu, M. (2018). Immobilization of horseradish peroxidase on electrospun poly (vinyl alcohol)–polyacrylamide blend nanofiber membrane and its use in the conversion of phenol. *Polymer Bulletin, 75*(5), 1843–1865.

Umamaheswari, J., & Shanthakumar, S. (2016). Efficacy of microalgae for industrial wastewater treatment: a review on operating conditions, treatment efficiency and biomass productivity. *Reviews in Environmental Science and Bio/Technology, 15*(2), 265–284.

Verma, P., & Madamwar, D. (2003). Decolourization of synthetic dyes by a newly isolated strain of Serratia marcescens. *World Journal of Microbiology and Biotechnology, 19*(6), 615–618.

Vineh, M. B., Saboury, A. A., Poostchi, A. A., & Mamani, L. (2018). Physical adsorption of horseradish peroxidase on reduced graphene oxide nanosheets functionalized by amine: A good system for biodegradation of high phenol concentration in wastewater. *International Journal of Environmental Research, 12*(1), 45–57.

Wesenberg, D., Kyriakides, I., & Agathos, S. N. (2003). White-rot fungi and their enzymes for the treatment of industrial dye effluents. *Biotechnology Advances, 22*(1–2), 161–187.

Xu, H. M., Sun, X. F., Wang, S. Y., Song, C., & Wang, S. G. (2018). Development of laccase/graphene oxide membrane for enhanced synthetic dyes separation and degradation. *Separation and Purification Technology, 204*, 255–260.

Yang, L., Li, X., Li, X., Su, Z., Zhang, C., Xu, M., & Zhang, H. (2015). Improved stability and enhanced efficiency to degrade chlorimuron-ethyl by the entrapment of esterase SulE in cross-linked poly (γ-glutamic acid)/gelatin hydrogel. *Journal of Hazardous Materials, 287*, 287–295.

Zhang, Z., & Balay, J. W. (2014). How much is too much? Challenges to water withdrawal and consumptive use management. American Society of Civil Engineers.

Zhang, H., Luo, J., Li, S., Wei, Y., & Wan, Y. (2018). Biocatalytic membrane based on polydopamine coating: a platform for studying immobilization mechanisms. *Langmuir, 34*(8), 2585–2594.

Zheng, Q., Dai, Y., & Han, X. (2016). Decolorization of azo dye CI Reactive Black 5 by ozonation in aqueous solution: influencing factors, degradation products, reaction pathway and toxicity assessment. *Water Science and Technology, 73*(7), 1500–1510.

Zheng, X., Zhang, Z., Yu, D., Chen, X., Cheng, R., Min, S., … Wang, J. (2015). Overview of membrane technology applications for industrial wastewater treatment in China to increase water supply. *Resources, Conservation and Recycling, 105*, 1–10.

18 Lipase of Lactic Acid Bacteria
Diversity and Application

Vrinda Ramakrishnan, Bhaskar Narayan,
and Prakash M. Halami
CSIR-Central Food Technological Research Institute

CONTENTS

18.1 INTRODUCTION

Lipases (triacylglycerol acyl hydrolases EC 3.1.1.3) are a class of serine hydrolases that catalyze the hydrolysis of triglycerides to glycerol and free fatty acids over the oil-water interface (Liu et al. 2006). Microbial lipases today occupy a place of prominence among biocatalysts owing to their ability to catalyze a wide variety of reactions in aqueous and nonaqueous media as well as their chemo-, regio-, and enantio-specific behavior that has caused a tremendous research interest among scientists and industrialists. Microbial lipases are generally preferred over plant and animal lipases in most of the industrial applications since they are diversified in their enzymatic properties and substrate specificity (Gupta et al. 2004). Lipases catalyze the hydrolysis and transesterification of triacylglycerols, enantioselective synthesis, and hydrolysis of a variety of esters (Uppada et al. 2017). These multifunctional enzymes are involved in several biological processes that are associated with the metabolism of dietary triglycerides to inflammation. Commercially, lipases are being exploited as cheap and versatile catalysts to degrade lipids in more modern applications. Most lipases act at a specific position on the glycerol backbone of a lipid substrate, thus liberating glycerol and free fatty acids sequentially (Figure 18.1). Lipases from lactic acid bacteria (LAB) are of immense importance as biocatalysts of industrial significance, therapeutic applications, and flavoring agents in the food industry. LAB strains have either intracellular or extracellular lipases. The ability of LAB to perform unique fatty acid transformation reactions including isomerization, hydration, dehydration, and saturation has several industrial relevances. Hence, there is a worldwide thrust area of research on LAB lipase (Vrinda et al. 2015, 2016).

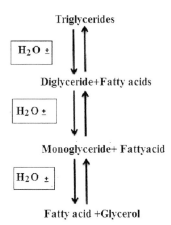

FIGURE 18.1 Diagrammatic representation of lipase-mediated reversible reaction.

An updated and extensive classification of bacterial esterases and lipases based mainly on a comparison of their amino acid sequences and some fundamental biological properties has been made, and these new insights result in the identification of eight different families with the largest being further divided into six subfamilies. This classification facilitated the prediction of important structural features such as residues forming the catalytic site or the presence of disulfide bonds, types of secretion mechanism and requirement for lipase-specific foldases, and the potential relationship to other enzyme families (Arpigny and Jaeger 1999). Lipase is the member of α/β hydrolase fold superfamily, having the conserved signature motive of Gly-Xaa-Ser-Xaa-Gly. Recently, Messaoudi et al. (2010) have proposed the classification of the family of bacteria through lipase based on the amino acid sequence, by the phylogenetic analysis of the identified 11 subfamilies of these lipases. Lipase of *Lactobacillus acidophilus* represents subfamily 1.7. Biochemically, lipase is hydrolyase that acts under the aqueous condition on the carboxyl bond of the triacylglycerides. Hence, natural substrate of lipase is long-chain triglycerides. A large number of LAB-producing lipases have been reported in the literature and thus (LAB) include *Lactobacillus breve, Lactobacillus casei, Lactobacillus fermentum, Lactobacillus plantarum, Pediococcus pentosaceus, Pediococcus cerevisiae,* and *Leuconostoc mesenteroides,* wherein lipase is found in cell-free extract of a disrupted cell (Katz et al. 2002, Jini et al. 2011, Vrinda 2013). In these cultures, the whole-cell extract of *Lactobacillus paracasei* and *Lactobacillus sake* is identified for lipase activity. Most of these lipases were evaluated for their substrate hydrolysis that includes oil from milk and dairy origin, tributyrin coconut oil, butter, fat, etc.

Most commonly, bacterial lipases of commercial importance are obtained from *Achromobacter, Alcaligenes, Bacillus, Bulkholderia, Chromobacterium,* and *Pseudomonas* (Gupta et al. 2004). However, due to their pathogenic nature, there is a safety concern for their food applications. LAB are generally considered to be weakly lipolytic, as compared with other groups of microorganisms. However, it is believed that lipolytic activity by LAB plays an important role in the determination of the special aroma of many different kinds of cheese, suggesting their unique property and action (Collins et al. 2003). Certain LAB, which are used as bulk starter cultures (*Lactococcus* and *Lactobacillus*), together with some other bacteria (*Leuconostoc, Enterococcus, Pediococcus,* and *Micrococcus*), which can survive pasteurization and/or contaminate cheese during maturation, are believed to significantly contribute to cheese fat hydrolysis (Collins et al. 2003). Most of the LAB are considered as generally recognized as safe (GRAS) and are used extensively as starter cultures in food and feed industries (Rai et al. 2010, Uppada et al. 2017). There are reports on lipase-producing LAB isolated from some fish products; however, lipase activity is less in comparison with other microorganisms such as *Bacillus* and *Pseudomonas* (Jini et al. 2011. Vrinda et al. 2012).

18.2 SCREENING OF LIPASE FROM LAB

Lipase production is known to be the basic property of most of the microorganisms wherein LAB are known to be weakly lipolytic. In general, screening by hydrolysis of tributyrin has been considered for obtaining lipase-positive cultures. Rhodamine B is another substrate used for screening lipase producer (Vrinda 2013). Different lipolytic LAB isolated from different food sources and environmental samples is described in Table 18.1. Most LAB associated with lipolytic property belong to *Lactococcus* and *Enterococcus*. LAB producing lipases are identified by using its ability to use triglyceride as a substrate or by the amount of fatty acid produced or clarification of emulsion in oil (Katz et al. 2002). The lipolytic activity of LAB on solid media is being identified based on the zone of hydrolysis of the substrate such as Victoria blue B, Spirit blue, Nile blue sulfate, and Night blue. Rhodamine B, a fluorescent dye, is being used in plate assay to detect the activity of several microbes including LAB (Figure 18.2).

For the purpose of in situ enrichment or for fermentation application, lipolytic LAB isolated from the same source is preferred in comparison with LAB from other sources. Those bacteria serve as a better starter culture for fermentation of similar or same products (Jini et al. 2011). Though, enzymatic hydrolysis is an alternative approach for both protein and lipid recovery and finds a wide range of applications in the food industry as it brings about products of high functionality and nutritive value (Rai et al. 2010). This is true in case of a fish waste, which is available in abundant and at low cost. In this direction, an attempt was made to isolate lipolytic LAB from fish processing waste (FPW), as these bacteria also play a major role in the degradation of lipid-rich fish waste through enzymatic hydrolysis of lipids present in FPW that could be exploited for in situ

TABLE 18.1
Lipolytic Lactic Acid Bacteria Isolated from Different Sources

Source	LAB	Substrate	Purpose	References
Fish processing waste	*Pediococcus acidilactici* NCIM5368, *Enterococcus faecalis* NCIM5367, *Pediococcus acidilactici* FM37, and *Pediococcus acidilactici* MW2	Tributyrin	Fermentative recovery of lipids and proteins	Rai et al. 2010
	Enterococcus faecalis NCIM 5367, *Pediococcus acidilactici* NCIM 5368	Tributyrin	Fermentative utilization of fish processing waste	Jini et al. 2011
	Enterococcus faecium NCIM5363 and *Enterococcus durans* NCIM5427	Tributyrin	Synergistic activity against gram-negative bacteria	Vrinda et al. 2012
Thai-fermented meat	*Lactobacillus pentosus, Lactobacillus* sp., *Pediococcus pentosaceus, Pediococcus lolii, Leuconostoc fallax, Weissella thailandensis, Weissella cibaria*, and *Weissella paramesenteroides*	Tween 20, Tween 40, Tween 60 or Tween 80	Aroma in meat	Somboon et al. 2015
Algerian traditional fermented milk products	*Lactobacillus curvatus, Leuconostoc mesenteroides* subsp. *mesenteroides, Lactobacillus plantarum, Lactobacillus brevis, Lactobacillus acidophilus*, and *Lactococcus lactis* subsp. *lactis*	Tributyrin	Flavor and aroma in milk products	Mechai et al. 2014
Fermented bamboo shoots	*Lactobacillus fermentum, Lactococcus* sp., *L. brevis, L. curvatus, Leuconostoc* sp., and *Lactobacillus xylosus*	Tributyrin	Technological attributes of native LAB	Sonar and Halami 2014

FIGURE 18.2 Screening and confirmation of lipase producing LAB. (a) Screening lipolytic LAB on MRS agar plate supplemented with 1% tributyrin. (b) Streaking of lipolytic LAB cultures on MRS agar tributyrin plate. (c) Rhodamine agar test for confirmation and differentiation of true lipase producing LAB isolates. Arrow indicates orange fluorescence of the colony on UV irradiation which is a characteristic feature of true lipase. (Courtesy of Vrinda, R., *Characterization of Lipase Producing Lactic Acid Bacteria Isolated from Fish Processing Waste*, PhD Thesis, University of Mysore, India, 2013.)

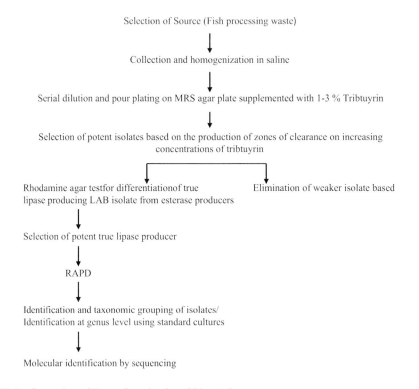

FIGURE 18.3 Screening of lipase from lactic acid bacteria.

enrichment of PUFA (Rai et al. 2010, Jini et al. 2011). Fish waste is known to be more in lipid, which increases the possibility of enrichment of lipolytic bacteria and thereby helps in faster and convenient screening for potent lipase-producing LAB. Importance of lipases from LAB unlike from other pathogenic sources possesses GRAS status, and thereby their metabolites can be directly used with minimal processing in the food industry (Vrinda 2013). Besides substrate-based screening, an efficient screening system is being developed for the identification of lipase-producing species/clones from a diverse population. There is a possibility of pinpointing enzyme variants that display the desired properties during molecular screening (Gupta et al. 2003).

With the advent of metagenomic approaches, these tools are being used worldwide for obtaining desirable lipase clone/s (Figure 18.3).

18.3 OPTIMIZATION OF LIPASE PRODUCTION

As stated before, LAB being poor in lipolytic activity, for their purification and application purpose required enzyme, have been produced in large scale. Literature is scanty in this regard for LAB lipase (Bhargavi et al. 2010). With respect to optimization of parameters, response surface modeling has been effectively used (Vrinda et al. 2015). In this study, for optimization of condition for enhanced lipase production, authors have employed statistical methodologies for maximal production by LAB isolates. Box-Behnken design was found to be suitable for increase in lipase production up to three-fold for both enterococci culture tested. The cost-effective medium designed by these authors was FPW, which was known to provide for growth and lipase production.

Being an inducible enzyme, lipase production is often the result of a synergistic combination of effective interactions between numerous parameters. It has been studied that lipase production often depends on several process variables, such as pH, temperature, incubation period, substrate concentration, inoculum level, and inducer concentration (Vrinda 2013, Vrinda et al. 2015). It is possible that lipase production can be stimulated in the presence of a specific substrate. It is known that conventional method of medium optimization that involves a variation of "one variable/one factor at a time" (OVAT/OFAT) is time-consuming, laborious, and expensive and may lead to misinterpretation of results. Unlike conventional methods, it is known that statistical experimental methods offer a simultaneous, systematic, and efficient variation of all the components economically (Vrinda et al. 2014). Predictive models provide informative tools for the rapid and cost-effective study of microbial growth, their products development, risk assessment, and scientific purposes. Utilization of FPW as a substrate for growth of LAB not only solves the problem of waste disposal but also helps in the conversion of the waste material for consequent utilization bringing economic advantages to the industry (Rai et al. 2010, Vrinda 2013). Lipase production under submerged and solid-state fermentation by *Lactobacillus acidophilus* NCIM 2909 was investigated by Bhargavi et al. (2010), wherein gingelly oil cake was used as a lipidic carbon source for higher enzyme production. These authors state that bioprocess usage of oil cakes is beneficial due to its availability and cheaper cost (Figure 18.4).

18.4 CHARACTERISTICS OF LIPASE PRODUCED BY LAB

Though lipase from several different bacterial species is well characterized, detailed information on lipase of LAB has not been documented. Literature suggests few reports on characteristics of lipase from LAB (El-Sawah et al. 1995, Esteban-Torres et al. 2015, Lopes et al. 2002). Upon purification, properties of lipases from LAB have been studied, and molecular weight has been determined.

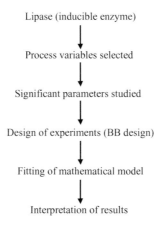

FIGURE 18.4 Optimization of lipase production.

TABLE 18.2
Characteristic of Lipase Produced by Lactic Acid Bacteria

LAB	Molecular Weight (kDa)	Properties	Reference
Lactobacillus delbrueckii subsp. *bulgaricus*	NA	pH 6.0 and temperature 35°C	El-Sawah et al. 1995
Lactobacillus plantarum MF32	75	pH 9.3 and temperature 37°C	Andersen et al. 1995
Lactobacillus plantarum 2739	65	pH 5.0 and temperature 15°C	Gobbetti et al. 1996
Lactococcus lactis subsp. *cremoris* E8	29 and a holoenzyme 109	Half-life of 1 h at 50°C	Holland and Coolbear 1996
Lactobacillus plantarum DSMZ 12028	4 bands, between 98 and 45 kDa	Extracellular lipase (temperature, pH not described)	Lopes et al. 2002
Lactobacillus plantarum	38	pH 6.0 and temperature 40°C	Brod et al. 2010
Lactobacillus spp.	NA	pH 9.0 and temperature 40°C	Padmapriya et al. 2011
Lactobacillus plantarum	30	pH 7.0 and temperature 40°C	Esteban-Torres et al. 2015
Enterococcus faecium MTCC5695	19.2	pH 10.8 and temperature 40°C	Vrinda et al. 2016
Lactobacillus plantarum 2739	65	pH 7.5 and temperature 35°C	Gobbetti et al. 1996

NA, not available.

As per the literature search, brief characteristics of lipase from the LAB have been compiled and are described in Table 18.2. Most of the lipases are found to be of acidic in nature; however, lipase with optimum activity at pH 10.8 is reported by Vrinda et al. (2016). Intracellular lipases obtained from LAB culture were purified by using an aqueous two-phase extraction system. The molecular weight of lipase produced by *Enterococcus faecium* was around 19.2 kDa, whereas lipase produced by *Enterococcus durans* was 21.6 kDa. The lipases were found to be stable in the presence of various organic solvents, suggesting their importance in industrial application. Most of the commercial applications of enzymes do not always need homogeneous preparation of the enzyme. However, a certain degree of purity is required, depending upon the final application, in industries such as fine chemicals, pharmaceuticals, and cosmetics (Gupta et al. 2004). For industrial purposes, the purification strategies employed should be inexpensive, rapid, high yielding, and amenable to large-scale operations. They should have the potential for continuous product recovery, with a relatively high capacity and selectivity for the desired product.

18.5 APPLICATION OF LAB LIPASES

Like many other enzymes, lipase is used in food processing, mostly for the modification and breakdown of lipid-polymer. Most commonly, commercial lipase is utilized for flavor development in the dairy product as well as in meat, baking, and alcoholic fermentation (Matthews et al. 2004). Some of the potential industries wherein lipase as an enzyme or lipase-producing LAB has been employed are described below.

18.5.1 DAIRY INDUSTRY

Lipase is very commonly used for the hydrolysis of milk fat. Especially for flavor development in cheese (Collins et al. 2003), Martínez-Cuesta et al. (2001) described *Lactobacillus casei* IFPL731 that had methionine aminotransferase activity resultant into the production of the typical cheese aroma. This strain showed multiple enzymatic systems that involve esterase, cell envelope proteinase, aminopeptidases, dipeptidases, specialized peptidases for proline-containing peptides, and amino acid–converting enzymes are found to be useful in flavor development in the cheese.

18.5.2 BAKING INDUSTRY AND MEAT PROCESSING

The lipolytic enzyme has been commonly used for the degradation of wheat lipid. Similarly, Uppada et al. (2017) studied the degradation property of lipase obtained from *L. plantarum* wherein complete degradation of meat was observed at 72 h. Padmapriya et al. (2011) carried out optimization of fermentation conditions such as temperature and pH for maximum lipase production from *Lactobacillus* sp. It was found that *Lactobacillus* sp. has more lipase activity when compared with another bacterial lipase. As a part of application study, it was evaluated for meat degradation, suggesting the industrial utility of lipase of *Lactobacillus* sp.

18.5.3 LIPASE BIOLOGY BIOSENSOR

For the quantitative determination of triglycerol, immobilized lipase is commonly used as a cost-effective sensor. These biosensing tools are very important in food industries as well as in clinical diagnostic, wherein the release of glycerol is being monitored. Due to the high diversity of lipase of LAB, they offer a range of molecules with different specificity that can be selectively employed for sensing purpose (Arpigny and Jaeger 1999, Messaoudi et al. 2010).

18.5.4 LIPASE IN WASTE MANAGEMENT

Lipases have wide-scale application in environmental waste treatment. Effluents, especially Abattoirs leather industries, poultry waste, etc., can be treated using lipase as an eco-friendly method. These lipid-rich waste treatments contain mostly triglyceride that can be efficiently hydrolyzed by lipase. Such hydrolysis will contribute to the reduction in the size of fat particles (Vrinda 2013). Wastewater of the poultry industry generally contained the high amount of oil and fat. These wastes possess low biodegradability. In the recent past, enzymatic biodegradation is suggested for fat degradation. Such biological process also takes care in preventing unpleasant odor from solid material accumulation as well as sedimentation. Though an eco-friendly process, lipase can be employed in waste treatment to hydrolyze and dissolve the fat material present in the wastewater and hence can reduce the usage of chemical solvents that are being known to affect microbial diversity (Vrinda 2013).

In a study by Dors et al. (2013), simultaneously enzymatic hydrolysis and anaerobic biodegradation of lipid-rich waste generated from the poultry industry were carried out. Such a biological process is used to minimize the environmental pollution caused by discharged lipid. Previously, it was found that the use of chemical substances for waste treatment showed a toxic effect to several methanogenic and acetogenic microorganisms (Dors et al. 2013). Some of LAB producing lipase or their enzyme preparation have been employed in different sectors, and detailed information pertaining to their applications has been provided in Table 18.3.

18.6 CLONING AND HET-EXPRESSION OF THE LIPASE OF LAB

Microbial diversity is a major resource for biotechnological products and processes (Gupta et al. 2004). An alternative approach is to use the genetic diversity of the microorganisms in a certain environment as a whole (also known as "metagenome") to obtain new or improved genes and gene products for biotechnological purposes (Henne et al. 2000). In the recent past, sequencing

TABLE 18.3
Application of Lipases Produced by Lactic Acid Bacteria (LAB)

Lipolytic LAB	Source	Application of Lipase Produced	Reference
Leuconostoc oenos (Oenococcus oeni)	Wine	Vinification	Matthews et al. 2004
Lactobacillus cuwatus and Lactobacillus sake	Fermented meat products	Starter cultures	Hammes et al. 1990
Lactobacillus casei subsp. casei IFPL73	Goat milk cheese	Adjunct culture for cheese preparation	Martínez-Cuesta et al. 2001
Lactobacillus acidophilus NCIM 2909	Standard culture	Application in oil cake	Bhargavi et al. 2010
Lactobacillus sp. G5	Goat curd	Hydrolysis of lipid	Rashmi and Gayathri 2014
Enterococcus faecium MTCC5695	Fish processing waste	Starter cultures for probiotic curds	Vrinda et al. 2014
Enterococcus durans NCIM 5427	Fish processing waste	Treatment of slaughter house waste	Vrinda et al. 2015
Lactobacillus plantarum MTCC4461	MTCC	Meat degradation and synthesis of flavor esters	Uppada et al. 2017

TABLE 18.4
Heterogeneous Expression of Lipase/Esterase of Lactic Acid Bacteria (LAB)

LAB	Source	Cloning and Characterization	Application	Reference
Lactococcus lactis	L. lactis NZ9700 standard culture	Cloning, characterization, controlled overexpression	Flavor formation in dairy products	Fernandez et al. 2000
Leuconostoc oenos (Oenococcus oeni)	Wine	A 912-bp ORF with GDSAG amino acid motif	Winemaking	Sumby et al. 2009
Lactobacillus plantarum	Fermented vegetables and meat products	39.075 kDa recombinant lipase/esterase of L. plantarum	Basic studies	Brod et al. 2010

of large metagenomic DNA fragments has fortuitously revealed numerous open reading frames, many of them encoding enzymes such as chitinase, lipase, esterase, protease, amylase, DNase, and xylanase. Limited studies are carried out in LAB lipase cloning and expression. Some of the lipase-like genes of LAB that have been expressed in heterogenous system has been described in Table 18.4.

18.7   GENOMIC VIEW OF LIPASE PRODUCED BY LAB

Lipase gene for LAB is poorly described in the public domains (NCBI GenBank). However, complete genome sequences show putative sequences as identified by similarity and conserved motif associated with GDSL amino acid sequences (Messaoudi et al. 2010). Based on that information, and P-BLAST search, some of the protein sequences that are identified as lipase-like are aligned and presented in Figure 18.5. This information provides clues for evolutionary relationship and diversity of lipase form LAB. It has been estimated that the diversity of lipase ranges from 30% to 90% in LAB.

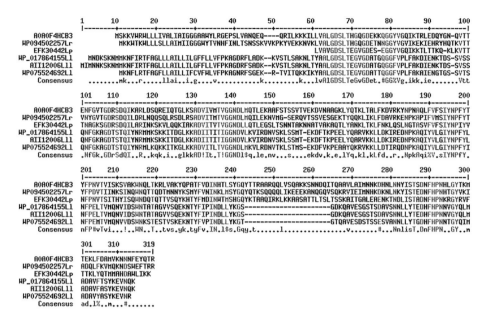

FIGURE 18.5 Multiple sequence alignment of lipase produced by lactic acid bacteria. WP075524692Ll: WP_075524692.1 lipase of *Lactococcus lactis;* WP_017864155Ll: WP_017864155.1 GDSL family lipase *Lactococcus lactis*; AII12006Lll: Lipase of *Lactococcus lactis* subsp. *lactis* NCDO 2118; EFK30442Lp: EFK30442.1 GDSL-like protein of *Lactobacillus plantarum* subsp. *plantarum* ATCC 14917; A0A0F4HCB3Lf: WP_046025709.1 lipase of *Lactobacillus fermentum*; and WP094502257Lr: WP_094502257.1, GDSL family lipase of *Lactobacillus reuteri.*

18.8 CONCLUSION

Lipases are one of the most versatile biocatalysts available in nature. The exponential increase in the use of lipases in biotechnology in comparison with other hydrolytic enzymes indicates the growing demand as well as application, especially in making fine chemicals. Lipases with improved properties are being produced by natural selection and protein engineering to further enhance the usefulness of these enzymes. The genomic view provided in this chapter supports this claim. Simultaneously, advances are being made in bioreactor and reaction technologies for effectively using the lipases. Purification and studying properties are needed to explore lipases of LAB for industrial application. More extensive research must be carried out in the field of lipases from LAB, as they are highly beneficial owing to their GRAS status and can be directly used in any industry with minimal processing. In addition, the supplementation of lipases during the anaerobic biodegradation of lipid-rich wastewaters represents an attractive solution for remediation of problems associated with biodegradation as well as an eco-friendly method. Despite its importance, studies on the mechanisms of production of LAB lipases and the role of lipidic substances used as inducers in lipase production are to be investigated. Recent advances in recombinant DNA technologies, high-throughput technologies, genomics, and proteomics can help in identifying LAB lipase with potential commercial relevance.

REFERENCES

Andersen HJ, Østdal H, Blom H (1995) Partial purification and characterization of a lipase from *Lactobacillus plantarum* MF32. *Food Chem*, 53:369–373.
Arpigny JL, Jaeger KE (1999) Bacterial lipolytic enzymes: classification and properties. *Biochem J*, 343:177–183.

Bhargavi PL, Manjushri R, Reddy PN (2010) Lipase production by lactic acid bacteria in submerged and solid state fermentation. *Biotechnol An Ind J*, 4:126–129.

Brod FC, Vernal J, Bertoldo JB, Terenzi H, Arisi AC (2010) Cloning expression, purification, and characterization of a novel esterase from *Lactobacillus plantarum*. *Mol Biotechnol*, 44:242–249.

Collins YF, McSweeney PLH, Wilkinson MG (2003) Lipolysis and free fatty acid catabolism in cheese: a review of the current knowledge. *Int Dairy J*, 13:841–866.

Dors G, Mendes AA, Pereira EB, de Castro HF, Furigo A (2013) Simultaneous enzymatic hydrolysis and anaerobic biodegradation of lipid-rich wastewater from poultry industry. *Appl Water Sci*, 3:343–349.

El-Sawah MMA, Sherief AA, Bayoumy SM (1995) Enzymatic properties of lipase and characteristics production by *Lactobacillus delbrueckii* subsp. bulgaricus. *Antonie van Leeuwenhoek*, 67:357–362.

Esteban-Torres M, Mancheño JM, de las Rivas B, Munoz R (2015) Characterization of a halotolerant lipase from the lactic acid bacteria *Lactobacillus plantarum* useful in food fermentations. *LWT-Food Sci Technol*, 60(1):246–252.

Fernandez L, Beerthuyzen MM, Brown J, Siezen RJ, Coolbear T, Holland R and Kuipers OP (2000) Cloning, characterization, controlled overexpression, and inactivation of the major tributyrin esterase gene of *Lactococcus lactis*. *Appl Environ Microbiol*, 66:1360–1368.

Gobbetti M, Fox PF, Smacchi E, Stepaniak L, Damiani P (1996) Purification and characterization of a lipase from *Lactobacillus plantarum* 2739. *J Food Biochem*, 20:227–246.

Gupta R, Gupta N, Rathi P (2004) Bacterial lipases: an overview of production, purification and biochemical properties. *Appl Microbiol Biotechnol*, 64:763–781.

Gupta R, Rathi P, Gupta N, Bradoo S (2003) Lipase assays for conventional and molecular screening: an overview. *Biotechnol Appl Biochem*, 37:63–71.

Hammes WP, Bantleon A, Min S (1990) Lactic acid bacteria in meat fermentation. *FEMS Microbiol. Rev.*, 7(1–2):165–173.

Henne A, Schmitez RA, Bomeke M, Gottschalk G, Daniel R (2000) Screening of environmental DNA libraries for the presence of genes conferring lipolytic activity on *Escherichia coli*. *Appl Environ Microbiol*, 66:3113–3116.

Holland R, Coolbear T (1996) Purification of tributyrin esterase from *Lactococcus lactis* subsp. *lactis* E8. *J Dairy Res*, 63:131–140.

Jini R, Swapna HC, Rai AK, Vrinda R, Halami PM, Sachindra NM, Bhaskar N (2011) Isolation and characterization of potential lactic acid bacteria (LAB) from freshwater fish processing wastes for application in fermentative utilisation of fish processing waste. *Brazilian J Microbiol*, 42:1516–1525.

Katz M, Medina R, Gonzalez S, Oliver G (2002) Esterolytic and Lipolytic Activities of Lactic Acid Bacteria Isolated from Ewe's Milk and Cheese. *J Food Protect*, 65(12):1997–2001.

Liu CH, Lu WB, Chang JS (2006) Optimizing lipase production of *Burkholderia* sp. by response surface methodology. *Process Biochem*, 41(9):1940–1944.

Lopes M de F, Leitao AL, Regalla M, Marques JJ, Carrondo MJ,Crespo MT (2002) Characterization of a highly thermostable extracellular lipase from *Lactobacillus plantarum*. *Int J Food Microbiol*, 76:107–115.

Martínez-Cuesta MC, Fernández de Palencia P, Requena T, Peláez C (2001) Enzymatic ability of *Lactobacillus casei subsp.casei* IFPL731 for flavour development in cheese. *Int Dairy J*, 11:577–585.

Matthews A, Grimaldi A, Walker M, Bartowsky E, Grbin P, Jiranek V (2004) Lactic acid bacteria as a potential source of enzymes for use in vinification. *Appl Environ Microbiol*, 70(10):5715–5731.

Mechai A, Debabza M, Kirane D (2014) Screening of technological and probiotic properties of lactic acid bacteria isolated from Algerian traditional fermented milk products. *Int Food Res J*, 21:2451–2457.

Messaoudi A, Belguith H, Gram I, Ben Hamida J (2010) Classification of EC 3.1.1.3 bacterial true lipases using phylogenetic analysis. *Afr J Biotech*, 9:8243–8247.

Padmapriya B, Rajeswari T, Noushida E, Sethupalan DG, Venil CK (2011) Production of lipase enzyme from *Lactobacillus* sp. and its application in the degradation of meat. *World Appl Sci J*, 12:1798–1802.

Rai AK, Swapna HC, Bhaskar N, Halami PM, Sachindra NM (2010) Effect of fermentation ensilaging on recovery of oil from fresh water fish viscera. *Enzyme Microb Technol*, 46:9–13.

Rashmi BS, Gayathri D (2014) Partial purification, characterization of *Lactobacillus* sp G5 lipase and their probiotic potential. *Int Food Res J*, 5:1737–1743.

Somboon T, Mukkharin P, Suwimon K (2015) Characterization and lipolytic activity of lactic acid bacteria isolated from Thai fermented meat. *J Appl Pharm Sci*, 5:6–12.

Sonar RN, Halami PM (2014) Phenotypic identification and technological attributes of native lactic acid bacteria present in fermented bamboo shoot products from North-East India. *J Food Sci Technol*, DOI:10.1007/s13197-014-1456-x.

Sumby KM, Matthews AH, Grbin PR, Jiranek V (2009) Cloning and characterization of an intracellular esterase from the wine associated lactic acid bacterium *Oenococcus oeni*. *Appl Environ Microbiol*, 75(21):6729–6735.

Uppada SR, Akula M, Bhattacharya A, Dutta JR (2017) Immobilized lipase from *Lactobacillus plantarum* in meat degradation and synthesis of flavor esters. *J Gen Eng Biotechnol*, DOI:10.1016/j.jgeb.2017.07.008.

Vrinda R (2013) Characterization of lipase producing lactic acid bacteria isolated from fish processing waste, PhD Thesis, University of Mysore, India.

Vrinda R, Bijinu B, Rai AK, Bhaskar N, Halami PM (2012) Concomitant production of lipase, protease and enterocin by *Enterococcus faecium* NCIM5363 and *Enterococcus durans* NCIM5427 isolated from fish processing waste. *Int Aqu Res*, 4:14.

Vrinda R, Louella CG, Halami PM, Bhaskar N (2014) Optimization of conditions for probiotic curd formulation by *Enterococcus faecium* MTCC 5695 with probiotic properties using response surface methodology. *J Food Sci Technol*, 51:3050–3060.

Vrinda R, Louella CG, Halami PM, Bhaskar N (2015) Kinetic modeling, production and characterization of an acidic lipase produced by *Enterococcus durans* NCIM5427 from fish waste. *J Food Sci Technol*, 52:1328.

Vrinda R, Louella CG, Niranjan S, Chandrashekar J, Halami PM, Bhaskar N (2016) Extraction and purification of lipase from *Enterococcus faecium* MTCC5695 by PEG/phosphate aqueous-two phase system (ATPS) and its biochemical characterization. *Biocatal Agric Biotechnol*, 6:19–27.

19 Role of Microbes in Environmental Sustainability and Food Preservation

En Huang
University of Arkansas for Medical Sciences

Ravi Kr. Gupta
Babasaheb Bhimrao Ambedkar (A Central) University

Fangfei Lou
The Ohio State University

Sun Hee Moon
University of Arkansas for Medical Sciences

CONTENTS

19.1 INTRODUCTION

Evolving spoilage and diseases-causing foodborne microorganisms are one of the biggest concerns that the food industry is facing. Spoilage is a complex process that renders foods undesirable for human consumption. Microbial spoilage usually changes food's sensorial characteristics; these include discoloration, textural changes, and formation of off-odors, off-flavors, and slime. Additionally, spoilage could lead to the deterioration of food's chemical and nutritional qualities. It is not surprising that food spoilage leads to considerable economic and environmental loss. It was estimated that over 10% of cereal grains and legumes and 50% of vegetables and fruits were lost due to spoilage microorganisms in the developing countries (Sperber, 2009). According to the US Department of Agriculture-Economic Research Service (ERS), a large proportion of US food supply is lost by retailers, foodservice workers, and consumers at these rates: fruits and vegetables (19.6%), fluid milk (18.1%), grain products (15.2%), caloric sweeteners (12.4%), processed fruits and vegetables (8.6%), meat, poultry and fish (8.5%), and fat and oils (7.1%) (Kantor, Lipton, Manchester, & Oliveira, 1997).

Foodborne illnesses have a significant economic and public health impact. The US Centers for Disease Control and Prevention (CDC) estimated that 31 major pathogens caused 9.4 million cases of foodborne illnesses, 55,961 hospitalizations, and 1,351 deaths annually in the United States (Scallan et al., 2011). Norovirus accounted for 58% of the foodborne illnesses, followed by *Salmonella* spp. (11%), *Clostridium perfringens* (10%), and *Campylobacter* spp. (9%). *Salmonella* spp. were the leading causative agents of hospitalizations (35%), followed by norovirus (26%), *Campylobacter* spp. (15%), and *Toxoplasma gondii* (8%). The leading causes of death were *Salmonella* spp. (28%), *T. gondii* (24%), *Listeria monocytogenes* (19%), and norovirus (11%) (Scallan et al., 2011). The health-related economic cost due to foodborne illnesses was as high as $77.7 billion annually in the United States (Scharff, 2012).

According to CDC-Foodborne Diseases Active Surveillance Network (FoodNet), the overall infections caused by six major bacterial pathogens (*Campylobacter, Escherichia coli* O157, *Listeria, Salmonella, Yersinia, and Vibrio*) decreased 23% in 2010 compared with that in 1996–1998 in the United States. *E. coli* O157: H7 is the most well-known Shiga toxin-producing *E. coli* (STEC). The O157 STEC is responsible for 63,153 foodborne illnesses, whereas non-O157 STEC causes 112,752 illnesses annually in the United States (Scallan et al., 2011). Infections due to *E. coli* O157 have declined 44% and reached the 2010 national health objective target (≤1 case per 100,000). However, the incidence of *Salmonella* remained high in the past 15 years. Furthermore, *Vibrio* infections increased by 115% since 1996 (CDC, 2011).

Food antimicrobials play important roles in preserving food quality and ensuring food safety. They are "substances used to preserve food by preventing the growth of microorganisms and subsequent spoilage, including fungistats, mold and rope inhibitors" (Codes of Federal Regulations, 21CFR170.3). The goal of using preservatives is to prevent or retard chemical or biological change in food (Davidson & Branen, 2005). However, adaptation and tolerance of spoilage and pathogenic microorganisms to current food preservatives pose serious problems to the food industry. Bacteria previously exposed to low concentrations of antimicrobial agent can adapt to the stress and become more resistant to the preservative at the levels used to control them. For example, an acid adaptation of Enterohemorrhagic *Escherichia coli* (EHEC) increased their survival in the organic acid preservative, lactic acid (Leyer, Wang, & Johnson, 1995). Similarly, the acid-tolerant EHEC strain, *E. coli* O157: H7 ATCC 43895, was capable of surviving for 21 days in apple cider in the presence of potassium sorbate or sodium benzoate (Miller & Kaspar, 1994). Furthermore, some yeast strains, such as *Saccharomyces cerevisiae* and *Zygosaccharomyces bailii*, have an innate resistance to the major weak acid preservatives, such as benzoic and sorbic acids. The mechanisms of acid resistance seem different for the two yeast species. *S. cerevisiae* actively extrudes the weak acid preservatives from the cell through efflux pumps, whereas *Z. bailii* limits the diffusion entry of weak acids and degrades these acids to nonantimicrobial products (Piper, Calderon, Hatzixanthis, & Mollapour, 2001).

In addition to the adaptive response, foodborne microorganisms may acquire resistance to antimicrobial agents through spontaneous mutations. Repeated use of an antimicrobial agent in food processing gradually selects for these resistant mutants. Enriching food processing environment with these new resistant strains has dire consequences. Food processors who have been successfully manufacturing products for years without a significant microbial challenge may gradually experience the buildup of spoilage or pathogenic microorganisms in their facility (Yousef, 2014). While addressing these crises, the processor may opt to use higher concentrations of the currently used preservative, if regulations permit it, or use alternative effective agent or treatment. Making processors' job even more challenging, consumers are increasingly demanding "clean labeling," and they are reluctant to accept products containing synthetic preservatives. In an attempt to address consumers' need, and to meet industry's responsibility to produce safe and acceptable products, food processors now are looking for new preservatives, including natural antimicrobials that can effectively extend product shelf life and maintain its safety.

19.2 CHEMICAL PRESERVATIVES: EFFICACY AND ACCEPTABILITY ISSUES

Food antimicrobial agents can be divided into traditional chemical preservatives and naturally occurring antimicrobials. Table 19.1 includes a noncomprehensive list of the regulatory-approved (US FDA) chemical food preservatives. Food processor's choice of antimicrobial agents depends on many factors such as antimicrobial spectrum, properties of antimicrobial agents (e.g., solubility and polarity), food-related factors (e.g., pH, lipid content), and food processing and storage conditions. In addition, sensory effect, cost of use, and product labeling should be considered when choosing proper antimicrobials (Davidson & Branen, 2005).

19.3 NATURAL ANTIMICROBIALS, THE ALTERNATIVE APPROACH

Natural antimicrobial agents, originating from animals, plants, and microorganisms, include promising alternatives to chemical food preservatives. Examples of these include ovotransferrin from egg-white, chitosan from crustaceans and arthropods, essential oils of spices and herbs, pediocin-like bacteriocin, and microbial fermentates of some starter cultures (Juneja, Dwivedi, & Yan, 2012). Table 19.2 lists some of the FDA-approved natural antimicrobial agents. Bacteria are the most

TABLE 19.1
Representative Regulatory-Approved (US FDA) Food Chemical Antimicrobials[a]

Antimicrobials	Spectrum	Examples of Use	CFR Designation
Weak lipophilic organic acids and esters			
Sorbic acid/sorbates	Yeasts, molds, bacteria	Dressings, wines, beverages, cheeses	182.3089; 182.3795; 182.3640; 182.3225
Parabens	Yeasts, molds, G+-bacteria	Bakery product, beverages, pickles, salad dressings	184.1490; 184.1670; 172.145
Benzoic acid/benzoates	Yeasts, molds	Beverages, dressings, margarine	184.1021; 184.1733
Other organic acids			
Acetic acid/acetates	Yeasts, bacteria	Baked goods, sources, dairy products	184.1005; 184.1721;
Diacetates	Yeasts, bacteria	Baked goods, soup, fats/oils, sauces, meat products	184.1754
Dehydroacetic acid	Yeasts, bacteria	Cut or peeled squash	172.130
Propionic acid/propionates	Molds	Bakery products, dairy products	184.1081; 184.1221
Lactic acid/lactates	Bacteria	Meats, fermented foods	184.1061; 184.1639; 184.1768; 184.1207
Inorganic anions			
Sulfites	Yeasts, molds	Dried fruits, wines	182.3739; 182.3766 182.3616; 182.3637
Nitrite/nitrate	*Clostridium* spp.	Cured meats	172.160;172.170; 172.175;172.177

[a] Modified from (Davidson & Branen, 2005; Jay, Loessner, & Golden, 2005; Kussendrager & Van Hooijdonk, 2000; Russell & Gould, 2003).

TABLE 19.2
Natural Antimicrobial Agents Approved by the US FDA

Antimicrobials	Spectrum	Examples of Use	CFR Designation
Lysozyme	*Clostridium* spp. and other bacteria	Cheeses, cooked meats	GRAS Notice No. GRN 000064
Lactoperoxidase	Bacteria	Yogurts	GRAS Notice No. 000196
Lactoferrin	Bacteria	Meats	GRAS Notice No. 000067
Nisin	*Clostridium* spp., other G$^+$ bacteria	Cheeses, canned meats	GRAS Notice No. GRN 000065
Natamycin	Molds	Cheeses, fermented meats	172.155

productive sources for antimicrobials. Additionally, most bacteria (>99%) are not readily cultured in laboratory media (Daniel, 2005); bacteria can enter the viable-but-nonculturable (VBNC) state under adverse conditions, such as low temperature, limited nutrient, and other stresses during food processing and storage (Giraffa & Neviani, 2001). The VBNC populations are mostly an untapped resource that may present great opportunities for researchers seeking novel antimicrobial agents. Bacteriocins of lactic acid bacteria, particularly, have been investigated extensively as potential alternatives to chemical preservatives.

19.3.1 BACTERIOCINS, WITH EMPHASIS ON LANTIBIOTICS

Bacteriocins are ribosomally synthesized antimicrobial peptides produced by bacteria. The majority of bacteriocins fall into two categories: lanthionine-containing peptides, i.e., lantibiotics (class I) and unmodified bacteriocins (class II). Nisin and paenibacillin (He, Yuan, Zhang, & Yousef, 2008) are examples of the lantibiotics, whereas pediocin and enterocin RM6 (Huang et al., 2013) are class II bacteriocins.

19.3.1.1 Structure and Classification of Lantibiotics

The term lantibiotics is the abbreviation of lanthionine-containing antibiotics. These are ribosomally synthesized peptides that contain lanthionines, which are formed during posttranslational modification. The characteristic residues in lantibiotics included dehydroalanine (Dha), dehydrobutyrine (Dhb), lanthionine (Lan), and methyllanthionine (MeLan). Dha and Dhb residues are unsaturated dehydroamino acids derived from serine and threonine, respectively. Dha or Dhb residue can form a thioether bond with the cysteine residue in the prepropeptide during biosynthesis, resulting in the unusual residues, lanthionine or methyllanthionine, respectively (Willey & Van der Donk, 2007). The chemical structure of paenibacillin (He et al., 2008) and the unusual residues are shown in Figure 19.1.

More than 60 lantibiotics, with different size, structure, or modes of action, have been discovered (Kuipers, Rink, & Moll, 2011). Lanthionine-containing peptides are divided into four classes (Table 19.3) based on their biosynthetic pathway and biological activity (Rea, Ross, Cotter, & Hill, 2011). Class I lantibiotics, such as nisin, have an elongated structure, whereas class II lantibiotic has a more compact and globular structure. For class I lantibiotics, two distinct enzymes, LanB and LanC, are responsible for dehydration of threonine/serine and formation of thioether linkages, respectively. For other classes, lantibiotics are modified by bifunctional enzymes (LanM in class II, RamC/LabKC in class III, and LanL in class IV) that exhibit dehydratase and cyclase activities (Rea et al., 2011; Willey & Van der Donk, 2007).

19.3.1.2 Biosynthesis of Lantibiotics

The genes for lantibiotic synthesis, export, and self-immunity are generally found in a gene cluster and are designated by the generic locus symbol *lan*. The gene cluster is either located on the chromosome (e.g., subtilin), the plasmid (e.g., epidermin), or transposon (e.g., nisin) (Kuipers et al., 2011).

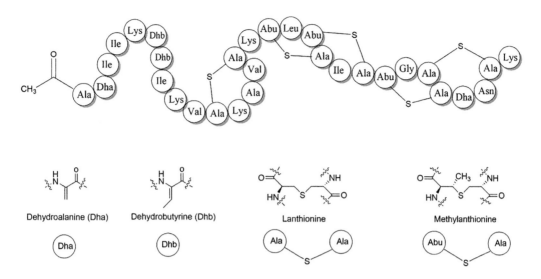

FIGURE 19.1 Structure of paenibacillin and its unusual residues.

TABLE 19.3
Classification of Lanthionine-Containing Peptides

Lantibiotics	Modification Enzymes	Export and Leader Cleavage	Conserved Motif in Leader Peptide	Other Features and Examples
Class I	LanB and LanC	LanT and LanP	"FDLD"	Linear structure, e.g., nisin
Class II	Bifunctional LanM	Bifunctional LanT(P)	"GC" or "GA"	Globular structure, e.g., mersacidin
Class III	RamC/LabKC	-	-	No antimicrobial activity, e.g., SapB
Class IV	LanL	-	-	No antimicrobial activity, e.g., venezuelin

Biosynthesis of lantibiotics involves several steps: (i) prepeptide (*lanA*) formation, (ii) posttranslational modifications including dehydration and cyclization (*lanB* and *lanC* for class I, *lanM* for class II), (iii) cleavage of the leader peptide (*lanP*), and (iv) secretion (*lanT*) of the mature peptides. In addition, genes involving in regulation (*lanRK*) and self-immunity (*lanEFG* and *lanI*) are often found in the same gene cluster (Chatterjee, Paul, Xie, & van der Donk, 2005; McAuliffe, Ross, & Hill, 2001). For example, the paenibacillin biosynthetic cluster consists of 11 open reading frames (ORFs) spanning 11.7-kb DNA sequence (Huang & Yousef, 2012). There are some unique features in paenibacillin gene cluster: a *lanN* gene for N-terminal acetylation of the lantibiotic, and an *agr* quorum sensing system for gene regulation.

19.3.1.2.1 Lantibiotic Precursor Peptide (LanA)

The structural gene (*lanA*) encodes the precursor peptide comprising a propeptide and an N-terminal leader sequence. The propeptide is the region corresponding to the mature lantibiotic. The propeptide region is rich in cysteine residues while the leader sequence usually lacks this amino acid (McAuliffe et al., 2001). The cysteine residue couples with Dha or Dhb residues, forming lanthionine or methyllanthionine, respectively. The leader peptide possesses a recognition motif for posttranslational modifications and serves as a signal for the secretion of the mature lantibiotics (Chatterjee et al., 2005).

19.3.1.2.2 Dehydratase (LanB) and Cyclase (LanC)

In class I lantibiotics, a dehydratase, LanB, catalyzes the selective dehydration of serine and threonine in the propeptide, forming Dha and Dhb residues, respectively. LanB comprises approximately 1,000 amino acids and is likely associated with the cytoplasmic membrane (Sahl & Bierbaum, 1998). Following dehydration, the cyclase, LanC, is responsible for the formation of the thioether bond between cysteine and Dha or Dhb. Koponen et al. (2002) provided the first direct evidence that NisB and NisC are required for dehydration and cyclization during its maturation. The deletion of *nisB* resulted in totally unmodified nisin precursor, whereas *nisC*-deficient mutant produced dehydrated precursor without lanthionine formation. NisC is a zinc protein where Zn^{2+} is ligated by two cysteine residues (Cys^{284}, Cys^{330}), one histidine (His^{331}) and a water molecule. Zinc ions activate cysteine residues for the nucleophilic attack on Dha or Dhb in the dehydrated nisin precursor (Li & van der Donk, 2007).

19.3.1.2.3 Bifunctional Enzyme (LanM)

In class II lantibiotics, a bifunctional enzyme, LanM, is responsible for dehydration and cyclization reactions. The LanM enzyme possesses 900–1,000 amino acids and shares 20%–27% similarity with the C-terminus of LanC proteins (Chatterjee et al., 2005). LanM requires ATP and Mg^{2+} for dehydration and correct cyclization (Xie et al., 2004). In a two-step dehydration model, the phosphoryl group of ATP is transferred to serine or threonine residue; then the resulting phosphor-Ser (pSer) or phosphor-Thr (pThr) undergoes β-elimination, forming the dehydrated Dha or Dhb (Chatterjee et al., 2005; You & van der Donk, 2007). Cyclization of the peptide is catalyzed by the C-terminal domain of LanM, the zinc-binding ligands essential for the catalytic activity of LanM (Paul, Patton, & van der Donk, 2007).

19.3.1.2.4 Proteases (LanP) and Transporters (LanT)

All lantibiotic precursors consist of an N-terminal leader peptide and a C-terminal propeptide. The modified precursor with the leader peptide is devoid of antimicrobial activity. Removal of the leader peptide gives rise to mature lantibiotics with activity. In class I lantibiotics, the cleavage of leader peptide is catalyzed by a serine protease, LanP, before or after the peptide is exported by the dedicated ABC-transporter, LanT. In class II lantibiotics, a multifunctional transporter, LanT (P), removes the leader peptide by its N-terminal peptidase domain during the export of the peptide (McAuliffe et al., 2001; Sahl & Bierbaum, 1998).

19.3.1.2.5 Regulation and Self-immunity

The production of lantibiotics is coordinately regulated by cellular events and the signal transduction pathway. For example, nisin biosynthesis is controlled by a typical two-component regulatory system (Chatterjee et al., 2005). The system comprises a histidine kinase (NisK) and a transcriptional response regulator (NisR). Nisin molecule is the signal for inducing the expression of the *nis* gene cluster. In the presence of nisin molecules, the membrane protein NisK transmits the signal to NisR. Then the activated NisR binds to *nisA* and *nisF* operators and thus triggers the transcription of the nisin gene cluster (Chatterjee et al., 2005). The production of other lantibiotics may be regulated by different mechanisms. For example, lacticin 481 production by *Lactococcus lactis* is induced through acidification resulting from lactic acid production (Hindré, Pennec, Haras, & Dufour, 2006).

Lantibiotic-producing strains protect themselves from killing by their own antibacterial products through specific immunity proteins. The self-immunity mechanisms involve two distant systems: the membrane-associated lipopeptide (LanI) and a dedicated ABC-transporter, LanFEG (Sahl & Bierbaum, 1998). LanI works as an intercepting molecule by binding to the lantibiotic peptides (Kuipers et al., 2011). The EpiFEG transporter in *Staphylococcus epidermidis* Tü3298 protects the producer by expelling the epidermin molecules from the cell membrane to the surrounding medium (Otto, Peschel, & Götz, 1998).

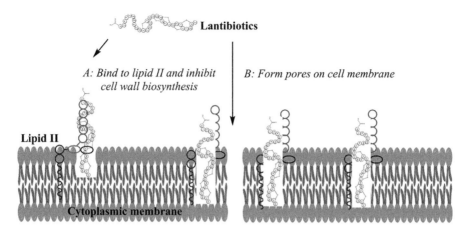

FIGURE 19.2 Schematic representation of mechanism of action of class I lantibiotics.

19.3.1.3 Mechanism of Action of Lantibiotics

Class I lantibiotics, such as nisin, disrupt the bacterial cytoplasmic membrane by pore formation and inhibit bacterial cell wall biosynthesis. Nisin molecules utilize lipid II, the precursor of cell wall synthesis, as a docking molecule for pore formation. When nisin molecule reaches cell membrane, the N-terminus of nisin (rings A and B) binds to pyrophosphate moiety in lipid II via intermolecular hydrogen bonds. The positively charged C-terminus of nisin inserts into the anion cell membrane, resulting in pore formation (Breukink & de Kruijff, 2006). The schematic representation mechanism of action of class I lantibiotics is shown in Figure 19.2. However, some class II lantibiotics (e.g., mersacidin) only inhibit cell wall synthesis by binding to lipid II; the interaction does not form pores on the cell membrane (Brötz, Bierbaum, Leopold, Reynolds, & Sahl, 1998).

19.4 LANTIBIOTIC BIOENGINEERING

Lantibiotic variants can be produced by manipulation of the structural gene *lanA*. In the whole-cell system, lantibiotics variants can be generated *in vivo* in the natural host organism or in a genetically well-characterized host (Cortés, Appleyard, & Dawson, 2009). To avoid production of the mixture of wild-type and mutated lantibiotics, a *lanA⁻* host is usually created by inactivation of the *lanA* gene. Inactivation can be achieved by deletion of *lanA*, gene replacement through recombination, or generating a frameshift mutation (Cortés et al., 2009). The *lanA⁻* gene can be replaced by a mutated *lanA* gene in the same locus (*in cis*) or complemented *in trans* with a separate mutated *lanA* on a plasmid. The *in cis* complementation system has been used to produce mutants such as subtilin, mutacin II, pep5, mersacidin, cinnamycin, and lacticin 3147 (Cortés et al., 2009). Example of lantibiotic variants generated by *in trans* complementation system included nisin, mersacidin, gallidermin/epidermin, actagardine, lacticin 3417, and nukacin ISK-1 (Cortés et al., 2009).

With the advance of lantibiotic biology, cell-free enzymatic synthesis of lantibiotics becomes possible by *in vitro* reconstitution of the bifunctional enzyme, LanM. Two class II lantibiotics, lacticin 481 and haloduracin, have been generated by this *in vitro* system (McClerren, Cooper, Quan, Thomas, Kelleher, & Van Der Donk, 2006; Xie et al., 2004). For the synthesis of lacticin 481, *lctA* and *lctM* were first cloned and overexpressed in *E. coli*. The His$_6$-tagged lanticin 481 prepeptide and LctM enzyme were isolated and purified by affinity chromatography, followed by reverse phase high-pressure liquid chromatography (RP-HPLC) and cationic exchange chromatography, respectively. In the presence of Mg^{2+} and ATP, the purified LctM enzyme catalyzes the formation of fully modified lanticin 481. Proteolytic removal of the leader peptide resulted in the production of active lacticin 481. Several lacticin 481 mutants or analogs containing nonproteinogenic amino acids were generated by the cell-free system (Levengood, Knerr, Oman, & van der Donk, 2009; Xie et al., 2004).

19.5 CHEMICAL SYNTHESIS OF LANTIBIOTICS

Chemical synthesis of lantibiotics is very challenging due to the complex thioether linkages. Total solution-phase synthesis of nisin was achieved through desulfurization reaction of disulfide peptides (Fukase et al. 1988). Recently, great progress has been made in the solid-phase synthesis of the lantibiotics (Tabor, 2011). The advancement of solid-phase peptide synthesis (SPPS) and the use of orthogonally protected building blocks allow for the on-resin synthesis of lantibiotics (Tabor, 2011). The first solid-supported synthesis of the lantibiotic, lactocin S, was reported recently. In addition, lactocin S analogs with improved oxidative stability were chemically synthesized by replacing the sulfur in lanthionine with a methylene unit (Ross, Liu, Pattabiraman, & Vederas, 2009; Ross, McKinnie, & Vederas, 2012). Some lantibiotics, such as nisin, epilancin 15X, lacticin 3147 A2, and paenibacillin, possess interlocking lanthionine or methyllanthionine rings, which represents another challenge in lantibiotic synthesis. The successful synthesis of two components of lacticin 3147 (Liu, Chan, Liu, Cochrane, & Vederas, 2011) and epilancin 15X (Knerr & van der Donk, 2012) provided a general methodology for the total synthesis of the lantibiotic family.

19.6 APPLICATION OF LANTIBIOTICS IN FOOD AND MEDICAL FIELDS

Lantibiotics have the potential to extend the shelf life and improve the safety of food. Lantibiotics can be applied to food products through three different routes: (i) direct additions of the purified or semipurified peptide, (ii) incorporation of powdered fermentate, or (iii) introduction of the lantibiotic-producing strains to fermented foods (Mills, Stanton, Hill, & Ross, 2011). Nisin is the most studied lantibiotic, and it has been approved as a food preservative by the US FDA in 1988. Nisin has no known toxicity to human and currently is allowed to be used in over 40 countries (Healy, O'Mahony, Hill, Cotter, & Ross, 2011).

In addition to food application, lantibiotics are promising candidates for treating bacterial infections. For example, nisin is active against drug-resistant pathogens, including methicillin-resistant *Staphylococcus aureus* (MRSA), vancomycin-resistant *Enterococcus* (VRE), and *Clostridium difficile* (Piper, Cotter, Ross, & Hill, 2009). Actagardine is a 19-amino-acid class II lantibiotic, which is produced by *Actinoplanes garbadinensis* (Boakes, Cortés, Appleyard, Rudd, & Dawson, 2009). The semisynthetic derivative of this lantibiotic(deoxyl-actagardine, NVB302) with selective activity against *C. difficile* has been tested in phase I clinical trial (Shi, Bueno, & van der Donk, 2012). Lantibiotic salivaricin A is produced by a probiotic strain *Streptococcus salivarius* K12. This lantibiotic-producing strain has been used in lozenges to suppress the oral bacteria implicated in halitosis (Burton, Wescombe, Moore, Chilcott, & Tagg, 2006).

19.7 BIOPRESERVATION USING BACTERIOPHAGES

Bacteriophage lytic enzymes, or lysins, are another category of effective antimicrobials for food preservation (Fischetti, 2008). Bacteriophages (or phages) are viruses that specifically infect and multiply in bacteria. They are ubiquitous in environments and have been isolated from a variety of foods, such as meat products, seafood, fermented dairy products, fresh produce, and mushrooms (Fieseler, Loessner, & Hagens, 2011). Bacteriophages rely on their obligate bacterial hosts for replication; therefore, they are harmless to human, animals, and plants (García, Rodríguez, Rodríguez, & Martínez, 2010).

Bacteriophages can be classified as lytic or lysogenic depending on their life cycles. Lytic phages strictly follow the lytic life cycle in which they reproduce themselves in the bacterial

TABLE 19.4

Regulatory-Approved Bacteriophage Preparations

Commercial Bacteriophages	Target Microorganisms	Uses	Regulatory Designation
AgriPhage™, (Omnilytics, Inc.)	*Xanthomonas campestris* pv. *vesicatoria* and *Pseudomonas syringae* pv. *tomato*	Biological control of diseases on tomatoes and peppers	EPA Reg. No. 67986-1
LISTEX™, (EBI Food Safety)	*Listeria monocytogenes*	Processing aid in all food products	FDA (GRAS No. 000198)
ListShield™, (Intralytix, Inc.)	*L. monocytogenes*	Direct application onto foods; surfaces in food facilities	21CFR172.785 EPA Reg. No. 74234-1
EcoShield™, (Intralytix, Inc.)	*Escherichia coli* O157:H7	Red meat parts and trim prior to grinding	FCN No. 1018

host and immediately release the phage progeny by destroying the host. Lysogenic phages, on the other hand, integrate their viral DNA into the bacterial chromosome and become the prophages. Prophages replicate as part of the bacterial host genome without immediately transcribing and making new phages. The prophages may be activated under adverse environmental conditions, multiplying in the bacteria and destroying the host cells (García, Martínez, & Rodríguez, 2011).

Lytic bacteriophages are promising biocontrol agents against pathogenic bacteria. The use of bacteriophages in food has received great attention in the past few years. Compared with other antimicrobial agents, bacteriophages have strict host specificity and self-replicating capacity. A single phage or mixed phage cocktail only kills specific pathogens without interfering with other desired bacteria (e.g., starter culture) in foods. Therefore, the consumption of bacteriophages in food products will not harm the commensal microbiota in the human gastrointestinal tract (Fieseler et al., 2011; García et al., 2010). The US Environmental Protection Agency (EPA) has approved a bacteriophage cocktail (AgriPhage) for biological control of plant bacterial diseases. In addition, the US FDA has approved the usage of bacteriophage preparations as a processing aid or food additives for human consumption (Table 19.4).

19.8   CONCLUSIONS AND FUTURE DIRECTIONS

The food industry is faced with increasing consumers' demand for "clean labeling" because of their reluctance to accept products containing synthetic preservatives. Antimicrobials, including bacteriocins from lactic acid bacteria and other nonpathogenic microorganisms, have the great potentials to be used as natural food preservatives. With the advance of lantibiotic bioengineering, lantibiotic variants were successfully produced by engineered native or heterologous hosts, or even by enzymatic synthesis *in vitro*. Moreover, great achievement has been made in the chemical synthesis of lantibiotics containing interlocking thioether rings; this also provides easy access to lantibiotic variants for structure-activity analyses.

The majority of bacteriocins were discovered by a screening of fermented food or other environmental samples. Considering that only 0.1%–1% of bacteria are culturable in laboratory media (Daniel, 2005), the dominant nonculturable microorganisms were overlooked by traditional screening approach. Recently, metagenomic-based methods have been successfully used for screening bioactive compounds. The metagenomic approach is a culture-independent strategy, which involves collecting the total genetic materials (metagenomic DNA) from a mixed community of organisms, cloning the extracted DNA in a suitable bacterial host, followed by activity-driven screening or sequenced-based selection (Brady, Simmons, Kim, & Schmidt, 2009). This metagenomic approach is very promising in finding new antimicrobial agents, including bacteriocins, in the VBNC microorganisms from food or environmental samples.

Although over 60 bacteriocins with different size, structure, or mechanisms of action have been discovered, nisin is the only FDA-approved lantibiotic widely used in the food industry. For some bacteriocins from lactic acid bacteria, they could be used in food as dairy fermentate. The use of other bacteriocins as food antimicrobials is regulated by the US FDA (FDA, 2008). Like other food additives, the US FDA recommends a very comprehensive toxicological study, including 13 different toxicity tests, to establish the safety of the proposed food antimicrobials (FDA, 2000). The high cost of the lengthy toxicity study becomes one of the major hurdles for the development of new antimicrobial agents.

Bacteriocin-based food preservation is not effective against Gram-negative foodborne pathogens, which usually cause serious problems with the consumption of meat and poultry products. The major pathogens of concern include STEC, drug-resistant *Salmonella* spp., and *Campylobacter jejuni*. Several bacteriophages with activity against Gram-negative foodborne pathogens are generally recognized as safe (GRAS) and approved by the US FDA. These approved phages can be used as processing aids and/or direct additives into food. However, like many other new technologies, consumer acceptance of bacteriophage usage may present a challenge to the food industry.

REFERENCES

Boakes, S., Cortés, J., Appleyard, A. N., Rudd, B. A. M., & Dawson, M. J. (2009). Organization of the genes encoding the biosynthesis of actagardine and engineering of a variant generation system. *Molecular Microbiology, 72*(5), 1126–1136.

Brady, S. F., Simmons, L., Kim, J. H., & Schmidt, E. W. (2009). Metagenomic approaches to natural products from free-living and symbiotic organisms. *Natural Product Reports, 26*(11), 1488–1503.

Breukink, E., & de Kruijff, B. (2006). Lipid II as a target for antibiotics. *Nature Reviews Drug Discovery, 5*(4), 321–323.

Brötz, H., Bierbaum, G., Leopold, K., Reynolds, P. E., & Sahl, H. G. (1998). The lantibiotic mersacidin inhibits peptidoglycan synthesis by targeting lipid II. *Antimicrobial Agents and Chemotherapy, 42*(1), 154–160.

Burton, J. P., Wescombe, P. A., Moore, C. J., Chilcott, C. N., & Tagg, J. R. (2006). Safety assessment of the oral cavity probiotic *Streptococcus salivarius* K12. *Applied and Environmental Microbiology, 72*(4), 3050–3053.

CDC (2011). Trends in Foodborne Illness, 1996–2010. https://www.cdc.gov/winnablebattles/FoodSafety/pdf/Trends_in_Foodborne_Illness.pdf (Accessed September 27, 2018).

Chatterjee, C., Paul, M., Xie, L., & van der Donk, W. A. (2005). Biosynthesis and mode of action of lantibiotics. *Chemical Reviews, 105*(2), 633–684.

Cortés, J., Appleyard, A. N., & Dawson, M. J. (2009). Whole-Cell generation of lantibiotic variants. In D. A. Hopwood (Ed.), *Complex Enzymes in Microbial Natural Product Biosynthesis, Part A: Overview Articles and Peptides* (pp. 559–574). San Diego, CA: Elsevier.

Daniel, R. (2005). The metagenomics of soil. *Nature Reviews Microbiology, 3*(6), 470–478.

Davidson, P. M., & Branen, A. L. (2005). Food antimicrobials-an introduction. In P. M. Davidson, J. N. Sofos, & A. L. Branen (Eds.), *Antimicrobials in Food* (pp. 1–10). Boca Raton, FL: CRC Press.

FDA, 2000. Guidance for Industry and Other Stakeholders: Toxicological Principles for the Safety Assessment of Food Ingredients. http://www.fda.gov/Food/GuidanceRegulation/GuidanceDocumentsRegulatoryInformation/IngredientsAdditivesGRASPackaging/ucm2006826.htm (Accessed September 27, 2018).

FDA, 2008. Guidance for Industry: Microbiological Considerations for Antimicrobial Food Additive Submissions. http://www.fda.gov/food/guidanceregulation/guidancedocumentsregulatoryinformation/ingredientsadditivesgraspackaging/ucm230417.htm (Accessed September 27, 2018).

Fieseler, L., Loessner, M. J., & Hagens, S. (2011). Bacteriophages and food safety. In C. Lacroix (Ed.), *Protective Cultures, Antimicrobial Metabolites and Bacteriophages for Food and Beverage Biopreservation* (pp. 161–178). Philadelphia, PA: Woodhead Publishing.

Fischetti, V.A. (2008). Bacteriophage lysins as effective antibacterials. Current Opinion in Microbiology, 11, 393–400.

Fukase, T., Kitazawa, M., Sano, A., Shimbo, K., Fujita, H., Horimoto, S., Wakamiya, T., & Shiba T. (1988). Total synthesis of peptide antibiotic nisin. Tetrahedron *Letters*, 29, 795–798.

García, P., Martínez, B., & Rodríguez, A. (2011). Bacteriophages and phage-encoded proteins: prospects in food quality and safety. In M. Rai & M. Chikindas (Eds.), *Natural Antimicrobials in Food Safety and Quality* (pp. 10–26). Cambridge, MA: CABI Publishing.

García, P., Rodríguez, L., Rodríguez, A., & Martínez, B. (2010). Food biopreservation: promising strategies using bacteriocins, bacteriophages and endolysins. *Trends in Food Science & Technology, 21*(8), 373–382.

Giraffa, G., & Neviani, E. (2001). DNA-based, culture-independent strategies for evaluating microbial communities in food-associated ecosystems. International Journal of Food Microbiology, 67(1–2), 19–34.

He, Z., Yuan, C., Zhang, L., & Yousef, A. E. (2008). N-terminal acetylation in paenibacillin, a novel lantibiotic. *FEBS Letters, 582*(18), 2787–2792.

Healy, B., O'Mahony, J., Hill, C., Cotter, P. D., & Ross, R. P. (2011). Lantibiotic-related research and the application thereof. In G. Wang (Ed.), *Antimicrobial Peptides: Discovery, Design, and Novel Therapeutic Strategies* (pp. 22–26). Oxfordshire: CAB International.

Hindré, T., Pennec, J. P., Haras, D., & Dufour, A. (2006). Regulation of lantibiotic lacticin 481 production at the transcriptional level by acid pH. *FEMS Microbiology Letters, 231*(2), 291–298.

Huang, E., & Yousef, A. E. (2012). Draft genome sequence of *Paenibacillus polymyxa* OSY-DF, which coproduces a lantibiotic, paenibacillin, and polymyxin E1. *Journal of Bacteriology, 194*(17), 4739–4740.

Huang, E., Zhang, L., Chung, Y. K., Zheng, Z., & Yousef, A. E. (2013). Characterization and application of enterocin RM6, a bacteriocin from Enterococcus faecalis. Biomed Research International, *2013*, 206917. doi: 10.1155/2013/206917.

Jay, J. M., Loessner, M. J., & Golden, D. A. (2005). *Modern Food Microbiology* (pp. 301–350). New York: Springer.

Juneja, V. K., Dwivedi, H. P., & Yan, X. (2012). Novel natural food antimicrobials. *Annual Review of Food Science and Technology, 3*, 381–403.

Kantor, L. S., Lipton, K., Manchester, A., & Oliveira, V. (1997). Estimating and addressing America's food losses. *Food Review, 20*(1), 2–12.

Knerr, P. J., & van der Donk, W. A. (2012). Chemical synthesis and biological activity of analogues of the lantibiotic epilancin 15X. *Journal of the American Chemical Society, 134*(18), 7648–7651.

Koponen, O., Tolonen, M., Qiao, M., Wahlstrom, G., Helin, J., & Saris, P. E. J. (2002). NisB is required for the dehydration and NisC for the lanthionine formation in the post-translational modification of nisin. *Microbiology, 148*(11), 3561–3568.

Kuipers, A., Rink, R., & Moll, G. N. (2011). Genetics, biosynthesis, structure, and mode of action of lantibiotics. In D. Drider & S. Rebuffat (Eds.), *Prokaryotic Antimicrobial Peptides: From Genes to Applications* (pp. 147–169). New York: Springer.

Kussendrager, K. D., & Van Hooijdonk, A. (2000). Lactoperoxidase: physico-chemical properties, occurrence, mechanism of action and applications. *British Journal of Nutrition, 84*, 19–25.

Levengood, M. R., Knerr, P. J., Oman, T. J., & van der Donk, W. A. (2009). In vitro mutasynthesis of lantibiotic analogues containing nonproteinogenic amino acids. *Journal of the American Chemical Society, 131*(34), 12024–12025.

Leyer, G. J., Wang, L. L., & Johnson, E. A. (1995). Acid adaptation of *Escherichia coli* O157:H7 increases survival in acidic foods. *Applied and Environmental Microbiology, 61*(10), 3752–3755.

Li, B., & van der Donk, W. A. (2007). Identification of essential catalytic residues of the cyclase NisC involved in the biosynthesis of nisin. *Journal of Biological Chemistry, 282*(29), 21169–21175.

Liu, W., Chan, A. S. H., Liu, H., Cochrane, S. A., & Vederas, J. C. (2011). Solid supported chemical syntheses of both components of the lantibiotic lacticin 3147. *Journal of the American Chemical Society, 133*(36), 14216–14219.

McAuliffe, O., Ross, R. P., & Hill, C. (2001). Lantibiotics: structure, biosynthesis and mode of action. *FEMS Microbiology Reviews, 25*(3), 285–308.

McClerren, A. L., Cooper, L. E., Quan, C., Thomas, P. M., Kelleher, N. L., & Van Der Donk, W. A. (2006). Discovery and in vitro biosynthesis of haloduracin, a two-component lantibiotic. *Proceedings of the National Academy of Sciences, 103*(46), 17243–17248.

Miller, L. G., & Kaspar, C. W. (1994). *Escherichia coli* O157: H7 acid tolerance and survival in apple cider. *Journal of Food Protection, 57*(6), 460–464.

Mills, S., Stanton, C., Hill, C., & Ross, R. (2011). New developments and applications of bacteriocins and peptides in foods. *Annual Review of Food Science and Technology, 2*, 299–329.

Otto, M., Peschel, A., & Götz, F. (1998). Producer self-protection against the lantibiotic epidermin by the ABC transporter EpiFEG of *Staphylococcus epidermidis* Tü3298. *FEMS Microbiology Letters, 166*(2), 203–211.

Paul, M., Patton, G. C., & van der Donk, W. A. (2007). Mutants of the zinc ligands of lacticin 481 synthetase retain dehydration activity but have impaired cyclization activity. *Biochemistry, 46*(21), 6268–6276.

Piper, C., Cotter, P. D., Ross, R. P., & Hill, C. (2009). Discovery of medically significant lantibiotics. *Current Drug Discovery Technologies, 6*(1), 1–18.

Piper, P., Calderon, C. O., Hatzixanthis, K., & Mollapour, M. (2001). Weak acid adaptation: the stress response that confers yeasts with resistance to organic acid food preservatives. *Microbiology, 147*, 2635–2642.

Rea, M. C., Ross, R. P., Cotter, P. D., & Hill, C. (2011). Classification of bacteriocins from gram-positive bacteria. In D. Drider & S. Rebuffat (Eds.), *Prokaryotic Antimicrobial Peptides: From Genes to Applications* (pp. 29–53). New York: Springer.

Ross, A. C., Liu, H., Pattabiraman, V. R., & Vederas, J. C. (2009). Synthesis of the lantibiotic lactocin S using peptide cyclizations on solid phase. *Journal of the American Chemical Society, 132*(2), 462–463.

Ross, A. C., McKinnie, S. M. K., & Vederas, J. C. (2012). The synthesis of active and stable diaminopimelate analogues of the lantibiotic peptide lactocin S. *Journal of the American Chemical Society, 134*(4), 2008–2011.

Russell, N. J., & Gould, G. W. (2003). Major preservative technologies. In N. J. Russell & G. W. Gould (Eds.), *Food Preservatives* (pp. 14–24). New York: Kluwer Academic/Plenum Publishers.

Sahl, H. G., & Bierbaum, G. (1998). Lantibiotics: biosynthesis and biological activities of uniquely modified peptides from gram-positive bacteria. *Annual Reviews in Microbiology, 52*(1), 41–79.

Scallan, E., Hoekstra, R. M., Angulo, F. J., Tauxe, R. V., Widdowson, M. A., Roy, S. L., Jones, J. L., & Griffin, P. M. (2011). Foodborne illness acquired in the United States-major pathogens. *Emerging Infectious Diseases, 17*(1), 7–15.

Scharff, R. L. (2012). Economic burden from health losses due to foodborne illness in the United States. *Journal of Food Protection, 75*(1), 123–131.

Shi, Y., Bueno, A., & van der Donk, W. A. (2012). Heterologous production of the lantibiotic Ala (0) actagardine in *Escherichia coli. Chemical Communications, 48*(89), 10966–10968.

Sperber, W. H. (2009). Introduction to the microbiological spoilage of foods and beverages. In W. H. Sperber & M. P. Doyle (Eds.), *Compendium of the Microbiological Spoilage of Foods and Beverages* (pp. 1–40). New York: Springer.

Tabor, A. B. (2011). The challenge of the lantibiotics: synthetic approaches to thioether-bridged peptides. *Organic & Biomolecular Chemistry, 9*(22), 7606–7628.

Willey, J. M., & Van der Donk, W. A. (2007). Lantibiotics: peptides of diverse structure and function. *Annual Review of Microbiology, 61*, 477–501.

Xie, L., Miller, L. M., Chatterjee, C., Averin, O., Kelleher, N. L., & van der Donk, W. A. (2004). Lacticin 481: in vitro reconstitution of lantibiotic synthetase activity. *Science, 303*(5658), 679–681.

You, Y. O., & van der Donk, W. A. (2007). Mechanistic investigations of the dehydration reaction of lacticin 481 synthetase using site-directed mutagenesis. *Biochemistry, 46*(20), 5991–6000.

Yousef, A. E. (2014). Resistance to processing. In Batt, C. A. & Tortorello, M.-L. (Eds.), *Encyclopedia of Food Microbiology*, vol. 3 (pp. 280–283). Amsterdam: Elsevier Ltd, Academic Press.

20 Bioinformatics and Applications in Biotechnology

Desh Deepak Singh
Panjab University

CONTENTS

20.1 INTRODUCTION

Bioinformatics can be defined as an interdisciplinary science in which one analyses the rapidly evolving biological data to gain meaningful insights into the processes of life. It has contributed in a major way in gene identification and annotation from sequenced genomes, Single nucleotide polymorphism (SNP) analysis and role in population heterogeneity, molecular basis of diseases, drug design, etc. The developments in the field have kept pace with the increasing computational and storage power in the area of information technology. With today's capability of cluster computing and big data, the field is poised for further exponential development and analytic capabilities.

20.2 HISTORICAL DEVELOPMENTS

With the sequencing of insulin sequence by Fredrick Sanger in 1955, the foundations for the development of computational tools in analysis of biological data were laid. Margaret Oakley Dayhoff at the Georgetown University Medical Center made pioneering efforts in application of mathematics and computational techniques to biological data by developing the one-letter code for amino acids and the scoring matrix PAM (point accepted mutations) in 1966. She also compiled an atlas of protein sequence and structures, and these developments resulted in the development of tools for sequence alignment and construction of phylogenetic trees. Again, Fredrick Sanger and his colleagues developed methods for DNA sequencing in 1977, and this resulted in a rapid availability of gene sequences. The bacteriophage PhiX174 was first sequenced by Sanger and his team in the mid-1970s, and the first bacterial genome, *Haemophilus influenza*, was sequenced in 1995 by Craig Venter and his team. The 1990 paper by David Altschul from National Center for Biotechnology Information on basic local alignment search tool (BAST) became a landmark tool for comparing protein sequences, and since then, many people have contributed toward the development of databases and tools for bioinformatics analysis. The 20.2 billion base human genome sequence was sequenced by human genome organization and Celera genomics in 2001 opening new vistas in biomedical research.

20.3   ENABLING TECHNOLOGIES FOR DEVELOPMENT OF BIOINFORMATICS TOOLS

20.3.1   NEXT-GENERATION SEQUENCING

The conventional Sanger sequencing of genomes is slow and has many rate-limiting steps, which have been overcome using the techniques of parallel or deep sequencing. Using next-generation sequencing (NGS), an entire human genome can be sequenced in almost a day, which otherwise took many years of hard labor using the Sanger sequencing (Behjati & Tarpey, 2013). In NGS, the DNA is fragmented to small readouts from 30–40 bp to about 500bp, denatured, and attached to beads with each bead in a nanowell; the DNA fragment is amplified usually by emulsion PCR and the fragments are sequenced. Extensive bioinformatics tools are then used to reassemble the genome. The first massively parallel signature sequencing incorporating parallel, adaptor/ligation mediated bead based sequencing technologies were made available by Lynx Therapeutics in 2000 (Brenner et al., 2000). Three technologies have revolutionized NGS:ion torrent personal genome machine (PGM), single-molecule real-time sequencing (SMRT) by PacBio, sequencing by synthesis by Illumina, among others. In the ion torrent PGM, the protons released as nucleotides are incorporated to make a double strand of single-stranded DNA fragment amplified in nanowells are captured by proton-sensing silicon wafers in the nanowells. As nucleotides are added to wells, signal emanates from only those wells in which the nucleotide is incorporated. The PacBio technique involves binding of DNA polymerase to the single-stranded amplified DNA strands, and when a γ-phosphate-labeled nucleotide is added to form the double strand, a distinct pulse of incorporated fluorophore is detected in real time. In illumina technique, the clonally amplified DNA immobilized on acrylamide coating on the surface of a glass flow cell sequencing by synthesis approach is used using fluorescently labeled nucleotides (Quail et al., 2012). Of late DNA, nanopore sequencing pioneered by Oxford Nanopore technologies offers real-time, scalable, and direct DNA sequencing. Single DNA/RNA molecules can be analyzed by passing them through a nano-orifice in a protein such asalpha hemolysin using electrophoresis. The system also bypasses the need for PCR amplification and is set to become the benchmark in NGS with substantially reduced operating costs.

20.3.2   HIGH-PERFORMANCE COMPUTING

With enormous amount of data being thrown up by powerful experimental and sequencing techniques, the enabling technologies to analyze them call for very-high-end computational capabilities. A computer cluster is assembled to work as one machine harnessing the power of each computer synergistically. High speed networks coupled with software for distributed computing have made it possible to link a large number of computers to work as one machine. As of November 2016, the Chinese Sunway TaihuLight is the world's most powerful supercomputer reaching 93.015 petaFLOPS (10^5 floating point operations per second). It consists of 40,960 processors with each processor containing 256 processing cores for a total of about 10 million CPU cores across the entire system (Fu et al., 2016). The Blue Gene high-performance computing system was developed by IBM in collaboration with Department of Energy's Lawrence Livermore National Laboratory in California. It was built to specifically observe the process of protein folding and gene development. Blue Gene/L uses 131,000 processors for 280 trillion operations every second. The power of the system can be gauged from the fact that, on a calculator, one has to work nonstop for 177,000 years to perform the operations that Blue Gene cando in 1s (http://www-03.ibm.com/ibm/history/ibm100/us/en/icons/bluegene/).

20.3.3 BIG DATA

The term refers to complex and large data that cannot be handled by traditional data processing software. This has become important in biology because of rapid explosion of genomes, which are sequenced and have to be analyzed. The traditional relational data management systems have serious limitations in handling such large datasets. The data growth challenges are 3D in terms of volume, velocity, and variety where volume refers to observing and tracking, velocity refers to availability of data in real time, and variety refers to the data in the form of text, images, audio, and video. One of the methods to tackle it is based on MapReduce technology developed by Google. Using this method, the queries are split and distributed over parallel nodes and processed in parallel, and results are gathered and delivered (Wickham, 2011). Some techniques being developed to handle big data are to represent data as tensors in which data can be represented as geometric objects and cloud-based infrastructures among a host of other techniques.

20.4 APPLICATIONS OF BIOINFORMATICS RESOURCES

20.4.1 GENOME ASSEMBLY AND ANNOTATION

A major application of bioinformatics software, tools, and storage systems start with data generated through genome sequencing. Since NGS involves generation of readout, which could be upto 30–40 bp, reassembling a genome with millions of such readouts as in humans is a mammoth task. Whole-genome shotgun sequencing (WGS) samples the genomic DNA by making small readouts and algorithm packages such as SSAKE, SHARCGS, VCAKE, and Velvet, are widely used in the reassembly of the fragments. A special problem arises in de novo WGS assembly of NGS data because there is no draft reference to fall back upon and another challenge is finding the fragments from the repetitive region of DNA. A wayout is oversampling of target genomes with readouts from different random positions and then finding out the overlap regions and reconstructing the genomesand their resolution. The notion of k-mers is used which consist of consecutive bases and reduces the assembly complexity and computational cost considerably. An overlap graph, e.g., the De Bruijn graph, is used in which the nodes represent the readouts and the edges represent the overlaps, is analyzed and results in contigs and finally sequences. SSAKE is one of the earliest assemblers and chooses the reads with end to end confirmation and then the candidates with multiple extensions. VCAKE is also an iterative extension algorithm and can also incorporate imperfect matches during contig extension. These and many other widely used genome assembly programs have revolutionized the NGS data compilation and their availability (Miller et al., 2010).

Once the genome has been sequenced and assembled properly, it needs annotation for identification of genes, mobile genetic elements, repeats, genome duplication, and diversity. Similarity search tools such as BLAST are used to identify obvious homologs, and more sophisticated tools such as GLIMMER (Gene Locator and Interpolated Markov ModelER), GENSCAN, and Artemis are some of the tools widely used for genome annotation. GLIMMER uses interpolated Markov models for microbial gene identification (Salzberg et al., 1998). The encyclopedia of DNA elements (ENCODE) project of National Human Genome Research Institute aims to identify all functional elements in human genome. Since only 1.5% of the genomic DNA is recognized as coding region, the role of the remaining component is not clear. The project aims to identify and fully characterize the regulome, which controls gene expression, and the findings will have major impact on understanding of the disease. Various bioinformatics tools are extensively used in the project to analyze and store the data so that it become useful during subsequent uses. A major outcome of the project is the FactorBook repository, which houses the database of more than 100 different transcription factors and their recognition sites in genomic DNA.

20.4.2 SNP IDENTIFICATION

SNP is a single nucleotide DNA variation on the genomes of different members of a species and occurring ay specific positions. They occur due to substitutions, deletions, and insertions and give a fundamental insight into allelic variations and individual, ethnic predispositions. A major effort is being made to correlate the SNP variations and their contributions to disease and health in different individuals. The international HapMap project aims to develop a halotype map of human genome to find genetic variations responsible for disease, response to drugs, environmental factor, etc. In phase III of the project, 11 global ancestry groups have been assembled: ASW (African ancestry in Southwest USA); CEU (Utah residents with Northern and Western European ancestry from the CEPH collection); CHB (Han Chinese in Beijing, China); CHD (Chinese in Metropolitan Denver, Colorado); GIH (Gujarati Indians in Houston, Texas); JPT (Japanese in Tokyo, Japan); LWK (Luhya in Webuye, Kenya); MEX (Mexican ancestry in Los Angeles, California); MKK (Maasai in Kinyawa, Kenya); TSI (Tuscans in Italy); YRI (Yoruba in Ibadan, Nigeria) (Altshuler et al., 2010). We have also undertaken a study on genetic variations in Arg5Pro and Leu6Pro, which modulate the structure and activity of GPX1 and increase genetic risk for vitiligo (Mansuri et al., 2016) Bioinformatics tools of sequence alignment, modeling, etc. were extensively used in this study.

20.4.3 SYSTEMS NETWORKS

Protein-protein interaction (PPI) mapping is an important event in mapping dynamic biological events. Protein interactions are important for signal transduction, gene expression, metabolic pathways regulation, muscle contraction, cancer development, etc. SH2 (Src homology region 2), and PTB (phosphotyrosine-binding domain) are small protein domains that mediate many signal transductions mediated through phosphotyrosines. This information flow through PPIs results in regulation of cell cycle, cell shape and movement, cell proliferation, differentiation, and survival (http://stke.sciencemag.org/content/2003/191/re12.long).

Many methods have been developed for rapid detection of PPIs, and they have become the basis for the development of insilico tools such as Strings database (https://string-db.org/) and Rosetta through which PPI can be identified. Experimentally protein affinity chromatography, immunoprecipitation, and library-based methods have been used for identifying PPIs. For example in one of the methods, a labeled protein is used as a probe for immobilized candidates expressed from a cDNA library on a nitrocellulose paper and the information thus generated is used for validation of in-silico tool Rosetta employs a low-resolution, rigid-body, Monte Carlo search followed by simultaneous optimization of backbone displacement and side-chain conformations using Monte Carlo minimization. The resulting interactions are ranked by an energy function, which takes into account Van der waals interactions, an implicit solvation model, and an orientation-dependent hydrogen bonding potential, and best ranking hits are clustered to select the final predictions (Gray et al., 2003).

20.4.4 DRUG DESIGN

A major application of bioinformatics tool is drug design when structure of candidate protein is known. Various tools are used to find out the free energy of binding of various conformers of a small molecule in the active site by calculating the Van der Waal's interactions, hydrogen bonding, solvation energy, and electrostatic interactions. A classic example is the design of the antitumor drug Gleevec. Gleevec has high specificity for the onco-protein BCR-Abl, which is a fusion protein resulting from reciprocal translocation between chromosome 9 and chromosome 22 (Philadelphia translocation), leading to a constitutively active kinase. Gleevec was approved by the FDA for clinical use in 2001 for treating chronic myeloid leukemia (CML) and gastrointestinal stromal tumors. The DFG-loop (Asp-Phe-Gly) in the kinase structure was taken as a structural feature that differs between kinases that bind Gleevec tightly or weakly. In the structure of Abl kinase, this loop adopts a so-called out conformation in both the apo and Gleevec-bound protein, while in the closest homolog and weak binder Src kinase, it occupies an in conformation in the apo protein that would have to move into the "out" position to accommodate the drug. It was hypothesized that the preferential occupancy of the DFG-out state by Abl but not Src is the primary source of Gleevec selectivity. A variety of approaches, both experimental and computational, were taken to quantify the free energy profile of the DFG-loop dynamics (Agafonov et al., 2015).

We have been working extensively in the area of lectins and leishmaniasis over the past many years and have some very encouraging leads from bioinformatics analysis, which we are developing further through experimental work. The structural studies on BanLec, a lectin from banana whose structure was solved by us (Singh et al., 2005), revealed some very useful insights into the basis for its unique sugar specificity. It has two sugar-binding sites per subunit in comparison with other members of the family studied till then. Bioinformatics analysis and modeling studies explained the basis for its recognition of 1-2, 1-3, and 1-6 linked mannosides. It is very encouraging to note that, based on these findings and further studies, BanLec has shown potential anti-HIV effects, which are mediated through its bidentate binding to specific viral glycans (Hopper et al., 2017). The mitogenic activity of some lectins contributes to inflammation, and in an interesting work based on rational analysis and modeling, BanLec mutant H84T resulted in an engineered product with decreased mitogenic activity while retaining the antiviral activity (Swanson et al., 2015). We have also extended our bioinformatics analysis of sugar-binding sites to other members of the jacalin-related family having the β-prism fold and found the preponderance of three sugar-binding sites per subunit in monocots, which could be related to their role in plant defense (Sharma et al., 2007).

We have used bioinformatics analysis to identify potential lectins in *Mycobacterium tuberculosis*H37Rv genome and have identified 11 candidate genes (Singh et al., 2007). The work formed the basis for cloning expression and structural analysis of these genes to identify their role in disease pathogenesis (Patra et al., 2010). Similarly, we have analyzed the genome of *Leishmania major* through a bioinformatics study and found 194 putative adhesion-like genes, which could be involved in host-pathogen interaction. Six of these have a well-defined adhesion domain, whereas the rest have been annotated as hypothetical proteins in available databases which require further experimental validation (Singh & Singh, 2008). Our study indicates a strong likelihood of these candidates in pathogenesis and we have extended this work to clone, express, purify, and characterize these proteins to understand their role in leishmaniasis (Singh et al., 2015; Kumar et al., 2017).

20.5 CONCLUSIONS

With the rapid elucidation of structures of many proteins implicated in diseases and the dynamics of protein interactions becoming clearer, rational drug design is gaining a confident foothold. The power of bioinformatics therefore spans the entire landscape from gene to disease modulation, and this is going to grow rapidly in the future and create many more useful products.

REFERENCES

Agafonov RV, Wilson C and Kern D (2015). Evolution and intelligent design in drug development. *Front MolBiosci* 2:1.

Altshuler DM et al. (2010). Integrating common and rare genetic variation in diverse human populations. *Nature* 467:52–58.

Behjati S, Tarpey PS (2013). What is next generation sequencing? *Arch Dis Child Educ Pract Ed* 98:236–238.

Brenner S et al. (2000). Gene expression analysis by massively parallel signature sequencing (MPSS) on microbead arrays. *Nature Biotechnol* 18:630–634.

Gray JJ, Moughon S, Wang C, Furman OC, Kuhlman B, Rohal CA, Baker D (2003). Docking with simultaneous optimization of rigid-body displacement and side-chain conformations. *J Mol Biol* 331:281–299.

Fu H et al. (2016). The Sunway TaihuLight supercomputer: system and applications. *Sci China Inf Sci* 59:072001.

Hopper JTS et al. (2017). The tetrameric plant lectin BanLec neutralized HIV through bidentate binding to specific viral glycans. *Structure* 25:773–782.

Kumar M, Ranjan K, Singh V, Pathak C, Pappachan A, Singh DD (2017). Hydrophilic surface protein A (HASPA) of Leishmania donovani: expression, purification and biophysico-chemical characterization. *Protein J* 36:343–351.

Miller JR, Koren S, Sutton G (2010). Assembly algorithms for next-generation sequencing data. *Genomics* 95:315–327.

Mansuri MS, Laddha NC, Dwivedi M, Patel D, Alex T, Singh M, Singh DD, Begum R (2016). Genetic variations (Arg5Pro and Leu6Pro) modulate the structure and activity of GPX1 and genetic risk for vitiligo. *Exp Dermatol* 25:654–657.

Patra D, Srikalaivani R, Misra A, Singh DD, Selvaraj M, Vijayan M (2010). Cloning, expression, purification, crystallization and preliminary X-ray studies of a secreted lectin (Rv1419) from *Mycobacterium tuberculosis*. *Acta Crystallogr Sect F Struct Biol CrystCommun* 66:1662–1665.

Quail MA, Smith M, Coupland P, Otto TD, Harris SR, Connor TR, Bertoni A, Swerdlow HP, Gu Y (2012). A tale of three next generation sequencing platforms: comparison of ion torrent, pacific biosciences and illumina MiSeq sequencers. *BMC Genomics* 24(13):341.

Salzberg SL, Delcher AL, Kasif S, White O (1998). Microbial gene identification using interpolated Markov models. *Nucleic Acids Res* 26:544–548.

Sharma A, Chandran D, Singh DD, Vijayan M (2007). Multiplicity of carbohydrate binding sites in β-prism fold lectins: occurrence and possible evolutionary implications. *J Bio Sci* 32:1089–1110.

Singh DD, Chandran D, Jeyakani J. Chandra NR (2007). Scanning the genome of *Mycobacterium tuberculosis* to identify potential lectins. *Protein Pep Lett* 14:683–691.

Singh V, Nair DN, Kaushal RS, Kumar M, Pappachan A, Singh DD (2015). Heterologous expression, purification and characterization of L-type lectin homologue from *Leishmania donovani*. *Biotech Rep* 8:81–87.

Singh DD, Saikrishnana K, Kumar P, Surolia A, Sekar K, Vijayan M (2005). Unusual sugar specificity of banana lectin from Musa paradisica and its probable evolutionary origin. Crystallographic and modelling studies. *Glycobiology* 15:1025–1032.

Singh V, Singh DD (2008). *Leishmania major*: genome analysis for identification of putative adhesin-like and other surface proteins. *ExpParasitol* 118:139–145.

Swanson MD et al. (2015). Engineering a therapeutic lectin by uncoupling mitogenicity from antiviral activity. *Cell* 163:746–758.

Wickham H (2011). The split-apply-combine strategy for data analysis. *J Statist Software* 1:1–29.

21 Microbial Bioinoculants for Sustainable Agriculture
Trends, Constraints, and Future Perspectives

Hillol Chakdar
ICAR–National Bureau of Agriculturally Important Microorganisms

Sunil Pabbi
ICAR–National Bureau of Agriculturally Important Microorganisms
Indian Agricultural Research Institute

CONTENTS

21.1 INTRODUCTION

The unprecedented increase in population is continuously exerting pressure on the limited natural resource base, making it difficult to produce more food and fiber. In quest for achieving enhanced levels of food production from the limited cultivable area, intensive agriculture practices were adopted, leading to extensive use of chemical fertilizers. But it resulted in numerous problems such as nutrient imbalance, micronutrients deficiency, deterioration of soil health, and stagnation of crop yields. The situation is becoming alarming day by day, forcing us to look for viable and sustainable alternatives. By the turn of the century, there has been a major shift where alternative farming, which includes integrated farming and organic farming, has been advocated and practiced globally so as to have eco-friendly, contamination-free agriculture that avoids/largely excludes chemical fertilizer, pesticides, etc. and mainly relies on organic inputs including crop rotation, legume, green manure, compost, and biological pest control. This new approach for farming is often referred to as "sustainable Agriculture" that promotes the use of renewable inputs such as biofertilizers, green manure, and vermicompost. This is also important from the view point of both environmentally safe technologies and providing some sort of fertilizer to the poor and marginal farmers, which form a major part of the farming system in Asia, in general, and India, in particular. Biofertilizers are low cost inputs that play a significant role in crop yields and enhance nutrient availability to the crop plants. The basic concept of biofertilizer is to domesticate some of the microorganisms in our agriculture production system, so that vast natural reservoir of nutrients in atmosphere (nitrogen) and soil can be trapped as an additional source to meet our requirements.

21.2 BIOFERTILIZERS IN AGRICULTURE

Biofertilizers are preparations containing live or latent cells of efficient strains of microorganisms (nitrogen fixing, phosphate solubilizing, cellulolytic, plant growth promoting) in sufficient numbers, used with the objective of increasing the number of such microorganisms and accelerating those microbial processes, which augment the availability of nutrient that can be easily assimilated by plants. In other words, these may be referred as "nutrient inputs of biological origin for enhanced plant growth and soil health improvement."

The basic aim of their use in agriculture is to increase their population in quantity and quality in soil or around the roots to allow them to carry out enhanced metabolic activities. These are also referred to as microbial inoculants as they contain living/latent cells of different type of micro-organisms. Most biofertilizers belong to one of the categories: nitrogen fixing, nutrient (P, K, Zn, etc.), solubilizing and/or mobilizing, and plant growth promoting. Nitrogen-fixing biofertilizers fix atmospheric nitrogen into forms that are readily taken up by plants. These include *Rhizobium*, *Azotobacter*, blue green algae (BGA), *Azospirillum*, and *Azolla*. Nutrient-solubilizing and nutrient-mobilizing microorganisms dissolve or mobilize inorganic and organic sources of specific nutrient for easy uptake by plants. Certain groups of bacteria enhance the growth of plants through different mechanisms by production of growth-promoting metabolites and are referred as plant growth–promoting rhizobacteria (PGPR). Nowadays, biofertilizers are available for almost all crops including horticulture and forestry. There are certain crop-specific biofertilizers, whereas others are common to many crops. Some of the major biofertilizers and target crops are given in Table 21.1. Commercial production of biofertilizers is not yet up to the mark, and only a few authentic biofertilizers are being produced (Table 21.2). To meet the demand of the biofertilizers in the country, policy changes and quality monitoring are urgently required.

TABLE 21.1
Some Crop-Specific Biofertilizers

Type of Biofertilizer	Function/Contribution	Target Crops
Nitrogen-fixing	20–30 kg N/ha/season	Pulse legumes such as gram, pea, and
Rhizobium	10%–35% increase in crop yield	lentil. Oil seed legumes such as soybean and groundnut; forage legumes such as lucerne and clover
Azotobacter	20–25 kg N ha/season	Wheat, cotton, maize, sugarcane, sorghum,
	10%–15% increase in yield	rice, vegetables, and many other crops
Azospirillum	10%–15% increase in yield	Cereals, sorghum, millets, etc.
	20–40 kg N/ha	
	Production of growth-promoting substances	

(Continued)

TABLE 21.1 (*Continued*)
Some Crop-Specific Biofertilizers

Type of Biofertilizer	Function/Contribution	Target Crops
Blue green algae	25–30 kg N/ha/season 10%–12% increase in yield Improvement in physicochemical and biological properties of soil	Mainly for rice (flooded rice)
Azolla	20–80 kg N ha/season 10%–20% increase in yield Improvement in physicochemical and biological properties of soil	Flooded rice
Phosphate solubilizing	Make 30–35 kg P_2O_5/ha available to plants Increase in grain yield Production of growth-promoting substances	For all agriculture, horticulture crops etc.
Phosphate mobilizing	Saving of 25%–30% of phosphatic fertilizer Help in uptake of water Protection from root pathogens	For agriculture, horticulture, and tropical forestry

Plant growth–promoting
  rhizobacteria
	Production of plant hormones
	Supplying nutrients such as biologically
	  fixed N or solubilizing P
	Suppression of pathogens
		For all agriculture, horticulture crops, etc.

TABLE 21.2
Some of the Commercially Available Biofertilizers/Bioinoculants Available in India and Other Countries

Class	Name of Biofertilizer/ Bioinoculant	Used Organism	Type of Formulation	Target Crop	Remarks
N fixers	Cell-Tech® Liquid (Novozyme)	Rhizobium	Liquid	Soybean	-

	Optimize® (Monsanto
	  BioAg Alliance)
		Bradyrhizobiumm japonicum
		  with lipochitooligosaccharide
			Liquid	Soybeans	-

	MicroAZ-ST Dry (TerraMax)	*Azospirillum brasiliense* *Azospirillum lipoferum*	Powder	Maize	-
	Symbion-N (Azos) (T Stanes & Co Ltd.)	*A. lipoferum*	Liquid	Multiple crops	-

	MicroAZ-IF Liquid™
	  (Terramax)
		Two improved strains of
		  Azospirillum
			Liquid	Maize	-

	Life Force Bio-N™ (*Nutri Tech Solutions*)	*Azotobacter chroococcum*	Liquid	Many crops	-
	NutriLife Bio-N (*Nutri Tech Solutions*)	*Azotobacter*	Liquid	Many crops	-
	Nodulator® (BASF)	*B. japonicum*	Liquid/Peat	Pea Lentil	-

	Nodulator® N/T
	  (BASF)
		Bacillus subtilis MBI 600
		  + *B. japonicum*
			Liquid	Soybean	-

	Nodulator® PRO (BASF Canada Inc.)	*B. japonicum*	Liquid	Soybean	-

	Nodulator® XL (BASF
	  Canada Inc.)
		Rhizobium leguminosarum
		  biovar *viceae* 1435
			Peat	Pea Lentil	3%–8% yield
					increase

(*Continued*)

TABLE 21.2 (*Continued*)

Some of the Commercially Available Biofertilizers/Bioinoculants Available in India and Other Countries

Class	Name of Biofertilizer/ Bioinoculant	Used Organism	Type of Formulation	Target Crop	Remarks
P solubilizers	*Symbion-P®* (*T Stanes & Co. Ltd.*)	*Bacillus megaterium*	Liquid	Multiple crops	Supplies up to 10–15 P/ha
	Jumpstart® (Monsanto)	*Penicillium bilaiae*	Granular	Maize	-
N fixer + P solubilizer	TagTeam® (Novozymes)	Rhizobium + *Penicillium bilaii*	Granular		9%–10% yield increase
	Vertex-IF	*Azospirillum Pseudomonas Pantoea*	Liquid	Maize	-
	Anubhav liquid biofertilizers (Anand Agricultural University)	*Azotobacter chroococcum, Azospirillum lipoferum*, and *Bacillus coagulans*	Liquid	Cotton, banana, potato, rose, turmeric, papaya, etc.	
N fixer + P solubilizer + K solubilizer	BioNPK (ICAR—National Bureau of Agriculturally Important Microorgnisms)	*Azotobacter Paenibacillus Bacillus*	Liquid	Multiple crops	Curtails use of chemical fertilizer by 25%–30% and increase yield by 10%–15%
N fixer + P solubilizer + Zn solubilizer + plant growth promoter	Arka Microbial Consortium (ICAR-Indian Institute of Horticultural Research)	-	Carrier based	Vegetables	Reduction in N and P fertilizer requirement by 25%–30% and yield increase by 5%–15 %
Plant growth promoter	QuickRoots® (Novozyme)	*Bacillus amyloliquefaciens Trichoderma virens*	Liquid	Canola, maize, field pea, lentil, small grains, and soybean	-

21.2.1 NITROGEN-FIXING BIOFERTILIZERS

The atmosphere around us contains about 78% nitrogen (by volume), but it cannot be utilized in free form by plants. Certain microorganisms have the ability to use this atmospheric nitrogen under normal conditions and convert to ammonia, which can then be easily taken up by plants. These are called nitrogen-fixing microorganisms. The conversion of N_2 into ammonia by microorganisms is done by the action of enzyme nitrogenase and thence into proteins, and the process is called biological nitrogen fixation (or dinitrogen fixation). All these nitrogen-fixing microorganisms (diazotrophs) are prokaryotic in nature and are broadly divided into three categories, viz., symbiotic, asymbiotic, or free living and associative (Table 21.3).

TABLE 21.3

Some Examples of Nitrogen-Fixing Bacteria Belonging to Different Categories

Category	Examples
Symbiotic	*Rhizobium*—legume symbiosis
	Rhizobium—*Parasponia* (nonlegume) symbiosis
	Frankia—Trees (e.g., *Alder, Casuarina*)
	Azolla—*Anabaena*
	Azotobacter paspali—*Paspalum notatum*
Free living	
1. Aerobic	*Azotobacter, Beijerinckia,* Cyanobacteria (e.g., *Nostoc, Anabaena, Tolypothrix, Aulosira*)
2. Facultative	*Klebsiella pneumonia, Bacillus polymyxa*
3. Anaerobic	*Clostridium, Desulfovibrio, Rhodospirillum*
	Rhodopseudomonas, Desulfotomaculum, Desulfovibrio
	Chromatium, Chlorobium
Associative	*Azospirillum, Herbaspirillum*
	Gluconobacter diazotrophicus, Azoarcus

21.2.1.1 Rhizobium

This is the most widely used among all nitrogen fixers, which colonizes the roots of specific leguminous plants to form root nodules. *Rhizobium* species such as *Bradyrhizobium, Mesorhizobium, Sinorhizobium,* and *Azorhizobium* in association with leguminous plants reduce atmospheric nitrogen and provide alternative to the use of energy-expensive ammonium fertilizer (urea). The legume-rhizobia symbiosis culminates in the formation of nitrogen-fixing root or stem nodules. It must be mentioned that not all legumes fix nitrogen. Of the three segregate families of legumes, the capacity to form nodules appears to be absent from the majority of species of Caesalpiniaceae. All members of family Mimosaceae and Fabaceae show formation of nodules with rhizobia. It is believed that legume-*Rhizobium* symbiosis can contribute at least 70 million metric tons nitrogen per year. Although the rhizobia are specific for leguminous crops, every *Rhizobium* cannot colonize every leguminous crop. It is crop specific, e.g., *Rhizobium trifolli* colonizes barseem, *Rhizobium meliloti* for Lucerne, *Bradyrhizobium japonicum* for soybean, *Rhizobium lupini* for chickpea, etc. All these different rhizobia have been developed as biofertilizer inoculants, and the farmer has to use the specific *Rhizobium* for a particular crop. Application of the inoculant to the seed surface prior to sowing is the traditional, most commonly used means of inoculation. An appropriate strain can increase the crop yield up to 10%–35%, and a nitrogen saving of 20–30 kg/ha/season.

Although inoculation of legume is a promising strategy for manipulation of rhizobial microflora and improving crop productivity and soil fertility, legume inoculation often fails where there is presence of adequate native rhizobia and high levels of mineral N especially in tropical soils. Thus, one needs to identify conditions where inoculation is needed and be beneficial. One may decide about inoculation provided, if population density of species-specific rhizobia is low, there is no immediate history of same or symbiotically related legume being grown in the area, legume follows a nonleguminous crop in a rotation, soils are acidic, and alkaline and saline or soil is poor in mineral nitrogen especially nitrate (Thilakarathna and Raizada, 2017; Salvagiottiet al., 2008).

21.2.1.2 Azotobacter

Azotobacter is a free living, heterotrophic nitrogen-fixing bacterium that occurs in the rhizosphere of a variety of plants. It possesses highest respiratory rate and can fix up to 25 kgN/ha under optimum conditions and increase yield up to 50% (Chandra et al., 2004). *Azotobacter* is found on neutral to alkaline soils, in aquatic environments, in the plant rhizosphere and phyllosphere. *Azotobacter* was initially discovered using a medium lacking combined nitrogen source. *Azotobacter chroococcum* is the most common species present in the soil; however, there are six species of *Azotobacter*, viz., *A. chroococcum, Azotobacter vinelandii, Azotobacter beijerinckii, Azotobacter nigricans, Azotobacter Armeniacus*, and *Azotobacter paspali*. Except *A. paspali*, which is a rhizoplane bacterium, the other members are largely soil borne and rhizospheric. Because of its free living nature, it can be used for a variety of crops, and its potential as a biofertilizer for various nonleguminous crops is well documented (Yazdani et al., 2009, Mrkovački et al., 2007, Hajnal-Jafari et al., 2012, Govedarica et al., 2002). It is known to influence plant growth through its ability to fix molecular nitrogen, production of growth-promoting substances such as IAA, gibberellin or gibberellin-like compounds, naphthalene acetic acid (NAA) and vitamins, excretion of ammonia, production of antifungal metabolites, and phosphate solubilization. Due to production of antifungal compounds, it also prevents the plants from different fungal diseases. Beneficial effects of *Azotobacter* inoculation have been reported in various cereals, vegetables, oil seeds, legumes, and cash crops also. Seed inoculation or seedling inoculation (by dipping the roots in microbial solution) is common practice. Seed inoculation of *A. chroococcum* increases the yield of field crops by about 10% and of cereals by about 15%–20%. The response to inoculation was increased by manuring or by fertilizer application. Coinoculation of *Azotobacter* with other bioinoculants such as *Rhizobium, Azospirillum*, P-solubilizers, and AM fungi has been reported to enhance the growth and yield of legumes, cereals, and vegetable crops (Murukumar et al., 2018; Arora et al., 2018). Since repeated application of *Azotobacter* is often needed for sustained effect and its requirement for large amounts of available carbon for its survival in soil, addition of FYM, compost, and other organic amendments to agricultural soils is also often recommended (Hindersah et al., 2018; Rodrigues et al., 2018).

21.2.1.3 Blue Green Algae

BGA or cyanobacteria are an important group of photosynthetic biofertilizers known since long to be responsible for the inherit and sustained fertility of rice fields. Rice field conditions are conducive for the growth of cyanobacteria, and improvement in the fertility status of rice field soils in tropics by utilizing these as nitrogen input has led to their agronomical potential. These are self-supporting, photoautotrophic microorganisms and can be produced in bulk with ease where water and carbon dioxide is available and temperature is suitable. It is an easily manageable, self-generating system, which not only provides valuable nutrients to plants in terms of nitrogen, amino acids, and growth-promoting substances but also improves soil health and texture. On an average, inoculation of rice field with BGA provides 25–30 kg N/ha and a progressive increase in yield by 10%–12%. The gradual buildup of BGA population results in soil fertility with a residual effect on succeeding crops also. A major role of BGA biofertilizer comprises organic matter accumulation and buildup in soil, leading to increase in exchange capacity of soil, water holding capacity, and buffering capacity of soil. BGA also produce humus on soil surface after death and dissolve certain soil minerals, maintaining a reservoir of elements in a semiavailable form for higher plants. Algalization also checks weeds proliferation by blocking nutrient supply and light and, on the other hand, induces early grain setting and maturity.

Algalization is an important input in paddy cultivation since it forms a perpetually renewable source of nutrients and improves soil health (Singh, 1961; Venkataraman, 1981). It was in the 1950s that Watanabe et al. (1951) initiated work on inoculation of rice fields with BGA, and it was further developed in India, Burma, Egypt, and China. In India, a considerable progress and development has been made in blue green algal biofertilizer technology. It has also been demonstrated that use of this technology results in soil fertility enrichment and improves rice crop yields (Mishra and Pabbi, 2004). The BGA are most often referred to as "paddy organisms" because of their abundance in the rice fields and play a significant role in rice culture. In addition to all these advantages of the BGA biofertilizer, it is also very easy to produce. Considering the fact that 87% of the rice area in our country accounts for holdings of 1–4 ha and 13% of the holdings even less than 1 ha, an inexpensive rural-oriented algal biofertilizer technology for rice was developed (Venkataraman, 1981) where farmers can produce this biofertilizer in their own field depending on their requirement with bare minimum inputs. The basic method involves mass production of a mixture of BGA along with a carrier, i.e., soil and then drying the mixture in sun. This technology was introduced as a package of practices in number of Indian states. The technology has further undergone several changes, and new protocols for its commercial production are now available, yielding quality product with higher titre value and longer shelf life (Pabbi, 2008). Extensive field trials conducted in different parts of the country with different chemical N levels have shown that algal supplementation results in saving of about one-third chemical N fertilizer to obtain similar crop yields. Algalization has always shown a perceptible gain in yield in presence of inorganic fertilizer; however, the gains are higher at lower fertilizer doses and marginal at optimum and higher doses. An extensive study has been done to assess the impact of BGA technology on farming practices and economic conditions of farming households in major rice-growing states of India, i.e., Punjab, Uttar Pradesh, and Haryana (Bhooshan et al., 2018). It was observed that use of BGA resulted in 25.2% reduction of chemical fertilizer (urea) input with an overall 3.8% increase in the yield and a marginal decrease in per acre cultivation cost. The study provided the comparative analysis on yield of paddy, urea consumption, and income with and without BGA application where it was observed that farmers earned about 3% greater income along with reduction in dosage of urea while reaping higher yield of paddy. There is no doubt that innumerable field trials conducted with Algal biofertilizer intervention have shown promising results in terms of both nitrogen saving and crop yield. The results may not be consistent always as the local ecological conditions differ and the inoculated organisms have to work under natural conditions which cannot be modified. Thus, at locations where the conditions are conducive for rapid growth and multiplication of inoculated algae, it gives good results, but at other places, the outcome is not as expected (Pabbi, 2015). This problem can be overcome to some extent by developing algal biofertilizer inoculants suitable for different agro climatic regions. This can be successfully achieved by isolating the local indigenous organisms from different areas or developing genetically stable strains, which can grow quickly under varied soil and climatic conditions and efficiently contribute by fixing atmospheric nitrogen. Such strains combined with capacity to withstand field doses of agrochemicals would be of greater importance for field application.

21.2.1.4 Azolla

Azolla, a water fern, also serves as a potential biofertilizer for rice and has a vast scope in Indian agriculture. The *Azolla* technology is most suited for irrigated rice areas having proper water management, which presently constitutes >20 million hectare. The successful use of this biofertilizer in just 1 million hectare rice area would result in a saving of >60,000 tons of urea. The biofertilizer potential of *Azolla* depends on total biomass production and N content. It grows very fast under favorable conditions, and its doubling time ranges between 2 and 10 days depending upon kind of species. It can fix as much as 3.6 N/ha/day. It can be used both as a green manure, before transplanting rice, and as a dual crop along with rice as biofertilizer. When grown once before or after planting of rice, it can produce up to 25 tons fresh biomass contributing up to 50 N/ha in about 25–30 days. During the crop duration of rice, *Azolla* can be applied thrice, i.e., once as a green manure before transplanting and twice as a dual crop along with rice and can contribute up to 80 N/ha. Application of *Azolla* has several other advantages like BGA. Besides contributing for nitrogen, it improves carbon status of soil, mobilizes mineral elements, improves soil physicochemical properties, and prevents weed growth. It also prevents water loss from the field by checking evaporation. Besides, it also serves as a raw material for biogas production and can be used as a feed for livestock, poultry, birds, and fishes.

The large-scale production of *Azolla* is again simple, and it can be easily multiplied in tanks, plastic trays, or simple earthen pots with negligible inputs or can be naturally grown in ponds, lakes, ditches, and canals. *Azolla* is susceptible to high temperature (>40°C) as well as low temperature and is prone to infection by pests and predators. The requirement of high inoculum to the tune of 0.5–1 ton/ha fresh *Azolla* biomass and problems of transportation has restricted its use.

21.2.1.5 Azospirillum

A unique nitrogen-fixing bacterium under the name *Spirillum lipoferum* was reported by Beijerinck in 1925 who observed its presence in enrichment cultures of *A. chroococcum*. Later, Day and Döbereiner (1976) had shown that *Azospirillum* could be isolated from the roots of tropical grass *Digitaria decumbens* using a semisolid N-free medium enriched with sodium malate. *Azospirillum* is recognized as a ubiquitous soil organism capable of colonizing the roots as well as the aboveground portions of a wide variety of plants including cereal crops. This bacterium is gram negative, motile with long polar flagellum, and generally vibroid in shape and contains poly-β-hydroxybutyrate granules. The cells change shape and size with culture age and produce cysts. They can grow under anaerobic with NO_3^- as electrons acceptor, microaerophilic in presence of N_2 or NH_3 as nitrogen sources, and fully aerobic conditions with combined nitrogen (NH_3, NO_3^-, amino acids) only. *Azospirillum* species also grow well on organic acids, such as malate, succinate, lactate, and pyruvate.

Azospirillum has been largely and extensively used for crop plants belonging to family Gramineae such as wheat, sorghum, pearlmillet, fingermillet, barley, and maize; however, its response has been found quite consistent in cropssuch as sorghum (*Sorghum bicolor*), pearlmillet (*Pennisetum americanum*) and fingermillet (*Eleusine coracana*). *Azospirillum* species inoculation can contribute 20–40 N/ha, resulting in an increase of crop yield by 10%–15%. The yield increase up to 11% in many crops has been reported (Wani, 1992). Despite their nitrogen-fixing capability, the increase in yield caused by *Azospirillum* inoculation is mainly attributed to an improvement in root development by the production of plant growth–promoting substances, such as auxins, cytokinins, and gibberellins. After inoculation onto plant roots, *Azospirillum* cells induce remarkable changes in the morphology and behavior of the entire root system. Hairs close to the root tip take on a more distinctive appearance, and the overall density and the length of the root system increase. Root hairs consist of expanded root epidermal cells, which play a role in water and nutrient exchanges and also help to anchor root to its surroundings. It also increases the diameter and length of both lateral and adventitious roots and thereby leads to additional branching of the lateral roots. These developments in the root system in turn increase absorptive area and volume of the soil substrate available to the plant, thereby resulting in increased uptake of soil nutrients. Like *Azotobacter*, it can also be used for treatment of seeds or seedlings. Another bacterium *Herbaspirillum*, which is taxonomically very closely related to *Azospirillum*, has also been found to be associated with grasses and contributes for nitrogen.

Strains of *Azospirillum* are known to produce siderophores, which form complexes with the metal ions, thereby contributing to improve the iron nutrition of plants and offer protection from minor pathogens. Similarly, azospirilla form antibiotic substances that suppress plant diseases due to their fungistatic activity against a wide range of phytopathogenic fungi, e.g., protecting cotton plants against *Thielaviopsis basicola* and *Fusarium oxysporum*. The field experiments in Israel have demonstrated that *Azospirillum*-inoculated sorghum plants made better use of moisture stored in soils from winter precipitation than did uninoculated plants (Sarig et al., 1988). In both green house and field experiments, inoculated plants were efficient at absorbing nitrogen, phosphorus, potassium, and other microelements from soil than uninoculated plants. Even the synergistic effects of *Azospirillum* with *Rhizobium* on different leguminous crops have been reported where stimulation of nodulation may be due to an increase in production of lateral roots and in root hair branching, and this, in turn, has been proposed to be due to production of phytohormones by *Azospirillum*. Phytohormone production and modulation of plant defense–related genes through seed and foliar application of *Azospirillum brasilense* have been reported to improve the growth of maize (Fukami et al., 2017). Fasciglione et al. (2015) reported that growth and quality of lettuce increases in *Azospirillum*-inoculated plants grown under salt stress.

21.2.1.6 Acetobacter

Gluconoacetobacter diazotrophicus is a gram-negative, microaerophilic, nitrogen-fixing bacterium, which was initially isolated from washed roots and stems of sugarcane plant. This is an endophytic diazotrophic bacterium, which is specifically found associated with sugar-rich non-leguminous plants such as sugarcane, sweet potato, sweet sorghum, cameroon grass, and coffee. It colonizes roots, stems, and leaves of host plants and is reported to fix 200 N/ha/year. In sugarcane, it is reported to contribute >50% of biologically fixed nitrogen and has the ability to grow and fix nitrogen at sugar concentration as high as 30%. Strains of *Gluconoacetobacter* have been shown to produce considerable amount of IAA. This can be used alone or in combination with AM fungi to produce synergistic effect on plant growth and yield of sugarcane, sweet potato, sweet sorghum, etc. (Meenakshisundaram and Santhaguru, 2011; Logeshwaran et al., 2011).

21.2.2 Nutrient (P, K, Zn)-Solubilizing and Nutrient-Mobilizing Biofertilizers

Phosphorus is only next to nitrogen in terms of nutritional elements required by plants. The "P" requirement of crop is often met by addition of phosphatic fertilizers, but only 15%–20% of added "P" fertilizer is available to the plants, as rest gets fixed in the soil as metal phosphates. There are certain microorganisms, which play a major role in the solubilization and uptake of native and applied phosphorus. These are otherwise referred to as phosphate-solubilizing microorganisms (PSMs). PSMs include several bacteria, fungi, actinomycetes, and cyanobacteria dominantly species of *Bacillus*, *Pseudomonas*, *Aspergillus*, *Penicillium*, *Anabaena*, etc., which bring about solubilization of bound phosphates in soil. These efficient cultures have been multiplied on mass scale and developed as inoculants. Their inoculation results in solubilizing insoluble inorganic phosphates and can also mineralize organic phosphatic compounds present in soil. Inoculation of PSMs to seed or seedlings makes 30–35 P_2O_5/ha available to plants and increases the grain yield of plants. It is also possible to use nonconventional sources of phosphorus like rock phosphate when using PSMs, which produce the same amount of yield as with super phosphate (SSP) due to their ability to solubilize rock phosphate (Tilak, 1993). PSMs have also been reported to produce growth-promoting substances.

Besides phosphate solubilization, there is another group of microorganisms that help in mobilizing nutrients for uptake by plants especially those less mobile in soil solutions like "P." They also help in uptake of water and protect roots from root pathogens. The microorganisms responsible for this are certain fungi called AM fungi. These form symbiotic association with the roots of plants important in agriculture, horticulture, and tropical forestry. Of all the AM fungi, endomycorrhizae are more important as biofertilizer. Several studies carried out on many important plants/crops such as tobacco, chillies, millet, tomato, citrus, and mango have shown that their inoculation improves seedling growth and vigor, which is mostly attributed to increase in uptake of nutrients, especially, phosphorus. Nearly 25%–30% of phosphatic fertilizer can be saved through the use of efficient AM fungi.

Almost 80%–90% of the mineral K in soil remains unavailable to the plants. Microorganisms capable of producing organic acids can make this mineral K available to plants in a way very similar to PSB. Production of organic acids directly dissolves either mineral K or chelate silicate ions to bring K into soil solution. A large number of bacteria such as *Bacillus*, *Paenibacillus*, *Sphingomonas*, *Burkholderia*, and *Arthrobacter* are known to solubilize K minerals. Various studies in the past have shown the benefits of K-solubilizing bacteria (KSB) such as *Bacillus mucilaginosus*, *Bacillus edaphicus*, and *Paenibacillus mucilaginosus* on plant growth and K uptake (Sheng, 2005, Sheng and He, 2006, Liu et al., 2012, Sangeeth et al., 2012).

In soil, Zn is present in the form of sulfates, oxides, carbonates, silicates, sulfides, etc. pH is an important factor that governs the proportion of plant-available zinc forms in soil. In general, Zn solubility increases with a reduction in soil pH. At alkaline pH, Zn becomes fixed in the form of insoluble carbonates, sulfides, phosphates, etc., and proportion of the bioavailable zinc pool reduces. Microbe-mediated solubilization of Zn can be accomplished by a range of mechanisms such as excretion of organic acids, proton extrusion, or production of chelators (Goteti et al., 2013). Ramesh et al. (2014) implicated organic acids for reduction of pH of soybean rhizosphere and enhanced uptake and accumulation of zinc by Zn- and P-solubilizing *Enterobacter cloaceae* MDSR9. Bacteria isolated from Zn hyperaccumulator *Sedum alfredii* increased DTPA-extractable Zn in the rice rhizosphere and enhanced Zn uptake and biofortification of rice grain (Yuyan et al., 2014). Zn solubilization by *Exiguobacterium aurantiacum* was associated with drop in pH of the medium due to organic acid production, and the strain could significantly enhance the Zn content in wheat grains as compared with uninoculated control (Shaikh and Saraf, 2017).

21.2.3 PLANT GROWTH–PROMOTING RHIZOBACTERIA

Rhizosphere, the region around root system of plants, is influenced by the living root and is the most happening region in the soil. The beneficial free living soil microorganisms isolated from this region having ability to improve plant health or increase yield are referred as PGPR. The effects of PGPR on plant growth can be mediated by direct or indirect mechanisms (Glick, 1995). The direct effects are attributed to the production of plant hormones such as auxins, cytokinins, and gibberellins; or by supplying nutrients like biologically fixed N or solubilizing P. The indirect mechanism include suppression of pathogens by production of siderophores, HCN, ammonia, antibiotics, and other metabolites by induced systemic resistance and/or by competing with the pathogen for nutrients or for colonization space. The well-known PGPR include bacteria belonging to genera *Azotobacter*, *Azospirillum*, *Klebsiella*, *Pseudomonas*, *Bacillus*, *Arthrobacter*, *Xanthomonas*, *Serratia*, *Pantoea*, and *Enterobacter*, and many more such microorganisms are yet to be isolated and identified (Verma et al., 2018; Chakdar et al., 2018).

21.3 GENETIC IMPROVEMENT OF BIOINOCULANT STRAINS

Genetic improvement of microbial strains is more common for industrial applications. Improved strains developed through mutation or genetic engineering have long been used for production of antibiotics, enzyme, or pharmaceuticals. Such improved strains are grown in controlled environment and practically do not face any competition as encountered by those released in natural environment. Therefore, the application and utility of the genetically improved strains for agriculture are limited, as they are eventually released in soil and face tough competition with indigenous microflora. Furthermore, horizontal transfer of genes also poses an environmental concern. Despite all these constraints, genetic improvement of bioinoculants can improve the efficiency of the inoculants several folds if the improvements are stable and ecologically sustainable (Figure 21.1).

Genetic improvement of bioinoculants through induced mutagenesis can also be achieved. Mutations in specific genes involved in synthesis of growth-promoting substances/secondary metabolites or organic acids/chelators involved in nutrient mobilization or deletion mutations or disruptions in repressors or negative regulators can improve the efficacy of the bioinoculants. Although developing mutant strains for bioinoculants is probably the easiest recourse for genetic improvement, their stability and competence in soil are the major constraints. Directed mutagenesis of *Pseudomonas fluorescens* F113 produced a triple mutant (mutant for *kin*B involved in swimming motility, *sad*B and *wsp*R encoding repressor of motility), which was hypermotile and had increased colonization efficiency and biocontrol activity as compared with wild-type strain (Barahona et al., 2011). Frederix et al. (2014) developed a *Pra*R (encoding a repressor associated with quorum sensing) mutant of *Rhizobium leguminosarum*, which showed enhanced root biofilm formation and nodulation competitiveness. Nitrous acid mutagenesis of *Burkholderia cepacia* strain RRE25 produced IAA overproducing mutants, which could significantly enhance the nitrogen, phosphorus, and potassium uptake in rice (Singh et al., 2013). Disruption of phlF (negative regulator of 2,4 DAPG biosynthesis) gene in *P. fluorescens* J2 resulted in higher production of 2,4 DAPG and increased inhibition of *Ralstonia solanacearum* (Zhou et al., 2014). Tripura et al. (2007) reported

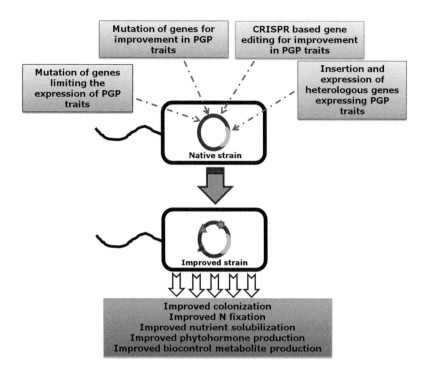

FIGURE 22.1   Genetic improvement of bioinoculants.

EMS-mutated strains of groundnut-associated *Serratia marcescens* GPS-5 with ~40% increase in mineral phosphate solubilization activity. Majority of these studies are confined to the labs, and only very few works have been evaluated under field trial. Zhang et al. (2002) developed UV mutants of *B. japonicum* and showed that there was an increase in nodule numbers, dry weight, and total nitrogen fixation in soybean grown in field.

Recombinant DNA technologies offer a very attractive means to genetically manipulate the potential microbial inoculants. Expression of heterologous genes helping in signaling, colonization, and improving the expression of traits of interests in the microbial inoculants can be achieved through genetic engineering. Furthermore, genetic engineering can help to develop strains, which can sense the physicochemical conditions in the microenvironment or presence of pathogen, and specifically induce a set of genes to help plant growth. Majority of the genetic modification studies have been carried out in rhizobia, whereas the other bioinoculants have received comparatively less attention. Orikasa et al. (2010) reported that introduction of vktA catalase gene from *Vibrio rumoensis* S1-T to *Rhizobium leguminosarum bv. phaseoli* USDA 2676 resulted in almost threefold and two fold increase in catalase activity and nitrogen fixing efficiency in the rhizobia. Rhizobia are known to produce leg hemoglobins, whereas hemoglobins are usually absent in rhizobia but have been reported in a number of bacteria. These hemoglobins are known to be synthesized under hypoxic conditions. Heterologous expression of such hemoglobins in rhizobia can improve the respiration rates and increase nitrogen fixation under low oxygen. In 1999, Ramírez et al. reported that *Rhizobium etli* expressing *Vitreoscilla* sp. hemoglobin (VHb) exhibited higher nitrogenase activity and total nitrogen content (68% and 14%–53%, respectively) in bean plants as compared with inoculation with *R. etli* wild type. Ethylene is a well-known inhibitor of nodulation in many legumes, and nodulation has been reported to increase with the treatment of inhibitors for ethylene biosynthesis or perception (Lee and LaRue, 1992; Nukui et al., 2000). Ma et al. (2004) reported that introduction of ACC deaminase gene from *R. leguminosarum* to *Sinorhizobium meliloti* resulted in 35%–40% more nodulation in alfa-alfa as compared with the wild-type *S. meliloti*.

Genetic modification of *Pseudomonas* strain with antibiotic-producing genes from other potential strains has been reported to improve the biocontrol activities significantly (Bainton et al., 2008). He et al. (2016) reported enhancement of antifungal activity of *Burkholderia pyrrocinia* through introduction of chitinase gene from *Bacillus subtilis*. Very similar to the mutant bioinoculant strains, the genetically modified strains have also been reported to compete with the indigenous microflora. McClung (1998) reported that a genetically modified strain of *S. meliloti* showed significant nodulation in alfa-alfa growing soils with low population of indigenous rhizobia while it failed in the soils where the number of rhizobia was higher. Ryder et al. (2012) showed that plant growth–promoting ability of *Trichoderma hamatum* could be enhanced by genetic manipulation, which otherwise rendered negative impacts on its biology and ecology as a saprotrophic competitor and antagonist of soil-borne pathogens.

21.4   NEXT-GENERATION BIOINOCULANTS

21.4.1   Nanobioinoculants—The New-Generation Bioinoculants

Nanoparticles (of sizes ≤100 nm) can find diverse application in plant nutrition and protection due to their size-dependent features, unique optical properties, and surface-to-volume ratio. Such applications may range from delivery biopesticides encapsulated in nanomaterials, stabilization of biopesticides with nanomaterials, sustained release of biofertilizers, and micronutrients for efficient use (Ghormade et al., 2011). In comparison with other formulation technologies, nanoparticles are advantageous as they can be used to deliver growth-promoting inoculants and/or their active chemical ingredients in a controlled way, for example, to a particular type of tissue/cells at a specific time (Timsuk et al., 2018). Qin et al. (2016) reported that silicon nanoparticle–immobilized chitinase enzyme of *Bacillus thuringiensis* had significant nematicidal activity with higher pH and thermotolerance. *B. thuringiensis*-coated zinc oxide nanoparticles showed significant reduction in fecundity and hatchability of *Callosobruchus maculatus* (Malaikozhundan et al., 2017). It has been suggested not only that nanoparticles can be used for delivery of bioinoculants, but that they are also known to improve the efficacy of inoculants. Shukla et al. (2015) reported that gold nanoparticles significantly increased the growth of *P. fluorescens*, *Paenibacillus elgii*, and *B. subtilis*. Population of growth-promoting bacterial population increased twofold when nanosilica was applied to the maize rhizosphere (Karunakaran et al., 2013). Timsuk et al. (2018) reported that growth of wheat seedlings increases due to enhancement in efficiency and attachment of growth-promoting bacteria upon treatment with TiO_2 nanoparticles. Despite having a tremendous potential for combining nanotechnology with microbial technology to develop smart technologies for agriculture, most of the researches are focused on development of nanoparticles through microbes or antimicrobial properties of nanoparticles.

21.4.2   Biofilm-Based Bioinoculants

Biofilm formation by microorganisms offers them adaptive advantage for colonization, stress tolerance, combating nutrient depletion, etc. Biofilms also help plant beneficial microorganisms to colonize different plant parts. Development of multispecies biofilms using *Trichoderma* as a matrix with *A. chroococcum*, *P. fluorescens*, and *B. subtilis* as partners showed positive effects on plant growth–promoting traits (Triveni et al., 2012, 2013). Experiments deploying *Trichoderma*- and *Anabaena*-based bacterial biofilms illustrated their role as plant growth–promoting and biocontrol agents in *Macrophomina phaseolina*-challenged cotton crop (Triveni et al., 2015). Babu et al. (2015) studied the varietal responses of *Anabaena*- and *Trichoderma*-based biofilms and their subsequent establishment in the rhizosphere of cotton. However, the works on biofilm-based bioinoculants are still in their initial stages and alternative matrices; interaction with a wide variety of plants needs further to be studied.

21.5 CONCLUSION

India has been the pioneer in the field of biofertilizers and has generated valuable information on their practical utilization as an important input in crop production. Number of on-field trials (OFTs) has demonstrated the superiority of these microbial products and have shown promising results both in terms of nutrient (N, P, K, etc.) input saving as well as crop yield. No doubt, the results may not be always consistent, as these are biological system and have to compete with both biotic and abiotic factors, resulting in varied effect, but at places where the conditions are appropriate, these microorganisms multiply and gradually establish to give desired results. Recent developments of new formulations such as liquid carrier materials, production under controlled conditions, development of growth media to attain rapid cell growth, and process design with effective quality control have led to higher "titre value" and longer shelf life of the product, leading to sustained performance. Popularization coupled with success of demonstration trials has led to wider acceptance and adoption at the farmers' level. The demand has increased, but the gap between demand and production is still wide. Although these new technologies hold promise, are economically viable, and can be exploited commercially considering the immense market potential, there is a need to create viable, remunerative, and enabling environment for participation of the more private sector in order to meet the demand and quality. Formulation of standards/regulations for quality control for ever-growing list of new biofertilizers/formulations is very essential and will go a long way in making high-quality biofertilizers available to the consumers and win their confidence, thereby sustaining productivity and fertility of soils. We also need to develop composite biofertilizer inoculants that can be useful for many crops and are multifunctional to further boost their role in agriculture. This will go a long way in not only providing economic benefits but also improving and maintaining fertility and sustainability in natural ecosystem.

REFERENCES

Arora, M., Saxena, P., Abdin, M.Z., Varma, A. Interaction between Piriformosporaindica and Azotobacterchroococcum governs better plant physiological and biochemical parameters in Artemisia annua L. plants grown under in vitro conditions. *Symbiosis* 75 no. 2 (2018): 103–112.

Babu, S., Bidyarani, N., Chopra, P., Monga, D. Evaluating microbe-plant interactions and varietal differences for enhancing biocontrol efficacy in root rot disease challenged cotton crop. European Journal *of* Plant Pathology 142 (2015): 345–362.

Bainton, N.J., Lynch, J.M., Naseby, J., Way, A. Survival and ecological fitness of *Pseudomonas fluorescens* genetically engineered with dual biocontrol mechanisms. *Microbial Ecology* 48 no. 3 (2008): 349–357.

Barahona, E., Navazo, A., Martinez-Granero, F., Bonilla, T.Z., Perez-Jimenez, R.M., Martin, M., Rivilla, R. A Pseudomonas fluorescens F113 mutant with enhanced competitive colonization ability shows improved biocontrol activity against fungal root pathogens. *Applied and Environmental Microbiology* (2011). doi:10.1128/AEM.00320-11

Bhooshan, N., Pabbi, S., Singh, A., Sharma, A., Chetan, Jaiswal, A., Kumar, A. Impact of blue green algae (BGA) technology: an empirical evidence from north western indo-gangetic plains. *3 Biotech* (2018). doi:10.1007/s13205-018-1345-5

Chakdar, H., Dastager, S.G., Khire, J.M., Rane, D., Dharne, M.S. Characterization of mineral phosphate solubilizing and plant growth promoting bacteria from termite soil of arid region.*3 Biotech* 8 no. 3 (2018): 463.

Chandra, R., Kumar, K., Singh, J. Impact of anaerobically treated and untreated (raw) distillery effluent irrigation on soil microflora, growth, total chlorophyll and protein contents of Phaseolus aureus L. *J. Environ. Biol.* 25 (2004): 381–385.

Day, J. M., Döbereiner, J. Physiological aspects of N2-fixation by a Spirillum from Digitaria roots. *Soil Biology and Biochemistry* 8 no. 1 (1976): 45–50.

Fasciglione, G., Casanovasa, E.M., Quillehauquya, V., YommiaMaría, A.K., Goñibc, G., et al. *Azospirillum* inoculation effects on growth, product quality and storage life of lettuce plants grown under salt stress. *ScientiaHorticulturae* 195 no. 12 (2015): 154–162.

Frederix, M., Edwards, A., Swiderska, A., Stanger, A., Karunakaran, R., Williams, A.,...Downie, J.A. Mutation of praR in R hizobiumleguminosarum enhances root biofilms, improving nodulation competitiveness by increased expression of attachment proteins. *Molecular Microbiology* 93 no. 3 (2014): 464–478.

Fukami, J., Ollero, F.J., Megias, M., Hungria, M. Phytohormones and induction of plant-stress tolerance and defense genes by seed and foliar inoculation with *Azospirillumbrasilense* cells and metabolites promote maize growth. *AMB Express* 7 (2017): 153.

Ghormade, V., Deshpande, M.V., Paknikar, K.M. Perspectives for nano-biotechnology enabled protection and nutrition of plants. Biotechnology *Advances* 29 no. 6 (2011): 792–803.

Glick, B.R. The enhancement of plant growth by free-living bacteria. *Canadian Journal of Microbiology* 41 no. 2 (1995): 109–117.

Goteti, P.K., Daniel, L., Emmanuel, A., Desai, S., Shaik, M.H.A. Prospective zinc solubilising bacteria for enhanced nutrient uptake and growth promotion in maize (Zea Mays L.). *International Journal of Microbiology* (2013). doi:10.1155/2013/869697.

Govedarica, M., Milošević, N., Jarak, M., Kuzevski, J., Krstanović, S., Krunić, V. Bakterizacijakaomeraborbe protivrizomaniješećernerepe. ZbornikradovaInstitutazaratarstvo i povrtarstvo 36 (2002): 33–42.

Hajnal-Jafari, T., Jarak, M., Djuric, S., Stamenov, D. Effect of co-inoculation with different groups of beneficial microorganisms on the microbiological properties of soil and yield of maize (Zea mays L.). *Ratarstvo i povrtarstvo* 49 no. 2 (2012): 183–188.

He, L.M., Ye, J.R., Ren, J.H., Huang, L., Wu, X.Q. Enhancement of antifungal activity of Burkholderiapyrrocinia JK-SH007 genetically modified with Bacillus subtilis Chi113 gene. Forest Pathology 46 no. 6 (2016): 632–646.

Hindersah, R., Handyman, Z., Indriani, F.N., Suryatmana, P., Nurlaeny, N. Azotobacter population, soil nitrogen and groundnut growth inmercury-contaminated tailing inoculated with Azotobacter. *Journal of Degraded and Mining Lands Management* 5 no. 3 (2018): 1269–1274.

Karunakaran, G., Suriyaprabha, R., Manivasakan, P., Yuvakkumar, R., Rajendran, V., Prabu, P., Kannan, N. Effect of nanosilica and silicon sources on plantgrowth promoting rhizobacteria, soil nutrients andmaize seed germination. *IET Nanobiotechnology* (2013). doi:10.1049/iet-nbt.2012.0048

Lee, K.H., Larue, T.A. Exogenous ethylene inhibits nodulation of Pisumsativum L. cv Sparkle. *Plant Physiology* 100 no. 4 (1992): 1759–1763.

Liu, D., Lian, B., Dong, H. Isolation of Paenibacillus sp. and assessment of its potential for enhancing mineral weathering. Geomicrobiology *Journal* 29 (2012): 413–421.

Logeshwaran, M., Thangaraju, K., Rajasundari, K. Antagonistic potential of Gluconacetobacterdiazotrophicus against Fusariumoxysporumin infecting sweet potato (Ipomeabatatus) P. *Archives of Phytopathology and Plant Protection* 44 no. 3 (2011): 216–222.

Ma, W., Charles, T.C., Glick, B.R. Expression of an exogenous 1-aminocyclopropane-1-carboxylate deaminase gene in Sinorhizobiummmeliloti increases its ability to nodulate alfalfa. *Applied and Environmental Microbiology* 100 no. 70 (2004): 5891–5897.

Malaikozhundan, B., Vaseeharan, B., Vijayakumar, S., Thangaraj, M.P. Bacillus thuringiensis coated zinc oxide nanoparticle and its biopesticidal effects on the pulse beetle, Callosobruchusmaculatus. Journal of *Photochemistry and Photobiology* B 174 (2017): 306–314.

McClung, G. Commercialization of a genetically modified symbiotic nitrogen-fixer, Sinorhizobiummmeliloti. The Biosafety Results of Field Tests of Genetically Modified Plants and Microorganisms – 5th Int. Symposium, Braunschweig, 6–10 September 1998.

Meenakshisundaram, M., Santhaguru, K. Studies on association of arbuscularmycorrhizal fungi with Gluconacetobacterdiazotrophicus and its effect on improvement of Sorghum bicolor (L.) ab. International Journal of *Current Research and Review* 1 no. 2 (2011): 23–30.

Mishra, U., Pabbi, S. Cyanobacteria: a potential biofertilizer for rice. *Resonance* 9 no. 6 (2004): 6–10.

Mrkovački, N., Mezei, S., Čačić, N., Kovačev, L. Effectiveness of different types of sugarbeet inoculation. ZbornikradovaInstitutazaratarstvo i povrtarstvo 43 no. 1 (2007): 201–207.

Murukumar, D.R., Indi, D.V., Amrutsagar, V.M. Synergistic effect of Azotobacter and phosphate solublizing bacteria (PSB) inoculation on yield of Rabi sunflower under dryland conditions. *Indian Journal of Dryland Agricultural Research and Development* 33 no.1 (2018): 41–44.

Nukui, N., Ezura, H., Yuhashi, K., Yasuta, T., Minamisawa, K. Effects of ethylene precursor and inhibitors for ethylene biosynthesis and perception on nodulation in Lotus japonicus and Macroptiliumatropurpureum. *Plant and Cell Physiology* 41 no. 7 (2000): 893–897.

Orikasa, Y., Nodasaka, Y., Ohyama, T., Okuyama, H., Ichise, N., Yumoto, I., Morita, N., Wei, M., Ohwada, T. Enhancement of the nitrogen fixation efficiency of genetically-engineered Rhizobium with high catalase activity. *Journal of Bioscience and Bioengineering* 110 no. 4 (2010): 397–402.

Pabbi, S. Cyanobacterial biofertilizers. *Journal of Eco-friendly Agriculture* 3 no.2 (2008): 95–111.

Pabbi, S. Blue green algae: a potential biofertilizer for rice. In: Sahoo D.B. and Seckbach J. (eds.), *The Algae World*. Springer India Pvt Ltd., New Delhi, 2015, pp.449–466.

Qin, X., Xiang, X, Sun, X., Ni, H, Li, L. Preparation of nanoscale Bacillus thuringiensis chitinases using silica nanoparticles for nematicide delivery. *International Journal of Biological Macromolecules* 82 (2016): 13–21.

Ramesh, A., Sharma, S.K., Sharma, M.P., Yadav, N., Joshi, O.P. Plant growth-promoting traits in Enterobacter cloacae subsp. dissolvens MDSR9 isolated from soybean rhizosphere and its impact on growth and nutrition of soybean and wheat upon inoculation. *Agricultural Research*, 3 no. 1 (2014): 53–66.

Ramírez, M., Brenda, V., Raúl, A.P., Soberón, M., Mora, J., Hernández, G. Rhizobium etli genetically engineered for the heterologous expression of Vitreoscilla sp. Hemoglobin: effects on free-living and symbiosis. *Molecular Plant-Microbe Interactions* 12 no. 11 (1999): 1008–1015.

Rodrigues, A.M., Ladeirab, L.C., Arrobas, M. Azotobacter-enriched organic manures to increase nitrogen fixation and crop productivity. *European Journal of Agronomy* 93 (2018): 88–94.

Ryder, L.S., Hariss, B.D., Soanes, D.M. Saprotrophic competitiveness and biocontrolfitness of a genetically modified strain of the plant-growth-promoting fungus Trichoderma hamatumGD12. *Microbiology* 158 (2012): 84–97.

Salvagiotti, F., Cassman, K.G., Specht, J.E., Walters, D.T., Weiss, A., Dobermann, A. Nitrogen uptake, fixation and response to fertilizer N in soybeans: a review. *Field Crops Research* 108 (2008): 1–13.

Sangeeth, K.P., Bhai, R.S., Srinivasan, V. Paenibacillusglucanolyticus, a promising potassium solubilizing bacterium isolated from black pepper (Piper nigrum L.) rhizosphere. Journal of Spices *and Aromatic* Crops 21 (2012): 334–336.

Sarig, S., Blum, A., Okon, Y. Improvement of the water status and yield of field-grown grain sorghum (Sorghum bicolor) by inoculation with Azospirillumbrasilense. Journal *of* Agricultural Science 110 (1988): 271–277.

Shaikh, S., Saraf, M. Biofortification of Triticumaestivum through the inoculation of Zinc solubilizing plant growth promoting rhizobacteria in field experiment. *Biocatalysis and Agricultural Biotechnology* (2017). doi:10.1016/j.bcab.2016.12.008.

Sheng, X.F. Growth promotion and increased potassium uptake of cotton and rape by a potassium releasing strain of Bacillus edaphicus. *Soil Biology and Biochemistry* 37 (2005): 1918–1922.

Sheng, X.F., He, L.Y. Solubilization of potassium-bearing minerals by a wild type strain of Bacillus edaphicus and its mutants and increased potassium uptake by wheat. *Canadian Journal of Microbiology* 52 (2006): 66–72.

Shukla, S.K., Kumar, R., Mishra, R.K., Pandey, A., Pathak, A., Zaidi, M.G.H., Srivastava, S.K., Dikshit, A. Prediction and validation of gold nanoparticles (GNPs) on plant growth promoting rhizobacteria (PGPR): a step toward development of nano-biofertilizers*Nanotechnology Reviews* (2015). doi:10.1515/ntrev–2015-0036.

Singh, R.N. (1961). Role of blue-green algae in nitrogen economy of Indian agriculture. Indian Council of Agricultural Research, New Delhi.

Singh, R.K., Malik, N., Singh, S. Improved nutrient use efficiency increases plant growth of rice with the use of IAA-overproducing strains of endophyticBurkholderiacepacia strain RRE25. *Microbial Ecology* 66 no. 2 (2013): 375–384.

Thilakarathna, M.S., Raizada, M.N. A meta-analysis of the effectiveness of diverse rhizobia inoculants on soybean traits under field conditions. *Soil Biology and Biochemistry* 105 (2017): 177–196.

Tilak, K.V.B.R. Associative effects of vesicular-arbuscular mycorhizae with nitrogen fixers. *Proceedings of the Indian National Science Academy* B59 no. 3 & 4 (1993): 325–332.

Timsuk, S., Seisenbaeva, G., Behers, L. Titania (TiO2) nanoparticles enhance the performance of growth-promoting rhizobacteria. *Scientific Reports* 8 (2018): 617.

Tripura, C., Sashidhar, B., Podile, A.R. Ethyl methanesulfonate mutagenesis–enhanced mineral phosphate solubilization by groundnut-associated Serratiamarcescens GPS-5. *Current Microbiology* 54 (2007): 79.

Triveni, S., Prasanna, R., Kumar, A., Bidyarani, N. Evaluating the promise of Trichoderma and Anabaena based biofilms as multifunctional agents in Macrophominaphaseolina infected cotton crop. *Biocontrol Science and Technology* 25 (2015): 656–670.

Triveni, S., Prasanna, R., Saxena, A.K. Optimization of conditions for in vitro development of Trichodermaviride-based biofilms as potential inoculants. *Folia* Microbiologica 57 (2012): 431–437.

Triveni, S., Prasanna, R., Shukla, L., Saxena, A.K. Evaluating the biochemical traits of novel Trichoderma-based biofilms for use as plant growth-promoting inoculants. *Annals of* Microbiology 63 (2013): 1147–1156.

Venkataraman, G.S. *Blue-green Algae for Rice Production: A Manual for its Promotion* (No. 46). Food & Agriculture Org, Rome, Italy, 1981.

Verma, R.K. et al. Role of PGPR in sustainable agriculture: molecular approach toward disease suppression and growth promotion. In: Meena V. (eds.) *Role of Rhizospheric Microbes in Soil.* Springer, Singapore, 2018.

Wani, S.P., Lee, K.K. Role of biofertilisers in upland crop production. In: Tandon H.L.S. (ed.) *Fertilisers of Organic Manures, Recycle Wastes and Biofertilisers.* Fertiliser Development and Consultation Organization, New Delhi, 1992, pp. 91–112.

Watanabe, A., Nishigaki, S., Konishi, C. Effect of nitrogen-fixing blue-green algae on the growth of rice plants. *Nature* 168 no. 4278 (1951): 748.

Yazdani, M., Bahmanyar, M.A., Pirdashti, H., Ali, M. Effect of phosphate solubilization microorganisms (PSM) and plant growth promoting rhizobacteria (PGPR) on yield and yield components of corn (Zea mays L.). World *Academy of Science, Engineering and Technology* 25 no. 1 (2009): 90–92.

Yuyan, W., Yang, X., Zhang, X., Dong, L., Zhang, J., Wei, Y., Feng, Y., Lu, L. Improved plant growth and Zn accumulation in grains of Rice (Oryza sativa L.) by inoculation of endophytic microbes isolated from a Zn hyperaccumulator, Sedum Alfredii H. *Journal of Agricultural and Food Chemistry* (2014). doi:10.1021/jf404152u.

Zhang, H., Daoust, F., Charles, T.C., Driscoll, T., Prithviraj, B., Smith, D.L. Bradyrhizobiumjaponicum mutants allowing improved nodulation and nitrogen fixation of field-grown soybean in a short season area. *The Journal of Agricultural Sciences* 138 no. 3 (2002): 293–300.

Zhou, T.T., Li, C.Y., Chen, D., Wu, K., Shen, Q.R., Shen, B. phlF−mutant of Pseudomonas fluorescens J2 improved 2,4-DAPG biosynthesis and biocontrol efficacy against tomato bacterial wilt. *Biological Control* 78 (2014): 1–8.

Index